Computers in Railways XVII

WIT*PRESS*

WIT Press publishes leading books in Science and Technology.
Visit our website for the current list of titles.
www.witpress.com

WIT*eLibrary*

Home of the Transactions of the Wessex Institute.
Papers contained in this volume are archived in the WIT eLibrary in volume 199 of WIT Transactions on the Built Environment (ISSN 1743-3509).
The WIT eLibrary provides the international scientific community with immediate and permanent access to individual papers presented at WIT conferences.
Visit the WIT eLibrary at www.witpress.com.

SEVENTEENTH INTERNATIONAL CONFERENCE ON RAILWAY ENGINEERING DESIGN & OPERATION

COMPRAIL 2020

Conference Chairmen

Giorgio Passerini
Marche Polytechnic University, Italy
Member of WIT Board of Directors

Jose Manuel Mera
Polytechnic University of Madrid, Spain

Ryo Takagi
Kogakuin University, Japan

International Scientific Advisory Committee

Organised by
Wessex Institute, UK
Polytechnic University of Madrid, Spain
Kogakuin University, Japan

Sponsored by
WIT Transactions on the Built Environment
International Journal of Transport Development and Integration

WIT Transactions

Wessex Institute
Ashurst Lodge, Ashurst
Southampton SO40 7AA, UK

Senior Editors

Editorial Board

Computers in Railways XVII

Railway Engineering Design and Operation

Editors

Giorgio Passerini
Marche Polytechnic University, Italy
Member of WIT Board of Directors

Jose Manuel Mera
Polytechnic University of Madrid, Spain

Ryo Takagi
Kogakuin University, Japan

WIT PRESS Southampton, Boston

Editors:
Giorgio Passerini
Marche Polytechnic University, Italy
Member of WIT Board of Directors

Jose Manuel Mera
Polytechnic University of Madrid, Spain

Ryo Takagi
Kogakuin University, Japan

Published by

WIT Press
Ashurst Lodge, Ashurst, Southampton, SO40 7AA, UK
Tel: 44 (0) 238 029 3223; Fax: 44 (0) 238 029 2853
E-Mail: witpress@witpress.com
http://www.witpress.com

For USA, Canada and Mexico

Computational Mechanics International Inc
25 Bridge Street, Billerica, MA 01821, USA
Tel: 978 667 5841; Fax: 978 667 7582
E-Mail: infousa@witpress.com
http://www.witpress.com

British Library Cataloguing-in-Publication Data

A Catalogue record for this book is available
from the British Library

ISBN: 978-1-78466-403-9
eISBN: 978-1-78466-404-6
ISSN: 1746-4498 (print)
ISSN: 1743-3509 (on-line)

The texts of the papers in this volume were set individually by the authors or under their supervision. Only minor corrections to the text may have been carried out by the publisher.

Preface

The present volume contains a selection of the papers presented at the 17th International Conference on Railway Engineering Design and Operation (COMPRAIL 2020) which was to take place in Berlin, organised by the Wessex Institute of Technology, and was instead held online due to SARS-Covid pandemic. The Meeting, which started in Frankfurt in 1987, has been successfully reconvened every two years in different locations around the world. Although this was the first online edition, the large number of delegates and the quality of papers and presentations testified the worldwide interest and the success of the conference series.

COMPRAIL Conference aims at presenting and discussing the use of advanced technologies in railway systems, promoting wider understanding of new developments in the management, design, manufacturing and operation of railways and related transportation systems. It also emphasises the use of digital technologies such as Information and Communication Technologies in advanced cutting-edge railway engineering. In such framework, modern railway engineering encompasses a variety of disciplines and involves managers, consultants, designers of train control systems and computer specialists, in addition to railway engineers. This is reflected in the contents of this Issue, which reveals the strongly multidisciplinary approach and the international character of modern railway engineering development.

The papers contained in this Issue, as well as all others presented at the Wessex Institute Conferences are archived in the Institute's eLibrary (witpress.com/elibrary) where they are permanently available to the international community.

The Editors wish to acknowledge the authors, the members of the Scientific Committee, the Referees, the Institutional Partners, who supported the Conference, and Marta Graczyk, of the Conference Secretariat, who did an outstanding job of coordination together with the WIT Press staff, Isabelle Rham, in particular.

Finally, the Editors, the Chairmen, the Members of Scientific Committee and of the Wessex Institute wish to remember the late Professor Carlos Brebbia, founder of Wessex Institute and of WIT Press, who foresaw and established this series of meetings.

The Editors

2020

Contents

Section 3: Timetabling and rescheduling

Section 4: Safety and security

Section 5: Modelling and simulation

Section 7: Environmental impact and CO_2 reduction

Section 8: Railway logistics

SECTION 1
ADVANCED TRAIN CONTROL

AUTONOMOUS TRAINS IN FREIGHT TRANSPORT: SHOULD THE RAILWAY NOT HAVE THE ADVANTAGE OVER THE TRUCKS?

STEPHAN MÜLLER
German Aerospace Center (DLR), Institute of Transport Research, Germany

ABSTRACT

Should the automation of trains not be easier than the automation of trucks? This paper analyses the structural barriers to innovation of "autonomous freight trains" for large-scale deployment. Four theoretical effects are analysed that characterize the mechanism for innovation activities in advanced markets: (1) the hierarchical and nested architecture of value networks; (2) path dependency in technological paradigms; (3) organizational dynamics; and (4) the stalemate in technology. These four effects culminate, and the higher the maturity level of the four effects is, the more they limit innovation activities. The analysis shows that the economic mechanism strongly limits innovation activity for rail freight transport in general and also in relation to autonomous train operation. For road freight transport, the degree of maturity is less given and thus the truck is at an advantage, although the complexity of train automation in general is less than that of truck automation. The results contribute to the challenge of train automation and are worth considering in a strategic innovation policy for rail.

Keywords: autonomous trains, automated vehicles, rail freight services, innovation in freight transport, RioTinto, AutoHaul.

1 INTRODUCTION

Efficient freight transport is essential for economic interaction and economic development. Efficiency is assessed by its users and determined by transport system's characteristics such as its costs, service quality, accessibility, transport times and other variables. Trucking is currently most efficient for its users considering that a large part of freight is transported on road. Just in time production, complex production and value adding chains, decreasing shipment sizes, third party logistics are part of the economic paradigm enabled by trucks [1]. Moreover, to the dominance of trucks in transport chains contributed the fact that environmental costs and other external costs are not internalized in transport prices [2].

However, deep structural trends culminate in a request of the established road freight transport. Bormann et al. [3] summarize four trends: (1) digitization; (2) environmental sustainability; (3) urbanization; and (4) individualization. It needs adaption in road freight transport to address these challenges. Hence, electrification and automated driving are prominent innovation streams that address those challenges. The electrification (by e.g. battery electric drives, hydrogen or catenary hybrid truck) is tackling the environmental costs of trucking [4]. Regarding automated driving it is expected that this technology will decrease energy consumption, mainly by platoon driving modes. Moreover, anticipated impacts are an increase of road's capacity, a decrease of transport cost and an enhanced reliability [5], [6]. Additionally, mobility-as-a-service (MaaS) solutions in freight transport by, for example, new level of connectivity, sophisticated platform models for freight allocation or ride sharing enter the market and promise to increase the efficiency of road transport (e.g. the utilization of loading capacity and less empty trips). Based on the current industrial and political efforts, a likely future pathway for road freight transport is that electrified trucks will drive automated in an economic setting of MaaS. However, if the scenario of electrified, automated MaaS-trucking becomes reality the railway's traditional performance parameter will be less

WIT Transactions on The Built Environment, Vol 199, © 2020 WIT Press
www.witpress.com, ISSN 1743-3509 (on-line)
doi:10.2495/CR200011

competitive. Trucking would be cheaper, faster and more flexible than today and, approaches environmental friendliness and mass capability of trains. It is not improbable that a shift from rail to road will be an effect then. However, considering the high efforts and public resources invested over the last decades to achieve a shift from road to rail, the innovation stream for trucking seems to be thus quite challenging for railway's future.

Although electrified and automated MaaS-trucking will likely result in a shift from rail to road it is a contestable policy option to deprive or hinder the technology for trucking due to protect rail services. The improvements for the trucking system result from socio-political and economic pressure and are thus needed to be addressed. Without such innovation pathway the future of trucking is endangered and this, maybe most important, is challenges the functioning of the economy which is aligned with the performance of trucking. Instead, an option for the intermodal competition in freight transport can be innovation, such as the increase of automation in rail freight transport.

Considering this option, an intuition might be that the automation of trains should be easier than the automation of trucks. This is because, firstly, trains operate in a nearly closed environment, on tracks, centrally managed compared to individual vehicles on road, an open system with direct interaction with other road users, and many external influences (in particular in urban areas). Secondly, there is some long-lasting experience in the sector to automate trains. One is the automation of the metro train system in Lille, France by 1983 and in some other cities since then [7]. A SNCF consortium currently tests a driverless train for passenger transport in France [8]. The CargoSprinter, developed for the Deutsche Bahn by 1996 [9], was a self-propelled single wagon which could transport up to 10 TEU containers. The German Aerospace Center (DLR) has conceptualized a Next Generation Train Concept for cargo with autonomous single wagons for the last mile [10]. In Australia, the first driverless freight train is operated to transport mining products since mid-2019 [11]. Hence, trains should have clear advantages against trucks in time, costs and complexity when it comes to roll out the system wide automation (and electrification). Instead, however, there are only few current activities in practice towards automation of rail freight services compared to plenty towards electrified and automated MaaS-trucking. Some non-obvious barriers do hinder the innovativeness of railways.

Against this background, this article analyses the structural barriers to innovate in rail freight using the introduced example of autonomous freight trains.

The analysis applies theories from the so called "attacker's advantage" and "stalemate in technology" which are developed in innovation economics to describe the market mechanism for innovation activities in matured compared to emerging markets. Please note that we use autonomous trains in freight transport to explain the innovative capacity of rail using a concise example. Those readers who expect autonomous driving to be discussed in detail, as this amendment to this law and those subsequent technologies and that operating concept is necessary etc., might be disappointed at first sight. But please understand the economic mechanism as a fundamental background for innovation in the railway market. It also applies to other innovations like automatic coupling, ETCS and others. Understanding the mechanism helps the strategic decisions of the companies and the innovation policy for railways and ultimately also helps to implement autonomous driving on the railways "faster" than on the road.

In Section 2 the theories are briefly introduced and applied on autonomous trains. Section 3 discusses RioTinto's achievement in the AutoHaul project from the perspective of the theories. The article closes with recommendations for a strategic innovation policy towards autonomous trains and other innovations in rail freight transport.

2 THE INNOVATIVENESS DISTRESS OF RAILWAY: WHAT HINDERS AUTONOMOUS TRAINS AND OTHER INNOVATIONS?

Most stakeholders likely agree that it needs innovation to improve the market offer and the market situation in rail freight services. Widespread arguments to understand why great innovations such as autonomous trains and automated coupling are not in application are: because innovations in railways are "system innovations" and there is a general "chicken-and-egg problem". These arguments are evident in many cases however it needs a critical look on why the issues of system innovations and chicken-and-egg problems for many innovations could not have been overcome in the past century? To answer the question on the why rail cannot operationalize its advantage for autonomous trains compared to trucks we have chosen an approach of theories which are developed to explain why incumbent firms are disadvantaged faced with radical technology market entrants. In both cases the established is contested by and has to deal with a drastic change power.

Christensen and Rosenbloom [12] investigated this situation as an "attacker's advantage" which is based on three inquiries for explanation: First, hierarchical and nested architecture of value networks, secondly) path-dependency in technological paradigms and, thirdly) organizational dynamics. We add the theory of the "stalemate in technology" by Mensch [13] as the fourth which explains the relation of market matureness and the decrease innovative activity (towards a stalemate situation) because, without doubt, rail is faced a matured market situation since decades. All four theories are first presented and then transferred to the example of autonomous trains. Taken together, this should lead to a deep understanding of the innovative capacity of mature transport systems such as the railway today.

2.1 The hierarchical and nested architecture of value networks

Due to continued optimization efforts, the rail has become a sophisticated and matured product system over the course of its life time. In this section, it should be made clear that this degree of maturity can be a barrier for major innovation steps such as autonomous trains.

2.1.1 Why innovation activity for matured transport systems is limited to incremental steps

Christensen and Rosenbloom [12] introduced the notion of hierarchical and nested architecture of value networks to describe that in most cases products are assembled under use of other products and are sub modules of higher level products. For example, a locomotive and a wagon are produced by many downstream parts such as brake and engine and those are also made out of other parts. The many locomotive and wagon can only be used on infrastructure with switches and so on. And upstream it needs train control systems and switch control etc. (see Fig. 1). Railway is a complex architecture which has nested levels and hierarchical levels. Those levels are connected in value networks to create a functioning railway system.

The point is that the alignment and plurality of the nested and hierarchical elements allow only coordinated and incremental steps of change. Some parts are nearly impossible to change on rail nowadays such as the width of tracks. To change others needs a lot of coordinated re-orientation such as changing the block interval system needs up- and downstream adaptations (locomotive equipment, control system, management system, laws etc.). Even smaller innovations are framed by the architecture such as the wagon can be re-invented only within the frame of the structural gauge, permissible axle load and so on. The more advanced an architecture of a transport system became over time, the more this system

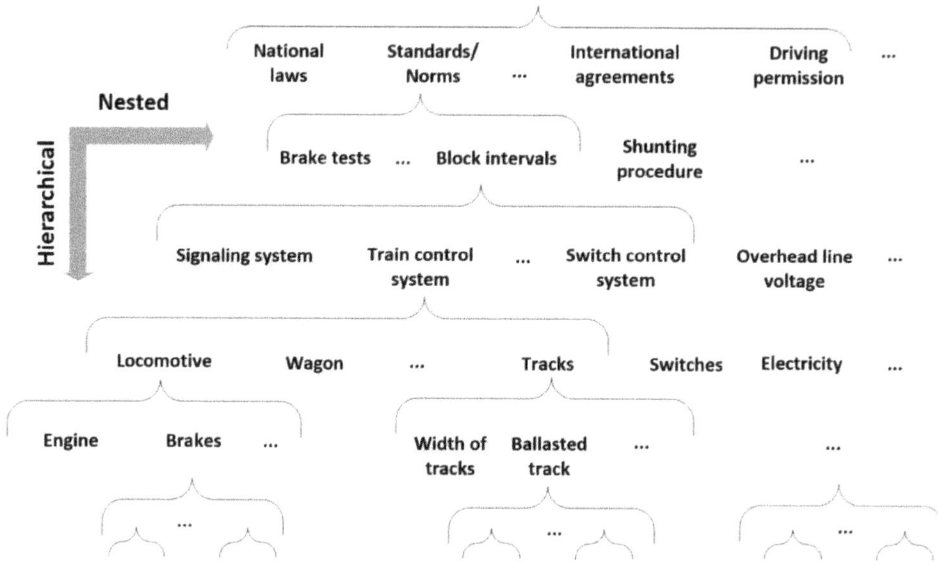

Figure 1: Example of the hierarchal and nested architecture of rail services.

is either limited to incremental innovations or it needs a system-wide, coordinated tremendously effort for the change (aligned with all interest parties what again makes incremental steps likely). Innovations are therefore blocked or in the best case enter the market as an incremental consensus solution.

2.1.2 The architecture of the value network as a barrier for autonomous trains

Looking at the architecture of the "rail freight transport" value chain, autonomous trains require adjustments at every hierarchical level and in almost every element, e.g. in infrastructure, train equipment, control systems, legislation, rotation planning. Moreover, sometimes several (competing) options for adaptations are possible. For example, train control for autonomous trains can be central with ETCS level 3 or decentral by connected sensors on trains. In the transition phase conventional trains and autonomous trains must be compatible somehow. The complexity further rises when considering that passenger and freight trains are operated on same tracks, are international and thus, adaptions or compatibility becomes a higher hierarchy issue. To achieve autonomous freight trains, almost the entire architecture has to be adapted drastically and adapted in a similar speed and intensity (horizontally and vertically aligned).

The advance of the architecture allows only small steps towards autonomous trains. To give an explicit example on this issue: The Siemens and Halske Telegraphenbauanstalt invented 1870 the relay boulder field which enables block interval train control. This innovation was a tremendously efficiency improvement for railways at that time [14]. This train control system however aligned and nested further locomotive development, train management, network management, network development, standardization, laws, rules, etc. – sustaining until today. To introduce a new train control technology like the European Train Control System (ETCS) level 2 needs enormous coordinated international efforts. Those efforts are taken by the European commission since 1996 (Directive 96/48/EC for interoperability of the trans-European high-speed rail system). Nearly 25 years later ETCS architecture is not deployed as the international standard. A difference between the innovations of 1870 and 1996 is that architecture has reached an enormously higher level of maturity, making the change in architecture more complex.

2.2 The path-dependency in technological paradigms

Almost every nation is home to a large railway company. One advantage is that this company has economic strength and robustness in a tough market environment. This section explains why incumbents however have a low innovation capacity. Especially in the case of drastic innovations like autonomous driving, there is a barrier to innovation.

2.2.1 Why radical innovations are rejected

With reference to a body of literature Christensen and Rosenbloom [12] argue that radical new technologies can be faithful to established companies. The notion of a radical technology is, that it is a completely new way of product architecture and functionality. Companies on market in a normal competitive mode innovate in a given technology paradigm what means that they are within an (as described before) architecture of value networks to provide a defined functionality of the product. They compete horizontal, e.g. making the best brake, or they compete by vertical integration to offer a complex product as national carriers do for example. Although radical technologies can bring about significant efficiency gains, incumbents are most likely to fight for what they have established. This strategy, however, repeatedly leads to its failure [15], [16]. The main reasons for this are:

- Radical technologies are competence-destroying: Once the architecture is created, each institution in the value network creates competences such as on technical issue, anticipation of customer needs or the development of a grounded business concept. This can be related to ownership of patents, protection such as by norms and standards, working skills within a company, production lines organization and so on. A radical technology is challenging many of those competences and devaluates them. For example, the innovation of the car devaluated massively the competence needed for horse carriages.
- The shift from product to process innovations: In a normal mode of innovation activity companies do first innovate on product level to enhance the architecture's functionality, reliability and convenience. However, they shift from product to process innovation what stabilizes or fixes the architecture from then on. The reason for the shift is that process innovations, e.g. higher efficient production, create margins when product innovations become difficult. The difficulty increases, for example, by the framing value network, technically high end, physical or economical limits (over-engineering). In a sense, the more advanced the architecture is, the more process innovations are lucrative compared to product innovations – which leads to the stabilization of the established.
- The strategic niche of radical innovations: Mass market products have high performance to satisfy the mass market demand. Radical technology normally underperforms in this market. They need market niches, where they provide new functionalities to users which are unsatisfied by mass market products. The niche is however of tremendous strategic relevance: the niche exists because the architecture of the mass market product cannot (technically, economically) satisfy this demand. When a radical technology focusses on this niche, the performance is specialized and unrivalled. Incumbents cannot or very hard serve this demand – they did not until the rise of the radical technology. Hence, the

underperformance of the challenger compared to products in the mass market becomes largely irrelevant. The performance in a strategic niche however is the key for growth and future market dominance. For example, first cars had the power of only 1 or 2 hp compared to more than 100 hp of a locomotive in 1890. This was however not relevant in the strategic niche where cars were deployed.

The point is, established companies in a technological paradigm have a) very low economic incentives and b) a strategic disadvantage to re-align the architecture of the value network from mass market products to niche market products (although some few companies can leap).

2.2.2 The path-dependency against autonomous trains

Path dependency implies that the actors manifest, defend and intensify their role in the architecture instead of pursuing alternatives. This has been the case with the railways for almost 200 years now.

Autonomous trains will destroy competences in the architecture. Concerned are for example, railway infrastructure companies. They made great investments in the past to enable non-autonomous trains with interlocking boxes, signaling and so on. Those elements provide a return on investment now but would become irrelevant in a world of autonomously communicating trains and operating companies. Many working skills are changing from today's planning and controlling to IT-based algorithmic, programming, computing, and hardware. Many of labor forces become obsolete which will be an issue of organizational change and labor unions. All those in the architecture, who are potential losers of the architectural re-orientation are potential brakeman of autonomous trains.

Rail freight service market is a very price sensitive market. The reason for the price sensitivity of the customs is because the competing services are considered to be very similar (homogenous products in functionality, reliability and convenience). Hence, the price becomes the selling point. However, in a price driven market margins are difficult (long term they are zero). If any actor would see the chance to compete by product innovations they would because this is the opportunity to leave the price competitive environment. However, they are captured in the architecture and margins are only possible with process innovations to higher efficiency in production. Although autonomous trains would be an efficiency increase by, for example, saving the driver costs, their rejection is likely.

In most countries normally one company dominates the national rail market. Competition increased since the market liberalization but is still on a low level. Private rail companies make their business, we can say, in niche markets which are opened by the market offer of the incumbent (often a former state owned national carrier). In a very price sensitive market, as the rail market is, those niches are not functional but private companies can operate them with a better price than the incumbent, for example, because of lower overhead costs. Maybe in many of those niches, autonomous trains could have a first application because it could decrease the production costs tremendously. However, the investments needed are very high and would decrease the margins for long time while at the same time the willingness to pay by the customs will likely decrease as well for the service with autonomous trains. The point is, it generally lacks in a functional niche for autonomous trains, where the investment is promising. Just to replace the production system for the same product will not ensure returns on investment. However, this lack is not a natural law but the path-dependency of the established players narrows them not to search for functional niches of new products.

2.3 Organizational dynamics

If an innovation is beneficial, it seems to make economic sense to introduce it to the market. In this section we will explain why an innovation that appears to increase efficiency to such an extent as autonomous trains still does not enter the market or will enter it only with a considerable delay.

2.3.1 Why breaking out of the architecture is difficult although willingness

Each company has a culture, technical and economic structure and develops knowledge and competences in its market over time. Christensen and Rosenbloom [12] explain that companies however tend to focus on their core business and core competences. Hence, they narrow the thinking and decisions for future innovations on their established value network and lose their capability to develop or think in alternative architecture. The decentral decisions in a company are guided by the same mind and are thus i) pre-determined and ii) aligned. "We've always done it this way". It becomes a collectively chosen pathway even if decentral in many aspects (economic structure, technical structure, cultural structure, even the sectoral structure).

The point is, on company level and on sectoral level stabilizing forces align structures and decisions. Thus, the willingness to break out of the architecture, which may emanate from one or the other decision-maker at any level of the architecture, is subordinated to organizational dynamics and its framing culture.

2.3.2 The slowing organizational dynamics against autonomous trains

Autonomous trains would have some advantages such as cost and reliability. Shifting train operations to autonomous trains should be a beneficial economic decision. However, high end automation of trains means a disruption of the existing railway system. From the perspective of the incumbents this means to open or rather to destabilize the established structures. No one knows whether incumbents will be the winners of this disruption or it might be a new entrant. Furthermore, it is clear that a suited and transferrable concept of autonomous trains has to be designed in a first stage. To find the concept which fits to market demand is costly and risky, likely linked to mis-investment and failure. It needs decades to find a general operated "design" and the more advanced the process is, the more the former established patents, processes, routines etc. are devaluated. The crux however is, that all these has to be done in market niches – there will not be an initial roll out of the new technology on national or international scale but to define the design to specific demand needs, for example, a company in rural condition. No one knows the transferability of a tested design in the niche to other customs towards a mass market.

National rail carriers for example are hence stabilized in this manner by many interest parties (politically, technically, economically) which makes it difficult to change the organizational dynamic towards autonomous trains.

2.4 The stalemate in technology

The railway market has been liberalized in most countries. As a result, established and new companies are competing for customers on the market. Surely it should be expected that innovations in a liberalized market environment will be used to make products better. This section explains the economic mechanism why innovations in a liberalized railway market are not a matter of course.

2.4.1 Why the market mechanism can block innovations

Normally, capital is invested if a return can be expected with a manageable risk. This also applies to investments in innovations. However, the market situation makes a difference on the return that can be expected. To explain the mechanism we will consider two extreme examples: When a radical technology enters the market, it is a "young technology". Major improvements in this technology can be achieved with relatively little capital input because the technology and their market are almost unexplored. Furthermore, significant innovation steps trigger purchases at high prices and thus market growth. The return on investment is correspondingly high and the risk low. Rapidly growing markets are therefore very attractive for investment in innovation. On the contrary, in a saturated market the technology is optimized and further innovations are very expensive but not significant. The lack of innovation does not trigger market growth. In the extreme, if the products are not innovative at all, the incentive to buy a new product is limited to replacement purchases. Thus, the investment in innovations is not attractive because such an investment is high, is risky and is, if given, little profitable. Therefore, growing markets or the expectation of market growth can be postulated as a prerequisite for investment in innovation. This unveils the mechanisms: no (expected) market growth means no innovation activity and, no innovation activity means no market growth. Mensch [13] described a situation as the stalemate in technology when a market is unable to innovate and to grow anymore – the market dynamic is paralyzed.

The point of the market mechanism is the nature of increasing market's matureness implies a stalemate in technology and the stalemate in technology is a strong barrier for innovation activity. Normally this market either is degenerating by market forces or subject to be disrupted from external technologies or is stabilized by subsidies.

2.4.2 The stalemate in technology of freight railways

In the Rail Freight Masterplan in Germany, the stakeholders postulated for the sectoral basic situation: "[…] in Germany, the largest rail freight market in Europe, only a few railway undertakings are recording appreciable growth and income that is just about in the black. However, these low rates of return are not sufficient for funding the investment required to safeguard the future of rail freight on a sustained basis" [17]. This statement indicates the technological stalemate. No market growth means no incentive (ability) to invest in innovations and in turn, no investment in innovations implies no market growth. Realistically, rail freight services have a production system which is not that much advanced compared to hundred years ago: centrally organized management, fixed slots, train composition with handmade coupling, frequent wagon checks, shunting and brake tests etc. Even the age of the rolling stock is aged, for example in Europe with an average about 30 years [18]. Although innovation to modernize rail freight are available such as automated coupling, dynamic pivoted bogies and the 5L wagon (low noise, lightweight, long-running, logistics capable, LCC-oriented) [19], [20], they are however not in operation – an important argument why they are not is the technological stalemate in the market which exists since decade already. Autonomous trains, which are a much larger technological and organizational transition, have thus very challenging market conditions as well. The market mechanism is a strong barrier to this innovation. It is difficult for market participants to provide it from their own economic power – it is probably necessary to subsidize innovation impulses.

2.5 Summary of the challenge

Four effects are discussed that are barriers for autonomous trains in large scale deployment:

1. The architecture of value networks: The hierarchical and nested architecture of the value-added network for rail freight transport, which has been developed for almost 200 years, only allows minor, aligned innovation steps. However, the innovation of autonomous rail freight transport means a massive change in the architecture; indeed, it means its disruption. To achieve this would mean a very long lasting coordinated undertaking.
2. The path-dependency: Some very important players would lose market power, profits and competences by autonomously operating trains. The potential losers of the technology are its potential brakeman.
3. Organisational dynamics: The incentive to develop autonomously running trains for an uncertain niche market, as expensive as the investment for the technology is, is not attractive. It makes economic sense to focus on the mass market.
4. The stalemate in technology: the market is not growing because there is a lack of innovation, and the lack of innovation is because the market is not growing. It is the inability of the market players to initiate positive market dynamics in this market environment. In the case of rail freight services this steady state exists since decades.

Although automation in rail freight might be strongly efficiency enhancing, might create new market growth from niches, profitability and social benefits, the four effects reflect why automation is not a common trend in the rail freight sector: The mechanism of innovation economics implies a deep innovativeness distress for rail freight.

In principle, the mechanism of innovation economics is relevant for the automation in road freight transport too. Briefly, the major difference from rail to road transport is, that road freight transport is not that much advanced as rail freight is. For example, the architecture is not that much matured in alignment and hierarchy even if they are aligned and nested but on a lower level. Another issue is that their market is not in the stalemate in technology – road freight logistics is a growing and (still) profitable business what triggers innovations activity.

Summing up, the state of rail freight do not support or only allows with considerable delay that autonomous trains become a sector-wide (cross-national) innovation stream.

3 THE DIFFERENCE OF THE AUTOHAUL PROJECT TO THE MAJORITY OF RAILWAY APPLICATIONS

The autonomous train in the AutoHaul project is without doubt a great achievement, from technical point of view as well as all the organizational aspects needed to implement the technology. This projects might contradict the discussion made before on the, let us say, innovativeness distress in railways. However, there are some important differences to the majority in the rail sector. Those can be pointed out as (based on [11]):

- The growing market: The Australian–British mining company RioTinto applied an autonomous train in a market where a) the mass capability is the functional core, b) where a long term growth is expected. Although the initially calculated costs of the project have been extravagated by far and needed nearly one billion in total – in long term growing condition the return on investment can be expected for the investor.
- The closed architecture: The mining company operated human driven rail services to transport the mining products close to an general architecture of railway which can be found everywhere in the world. However, they are exclusive users of the infrastructure and could re-create a disruptive architecture by a single company's decision. The point is not all to devaluate the achievement but to point out, that normally many (hierarchical and nested) companies have to align to such a development step.

- It was a mining company, not a rail company: What stereotypical for disruption is, is that it is rarely introduced by incumbents. Also in the case of the AutoHaul project, it was not a railway company which introduced the autonomous train but a mining company. Rail transport is "only" a part to fulfill the function of the product of RioTinto: mining. The company its organizational dynamic was not narrowed or rather path-dependent of the established rail system when rail is not the core product.
- The specific framework conditions: It is not the point to say that it seems to be obviously easier to run driverless trains in the Australian outback than in, let us say, densely populated central Europe. Solutions can be found to any circumstances. The point is, that it needed collaboration between public authorities and the project team to realize driverless trains. Many of existing rules, norms and standards needed to be overthought and permitted for this achievement. It needed politicians, officials and administrators that where open to leave the safety zone of the established framework for rail operations. In the Australian case there was this supportive attitude and this was one of the decisive factors for the success of the project – they overcame the path-dependency and the established organizational dynamics.

Considering the AutoHaul project on a closer look we see that the theoretical aspects where overcome. However we also see, that AutoHaul is distinct to the general rail freight service sector. In the next section we address the learnings into policy implications.

4 POLICY IMPLICATIONS

Without doubt, innovation is the key to trigger competitiveness of railways against trucking, new market growth and a shift from road to rail. To bring autonomous trains into practice is possible as the mining company RioTinto has demonstrated with the AutoHaul project. However, it needs to address the issues of the innovativeness distress of railways in the policy strategy to achieve it in the rail freight market. From the theories applied, for strategic innovation policy in rail freight following cornerstones can be provided:

- *The state is often needed as a donor*: In the situation of the technological stalemate, market players are generally not able to invest in innovations. This means that public funds or correspondingly state-subsidized loans are indispensable to trigger innovation in the rail freight services.
- *Accepting the "impending" radical change – shaping a new role for rail freight services*: a reversal trend for the market situation cannot be achieved with the incremental improvement of the market supply. It needs radical new technologies and business concepts to shape a new role of rail freight services in transport markets and, in new transport markets.
- *Growth from relevant market niches*: Instead of designing and envision a possible future system on the drawing board, it is needed to test the market demand for new products in small areas of application. It needs strategic niches were a new architecture of rail provides new functionality and thus willingness to pay. This applies both to the geographic scope and to the demand group. If the concept is successful in the market niche and transferable for further demand groups, the market supply will gradually grow in line with the market needs.
- *Involvement of new actors*: Established market players can incorporate radical innovations into their production processes. To this end, there must be a willingness to grow out of the niche and to revise the market offering. Since new players are not captured in architecture of a value network, they naturally have a higher willingness to

create a new one. In case the incumbent architecture is a kind of closed society, this would be a requirement. It is therefore particularly important to attribute a high potential to new players to give impetus to innovativeness in rail freight transport.

- *Radical innovations need adapted framework conditions*: A radical innovation, which revolutionizes the previous technological basis, is naturally not suitable for established production processes, standards, institutions, etc. They therefore need test fields, progressive thinking and acting authorities, ideal protection against resentment as well as capital/subsidies in order to gain access to established practice.
- *Not the state selects the winner but the winner selects the support of the state*: It makes no economic sense to select a new winning technology centrally (and possibly even on the basis of proposals from established market players). Many experiments and funding that is as open to technology as possible create a situation in which an effective market selection takes place and a really suitable solution, the winning technology, is created. Even if many ideas will fail, which means lost public funds in the first place, this helps to find a winning technology by learning from failed experiments. It is also more decisive to what extent society would benefit from a sustained, self-driven growth in rail freight transport.

The current flow of knowledge towards automated driving can be seen as an opportunity to break up established structures in rail freight transport and to re-launch the system's path to new market growth, profitability and social benefits. However, an innovation policy is needed that takes into account the specific characteristics of the current innovativeness distress in rail freight and the corresponding policy concepts.

REFERENCES

[1] Blanquart, C. & Burmeister, A., Evaluating the performance of freight transport: A service approach. *European Transport Research Review*, **1**, pp. 135–145, 2009.

[2] Union Inter Des Chemins Fer (UIC), External costs, 2015. https://uic.org/support-activities/economics/external-costs. Accessed on: 19 Sep. 2019.

[3] Bormann, R., Fink, P., Holzapfel, H., Rammler, S., Sauter-Servaes, T., Tiemann, H., Waschke, T. & Weirauch, B., The future of the German automotive industry: Transformation by disaster or by design? *Wiso Diskurs*, Friedrich Ebert Stiftung, 2018.

[4] Moultak, M., Lutsey, N. & Hall, D., Transitioning to zero-emission heavy-duty freight vehicles. White Paper, Washington, DC, 2017.

[5] Kunze, R., Haberstroh, M., Ramakers, R., Henning, K. & Jeschke, S., Automated truck platoons on motorways: A contribution to the safety on roads. *Automation, Communication and Cybernetics in Science and Engineering 2009/2010*, eds S. Jeschke, I. Isenhardt & K. Henning, Springer: Berlin and Heidelberg, 2011.

[6] Müller, S. & Voigtländer, F., Automated trucks in road freight logistics: The user perspective. *Advances in Production, Logistics and Traffic*, eds U. Clausen, S. Langkau & C. Thaller, Springer, 2019.

[7] Wang, Y., Zhang, M., Ma, J. & Zhou, X., Survey on driverless train operation for urban rail transit systems. *Urban Rail Transit*, **2**, 106–113, 2016. DOI: 10.1007/s40864-016-0047-8.

[8] Zasiadko, M., SNCF tests its first autonomous trains, 2019. www.railtech.com.

[9] Forschungsinformationssystem (FIS), CargoSprinter, 2019. www.forschungsinformationssystem.de/servlet/is/402123/ Accessed on: 26 Feb. 2019.

[10] German Aerospace Center (DLR), NGT cargo: What freight trains of the future will look like, 2017. www.dlr.de/dlr/en/desktopdefault.aspx/tabid-10280/385_read-21934/year-all/#/gallery/26733. Accessed on: 19 Sep. 2019.
[11] RioTino, Successful rollout of AutoHaul™ positions WA as global rail technology leader, 2019. www.riotinto.com/en/news/releases/AutoHaul-successfully-deployed.
[12] Christensen, C.M. & Rosenbloom R.S., Explaining the attacker's advantage: Technological paradigms, organizational dynamics, and the value network. *Research Policy*, **24**, pp. 233–257, 1995.
[13] Mensch, G., *Das technologische Patt: Innovationen überwinden die Depression.* Umschau Verlag Breidenstein KG, 1975.
[14] Kirchberg, P., *Die Entwicklung des Eisenbahnnetzes in Deutschland von der Reichsgründung bis zur Gegenwart*, 1993. DOI: 10.11588/ih.1993.0.22599.
[15] Porter, M.E., The competitive advantage of nations. *Harvard Business Review*, **March–April**, 1990.
[16] Christensen, C.M., *The Innovator's Dilemma: When New Technologies Cause Great Firms to Fail*, Harvard Business School Press: Boston, MA, 1997.
[17] Federal Ministry of Transport and Digital Infrastructure (BMVI), Rail freight masterplan, 2017. www.bmvi.de/SharedDocs/EN/publications/rail-freight-masterplan.html.
[18] Scordamaglia, D. & Katsarova, I., The fourth railway package: Another step towards a single European railway area. European Parliamentary Research Service, PE 579.088, March 2016
[19] König, R. & Hecht, M., White paper innovative rail freight wagon 2030: The "5L" future innovative initiative as a basis for growth in rail freight transportation, 2012.
[20] Thunder, T. & Islam, D., Assessment of existing and future rail freight services and technologies for low density high value goods in Europe. *European Transport Research Review*, **10**(9), 2017.

TERMINOLOGY, DIFFERENCES, AND CHALLENGES OF COMMUNICATIONS-BASED TRAIN CONTROL AND EUROPEAN TRAIN CONTROL SYSTEMS

LAURA MARTINEZ & ULLRICH MARTIN
Institute of Railway and Transportation Engineering (IEV), University of Stuttgart, Germany

ABSTRACT

Control systems guarantee safe train operations by (1) monitoring and supervising train speed, (2) defining the next allowable target position, and (3) activating automatic braking systems. Two different train control systems, (1) Communications-Based Train Control (CBTC) and (2) European Train Control Systems (ETCS), are usually implemented for railway operations. Among those systems, the CBTC is applied for light railway and urban mass-transit systems, whereas ETCS is used for heavy railway systems. These two train control systems have been used worldwide for more than 20 years. Although both systems share a common field of application, their scope, functionalities, and automation aims differ. This paper reviews the CBTC and ETCS systems, and their underlying moving principles. The applicability, terminology, differences, and challenges with regard to levels of implementation, radio communication systems, train integrity, and interoperability are analysed in detail. Based on this analysis, potential characteristics are identified that could be adopted from these systems to improve railway operation.
Keywords: Fixed Block, Moving Block, Communications-Based Train Control (CBTC), European Train Control Systems (ETCS), Train Control, Automatic Train Operation (ATO), Automatic Train Protection (ATP), Automatic Train Supervision (ATS).

1 INTRODUCTION

Communications-Based Train Control (CBTC) and European Train Control System (ETCS) are recognised in the railway sector – as their names suggest – as *train control systems*; however, ETCS sometimes is also referred as a *train protection system*. Despite their apparent similarity, their development and application areas have gone different ways.

ETCS was designed as an interoperable and fail-safe solution (i.e. in case of equipment or software failure, it will not endanger lives or property) initially just for European Union (EU) countries to supervise train movement depending on the speed limit. However, its popularity is far-reaching, especially in continents such as Asia and Africa [1]. To achieve the goal of interoperability, many EU stakeholders were involved in the development of specifications and further implementation, this degree of complexity slowed down the specification process. Currently, ERTMS has defined three ETCS levels: Level 1 and Level 2 have already been implemented and are detailed in the main versions of the system requirement specification, and the Level 3 specifications are currently under development (Section 4.1). Level 1 and Level 2 operate under the logic of Fixed Block Operation, while Level 3 under the logic of Moving Block Operation. These two separation principles are covered in the next section.

CBTC systems operate under the logic of the moving block separation principle. It has the following main characteristics: (1) precise train position determination at any given time independent of using track circuits or axle counters; (2) continuous track-to-train communication; and (3) an implementation of functions related to safe train operation as well as optional functions related to driverless operation (e.g. speed regulation), and to dispatching rules (Section 3.2). CBTC systems, tailor-made solutions, are successful in implementing many automation functionalities faster than ETCS for urban mass-transit applications in

WIT Transactions on The Built Environment, Vol 199, © 2020 WIT Press
www.witpress.com, ISSN 1743-3509 (on-line)
doi:10.2495/CR200021

cities such as Vancouver, Canada and Beijing, China. Conversely, CBTC is manufacturer dependent, i.e. a detailed standard is not fully specified for suppliers to achieve an interoperable cross-frontier application [2]–[4].

In this paper, the authors analysed the reasons behind these differences, paying special attention to why CBTC has achieved driverless operation while ETCS outperforms CBTC regarding interoperability. To accomplish this, the authors first describe the separation distance principle used in each train control system, their terminology, and their main differences and challenges in terms of levels of implementation, radio communication systems, train integrity, and interoperability. Furthermore, this review gives insight into why CBTC and ETCS – despite what their names could indicate – should not be seen as being part of the same category, but rather two systems comprised of different functionalities.

2 SEPARATION DISTANCE PRINCIPLES

Before going into more detail, it is important to understand the two main separation distance principles used in railway operation, Fixed and Moving Block Operation, because they influence the type of infrastructure and software requirements needed. In Fixed Block Operation (also known as a conventional operation), the railway tracks are divided into block sections of around 1000 m and are often protected by lineside signals. In this type of operation, track circuits or axle counters are mounted on the tracks and indicate whether a block is free or not. Only one train is allowed to move in a block section at one time (Fig. 1). Examples of this type of operation are the ETCS Level 1 and Level 2, which are specified by the European Rail Traffic Management System (ERMTS) [5].

In contrast, Moving Block Operation (also known as absolute braking distance) works under the principle that the next danger point location, either the rear of the train ahead or movable track elements (e.g. switches), is known precisely. Under this principle, the minimum separation between trains is determined by the safe braking distance of the train plus an additional safety distance (Fig. 2). In other words, the separation distance between trains is not fixed, rather it changes continuously depending on the trains' speeds or movable infrastructure elements. Examples of this operation are ETCS Level 3, which has not yet been specified by ERTMS, and the CBTC, which has been specified to some extent by the Institute of Electrical and Electronics Engineers (IEEE) [3], [5], [6]. Moving Block Operation has the potential to increase the network capacity (i.e. more trains per hour) without major investment in the current infrastructure. However, its implementation could lead to large investments in train equipment, telecommunication infrastructure, and on-board equipment maintenance [7].

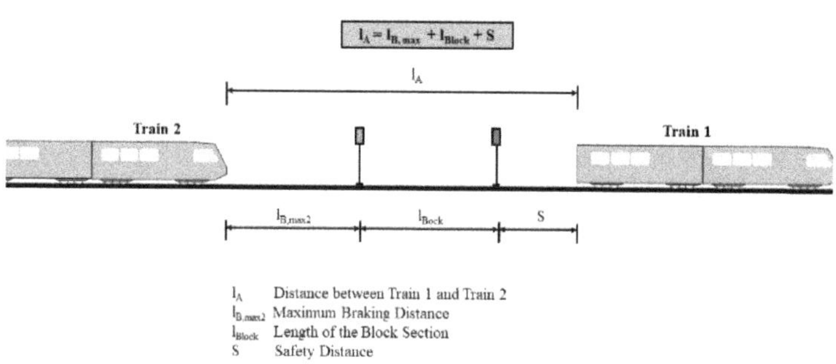

Figure 1: Fixed distance separation principle between consecutive trains – fixed block [5].

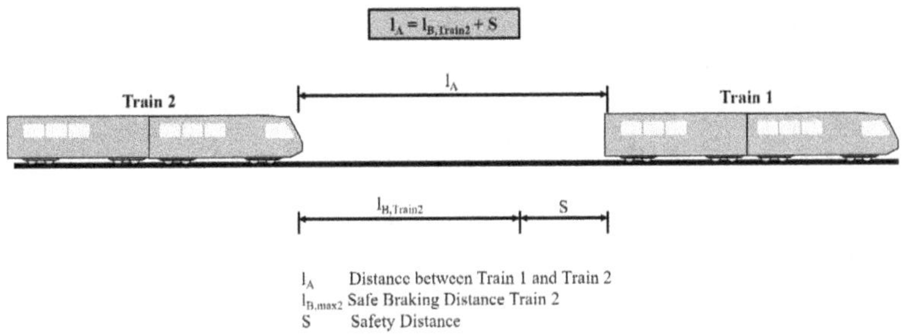

Figure 2: Absolute braking distance separation principle between consecutive trains – moving block [5].

3 TRAIN CONTROL SYSTEM TERMINOLOGY

3.1 The myth of train control systems applications

CBTC systems are mostly implemented in closed urban mass-transit systems because they require a higher line capacity (i.e. more trains per hour), shorter headways (i.e. the space or time interval between two successive trains), lower speed limits, and simpler and shorter networks (in contrast to heavy railway networks). On the other hand, ETCS is mostly implemented in heavy railway networks for the opposite reasons [3], [8]. Nevertheless, the trend of implementing a specific train control system for a specific railway network is slowly changing. For example, some countries plan to integrate ETCS Level 2 with additional Automatic Train Operation (ATO) functionalities into their suburban mass-transit railway networks (e.g. in Germany) [9]. Meanwhile, CBTC systems are been adopted in heavy passenger rail and heavy-haul applications (e.g. in Brazil) [10]. In addition, other initiatives combine both systems to integrate the heavy and light railway systems (e.g. in the UK) [11]. These deployments of CBTC and ETCS shatter the myth that each system is used specifically in just one type of railway network (heavy or light), proving that they can be adapted to different projects.

3.2 Scope and functionalities

Both CBTC and ETCS systems have acknowledged the definition of Grades of Automation (GoA) proposed by the International Association of Public Transport (UITP). The GoA (Table 1) is a classification – by functionalities and responsible – of the levels that make the automation of railway systems possible. The lowest level of automation is GoA1 where the driver performs all the main functionalities (setting train in motion, stopping the train, door closure, operation in event of a disruption), while in the highest level, GoA 4, the train control system regulates all the main functionalities. Nevertheless, independent of the driver's presence, the railway operator should assign a responsible person for the correct system functioning e.g. in the control centre.

CBTC and ETCS applications continuously use abbreviations to explain system functionalities. However, for each of these two control systems, the meaning of such abbreviations might vary. This could lead to misleading interpretations of their functionalities and scopes. Table 2 summarizes the perspectives of the main functionalities of train control systems for CBTC standards, ETCS standards, and for the UITP. The following paragraphs give a more detailed explanation of these perspectives.

Table 1: Grades of automation and examples. *(Source: Farooq and Soler [3], Pachl [5], and UITP [12].)*

Grade of Automa -tion	Type of Train operation	Setting Train in Motion	Stopping Train	Door Closure	Operation in event of Disruption	Examples
GoA 1	ATP with driver	Driver	Driver	Driver	Driver	ETCS Level 1
GoA 2	ATP and ATO with driver	Automatic	Automatic	Driver	Driver	ETCS Level 2 ETCS Level 3
GoA 3	Driverless	Automatic	Automatic	Train attendant	Train attendant	
GoA 4	Unattended Train Operation	Automatic	Automatic	Automatic	Automatic	CBTC with all functionalities

Table 2: The perspectives of railway system functionalities by type of system.

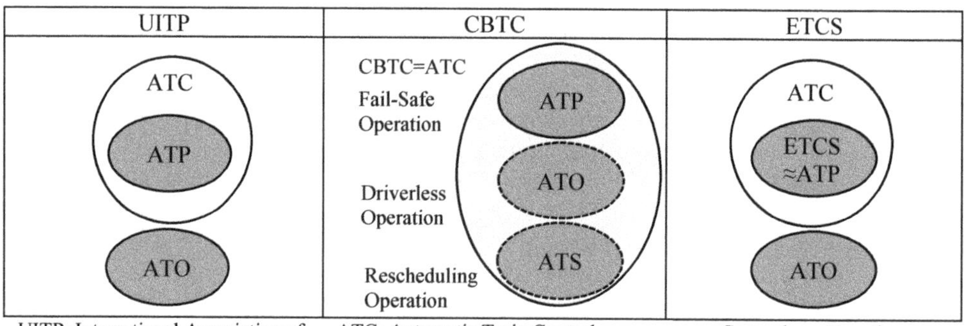

UITP: International Association of Public Transport
CBTC: Communications-Based Train Control
ETCS: European Train Control System

ATC: Automatic Train Control
ATP: Automatic Train Protection
ATO: Automatic Train Operation
ATS. Automatic Train Supervision

——Compulsory Function
-----Optional Function

For UITP, the Automatic Train Protection (ATP) has to ensure basic safety (e.g. avoiding collisions or exceeding speed limits) whereas the ATO assumes the functions of the driver except for door closing. UITP defines Automatic Train Control (ATC) as a general class of ATP that is in charge of route setting and train regulation [12].

CBTC standards additionally to the terms used by UITP describe the term Automatic Train Supervision (ATS) that monitors trains and adjusts their performance to maintain schedules. Regarding ATP and ATO functionalities, CBTC systems have similar meanings to UITP. CBTC Standard distinguishes ATP, ATO, and ATS as different subsystems inside of the ATC. The CTBC System must include ATP and may include ATO and/or ATS, but it needs to have the three functionalities to achieve full automation (GoA 4) [2].

Contrary to what one might think, ETCS as a Train Control System is closer related to the definition of ATP of CBTC Standards and UITP. Contrary to IEEE for CBTC, ERTMS does not contemplate the full automation of ETCS as an objective, ETCS is seen as a fail-safe component of signalling primarily independent of other optimization objectives (e.g. improve punctuality). Moreover, functionalities of CBTC, such as ATO and ATS are considered as external systems in ETCS but can be compatible with them e.g. ETCS over ATO is an alternative to integrate this functionality to the ETCS operation [9]. Similarly, the train dispatching operation (or ATS) is not a part of the ETCS specification. Initially, it was conceived by ERTMS as the European Train Management Layer (ETML), however, due to the complexity to integrate different management systems, its scope was limited to the development of an interface [13]. Currently, Shift2Rail, a European initiative made up of railway operators and manufacturers, develops an automated traffic management system interoperable with ETCS [14]. This Traffic Management System uses as inputs the rules and performance of fail-safe systems e.g. movement rules to reschedule trains more efficiently in case of delay. Moreover, we believe that if this Traffic Management System can coordinate both urban mass transit and heavy railway systems, independent of which train control system (ETCS or CBTC) is used, rescheduling of trains can be done more efficiently. That ultimately results in better use of resources, and hence a greater profit to end-users, railway operators, and infrastructure managers (e.g. reducing operational waiting time, operation costs, infrastructure optimization).

4 DIFFERENCES AND CHALLENGES BETWEEN ETCS AND CBTC SYSTEMS

4.1 Evolution and levels

In ETCS, the trains need to know three types of information: current position, the maximum allowed speed at a certain track section, and the message containing the next safe allowable position – called movement authority (MA). The rough current position is sent to the train via fixed Eurobalises mounted on the tracks, and between them, the train estimates on-board its fine position using different sensors (e.g. odometer). All ETCS levels use the same principle to estimate the current position; additionally, in ETCS Level 3 (not yet specified) satellite-supported systems are being considered for transmission of vital and non-vital information [15], although a question remains open regarding its suitability in adverse environments such as inside tunnels. For ETCS Level 2 and 3 thanks to the track-to-train bidirectional communication, trains also can transmit their position to the control centre (known as Radio Block Centre (RBC)). However, for the movement authority and maximum track allowable speed, the type transmission, and the logic behind it differ for each ETCS Level (Section 4.1).

For ETCS Level 1, the message of the movement authority and maximum track allowable speed for each block is transmitted from the lineside signals to the Lineside Electronic Unit (LEU), which sends the correct telegram to the controlled Eurobalises mounted on the tracks and finally from there to the train. If required, Euroloops or radio infill transmission could be integrated into the system to transfer the change of the lineside signal (e.g. from red to green) to the train early and avoid unnecessary braking of the train. In this level, track circuits or axle counters are required for the train integrity supervision (Fig. 3) [16]. In this level, an intermediate and affordable solution was conceived by ERTMS, called Limited Supervision. In this solution, the supervision of the driver following the braking curve is done only at the stop lineside signal and not along the tracks as it happens in the Level 1 Full Supervision (ETCS Level 1 LS).

For ETCS Level 2, instead of the lineside signals, LEU, and controlled Eurobalises, (which are optional), the vital information is transmitted by the RBC via Global System for Mobile Communications-Railway (GSM-R); and hence, a bidirectional continuous data transmission track-to-train can be achieved. The logic behind the movement authority estimation continues to be the Fixed Block Operation, and track circuits or axle clearance detection is also needed for train integrity supervision (Fig. 4) [16], [17].

For ETCS Level 3, the vital information is transmitted through radio communication. The logic behind the movement authority in this level is the Moving Block Operation [16], [18]. The latter means that the trains can travel closer to each other and the infrastructure can be optimized. For the real Moving Block Operation, track circuits or axle clearance detection is not needed for train integrity. So far, the most critical issues to complete the specifications of Level 3 are the definition of the radio communication technology (Section 4.2), and the assurance of the continuous supervision of train integrity (Section 4.3) (Fig. 5) [6], [16].

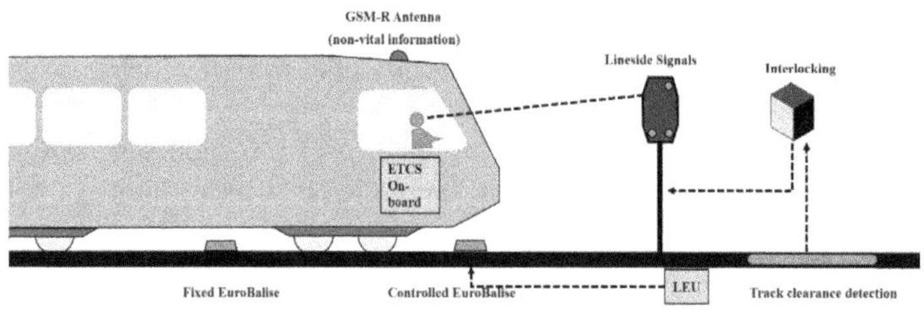

Figure 3: Equipment for ETCS – level 1 [16].

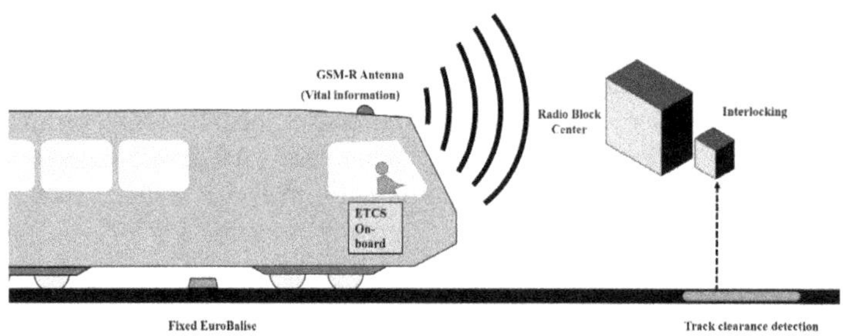

* optional: lineside signals, LEU, and controlled Eurobalises

Figure 4: Equipment for ETCS – level 2 [16].

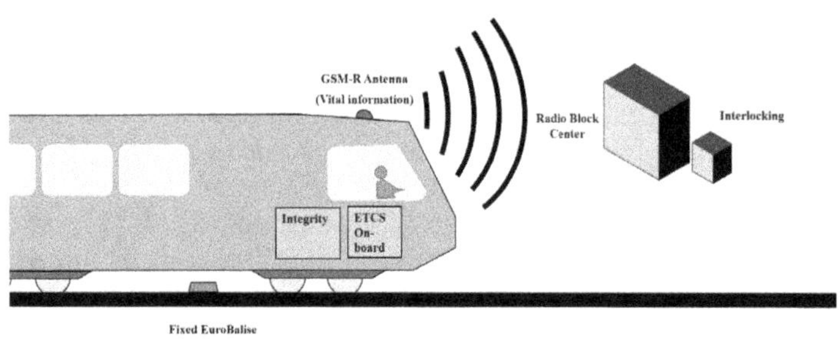

Figure 5: Equipment for ETCS – level 3 [16].

ETCS Level 3 is comprised of four different variants: overlay, hybrid, virtual, and moving block. Overlay variant integrates ETCS Level 1 or Level 2 systems plus the hybrid solution or plus the virtual block variants. The hybrid variant uses a limited amount of trackside circuits to guarantee train integrity detection in case of radio communication centre shut down. The virtual block solution uses virtual blocks, defined by the software and database, for managing track occupancy and train separation. Fig. 6 shows an example of the hybrid solution using virtual block sections. Trains report their position to the radio control centre, and it assigns them to each virtual block. In this variant, track circuits are not needed. To achieve the hybrid and virtual block solutions, only software and database updates are required. True moving block variant (Section 2) calculates the moving authority based on real-time train speed and position, and the state of other infrastructure elements; track circuits are not needed. Therefore, block size changes with the train speed, thus increasing network capacity [6]. Although, the overall capacity is restricted not only by the headway between trains, but also by the infrastructure itself (e.g. time to set up switches), and most significantly by the passenger boarding and alighting time at the station.

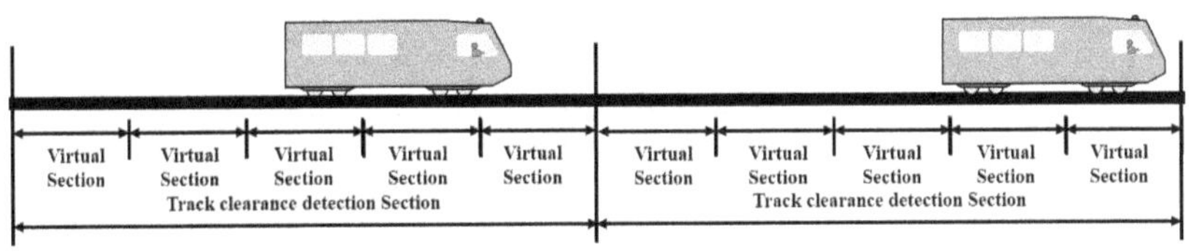

Figure 6: ETCS level 3 variant – hybrid with virtual block [6].

There are differing opinions about whether to build a new system or to migrate legacy systems to either ETCS Level 2 or Level 3. Theoretical studies have demonstrated that the headway for high-speed lines can be improved compared with ETCS Level 1 from 14% to 52% using ETCS Level 2, and up to 62% using ETCS Level 3 [16], [19], [20]. However, a particular case in Level 2 can be achieved by splitting the network into very small sections, smaller than the safety distance required for Moving Block discussed in Section 2 (Fig. 2). In this case, the network capacity with ETCS Level 2 might similar to ETCS Level 3. Some argue that because ETCS Level 2 does not need lineside signals, and technically, it could achieve a similar capacity performance as ETCS Level 3 (i.e. Hybrid ETCS Level 3), this solution could be a more profitable option. While others argue that ETCS Level 2 still needs track circuits or axle counters for integrity detection, and each additional piece of equipment requires extra and costly wayside hardware and maintenance. Either way, the investment of each level might be also dependent on the transition cost to the future railway communication technology (Section 4.2) [3].

CBTC systems have had their own evolution before becoming what they are today. Morar outlines four generations of train control systems, in which CBTC Systems is the fourth generation. Here, track circuits or axle counters are no longer needed, trains move according to Moving Block or Virtual Block principles, and the logic of estimating the moving authority depends on the train-borne, and the bidirectional and continuous train-to-wayside communication [4].

CBTC systems components include (Fig. 7): (1) integrated networks: comprised of train on-board network, train-to-trackside radio network, trackside backbone network; (2) wayside components: Zone Controller (ZC), Computer-based Interlocking (CI), and Access Points

(AP); and (3) on-board components: comprise Vehicle on-board controller-computer (VOBC), on-board ATP and ATO, Radio Communication System (RCS) [3]. CBTC systems estimate trains rough position through beacons installed on the tracks, and fine position through sensors installed on-board. AP serve specific Wi-Fi cells along the track. Because Wi-Fi has short-range coverage, AP are placed close to each other, such that their coverage areas overlap. ZC estimates the movement authority, and sends it via Wi-Fi to each AP, and finally, from there to each train (Section 4.2) [2], [3].

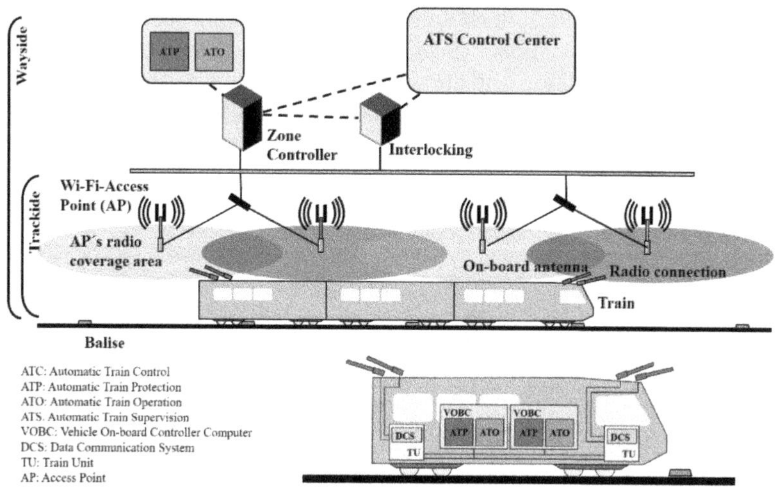

Figure 7: Equipment for CBTC systems [3].

Furthermore, CBTC is characterized by the spatial, frequency, temporal and power redundancy e.g. redundant AP coverage; as well as its remote diagnostic features that ensure its high availability. CBTC systems perform: (1) Service Affecting Diagnosis e.g. perform periodic self-testing of radio hardware, (2) Corrective Maintenance Diagnostics e.g. define the reasons of radio communication failure, and (3) Predictive Maintenance Diagnostics e.g. predict radio communication failure before it happens [21].

4.2 Radio communication

In CBTC systems, the train-to-trackside radio network is mostly Wi-Fi. Despite its susceptibility to high adjacent-channel interferences, network congestion, and its short-range capabilities at a high-frequency band, Wi-Fi radio communication in the 2,4 GHz and 5 GHz frequencies ranges is highly popular, because: (1) its unlicensed frequency band, (2) its high availability of cost-effective radio equipment, (3) ease of installation for indoor application, and (4) external-authority independent operation. In some cases, the Global Navigation Satellite System (GNSS), e.g. Global Positioning System (GPS) may work as a supplementary system of CBTC systems [3]. For heavy railway applications and outdoor applications, Wi-Fi radio communication is not seen as a viable solution due to its high latency. Heavy railway operations require low latencies and high reliability e.g. for virtual coupling [22].

For ETCS applications, the GSM-R is applied as the standard of radio communication technology since 1994. In Level 1, it might be used only for voice communication. In Level

2, for train-to-track communication of both voice and vital communication. In Level 3, it is no longer clear if GSM-R will be used, because GSM-R is expected to become financially unsustainable to maintain due to obsolescence of the underlying 2G mobile technology and increase of all future needs. The International Union of Railways (UIC) launched the Future Railway Mobile Communication System (FRMCS) project in 2012 to specify the user requirements of the new radio communication technology for heavy and light railway systems [23]. The fast evolution of the telecommunication sector (like the earlier launch of 5G for the mobile network market) has increased the complexity of FRMCS specification, and some previous efforts (i.e. LTE-R specification) were lost. Even though FRMCS technology is not clear yet, it is more likely to be 5G. It is a promising alternative that is being researched, although not yet implemented.

4.3 Train integrity

Train integrity refers to the system's ability to detect the train completeness with respect to its length at all times (i.e. to confirm the train has not lost any wagons), even when if the radio communication stops working. The importance of detecting integrity lies in the safe operation of the whole system, as well as avoiding reducing the network capacity due to the sudden stop of trains [6]. In CBTC systems, this is achieved through a periodic self-testing of radio hardware. Continuous messages of train completeness are sent from the antennas in the rear and head of the train to the wayside infrastructure by radio when the message is not sent the degraded mode (or failure mode) is activated. Under this mode, CBTC systems operate in a safe manner and minimize the operational impact by either using auxiliary wayside systems and/or strict operating rules. In ETCS Level 3 is not yet clear how to implement this principle. Specifically, for freight transport, the number and diverse wagons typologies make it harder and expensive to equip all of them with this technology. For this reason, hybrid ETCS Level 3, which still uses some track detection devices to guarantee train integrity, is seen by the industry and research groups as the most suitable option to support ETCS Level 3 in freight transport [3]. A second initiative, suitable for passenger and freight trains equipped with ETCS Level 3, it is to mount at each end of each wagon a Coupler Monitor Device (CMD). These devices report train integrity to the Head End Device (HED), located at the front of the train, and it, in turn, reports to the RBC [24]. These initiatives are still under development and are not yet included in the specification of ETCS Level 3.

4.4 Interoperability and interchangeability

Interoperability is the ability to operate a train equipped with an on-board system provided by one supplier on lines with wayside equipment provided by another supplier. Interchangeability is the ability to exchange elements of the systems with components provided by a supplier different from the original supplier [4]. Interoperability and interchangeability transcend beyond allowing the operation across-frontiers, but they are key factors in keeping system prices low for both heavy and light railway systems worldwide. Railway operators are not bound to one supplier during the system life cycle, and benefit from future developments of different suppliers (e.g. hardware and software updates). In this respect, IEEE needs to work more to describe the CBTC systems standards precisely, such that more suppliers could adhere to them [3]; meanwhile, ETCS has successfully achieved this for Level 1 and 2.

Regarding interoperability, the project Next Generation Train Control Systems (NGTC) carried out a detailed analysis of CBTC and ETCS functionalities, and identified common core and domain-specific functions for light and heavy railway applications. The aim of the project is to design the specifications of the future train control systems (the evolution of ETCS) with higher compatibility to CBTC systems. NGTC concentrates on the common core functions between ETCS and CBTC systems that correspond principally to the ATP functions [25]. The first step in identifying commonalities between systems has been accomplished, but it is necessary to take this initiative a step further, design and implement the specification compatible with both systems.

Table 3 summarizes the characteristics of CBTC and ETCS regarding the different characteristics explained above.

Table 3: ETCS and CBTC comparison. *(Source: Farooq and Soler [3], and Stanley [16].)*

Level	Level 1	Level 2	Level 3	CBTC
Application	(mostly)Heavy	(mostly)Heavy	(mostly)Heavy	(mostly)Light
Operation	Fixed Block	Fixed Block	Moving Block	Moving Block
Lineside Signals	Yes	Not needed	Not needed*	Not needed*
Interoperable and Interchangeability	Yes	Yes	Yes	No
Functionalities	ATP	ATP compatible with ATO	ATP compatible with ATO	ATP, ATO, ATS, Diagnosis
Standards by	ERTMS	ERTMS	Not yet specified	Partially by IEEE
Route Interlocking	Yes	Yes	Not yet specified	Yes
Type Data Transmission	Spot (Semicontinuous), Unidirectional (from track to train)	Continuous Bidirectional Track-to-train	Continuous Bidirectional Track-to-train	Continuous Bidirectional Track-to-train
Vital information is transmitted through	Lineside signals, and Electronic Unit, controlled Eurobalises (Euro loops and Radio Infill)	Radio communication via GSM-R	Radio communication (To be defined, probably 5G)	Radio communication via WiFi
Track clearance detection	Track circuits/ axle counter	Track circuits/ axle counter	On board (To be defined)	Continuous message

* In case of system failure, wayside signals might be required.

5 CONCLUSION

In this paper, the terminology, differences, and challenges of the CBTC and ETCS were discussed. One of the most important – and usually overlooked – differences between them is associated with their scope. ETCS is conceived as a standardized, fail-safe system for train operation, whereas CBTC systems are not fully specified and have more integrated functionalities that allow full automation such as driverless functionalities related to ATO, and train rescheduling related to ATS, to keep punctuality. ETCS does not aim to fully obtain automated operation, and ATO and ATS functionalities are not part of the scope of ETCS. However, those functionalities are developed separately as external systems compatible with ETCS for heavy and light railway applications (e.g. Traffic Management Systems). Nevertheless, the question remains if it is reasonable to integrate these functionalities, considering the fact that they are interdependent. For example, the outputs of safe operation (e.g. movement authority) are inputs of the Traffic Management System. By doing that, better service for railway users and lower operating costs for railway operators and infrastructure managers could be achieved.

Additionally, CBTC systems work under the moving block principle using mostly a continuous train-to-track radio communication via Wi-Fi, while ETCS could use either Fixed Block or Moving Block Operation. Fixed Block Operation has already been specified for ETCS Level 1, which combines lineside signals, controlled Eurobalises and track circuits or axle counters for a spot track-to-train communication, and Level 2 which combines GSM-R radio communication, and track circuits or axle counters to perform a continuous train-to-track communication. The Moving Block Operation is expected to be included in ETCS Level 3 specification that is under development. The success of the ETCS Level 3 specification process is highly dependent on the definition of the next radio communication for railway applications and how to achieve train integrity.

Regarding the advantages, CBTC was able to achieve a higher degree of automation, implement several functions including several types of self-diagnosis, and apply the Moving Block Operation faster than ETCS. Indeed, the radio communication technology used for CBTC systems (Wi-Fi) might not be optimal for heavy railway applications; its fast evolution can help as a lesson learned for ETCS Level 3; for example, to identify key factors during the planning phase that guarantee a smooth train-to-track communication in complex environments. Similarly, the ERMTS could learn from the good practices regarding moving block logic and train integrity already implemented in CBTC systems to specify ETCS Level 3. Finally, CBTC systems could learn from ETCS regarding interoperability and interchangeability. ERMTS has already accomplished these features for ETCS Level 1 and 2, and it is working towards the same goal for the specification of ETCS Level 3.

ACKNOWLEDGEMENTS

We gratefully acknowledge Maureen Kösters, Aditi Kumawat, Ijeoma Osakwe for the valued suggestions regarding this paper.

REFERENCES

[1] European Rail Traffic Management System, ERTMS Deployment Statistics. http://www.ertms.net/?page_id=58. Accessed on: 18 Jan. 2020.

[2] IEEE Standard for Communications-Based Train Control (CBTC) Performance and Functional Requirements, IEEE: Piscataway, 1999.

[3] Farooq, J. & Soler, J., Radio communication for communications-based train control (CBTC): A tutorial and survey. *IEEE Communications Surveys and Tutorials*, **19**, pp. 1377–1402, 2017.

[4] Morar, S., Evolution of communication based train control worldwide. *IET Professional Development Course on Railway Signalling and Control Systems (RSCS 2012) Proceedings*, London, UK, 2012.

[5] Pachl, J., *Systemtechnik des Schienenverkehrs*, Springer Fachmedien Wiesbaden: Wiesbaden, 2018.

[6] Furness, N. et al., *ERTMS Level 3-The Game Changer*, IRSE NEWS: 2–9.

[7] Schneider, L., Zugbeeinflussungssysteme für Stadtbahnen im Vergleich. *Der Eisenbahn Ingenieur (EI)*, pp. 31–34, 2019.

[8] Hansen, I.A., Pachl, J. & Albrecht, T., *Railway Timetabling & Operations: Analysis, Modelling, Optimisation, Simulation, Performance Evaluation*, 2nd ed., Eurailpress: Hamburg, 2014.

[9] Untersuchung zur Einführung von ETCS im Kernnetz der S-Bahn Stuttgart: Abschlussbericht (17FEI27440), January 2019.
[10] Vieira, P. et al., *Deploying a CBTC System in a Heavy-Haul Railroad*, 2008.
[11] Arpaci, M.E., *Case Study: Coexistence of CBTC and ETCS on Crossrail Project in London*, 2016.
[12] UITP, *Press Kit Metro Automation Facts, Figures and Trends*, Belgium.
[13] UIC, ERTMS: International Union of Railways. https://uic.org/rail-system/ertms/.
[14] Shift2Rail, Innovation Programme 2. https://shift2rail.org/research-development/key-documents/. Accessed on: 4 Feb. 2020.
[15] Zweigel, R., Gehrt, J.-J. & Abel, D., *Scientific Railway Signalling Symposium 2018*, Darmstadt, Germany, 2018.
[16] Stanley, P., *ETCS for Engineers*, DVV Media Group Eurail Press: Hamburg, 2011.
[17] Railway Signalling. The ERTMS/ETCS signalling system, 2013–2014.
[18] Damy, S., A novel GNSS-based positioning system to support railway operations, Doctoral thesis, Imperial College London, 2017.
[19] VÄYLÄ. ERTMS/ETCS Level 2 Capacity Benefits on the City Lines of the Helsinki Region, 2019.
[20] Winter, P., *Compendium on ERTMS, European Rail Traffic Management System*, DVV Media Group Eurailpress: Hamburg, 2009.
[21] Ali, N., *7 Key CBTC Functions Transit Operator Must Understand*, CBTC Solutions, 2017.
[22] Briso-Rodríguez, C., García-Loygorri, J.M. & Zhang, L., Transmission-based signaling systems. *Modern Railway Engineering*, ed. A. Hessami, InTech: Rijeka, 2018.
[23] SYSTRA, Study on the architecture of on-board radio communication, 2018.
[24] Martin, U., Körner, M. & Beck, R., Funktionale Sicherheitsanforderungen an eine ETCS L3 kompatible Mittelpufferkupplung. *Eurail*, **107**, pp. 15–19, 2015.
[25] Gurník, P., Next generation train control (NGTC): More effective railways through the convergence of main-line and urban train control systems. *Transportation Research Procedia*, **14**, pp. 1855–1864, 2016.

SECTION 2
SIGNALLING AND TRAIN CONTROL SYSTEMS

DEVELOPMENT OF THE SIGNAL ARRANGEMENT EVALUATION SYSTEM

KAZUMASA KUMAZAWA, TAKAYASU KITANO, TAKASHI SAKAGUCHI, SHUNICHI TANAKA & SATOSHI KATO
Signalling and Transport Information Technology Division, Railway Technical Research Institute, Japan

ABSTRACT
In order to construct new lines or to change track layouts in stations, it is necessary to determine the position of each signal, called an examination of the signal arrangement, in consideration of train operation and so on. To obtain the signal arrangement that satisfies the desired train operation, it is essential to consider headways. Furthermore, in the automatic block section, the signal aspect transition between adjacent signals must be determined, taking account of the position of each signal and so on. Therefore, to determine a signal arrangement requires a lot of time and a person who has the knowledge of both signalling system and train operation. To solve these problems, we developed a signal arrangement evaluation system. The system has two main functions. One function is the creation of the signal aspect transitions. In this function, this system calculates a braking distance corresponding to a signal aspect transition and checks whether the braking distance is lower than a distance between two signals or not. The other function is the creation of a headway diagram. In this function, the system calculates headways and checks whether the headways of all signals satisfy the desired train operation or not. In this paper, we explain these calculation logics and the structure of the developed system. In addition, we introduce a result of evaluating a signal arrangement in a model line by using the developed system.
Keywords: signalling system, train performance curve, train headway, automatic block system, signal aspect, examination of new railway lines, track layouts improvement.

1 INTRODUCTION

When planning new line construction and track layouts improvement, running time between stations and time interval between trains are important. Running time can be calculated from the characteristics of the rolling stock and infrastructure conditions such as gradients of a track, speed limits and the position of station. In addition, a signal arrangement is required to calculate time interval between trains. The task of deciding signal positions based on a target time interval is called an examination of a signal arrangement. The examination of the signal arrangement consists of four tasks. First, a signal aspect transitions that represents transitions between signal aspects is created based on the signal arrangement plan. Secondly, the train headway (hereinafter called the headway) that indicates time interval between trains is calculated based on the signal aspect transitions. Thirdly, headways at each signal are conformed whether it satisfies the target value or not. Finally, the above tasks are repeated until the target value of the headway is satisfied. Furthermore, in creating the signal aspect transitions, it is necessary to perform deceleration distance calculation for all transitions between signal aspects, and repeat the adjustment of the signal arrangement plan based on the result. The examination of the signal arrangement requires a lot of time and a skilled person who has the knowledge of signalling system and train operation. For that reason, system assistance is required to carry out the examination of the signal arrangement.

In order to solve these problems, we developed a signal arrangement evaluation system. First, we conducted hearing investigations with the person in charge of the examination of the signal arrangement to understand the overview of it. Based on the knowledge we have obtained from these investigations, we developed a system that can evaluate the signal

WIT Transactions on The Built Environment, Vol 199, © 2020 WIT Press
www.witpress.com, ISSN 1743-3509 (on-line)
doi:10.2495/CR200031

transitions and the headway at each signal. Then, we applied the developed system to a model line and confirmed its usefulness.

2 OVERVIEW OF AN EXAMINATION OF THE SIGNAL ARRANGEMENT

As shown in Fig. 1, the general procedure for an examination of the signal arrangement on several Japanese suburban lines is as follows: Step 1: creating a signal arrangement plan according to the examination conditions. Step 2: creating signal aspect transitions based on the signal arrangement plan. Step 3: calculation headways at each signals based on the signal aspect transitions and train performance curves, and Step 4: evaluating the signal arrangement. In the examination of the signal arrangement, the above Steps 1 to 4 are repeated to determine signal arrangement in which t the value of headways at all signals satisfies that of the target headway.

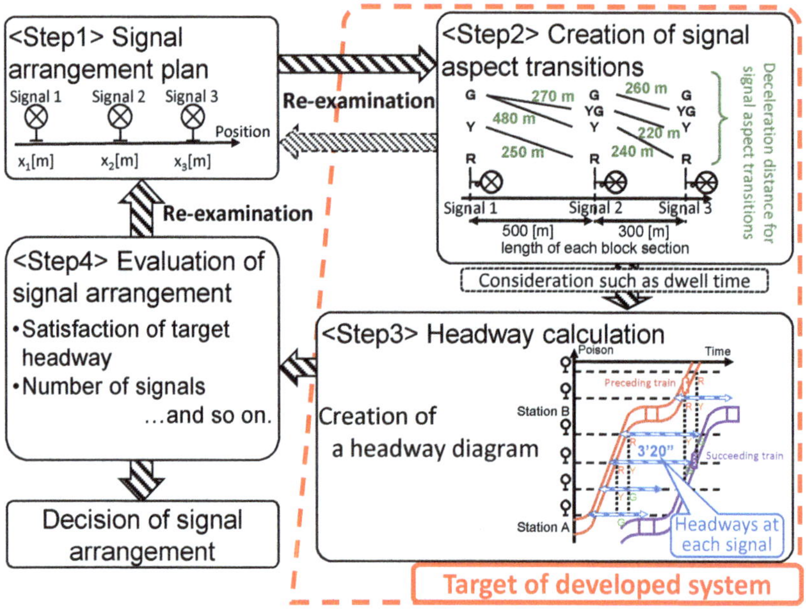

Figure 1: Overview of an examination of the signal arrangement.

In this paper, we focus on the creation of the signal aspect transitions (Step 2) and the calculation of headways (Step 3), which are important and time-consuming tasks in the examination of the signal arrangement. In order to develop the signal arrangement evaluation system, it is necessary to conduct hearing investigations with the person in charge of the examination of the signal arrangement and obtain their knowledge. This section summarizes the knowledge obtained from the investigation on Step 2 and Step 3.

2.1 Target signalling system

We focus on the Japanese automatic block system. In the automatic block section, main signals divide the line into several sections, and each section is set as a block [1]. Interval between trains is controlled by the signal aspect of the main signal. The signal aspect has the

Table 1: An example of directed speed by a signal [2].

Signal indication	Signal aspect	Directed speed
Proceed signal	Green (G)	Line speed
Reduced speed signal	Yellow and Green (YG)	65 km/h
Caution signal	Yellow (Y)	45 km/h
Restricted speed signal	Two Yellows (YY)	25 km/h
Stop signal	Red (R)	0 km/h

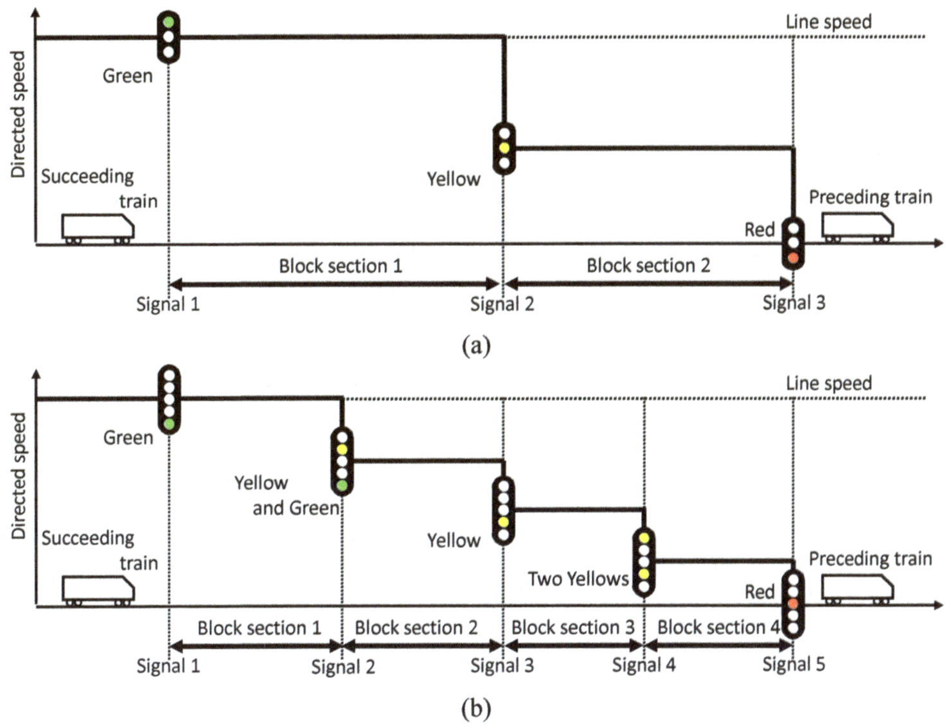

Figure 2: An example of signal aspect transitions. (a) Signal aspect transitions on general lines; and (b) Signal aspect transitions on high capacity lines.

directed speed (directed speed by a signal) shown in Table 1. The signal aspect of the main signal changes depending on the existence of the preceding train and the signal aspect transitions, and the driver operates the train based on this signal aspect (Fig. 2).

2.2 Creation of a signal aspect transitions

The signal aspect transitions shows transitions between each signal aspects [3]. The signal aspect changes based on the signal aspect transition system shown in Fig. 3. The transition between each signal aspect is basically composed of a proceed signal (G), a caution signal (Y), and a stop signal (R) shown by the solid lines in Fig. 3. However, in order to shorten the

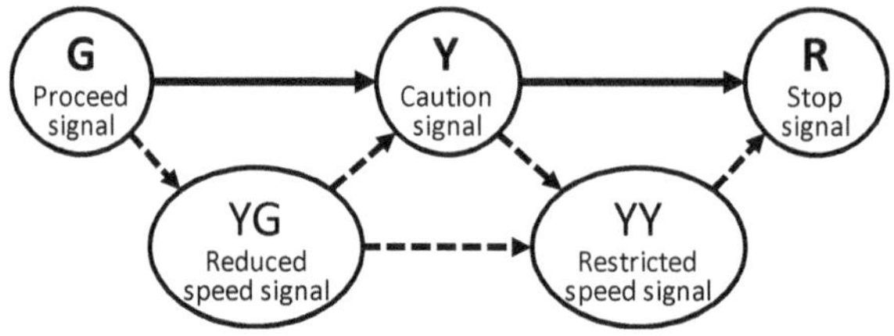

Figure 3: Signal aspect transition system.

headway or to compensate a shortage of deceleration distance, a transition (broken line) of a signal aspect routing through a reduced speed signal (YG) or a restricted speed signal (YY) may be set. In this section, we describe the creation of the signal aspect transitions corresponding to Step 2 of Fig. 1.

2.2.1 Calculation of running speed at each block section

In creating the signal aspect transitions, it is necessary to calculate the deceleration distance at each block section (see Section 2.2.2). The deceleration distance can be calculated based on the directed speed set for each signal (see Table 1), with the directed speed as the deceleration initial speed and deceleration final speed. For example, the deceleration distance from Y to R can be calculated as the deceleration distance from 45 km/h, which is the directed speed of Y, to 0 km/h, which is the directed speed of R. However, since G is directed to drive within a range not exceeding the line speed, the deceleration initial speed can't be determined. Therefore, in many cases, a deceleration initial speed in the case of G is calculated by a speed profile (a part of a train performance curve) created to examine signal aspect. The deceleration initial speed may be calculated by setting the line speed in the target block section (length of block section) as the deceleration initial speed or by setting the speed at the signal sighting position as the deceleration initial speed. In addition, since the line speed and the stopping station are different depending on the train type, it is necessary to create the speed profile for each train type and to calculate the deceleration initial speed at G based on each profile.

2.2.2 Calculation of the deceleration distance for each signal aspect transition

In cases where the train decelerates according to the signal aspect transition setting in each block section, the train must be able to decelerate to the speed directed by the signal aspect within the length of block section or within the signal sighting distance. Therefore, as shown in Section 2.2.1, it is necessary to calculate the deceleration distance corresponding to the signal aspect transition. The visibility of signals is not taken into consideration at the stage of examining the signal arrangement, therefore, the length of the block section is often used as a criterion for evaluation.

Eqn (1) is applied to the calculation of the deceleration distance S m for each signal aspect transition. For the gradient of eqn (1), either the minimum gradient value or the average gradient value within each block section is applied. If necessary, the idle running time until the start of deceleration may be added to eqn (1)

$$S = \frac{V_0^2 - V_1^2}{7.2 \times \left(\beta + \frac{\theta}{K}\right)}, \tag{1}$$

where

V_0 is the deceleration initial speed expressed in km/h,
V_1 is the deceleration finial speed expressed in km/h,
β is the braking deceleration rate expressed in km/h/s,
θ is the gradient in ‰, * Negative value if the gradient is downslope.
K is the inertia coefficient. * 31 for electric multiple-unit trains/30 for others.

In the examination of the signal arrangement, the braking deceleration rate of eqn (1) and the deceleration initial speed (V_0) for G are different depending on the train characteristic. For that reason, the deceleration distance is calculated for multiple train types and the maximum value is used. Therefore, in this examination, it is necessary to select multiple train types that may increase the deceleration distance and calculate the maximum value of the deceleration distance.

2.2.3 Setting of the signal aspect transition

Based on the length of block and the deceleration distance calculated for each signal transition, the signal transition at each block section is set [3]. For example, as shown in Fig. 4, the deceleration distance from G to Y is 520 m and the deceleration distance from G to YG is 280 m in the block section from signal 1 to signal 2. In this case, since the length of block section is 520 m or more, the signal aspect transition from G to Y can be set for signal 1 and signal 2. The signal aspect transition of each signal affects headways. From the viewpoint of headways, it is desirable that the signal aspect of each signal returns to G as soon as possible after the preceding train passed the section. Therefore, considering the signal aspect transition system shown in Fig. 3, it is desirable that the signal aspect transition consists of only G, Y and R, without YG or YY.

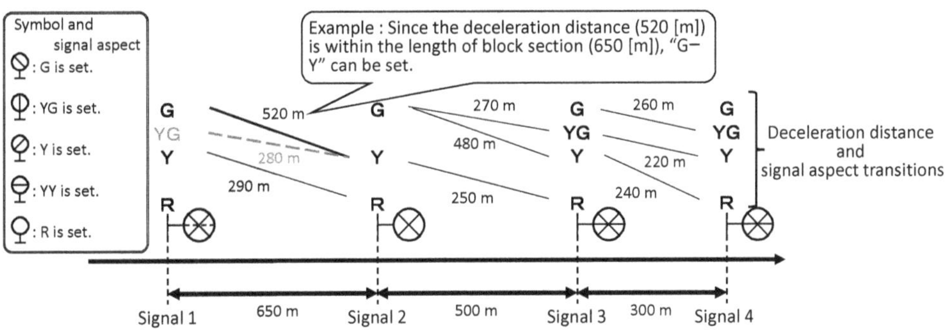

Figure 4: Setting of the signal aspect transition.

2.3 Calculation of headways

Two trains running successively must secure a sufficient time interval for safety. This time interval is called the headway [4] and is calculated from the train performance curves of two trains, the signal aspect transitions, and the dwell time. The train performance curve applied at this calculation takes the driving operation into account. Headways are calculated at each signal. As an example, we use Fig. 5 to explain how to calculate the headway at signal 1. The time from when a preceding train passes through signal 1 until when the train passes through the position (signal 3) necessary for signal 1 to show G again is defined as T_1. Furthermore,

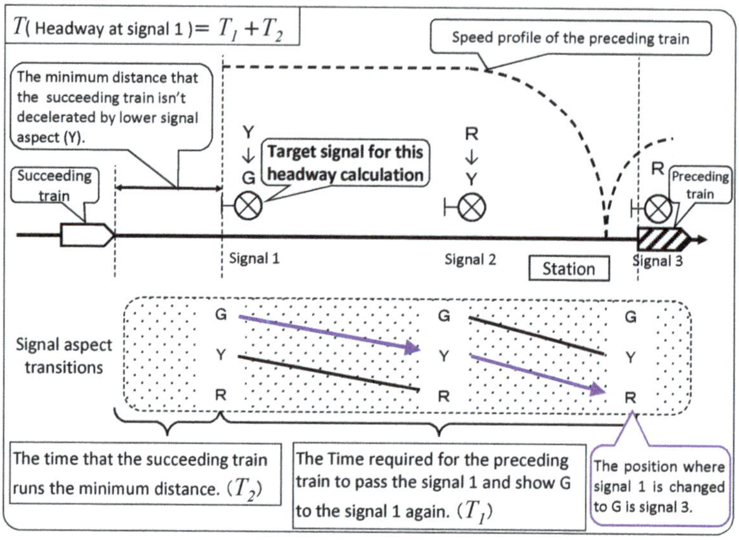

Figure 5: Calculation of a headway [5].

a certain time that a succeeding train runs the minimum distance which the train isn't decelerated by lower signal aspect (Y in Fig. 5) is defined as T_2. Basically, the sum of these T_1 and T_2 is the headway T at signal 1. In some cases, it is necessary for this calculation to consider the delay time caused by the driver's braking operation, the signalling system, and so on.

Headways are expressed by using a headway diagram shown in Fig. 6. The headway diagram is created based on the train movement obtained from train performance curves (time curve) and the signal aspect transitions. The vertical axis of the headway diagram represents position, and the horizontal axis represents time. Furthermore, headways calculated at each signal are displayed on the headway diagram. In the examination of the signal arrangement, the headways at all signals must be smaller than the target headway.

3 SIGNAL ARRANGEMENT EVALUATION SYSTEM

The examination of the signal aspect transitions and headways based on the signal arrangement plan (corresponding to Step 2 and Step 3 in Fig. 1) requires a lot of time in the examination of the signal arrangement. In order to support this evaluation (the part surrounded by the red dashed line in Fig. 1), we developed a signal arrangement evaluation system based on the knowledge described in Section 2.

3.1 Key points of evaluating the signal arrangement

The results from the investigation shown in Section 2, it was found out that there are two points that must be satisfied in order to decide the signal arrangement.

a) Conditions to be satisfied in the signal aspect transitions
   As shown in 2.2, in cases where the train decelerates according to the signal aspect transition setting in each block section, the train is required to decelerate to the speed directed by the signal aspect within the length of the block section. Furthermore, in the calculation of the deceleration distance, it is necessary to select multiple train types that

may increase the deceleration distance and calculate the maximum value of the deceleration distance.

b) Conditions to be satisfied in headways
Satisfaction of the target headway is very important in the examination of the signal arrangement. Therefore, headways at all signals must be smaller than the target headway.

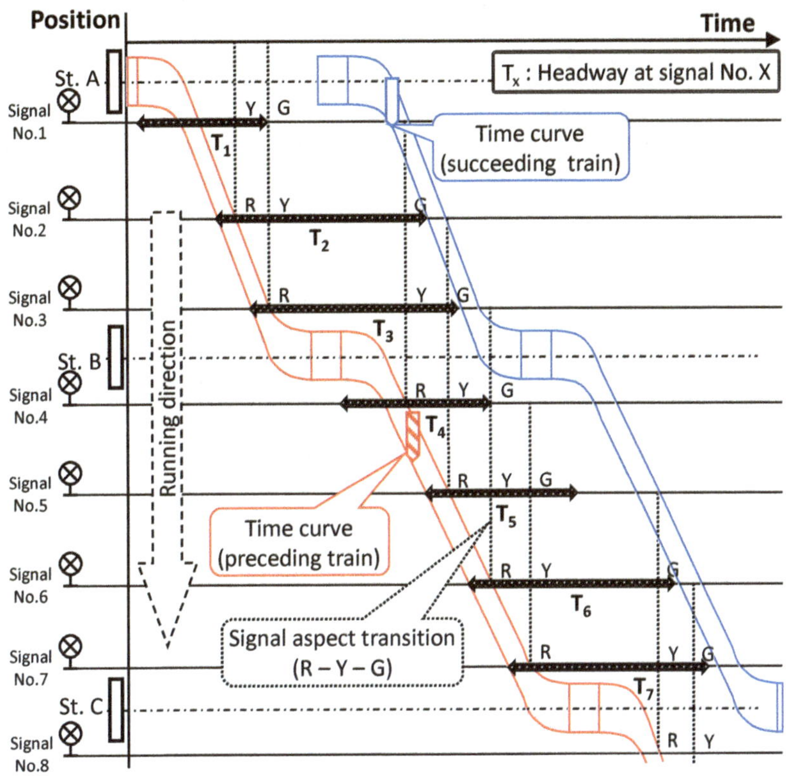

Figure 6: Headway diagram.

In the examination of the signal arrangement, after satisfying the above conditions, the signal arrangement is determined in consideration of the number of traffic lights, the margin for the target headway, and so on. Therefore, the signal arrangement evaluation system to be developed is required to evaluate the signal aspect transitions and the headway. The functions corresponding to each evaluation are summarized below.

a) Evaluation function of the signal aspect transitions

- Display the signal aspect transition.
- Calculate deceleration distance for each signal aspect transition.
  - Create speed profiles for multiple train types.
  - Calculate the deceleration distance for multiple vehicle types and output the maximum value.
- Display the length of block section.

- Confirm whether the deceleration distance for each signal aspect transition is no more than the length of each block section or not.

b) Evaluation function of the headway

- Calculate headways at each signal.
- Confirm that headways at each signal satisfy the target headway.

3.2 Development of the signal arrangement evaluation system

As shown in Section 3.1, the signal arrangement evaluation system requires the function of evaluating the signal aspect transitions and the function of evaluating the headway. Therefore, as shown in Fig. 7, we developed the system that integrates the evaluation function of the signal aspect transition and the headway calculation function.

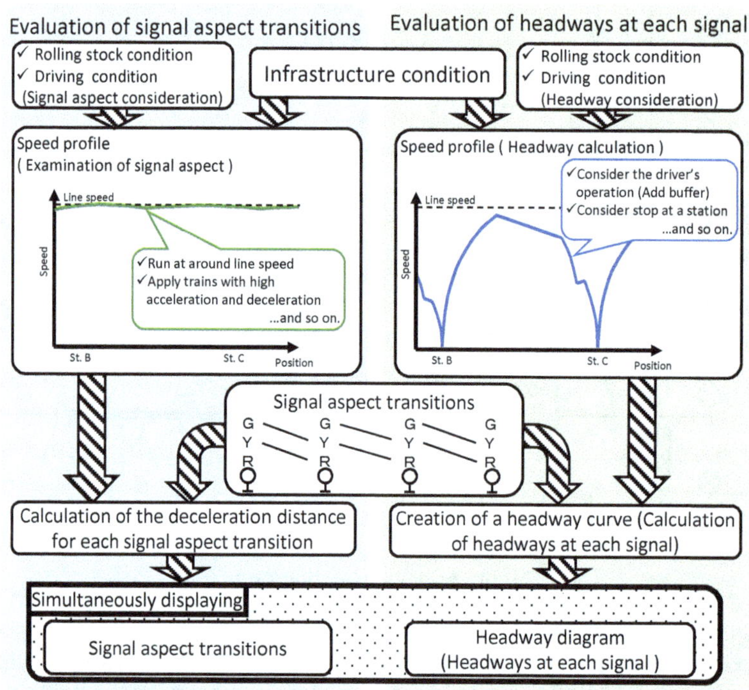

Figure 7: Configuration of the signal arrangement evaluation system.

We have developed a running time and headway calculation system [6], [7] that has a function of calculating headways. The function can calculate headways at each signal. Furthermore, this function has equivalence to the evaluation function of the headway shown in Section 3.1 because headways at each signal can be displayed on the headway diagram. Therefore, we decided to develop the signal arrangement evaluation system based on the system. On the other hand, we developed new functions for the evaluation of the signal aspect transition based on the contents summarized in Section 3.1.

3.2.1 Creation of a speed profile for examination of signal aspects

In order to evaluate the signal aspect transitions, it is necessary to calculate running speed at each block section shown in Section 2.2.1. In order to calculate running speed, the speed profile for examination of signal aspects is required. As a result of the investigation in Section 2, it is confirmed that this speed profile is different from that applied when calculating headways. The conditions for creating each speed curve are summarized as follows.

a) Speed profile for examination of signal aspects
   In the evaluation of the signal aspect transitions, the speed profile is required, which is created under the condition that the deceleration distance is maximum. The following conditions must be satisfied to create the speed profile that maximizes the deceleration distance.

   - The buffer time caused by the driving operation is not considered.
   - At each station, trains mainly pass through.
   - Applied trains are selected in consideration of the following.
     - Trains with low deceleration rate and long idling time before braking is activated
       ex) Freight trains and electric multiple-unit trains driven by locomotives
     - Trains with high deceleration initial speed
       ex) Trains mainly pass and have a high maximum speed or trains with a high acceleration performance

b) Speed profile for the headway calculation
   In the headway calculation, the speed profile used for creating the transport planning is applied. The creation conditions are as shown below.

   - The buffer time caused by the driving operation is considered.
   - Stops and passages at each station are set based on the transport planning.
   - Trains with low rolling stock performance are applied to avoid running time shortage.

3.2.2 Calculation of the deceleration distance

In order to evaluate the signal aspect transitions, it is necessary to calculate the deceleration distance corresponding to a signal aspect transition setting in each block section based on the speed profile for examination of signal aspects. The method of calculating the deceleration distance S m is as shown in Section 2.2.2, but we considered the following points as more realistic conditions.

- Considering idling time until braking is activated, the eqn (2) including idling time is applied to the deceleration distance calculation. The average gradient is used as the gradient value applied to the eqn (2)

$$S = \frac{V_0^2 - V_1^2}{7.2 \times \left(\beta + {}^{\theta}/_{K}\right)} + \frac{V_0}{3.6} \times t, \tag{2}$$

where
V_0 is the initial deceleration speed expressed in km/h,
V_1 is the final deceleration speed expressed in km/h,
B is the braking deceleration rate expressed in km/h/s,
θ is the gradient expressed in ‰, * Negative value if the gradient is downslope.

K is the inertia coefficient * 31 for electric multiple-unit trains/30 for others
t is idling time expressed in second.

- It is necessary to consider the deceleration distance of trains with different train performance. Therefore, it enables the calculation of two speed profiles for examination of signal aspects under different train conditions.

3.2.3 Display of signal aspect transitions and headway diagram

In the examination of the signal arrangement, it is necessary for the person in charge to confirm the signal aspect transitions and the headway diagram at the same time. Therefore, it is necessary for the system to be developed to simultaneously display the signal aspect transitions and the headway diagram. As described above, this system utilizes the function of calculating headways of the running time and headway calculation system. Therefore, in this system, it is possible for the signal aspect transitions that is the basis of the headway calculation to be displayed together with the headway diagram. Furthermore, the length of block sections and the deceleration distance for each signal aspect transition are displayed based on the deceleration distance calculation results shown in Section 3.2.2. If the deceleration distance is longer than the length of block section, the difference is displayed as the deceleration distance shortage. As a result, it is possible for the person in charge to confirm the satisfaction of the signal aspect transitions and the headways at all signals, which are indispensable for the examination of the signal arrangement.

4 EVALUATION OF THE DEVELOPED SYSTEM

In order to evaluate the developed system, we will carry out an examination of the signal arrangement on the model line.

4.1 Overview of the model line

The line conditions that require a lot of time for the examination of the signal arrangement are lines with short headway and lines where run electric multiple-unit trains and freight trains with different deceleration distances. Considering these conditions, the line shown in Fig. 8 is used as the model line.

- Ground equipment conditions: The line length is approximately 15 km, and there are 20 signals. There are many downward slopes toward the end point.
- Headway calculation conditions: Headways at all signals from station E to station A is calculated based on the train performance curve shown in Fig. 8. In addition, the target headway for this section is assumed to be 3 minutes.
- Evaluation conditions of the signal aspect transitions: The deceleration distance is calculated for the following two types with different train performance.
    - Train type 1: This train stops at station E and station A. This train is a freight train with a maximum speed of 75 km/h, the braking deceleration rate of 1.7 km/h/s, and idling time of 8 seconds.
    - Train type 2: This train stops at station E and station A stop. This train is an electric multiple-unit train with a maximum speed of 100 km/h, the braking deceleration rate of 2.5 km/h/s, and idling time of a second.

4.2 The results of evaluation

As shown in Fig. 9, we confirmed that the satisfaction of the signal aspect transitions and headways at each signal, which are indispensable for the examination of the signal arrangement, can be evaluated at the same time. Moreover, the time required to calculate the evaluation result is several seconds.

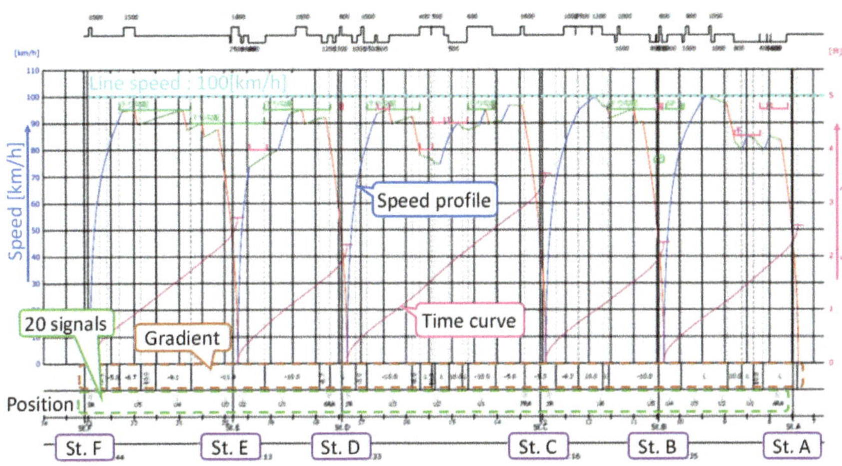

Figure 8: Train performance curve of the model line.

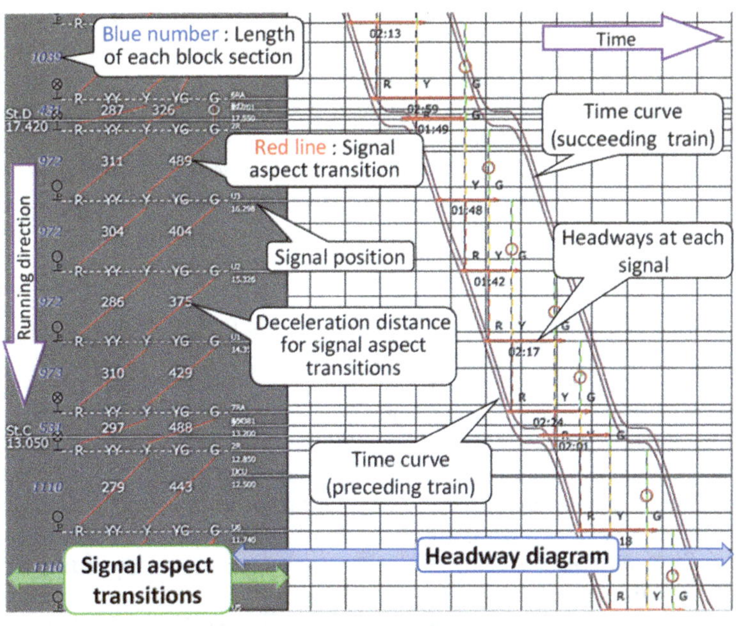

Figure 9: Evaluation results from the developed system.

a) The evaluation result of conditions to be satisfied in the signal aspect transitions
From the signal aspect transitions display section of Fig. 8, whether the deceleration distance is sufficient can be confirmed by displaying the length of block sections and the deceleration distance for each signal aspect transition. For the deceleration distance, the larger value of the deceleration distance calculation results of two train models is applied.
b) The evaluation result of conditions to be satisfied in headways
Satisfaction of the target operation time interval can be confirmed from headways at all signals displayed on the headway diagram.

It takes a lot of time for the person in charge to perform the same work without developed system. This system is considered effective in supporting the examination of the signal arrangement, because the evaluation result of a signal arrangement can be obtained within a few seconds.

5 CONCLUSION AND FUTURE WORKS

When planning new line construction and track layouts improvement, it is necessary to determine the position of each signal, called an examination of the signal arrangement, in consideration of train operation and construction costs. The examination of the signal arrangement requires a lot of time and a person who has the knowledge of signalling system and train operation. For that reason, system assistance is required to carry out the examination of the signal arrangement.

Therefore, in order to solve these problems, we have developed a signal arrangement evaluation system. First, we have conducted hearing investigations with the person in charge of the examination of the signal arrangement to understand the overview of its. Based on the knowledge we have obtained from these investigations, we have developed the system that can evaluate the signal transitions and headways at each signal. In addition, using the developed system, we have evaluated a signal arrangement of a model line, which requires consideration of two train types with different deceleration performance and has 20 signals. As a result, we have confirmed that it is possible to evaluate the signal arrangement in a few seconds (Fig. 9). We conclude that the developed system can shorten the time required for the examination of the signal arrangement.

The developed system introduced in this paper is planned to be developed further for practical use. In addition, as future works, we plan to construct a method that proposes a signal arrangement based on the target headway, instead of evaluating the signal arrangement.

REFERENCES

[1] Theeg, G. & Vlasenko, S. (eds), *Railway Signalling and Interlocking*, 2nd ed., PMC Media House: Hamburg, 2018.
[2] Ugajin, H., Shiroto, H., Fujinami, K. & Omino, K., Two greens: A new signal aspect for high speed train on wayside signaling. *Quarterly Report of RTRI*, **40**(1), pp. 13–17, 1999.
[3] Yoshitake, I., *Tetsudo no Untenhoan* (*Train Protection Facilities on Railway*), Japan Train Operation Association: Tokyo, 2006. (In Japanese.)
[4] Kuramoto, K., Kotani, T., Hirano J. & Akatsu N., *Untenriron – Saikaiteiban* (*Theory for Efficient Train Operation – Revised Edition*), Japan Train Operation Association: Tokyo, 2010. (In Japanese.)
[5] Kumazawa, K. & Takeuchi, Y., Construction of a comparative evaluation method of the driving operation form multiple viewpoints. *RTRI Report*, **31**(10), pp. 47–52, 2017. (In Japanese.)

[6] Hirano, J., Tomii, N., Yamashita, O. & Kimura, Y., Development a train performance computation system on an engineering workstation. *Computer in Railway III*, pp. 127–136, 1992.
[7] Railway Technical Research Institute, Signalling and Transport Information Technology Division, Running time and headway calculation system – SPEEDY. www.rtri.or.jp/rd/division/rd47/rd4740/rd47400101.html. Accessed on: 17 Apr. 2020. (In Japanese.)

ENHANCEMENT OF DETECTION FUNCTIONS OF A 3D-LASER-RADAR-TYPE OBSTACLE DETECTION SYSTEM AT A LEVEL CROSSING

YUKI YAO[1], SATORU KITAMURA[1], REIJI ISHIMA[1], TORU MURAKAMI[1],
TOSHIHIRO HAYASHI[2], TAKESHI TAKAHASHI[2], YOSHIYA KOWASHI[2] & YOSHIKI SEGAWA[3]
[1]Research and Development Center of JR East Group, East Japan Railway Company, Japan
[2]Security Project Dept. Intelligent Information Management Headquarters, IHI Corporation, Japan
[3]Platform Center Technology & Intelligence Integration, IHI Corporation, Japan

ABSTRACT

East Japan Railway Company (JR-East) widely uses obstacle detection systems at level crossings in order to prevent collision accidents between running trains and passing automobiles. One type is the Three Dimension Laser Radar (3DLR), which uses LiDAR technology and has been installed into more than 800 level crossings in JR-East. As the current specification, 3DLR has a lower threshold of detection height. 3DLR do not detect obstacles from the ground surface to the lower threshold height in order to avoid wrong detection, which comes from surface irregularity such as grass-growing. However, it may also happen that a falling down person at level crossing cannot be detected due to the lower threshold. With an increasing demand for safety at level crossing, even a falling down person needs to be detected. Therefore, we have enhanced the detection functions of 3DLR by implementing two algorithms for the lower threshold. One is to follow up each object even below the threshold height so that a falling down person can be detected. The other is to vary the threshold height along with the shape of the ground surface of level crossing so that every object with more-than-threshold-height size can be detected. We have done the field test for about two years at a level crossing in Nambu Line and confirmed that the enhanced detection function with two algorithms work very well with the ability of detection of a falling-down object. JR-East has plan to start using the enhanced detection function of 3DLR in 2020.

Keywords: 3DLR, LiDAR, obstacle detection, level crossing.

1 INTRODUCTION

Prevention of railroad level crossing accidents is a major challenge for railway operators. In Japan, due to various measures to prevent level crossing accidents, the number of accidents has decreased to about 26% compared to 1988 [1], [2]. However, over the past few years, the number of fatalities in level crossing accidents has remained unchanged at around 100. In these, pedestrians account for the majority of fatalities (73%), so ensuring the safety of pedestrians at the level crossing is important. In addition, 40% of the fatal pedestrians are elderly people aged 65 and over, and there has been a high level of public concern [3].

At the same time, railway operators are introducing obstacle detection devices to prevent accidents at level crossings. Several types of obstacle detection devices have been introduced, such as lasers, loop coils [4], and millimeter-wave devices [5]. However, conventional obstacle detection devices are designed to prevent collisions between vehicles and trains, not to detect pedestrians.

In this study, we developed an enhanced obstacle detection system to improve the detection function of pedestrians at the level crossing.

2 OBSTACLE DETECTION DEVICES

In JR-East, there are three types of obstacle detection devices for level crossings as shown in Table 1 and Fig. 1. Laser type, Loop coil type, and Three Dimension Laser Radar (3DLR)

WIT Transactions on The Built Environment, Vol 199, © 2020 WIT Press
www.witpress.com, ISSN 1743-3509 (on-line)
doi:10.2495/CR200041

type. JR-East mainly install 3DLR type because it can detect more obstacles and therefore it is safer than Laser type and Loop coil type.

2.1 Principle of operation of 3DLR type obstacle detection system

Fig. 2 shows the principles of measurement of this 3DLR. 3DLR is one of the LiDAR system. 3DLR device emits a laser pulse to an object, and measures the time that it takes for reflected laser to return to the radar (time-of-flight method) to acquire a distance to that object [6]. Fig. 3 shows the object detection method used by the 3DLR. A laser pulse is emitted in a way that scans the entire area of a level crossing in the horizontal and vertical directions, and 3D coordinate values of each point are measured based on the laser reflected and returning to the 3DLR. The coordinates higher than the road surface are then extracted based on coordinate values of each point, and points distributed in close proximity to each other are recognized as "one group" [7]. Data on these groups of points are processed to calculate the positions and sizes of objects. By repeatedly executing this measurement process and performing related signal processing tasks, it becomes possible to recognize objects and identify the types of objects. Furthermore, speeds and moving directions are calculated based on the amount of temporal change in their positions.

Table 1: Types of crossing obstacle detection.

	Laser type	Loop coil type	3DLR type (3D laser radar)
Detection target	Ordinary vehicles	Larger vehicles than trucks, buses, etc.	Larger objects than 1 m × 1 m × 1 m cubic
Advantages	Lower implementation costs than other methods	Little exposed equipment	Easier to install than other methods
Disadvantages	Have some gaps which cannot be detected	Cannot detect small objects and pedestrians	Cannot detect objects in a bad weather such as heavy rain or snowfall. Higher equipment cost
Number of installations. (In JR-East)	1,460	570	800

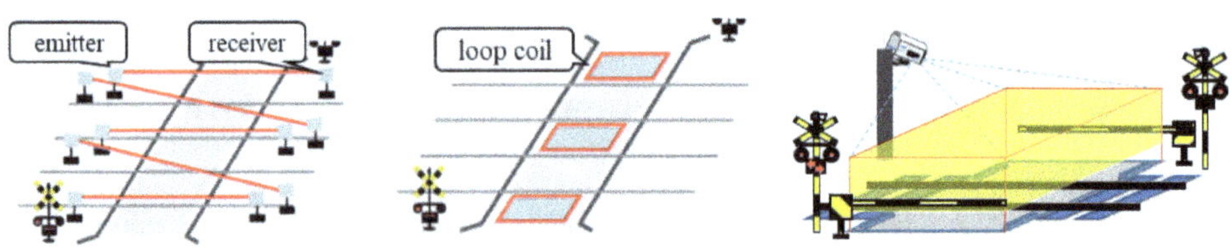

(a) (b) (c)

Figure 1: Overview of obstacle detection system at level crossing. (a) Laser type; (b) Loop coil type; and (c) 3DLR type.

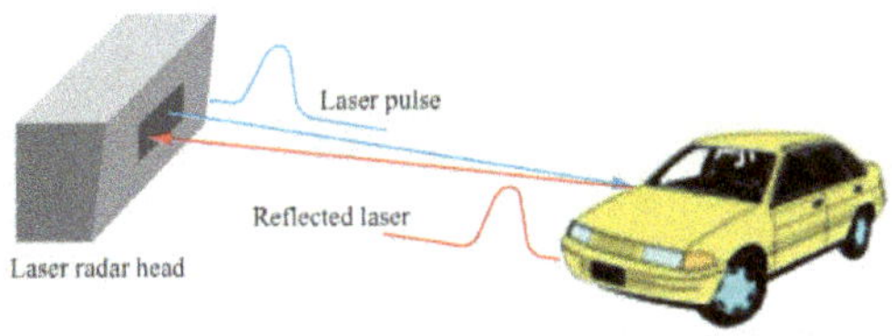

Figure 2: Measurement principle of 3DLR [6].

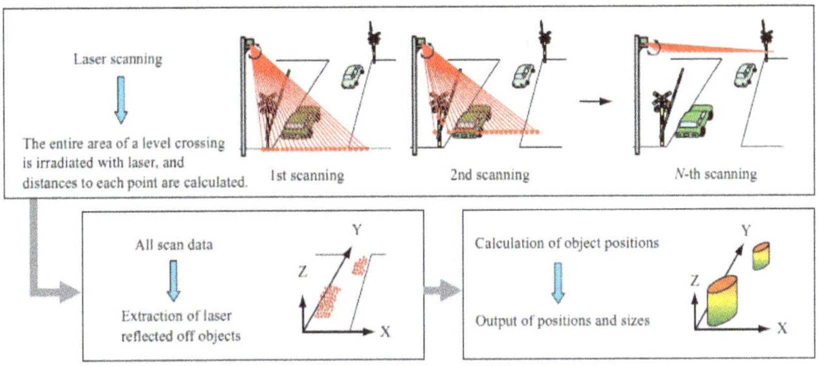

Figure 3: Object detection method [6].

3 PROBLEMS

3DLR can detect more obstacles and therefore it is safer than conventional Laser type and Loop coil type, but there are still some cases which are difficult to detect.

As shown in Fig. 4, we consider the following three cases:

- Case 1: Falling down while crossing. There have been cases where pedestrians have fallen down over due to unevenness near the rails in the level crossing as shown in Fig. 4(a). A falling down person, especially of an elderly person or in a wheelchair, is likely to lead directly to an accident because he/she takes much time to get up. Therefore, detecting a person who is about to fall down is considered to be important for accident prevention. With current 3DLR, detection after a fall down is difficult because the height of the fallen person is lower than the detection area.
- Case 2: Entering the track area. In some cases, moving from within the level crossing into the direction of the railroad tracks can lead to accidents because it is out of the detection area of the obstacle detection device as shown in Fig. 4(b). As it is very dangerous to go outside the railroad tracks, especially during warning at a level crossing, it needs to detect an object and output stop signal as soon as possible.
- Case 3: Stopping in front of the crossing bar. There is a case that the pedestrian was not able to cross the level crossing before the crossing bar closed and was left in front of the crossing bar and collided with the train as shown in Fig. 4(c). We assume that a vulnerable person such as an elderly person is unable to lift the crossing bar by themselves, or unable to reach the end of the level crossing and stops in front of the crossing bar. Currently, in many cases, the detection area is located 0.4 m away from the crossing bar, so that a person nearby the crossing bar cannot be detected.

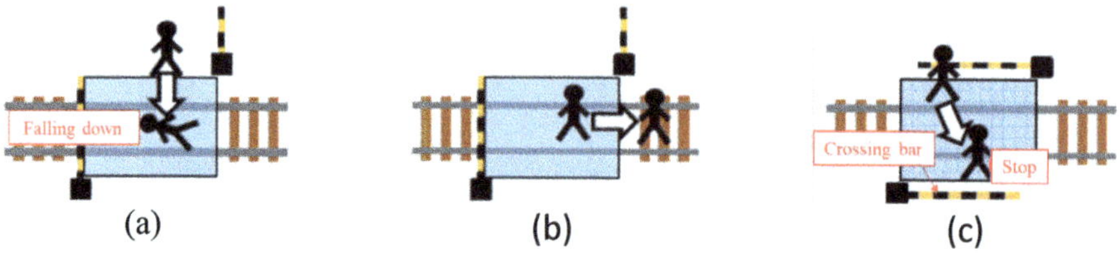

Figure 4: Pedestrian action pattern. (a) Case 1; (b) Case 2; and (c) Case 3.

In this study, we added a new algorithm to the object detection logic to improve the detection of pedestrian, which has not been possible to detect as shown the above three cases. For the development, we used the existing hardware and modified the software to improve the detection function at a lower cost.

4 EXAMINATION OF ALGORITHMS

Based on the action patterns of Section 3, we developed the functional specifications to be added in this study.

4.1 Function for Case 1 (falling down object detection)

A falling down object detection function was developed as a countermeasure for Case 1 in Section 3. In the current 3DLR, in order to prevent false detection due to grass-growing or bumps of the ground surface of the level crossing, an area is set up where an obstacle is not detected up to a certain height. Therefore, as shown in Fig. 5, it was difficult to detect the fall down of a person if the height was lower than the lower threshold height (0.5 m). Then, the lower threshold was set to 15 cm for the area only around the object being tracked, so that a person who has fallen down could be detected. Since the detection height is lowered only around the object, false detection of grasses and rain can be prevented.

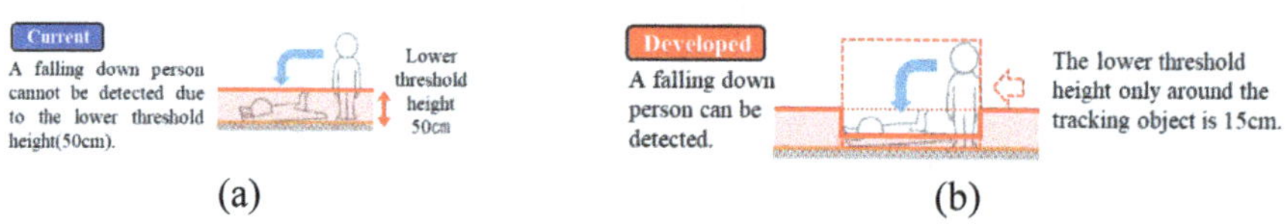

Figure 5: Falling down object detection function. (a) Current; and (b) Developed.

4.2 Additional function for Case 1 (map of ground surface of the level crossing)

In the current 3DLR, the lower threshold height was set at a certain level from the reference plane of the detection area, but it was treated as a constant value regardless of the shape of the ground surface of the level crossing. Therefore, as shown in Fig. 6, with the current function, it cannot detect pedestrians who have fallen down in the recesses. In this study, a map of the height corresponding to the ground surface of the level crossing was created and the lower threshold height was corrected by the height of a map along with the shape of the ground surface.

The map of ground surface of the level crossing is automatically created based on the data measured by the 3DLR. The detection area is divided into multiple square areas, and the height correction value of the ground surface is calculated for each area.

Figure 6: Map of ground surface of the level crossing. (a) Current; and (b) Developed.

4.3 Function for Case 2 and Case 3 (obstacle detection process for behavior judgment)

In addition to the current detection function, the following behavior judgment function was considered.

1. In order to detect pedestrians for Case 2 (Entering the track area), we set up a "detection area for track intrusion" in the direction of the track (refer to Fig. 7), and determine a "track intrusion" when an object moves from the detection area into the detection area for track intrusion.

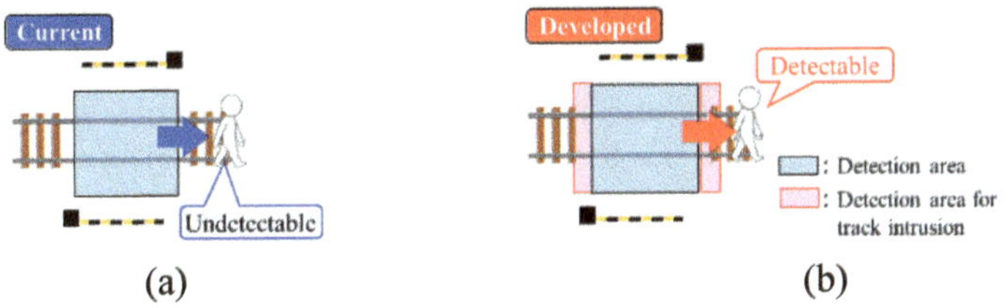

Figure 7: Detection of an object entering the track area from a level crossing. (a) Current; and (b) Developed.

2. In order to detect pedestrians for Case 3 (Stopping in front of the crossing bar), we set up a "detection area in front of crossing bar" (refer to Fig. 8). When an object moves from the detection area into this area, and it stays for more than the threshold time (5 seconds) during the warning of the level crossing, the function judges that it has stopped in front of the crossing bar. The reason why the behavior history is used for this judgment is to prevent false detection of some object such as reflectors of the crossing bar, and to exclude objects that exist in the area in front of crossing bar from the beginning.

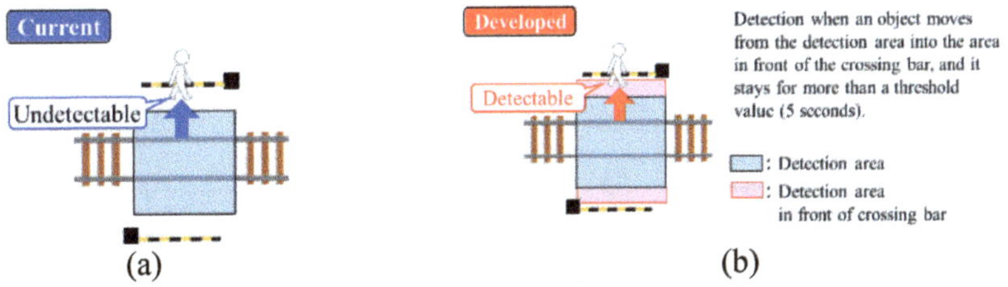

Figure 8: Detection of an object stopping in front of the crossing bar. (a) Current; and (b) Developed.

However, with regard to "Case 2 Detection of an object entering the railroad tracks from a level crossing", we wish to exclude the cases in which maintenance workers and vehicles enter the track area. But it is difficult to distinguish between them and other objects. In addition, the area that can be detected by a 3DLR is only in the vicinity of the level crossing, and it is difficult to continue tracking objects that have entered beyond the detection area. Therefore, it is necessary to keep issuing the "entering the railroad tracks" judgment for safety when an object is detected in the direction of the tracks. However, it is not possible to distinguish whether an object entering the railroad tracks continues to stay on the track or whether it has gone out of the tracks, so the judgment is continued. Therefore, it was decided to exclude Case 2 from this additional function.

4.4 Flow of detection and processing

Fig. 9 shows the software structure reinforced with the developed functions mentioned in 4.1 to 4.3. The height from the ground surface is corrected by "map of ground surface" as pretreatment. As a result, a pedestrian who has fallen down in the recesses can be detected in the later object detection process. In the object detection process, we added the "detection of falling down objects" function so that it can detect falling down objects continuously, and improved the current "object detection" and "object tracking" functions. In the last stage, we added a behavior judgment process (including "inter-area movement judgement" and "obstacle detection process for behavior judgement" processes) in parallel with the current obstacle detection process, and it has become able to detect of an object stopping in front of the crossing bar.

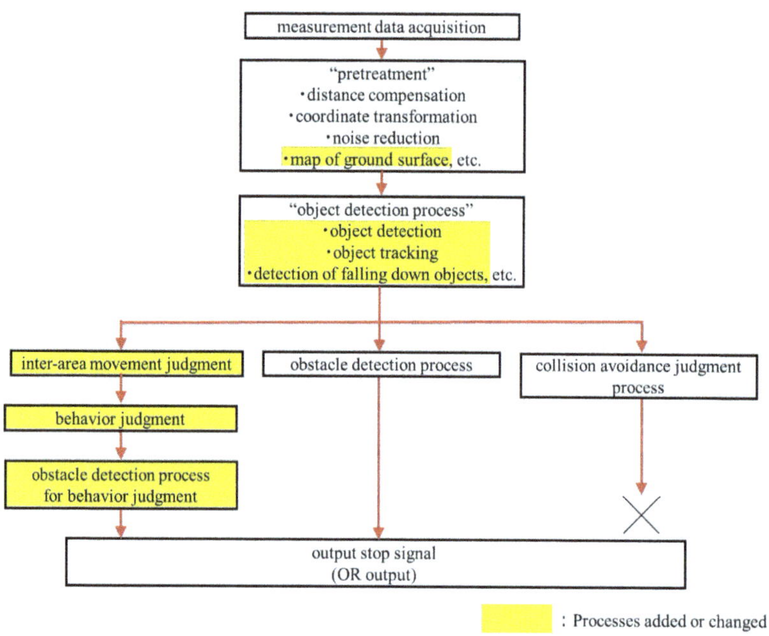

Figure 9: Software structure of 3DLR with developed functions.

5 FIELD TESTING AND EVALUATION

After confirming the operation of the additional functions mentioned in Section 4 in the factory, we conducted a field test at the Kuji level crossing next to Kuji station on the Nambu line in the Tokyo metropolitan area from March 2018 to February 2020. As shown in Fig. 10, this crossing is very busy with 6,000 pedestrians a day. As a result, even current obstacle detection device activates stop signals about 10 times a day, which affects normal railway operation.

Figure 10: Kuji level crossing at the rush hour.

5.1 Results of the falling down object detection function

The results of the number of obstructions detected in the one-week field test are shown in Table 2. Obstacle detection device activates stop signal to approaching train if an object stays in the detection area for more than 5 seconds. We defined correct detections as the total number of obstacle detection that stay in the detection area for more than 5 seconds by checking the camera image visually. The number of correct detections and false detections increased when the falling down object detection function was enabled compared to the current function. This is because the lower threshold of the height of the detection area has been changed from 50 cm into 15 cm, thus spreading the detection area. This is especially true in a level crossing with many pedestrians, such as at the Kuji level crossing, where the detection time of multiple people as a single detection group increases, resulting in an increase in false detection. Fig. 11 shows an example of the test to compare the ability to detect a falling down object. Fig. 11(b) shows the detection screen of the current function and (c) shows the detection screen of developed function.

Table 2: Field test detection results.

No.	Effective function	Total number of detections	Correct detections	False detections
1	Current function	68	6	62
2	Current function + Falling down object detection function	79	7	72

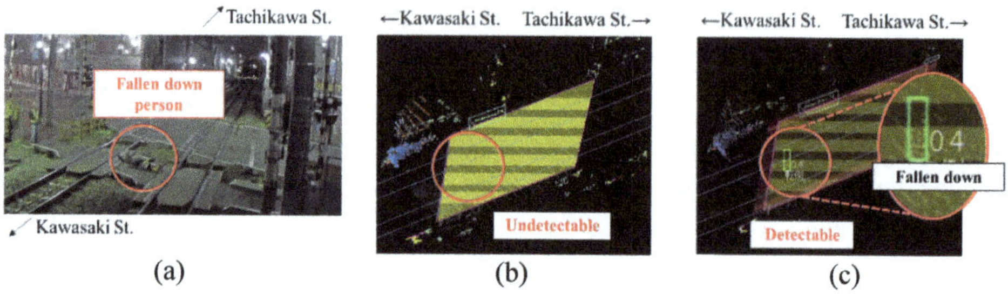

Figure 11: Falling down object detection status. (a) Camera image of a level crossing; (b) Detection screen-Plane view (current); and (c) detection screen-Plane view (developed).

5.2 Map of ground surface of the level crossing

In order to get the correct value of the height, the height data of each area for a certain period were aggregated, and then discrete data by vehicles, pedestrians, etc. were removed from those data. In Fig. 12(b), the height of the ground surface was corrected higher than the actual height due to vehicles and pedestrians. Fig. 13 shows the histogram of the measured height at this case. This shows the discrete data of vehicles and pedestrians apart from the data of the surface. On the other hand, in Fig. 12(c), by removing these discrete data as noise, it is confirmed a map of ground surface of the level crossing is produced correctly.

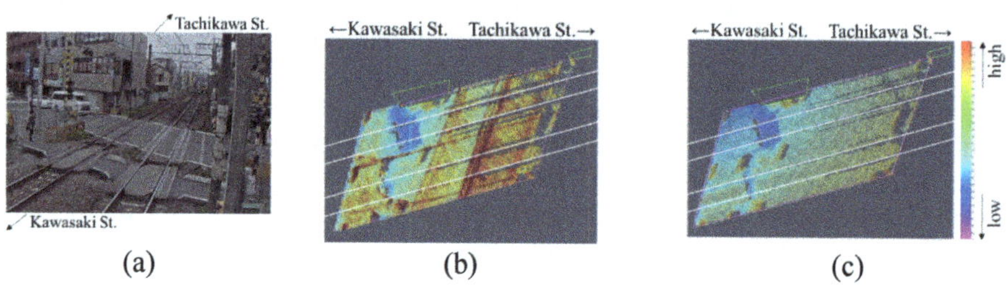

Figure 12: Map of ground surface of the level crossing (a) Capture of a level crossing; (b) Height map (w/o noise reduction); and (c) Height map (with noise reduction).

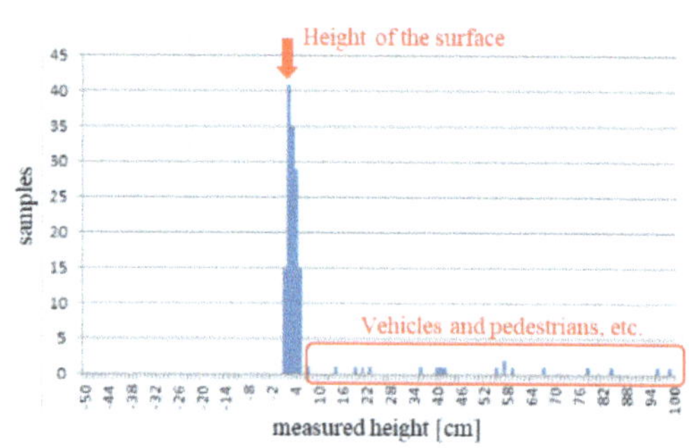

Figure 13: Histogram of measured height.

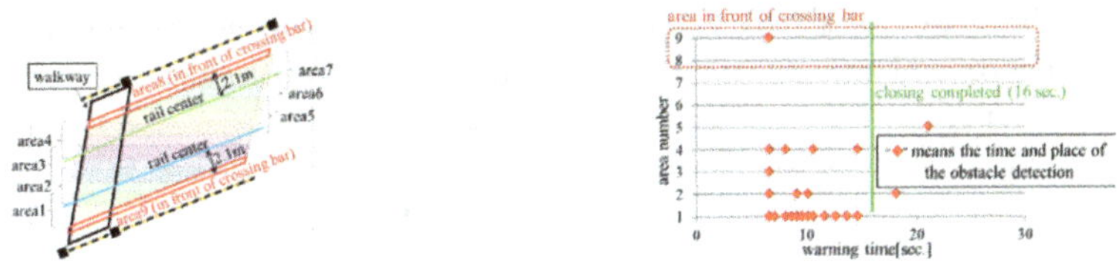

(a) (b)

Figure 14: Detection timing by area. (a) Detection area; and (b) Time and area of detection obstacle.

5.3 Obstacle detection process for behavior judgment

For Case 3(detection of an object stopped in front of the crossing), during the field test, the obstacle detection device sometimes misidentified that a detected object separated into two at the boundary between the detection area and the area in front of the crossing bar, and the timer of counting staying time was reset. This may be due to the unsuccessful handling of the object in the detection area and the area in front of the crossing bar, and further verification is required. Fig. 14 shows an example graph of the detection area and the time after level crossing start warning. This graph shows that there are no examples of pedestrians being detected in the area in front of the crossing bar (area 8 and 9) before and after the crossing bar is closed. Therefore, this function was found to be less effective in detecting pedestrians.

5.4 Evaluation

A field test was conducted at Kuji level crossing as described in Sections 5.1 to 5.3. It was confirmed that the falling down object detection function and the map of ground surface of the level crossing worked correctly, although the number of false detections slightly increased. We decided to adopt this function for operational use. On the other hand, it was found that obstacle detection process for behavior judgment of pedestrians stopping in front of crossing bar was not successful in taking over objects between the detection area and the area in front of crossing bar. In addition, there are few cases of detecting people staying in front of the crossing bar. For those two reasons, we have decided not to adopt this function for operational use.

6 CONCLUSION

In this study, we developed a new function of a 3DLR type obstacle detection system to improve the detection function of pedestrians at the level crossing. Three types of pedestrian action patterns that lead to level crossing accidents were extracted and countermeasure functions were examined. As a result, it was decided to implement two functions. One is to detect falling down objects by detection height being lowered only around the object. The other is to correct lower threshold height along with the shape of the ground surface by a map of the height corresponding to the ground surface of the level crossing. Regarding, the countermeasure against to entering in the direction of the track shown in Case 2, we decided not to implement this function due to difficulty of distinguishing maintenance activities. As for Case 3, we decided not to implement this function, either, due to less effectiveness of implementation.

Through the field test at Kuji level crossing, we confirmed that the falling down object detection function and the map of ground surface of the level crossing worked well and decided these functions to be put into practical use.

Based on the above results, JR-East plans to install these two functions to the existing 3DLR obstacle detection system for operational use from 2020.

REFERENCES

[1] Cabinet Office, Government of Japan, White Paper on Traffic Safety in Japan 2014, FY2013 Situation of Traffic Accidents and Current Situation of Measures for Traffic Safety. www8.cao.go.jp/koutu/taisaku/h26kou_haku/english/pdf/1_t1.pdf. Accessed on: 20 Feb. 2020.

[2] Cabinet Office, Government of Japan, White Paper on Traffic Safety, FY2018 Situation of Traffic Accidents and Current Situation of Measures for Traffic Safety. www8.cao.go.jp/koutu/taisaku/r01kou_haku/english/pdf/wp2019.pdf. Accessed on: 20 Feb. 2020.

[3] About measures to prevent level crossing accidents such as the elderly, Meeting for measures to prevent level crossing accidents such as the elderly, Japan. www.mlit.go.jp/common/001105649.pdf. Accessed on: Oct. 2015.

[4] Japan Railway Electrical Engineering Association, *An Introduction of Signal "Level Crossing Protection Device" for the Railroad Engineer*. (In Japanese.)

[5] Ujiie, T., Watanabe, T., Sonoda, N., Asano, A., Kenmoku, M. & Makino, S., Development of the next-generation crossing obstacle detection device using millimeter wave. *IEEJ Transactions on Industry Applications*, **127**(8), pp. 906–911, 2007.

[6] Hisamitsu, Y., Sekimoto, K., Nagata, K., Uehara, M. & Ota, E., 3-D laser radar level crossing obstacle detection system. *IHI Engineering Review*, **41**(2), pp. 51–57, 2008.

[7] Sekimoto, K., Nagata, K. & Hisamitsu, Y., 3D laser radar for traffic safety system. *Industrial Applications of Laser Remote Sensing*, pp. 143–152, 2012.

CHOICE OF SENSOR FUSION FRAMEWORK FOR TRAIN POSITIONING SYSTEM

MONISH SENGUPTA
Civil, Environmental and Geomatic Engineering, University College London, United Kingdom

ABSTRACT
The general trend in railway length is in overall decline across the country while the passenger demand is growing. To meet this growing demand as well as the capacity to deliver a safe and secure railway, there is a need for use of novel technology to cope with it. Hence the rise of ERTMS and CBTC systems where there is a shift from the conventional fixed block signalling system to moving block signalling system. Sensor fusion technique and sensors sit in the heart of these new systems. Literature review has been carried out to understand the fundamentals of these various sensors currently used on railways. With a combination of literature review and simulation, a study has been carried out via modelling and analysis to explore a better sensor fusion architecture applicable to railway environment which can adapt itself to the various sensor drawbacks. This report summarises the research undertaken and its significant findings in the filter performance with varying measurement and state noise, highlighting the potential for introducing alternative approaches to train positioning that would enable removal of all trackside location correction balises, leading to huge reduction in maintenance cost and more flexibility in future track design.
Keywords: ERTMS, CBTC, sensor fusion, signalling, train control, rail sensors.

1 INTRODUCTION

Traditional train navigation and control systems use fixed train detection systems such as track circuits and axle counters. These fixed train detection systems form the vital element of the traditional fixed block signalling system.

The capacity of railway is constrained by the fixed train detection system. The fixed train detection system formed by either track circuit or axle counter creates a fixed block. Only one train is allowed within this block which in turn affects the number of trains that can be driven within a given piece of railway. The new digital rail ERTMS level 1 and level 2 or CBTC although uses continuous supervision and automatic train protection, they are still constrained by the use of fixed block signalling system. ERTMS level 3 although uses moving block signalling system, it is constrained by the use of balises on track which restricts the train to fixed infrastructure along the track.

Existing railway on-board sensors used for ERTMS and CBTC and the data fusion technology do not restrict the location uncertainty within the location state of the train. Occasional missing of balises, wheel slip and slide and sensor misalignment leads to unbounded growth in location uncertainty.

As a result, the capacity and hence performance can get affected. Therefore, the railway signalling technology needs to come up with a framework of sensor fusion technology and control mechanism which can be used to reduce the state uncertainty and hence provide an optimal location estimate.

2 LITERATURE REVIEW

The sensors currently used for train navigation are tachometers, doppler radar and balise.

WIT Transactions on The Built Environment, Vol 199, © 2020 WIT Press
www.witpress.com, ISSN 1743-3509 (on-line)
doi:10.2495/CR200051

2.1 Tachometers

Tachometers are wheel angular speed sensors often used as a primary form of speed measurement device. A whole range of different types of analogue and digital sensors such as capacitative, active and passive electromagnetic tachometers, have been used for some time in rail vehicle applications [1]. They are frequently used due to their robustness and reliability [2].

Discussing about the limitations of a tachometer, Mirabadi et al. [1] suggested parameters limiting the tachometer output accuracy include resolution of the tachometer, sampling frequency, electrical noise, mechanical imperfections and wheel slip and slide.

Wheel slip and slide occurs when the tractive effort exceeds adhesive weight or braking effort exceeds adhesive weight respectively. The adhesive weight is the result of wheel rail friction force. Therefore, during heavy traction, the wheel can rotate freely in an event of wheel slip giving rise to excessive rotational speed. On the contrary, wheel can get locked in the event of wheel slide during heavy braking leading to zero wheel rotation. Geistler and Bohringer [3] while discussing about robust rail speed measurement mentioned that the use of wheel tachometers may lead to degraded operation due to the presence of wheel slip and slide. On the topic of wheel slip and slide, STARS (Satellite Technology for Advanced Railway Signalling) project [4] mentioned that tachometers are the most unreliable of all railway sensors, especially if mounted on a powered axle. The Transit Cooperative Research Programme (TCRP) report [5] on the other hand suggested that errors caused by wheel slip and slide can be minimized by the use of current sophisticated slip-slide control or, where feasible, placing the tachometer on an unmotored axle. Fitting the tachometer on to the unmotored axle however can only restrict wheel slip, but not slide as the braking effort can still lead to measurement inaccuracies.

2.2 Doppler Radar

Speed of the train can be measured directly by applying the Doppler principle [1]. The frequency shift as mentioned by Mirabadi et al. [1] is related to the speed by:

$$F_d = (2VCos\gamma)/\lambda, \tag{1}$$

where
F_d is the Doppler frequency shift
V is the train speed relative to the ground
γ is the radiation angle
λ is the transmitted carrier wavelength.

Doppler radar is a non-contact sensor with two microwave antennas which provides good accuracy and reliable output independent of the reflecting surface and the vibration [4]. Mirabadi et al. [6] while discussing about fault tolerant train navigation system identified the Doppler radar as the preferred sensor due to its low amplitude of white noise. Compared to tachometer, they identified Doppler radar to provide higher accuracy data. Heide et al. [7] suggested that use of 24GHz coded Doppler radar can provide high precision data (within 20db) for vehicle position and speed measurement.

Malvezzi et al. [8] on the other hand while discussing odometric estimation for train protection system stated that Doppler radar output is often affected by noise and systematic errors. This has been echoed by Mirabadi et al. [1] who identified the sources of errors due to very smooth reflective surface, change in radiation angle due to acceleration and braking action, vibration and radar noise and bias error.

While discussing about the other areas where Doppler effect has been used for speed sensing, Wrobel [9] stated in his research about LIght Detection And Ranging (LIDAR) sensor which also uses Doppler principle to measure the train speed. The doppler shift received by this sensor he suggested can accurately determine the speed by applying eqn (1). Doppler radar is also used in aircraft speed measurement [10].

2.3 Accelerometers

Accelerometers are sensors which can measure acceleration. They are not a common sensor that are used on existing speed measurement sensor framework for railways. Some CBTC system uses accelerometer as speed sensing device, but ERTMS in Europe do not prefer the use of accelerometers onboard.

It is a device that can measure acceleration generated by the movement of an object along an axis [11]. Linear accelerometers can be used to measure the acceleration of the vehicle in different dimensions [1].

An important element of this sensor as highlighted by Vettori et al. [12] is that if the sensor axis is not perfectly perpendicular to the gravity vector, the output signal is not null, even if the accelerometer is perfectly still. Mirabadi et al. [6] identified this error as bias error apart from other probable accelerometer sensor errors such as hysteresis error, dead zone, axis alignment error and pivot alignment error.

An important relation between the bandwidth and sensor accuracy has also been identified by Malvezzi et al. [11]. Faster we acquire the accelerometer output, worse the accuracy we get. This is explained by the fact that high sensing frequency requires a high ratio of stiffness to mass and therefore comes at the cost of having poorer sensitivity and resolution [13]. In this context, Endevco [14] in their study on the steps leading to identifying the right accelerometer stated that the frequency response is an important parameter when considering any accelerometer.

The Integrated Electronic Piezoelectric (IEPE) is the most commonly used accelerometer that does not need any external charge amplifier and uses ordinary low-cost cable [14].

2.4 Balise

Balise is a transponder fixed on the track by ERTMS and CBTC system to correct the position uncertainty which builds up within the train location subsystem over the time. It provides a position reference and direction information to the train. Balises are magnetically coupled transponders and do not require a constant energy source on their own [15], hence can be considered as passive. However, for ERTMS Level 1 system, balises can be active where they are connected to the signal via a Lineside Electronic Unit. Here the balises are continuously powered.

As pointed out by Mendizabal et al. [16], the balises are placed at approximately regular intervals. The distance between the balises are determined by the speed and complexity of the railway. As it is important for the train to stop precisely at stations, point zones and buffer stops, there can be up to 8 balises required at a minimum interval defined by the design [17]. The main reason of having more than one balise in critical stopping locations is the possibility of train missing to read a balise. Geistler and Bohringer [3] hence suggested a sensor fusion for position estimate in order to reduce the reliance on wayside balise.

2.5 Other sensors

This section deals with those sensors which do not currently form the sensor fusion framework for ERTMS or CBTC train positioning system.

2.5.1 Eddy current sensor

Acharya et al. [18] suggested eddy current sensors as an auxiliary system for speed sensing device on railways. Lion Precision [19] identified that eddy current sensor is appropriate where there is a presence of liquid and a risk of contamination. In this context, Geistler and Bohringer [3] mentioned that eddy current sensors are more robust than tachometers in terms of weather and environmental conditions. In addition to non-destructive testing Geistler [20] pointed out that eddy current sensors are also be able to detect inhomogeneities in magnetic resistance along the track. Magnetic resistance is an important element to consider when designing track circuits. As magnetic resistance affects the current flow, it can jeopardise the detection of a train presence within a given track circuit.

2.5.2 Global Positioning System (GPS)

GPS is a satellite-based radio navigation system developed by the United States Department of Defense to provide accurate absolute positioning information over extended periods of time worldwide under all weather conditions [21]. GPS is currently used by the UK rolling stock maintenance [22] to identify the number of trips that a train is doing on a daily basis. This data then feeds into their regular maintenance regime which then builds up its maintenance schedule for all available rolling stock. Bajaj et al. [23] mentioned that GPS is rapidly becoming a universal utility as the cost of integrating the technology into vehicles, machinery, computers, and cellular phones is decreasing. Yao et al. [24] mentioned that the GPS solution is relatively stable and can maintain a consistent level of navigation functionality for a long period, but that it also suffers from multipath effects and its signal may easily be blocked or lost under certain environment.

Bhatt et al. [21] mentioned that although GPS has been widely used in land vehicle navigation systems, standalone GPS are unable to provide continuous and reliable navigation solutions in the presence of signal fading and/or blockage such as in urban areas. Hu et al. [25] studied real time adaptive algorithm on GPS satellite data processing for vehicle navigation where they mentioned that in practice, the number of measurements may vary from epoch to epoch timing frequency in a sensor fusion environment due to GPS signal being blocked off by obstacles or simply due to satellite unavailability. While discussing about the errors in GPS signal, Bajaj et al. [23] mentioned that a major source of error arises from the fact that radio signal speed is constant only in a vacuum. Errors due to multipath fading, which occurs when a signal bounces off a building or terrain before reaching the receiver's antenna, also can reduce the accuracy [23].

Differential GPS (DGPS) employs both roving receivers that make satellite position measurements and stationary receivers that use their position to compute signal timing [23]. While comparing GPS, DGPS and Assisted GPS (AGPS), Bajaj et al. [23] mentioned that AGPS provides a better accuracy typically less than 15 m and DGPS as less than 0.9 m.

2.5.3 Other Global Navigation Satellite System (GNSS) sensors

GPS discussed in previous section is one form of GNSS. The other forms of GNSS are:

- GLONASS (Russia)
- Galileo (EU)
- Compass (China)

Moreno et al. [26] mentioned that there are two major railway projects which are trying to validate satellite technology for railway safety services. The first one is project 3InSAT, led by Ansaldo STS and the other one is led by European Space Agency. They are now being tested in Sardinia with an intention to remove track circuits and GSM-R infrastructure using satellites for both location and communication.

While discussing about precise and reliable train positioning system, Albrecht et al. [27] stated that a modular on-board positioning system based on GNSS (Global Navigation Satellite System) is being developed in the German national research project PiLoNav. Jiang et al. [31] suggested that the competition between various means of transportation pushed the railways to look at the GNSS positioning as a more cost effective solution as compared to the use of trackside equipment. STARS project [4] also put forward a framework of train positioning system using GNSS.

The Indian Regional Navigation Satellite System (IRNSS) is an independent navigation system to be a constellation of seven satellites. Its primary service area will cover the Indian sub-continent extended by about 1500 km. While carrying out a study on the effects of various environment factors on this system, Manjula and Raju [28] mentioned that under harsh conditions such as ionosphere scintillation and multipath effects, the performance of GNSS deteriorates leading to loss of phase lock and signal outages.

3 FUSION FRAMEWORK

From the above literature review, it is clear that a single sensor or a type of sensor is not good enough to carry out a train position estimate. Combination of sensors is the only way forward where the various noise profiles and sensor errors can be accommodated and at the same time provide an optimum train position estimate.

Kalman filter algorithm was chosen for fusing data from various sensors in order to get an acceptable state estimate. It has observation noise covariance R, state error covariance Q and estimate error covariance P.

The simulation run that has been programmed in MATLAB is:

Run 1: The train starts from stationary to reach a speed of 350km/hr at an acceleration of $1.2 m/s^2$.
Run 2: The train then coasts with acceleration $0 m/s^2$ and reaches a distance of 6.5 km.
Run 3: The train then brakes at a deceleration rate of $1.1 m/s^2$ till it comes to a halt, hence speed reaches 0 m/s.

Standard deviation for the sensor noise as chosen from the manufacturer's catalogue are shown in Table 1.

Table 1: Sensor error standard deviation table.

GPS	0.2 m
Doppler radar	0.027 m/s for speed < than 100 km/hr 0.01% of true speed for speed >than 100 km/hr
Tachometer	1% of true speed
Accelerometer	3% of true acceleration
Balise	0:25 m

Nine sensor frameworks are chosen as shown in Table 2.

Table 2: Sensor fusion framework.

Fusion framework	GPS	Doppler	Accelerometer	Tacho	Balise
Frm 1	√	√			
Frm 2	√		√		
Frm 3		√		√	√
Frm 4				√	√
Frm 5		√			√
Frm 6	√			√	
Frm 7		√		√	
Frm 8	√	√		√	
Frm 9	√	√	√	√	

The rationale for considering the sensors as shown in Table 2 is driven by the literature review carried out in Section 2.

Drawing the rationale for tachometer being the robust and reliable sensor [11] used on railways and currently been within the sensor fusion framework, we have considered the tachometer to be within the framework for most experiments. The same rationale applies to the doppler sensor also.

Accelerometer on the other hand is currently only used on some CBTC systems and given the bias errors [6] and the issue of measurement frequency [8], we do not consider it as our preferred sensor.

GPS on the other hand, although do not form the part of the existing rail sensor fusion framework, it is used for rolling stock maintenance as identified in Section 2.5.2. It provides relatively stable data [24] and is currently extensively used in road vehicle navigation.

Balise being an external sensor do not form part of the sensor fusion framework which sits within the train. Therefore, they provide location information periodically as set by the designer and not by the system. As identified, there is a high chance that the train at a high speed might miss the balise which doesn't help in estimating the train location which is the primary aim of this research. Hence use of balise in this framework is kept minimal.

Using the above sensor framework, we ran traditional Kalman filter assuming all the a priori assumptions and filter properties are correct using MATLAB software for each of the sensor combination.

4 ANALYSIS

4.1 Analysis of filter response in terms of variance covariance

The filter error variance covariance P is the measure of the filter stability. As pointed out by Groves [29], it records the correlation between the state estimates and the uncertainties. In a way it provides the best knowledge of the system dynamics at any instant t. As the P approaches zero, the filter puts less emphasis on the measurement [30].

The diagonal matrices of P are the variance covariances of "location", "speed" and "acceleration" respectively. Analysing the results of P11, P22 and P33 we find the following:

Doppler–Tachometer combination is not the best of the sensor framework for any of the state estimates. While it is worse for location state estimate, it is not providing the best estimate for speed and acceleration either.

GPS–Tachometer–Doppler–Accelerometer produces the best estimate for position, but with an increasing speed, the variance covariance matrix is increasing which indicates that with faster train speed and high acceleration, the filter might tend to diverge. A similar property is shown for the speed estimate also. However, for the acceleration estimate, the sensor framework exhibits better result.

The GPS–Doppler combination shows a similar performance for the location estimate to that of GPS–Tachometer–Doppler–Accelerometer combination, however it doesn't show any sign of dipping during the "zero" acceleration phase. This may due to the fact that it is not getting acceleration observation data and hence is solely relying on the filter. The sensor framework however is not exhibiting the best of the results for the speed and acceleration state estimate.

GPS–Accelerometer fusion framework exhibits the most stable filter for the location estimate. It doesn't show any change due to growing speed. The filter also shows stable behaviour for the speed and acceleration state estimates. It is the best sensor combination for the acceleration state estimate.

Doppler–Tachometer–balise is the current sensor fusion framework which is used ERTMS and in most CBTC systems. It exhibits a stable position estimates only once it receives the balise reading. However, it doesn't exhibit a good performance for speed estimate and shows a tendency to a growing variance covariance. This suggests it might not be the best sensor fusion framework for high speed rail network and high performing, high accelerating trains. Therefore, for mass transit network, where the trains need to accelerate and brake fast, this sensor framework may not be the best combination. The acceleration estimate is not impressive either.

The Tachometer–balise sensor framework exhibits a high degree of dependency on the balise reading, particularly for the location estimate. For the speed and acceleration estimate, this is the worst sensor combination as evident from the three graphs.

The Doppler–balise combination exhibits similar dependency on the balise reading as that of Tachometer–balise framework. But it is a better fusion combination than the Tachometer–balise framework. The speed estimate is not what expected given it has got a speed sensor. The speed estimate exhibits a trend towards growing uncertainty as the speed increases. While it is not the worst sensor combination for the acceleration state estimate, it is not the best either.

GPS–Tachometer combination does not exhibit the best estimate in any of the state estimates. While it is not the worst, it is not good either.

GPS–Tachometer–Doppler combination is very similar to GPS–Doppler in terms of position state estimate. However, it is not a good sensor combination for speed and acceleration state estimate.

4.2 Analysis in terms of error recording

Analysing the filter from the error perspective, we choose the best filter which gives least error.

Out of the nine sensor combinations as identified in Table 2, three sensor combinations give the best estimate in terms of least location estimate error. They are GPS–Doppler, GPS–Tachometer–Doppler and GPS–Tachometer–Doppler–Accelerometer.

At the beginning of the filter run, the GPS–Doppler is the most stable framework. But all the three sensor frameworks stabilise within ten seconds of the filter run and produces almost same location estimate error.

The change of acceleration occurs at 81 second and 107 second. Fig. 1 suggests that GPS–Doppler and GPS–Tachometer–Doppler sensor combination provides better estimate than the GPS–Tachometer–Doppler–Accelerometer sensor combination.

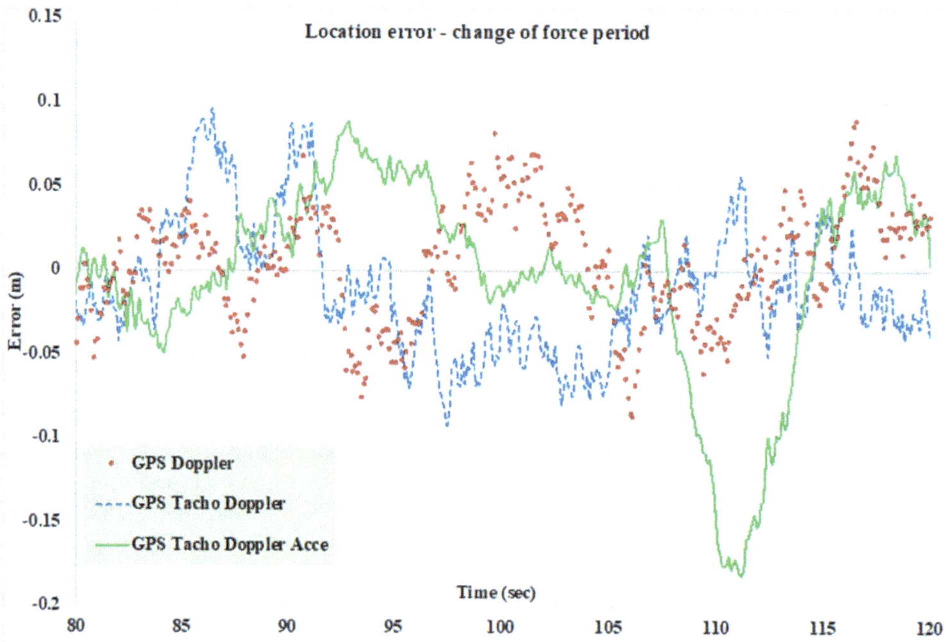

Figure 1: Comparison of location error during force change.

At the end of the mission, when the filter parameters are quite established and the train is coming to a halt, the three sensors are performing better than any other sensor combinations.

Similar comparison was carried out for speed and acceleration. For the speed estimate (Fig. 2), it seems that the best sensor fusion frameworks are GPS–Accelerometer, Doppler–Tachometer and GPS–Doppler–Tachometer–Accelerometer. During the change of acceleration phase, the best estimate is shown by GPS–Accelerometer and also by Doppler–Tachometer.

Fig. 2 and Fig. 3 show the speed state estimate error and acceleration state estimate error as recorded for the best combination of sensors. The best sensor framework that seems to produce the best acceleration estimate (Fig. 3) are GPS–Accelerometer and GPS–Tachometer–Doppler–Accelerometer sensor framework. While the GPS–Accelerometer framework produces better estimate during the initial phase of the filter run, they produce almost similar estimate during the final phase of the filter when the system parameters are well established. During the acceleration change phase, while both the framework produces similar error estimate, both exhibits zero error during zero acceleration phase. This suggests that including a sensor which can measure acceleration within the sensor fusion framework provides better acceleration estimate. Both exhibits sudden change of error at the point when there is a change of acceleration from 0 to -1.1 m/s^2.

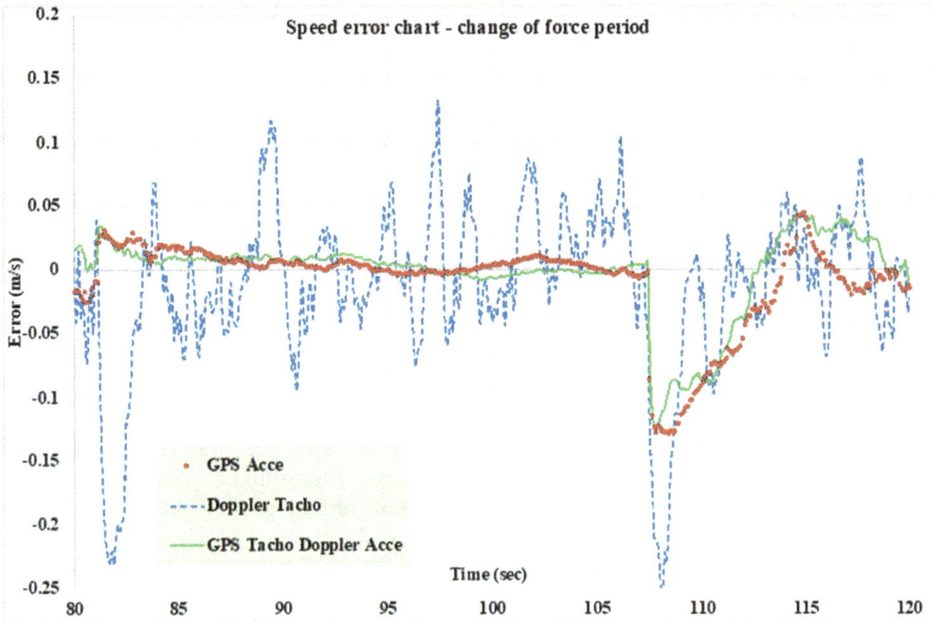

Figure 2: Comparison of speed error during force change.

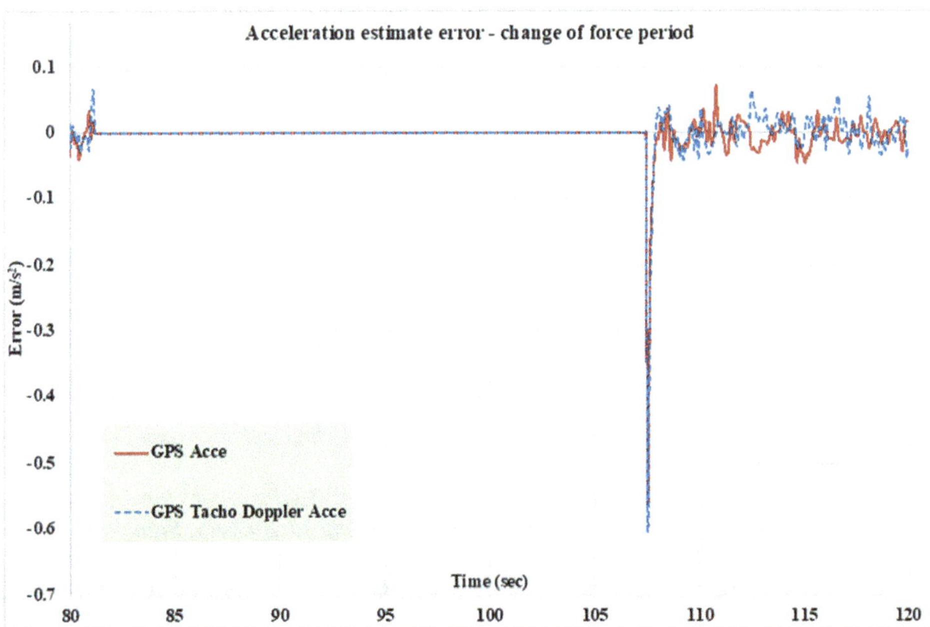

Figure 3: Comparison of acceleration during force change.

5 CONCLUSION

Judging the behaviour of the filter with various sensor combinations, we can draw the following conclusion: GPS is the chosen location sensor than that of balise. The rationale is wherever GPS has been fused with other sensor, the filter provides a better position estimate. Fusing an accelerometer with GPS always provides a better state estimate for all location, speed and acceleration.

Use of tachometers in a sensor framework effects the overall filter accuracy including that of speed estimate. None of the sensor fusion framework with tachometer provides a better speed estimate unless it is fused with GPS and accelerometer. While the traditional Doppler–Tachometer–balise combination provides a fairly good location estimate, it is not good as the estimate that we get by fusing accelerometer or GPS with it. Interestingly the combination of Doppler–balise and Doppler–Tachometer–balise provides the same state estimate for all location, speed and acceleration. This proves that the tachometer has no impact on the estimate even on speed. It is not adding any value to the filter estimate.

The accelerometer sensor which is currently scarcely used due to the reason explained in Section 2.3 turned out to provide fairly good state estimate. Therefore, further experiments will be carried out including accelerometer sensors.

None of the experiments involving balise provides any better state estimate. The indication is where balise is solely used for the purpose of location observation along with Doppler and Tachometer, the filter tends to increase its uncertainties till it gets a reading from the balise. Given the fact that the balise are mounted on the track at a fairly specific interval, the system is only constrained to receive balise information every specific distance of travel. This suggests that any incident of missing balise will lead to growing uncertainties within the location state estimate.

Therefore, the conclusions are:

1. GPS is the chosen sensor for location observation;
2. Doppler is the chosen sensor for speed observation;
3. Accelerometer is the chosen sensor for acceleration observation;
4. More simulations incorporating failure modelling within the sensor fusion framework; and
5. Standardise sensors and the fusion architecture for any ERTMS solution in order to achieve interoperability.

REFERENCES

[1] Mirabadi, A., Mort, N. & Schmid, F., Application of sensor fusion to railway systems. *International Conference on Multisensor Fusion and Integration for Intelligent Systems*, pp. 185–192, 1996.

[2] Malvezzi, M. et al., Train position and speed estimation by integration of odometers and IMUs. *9th World Congress on Railway Research*, pp. 22–26, 2011.

[3] Geistler, A. & Bohringer, F., Robust velocity measurement for railway applications by fusing eddy current sensor signals. *IEEE Intelligent Vehicles Symposium Proceedings*, pp. 664–669, 2004.

[4] Bazant, L., Toma, A. & Mocek, H., *Description of Methodology for Data Record Sorting and Saving Satellite Technology for Advanced Railway Signalling*. Tech. report 687414, p. 114, 2017.

[5] Transit Cooperative Research Programme (TCRP), Tech. report, August. pp. 17–64, 1989.

[6] Mirabadi, A., Mort, N. & Schmid, F., Design of fault tolerant train navigation systems. *Proceedings of the American Control Conference, San Diego, California*, June, pp. 104–108, 1999.
[7] Heide, P., Magori, V. & Schwarte, R., Coded 24 GHz Doppler radar sensors: A new approach to high-precision vehicle position and groundspeed sensing in railway and automobile applications. *IEEE NTC 1995 – Microwave Systems Conference, Conference Proceedings*, pp. 101–104, 1995.
[8] Malvezzi, M., Allotta, B. & Rinchi, M., Odometric estimation for automatic train protection and control systems. *Vehicle System Dynamics 49.5*, pp. 723–739, 2011.
[9] Wrobel, S. Multifunction LIDAR sensors for non contact speed and track geometry measurement in rail vehicles. PhD thesis, pp. 1689–1699, 2013.
[10] Maybeck, P.S., Stochastic models, estimation, and control. *New York*, vol. 1, pp. 1–16, 1979. http://www.cs.unc.edu/~welch/media/pdf/.
[11] Malvezzi, M. et al., Train position and speed estimation by integration of odometers and IMUs. *9th World Congress on Railway Research*, May, pp. 22–26, 2011.
[12] Vettori, G. et al., Innovative management of wheel-rail adhesion conditions in localization algorithms for the automatic train protection. *Chemical Engineering Transactions*, 33.2, p. 685, 2013.
[13] Shieh, J., Huber, J.A., Fleck, N.A. & Ashby, M.F., The selection of sensors. *Progress in Materials Science*, **46**(3–4), pp. 461–504, 2001.
[14] Endevco, *Steps to selecting the right accelerometer.*
[15] Sharma, R. & Lourde, R.M., Crosstalk reduction in balise and infill loops in automatic train control. *INES 2007 – 11th International Conference on Intelligent Engineering Systems Proceedings*, pp. 39–44, 2007.
[16] Mendizabal, J. et al., *Communication Technologies for Vehicles*, vol. 7865, pp. 27–39, 2013.
[17] UNISIG, Principles. *System Requirements Specification*, Chapter 3, 2012.
[18] Acharya, A., Sadhu, S. & Ghoshal, T.K., Train localization and parting detection using data fusion. *Transportation Research Part C: Emerging Technologies*, **19**(1), pp. 75–84, 2011.
[19] Lion Precision, Comparing non – contact sensors: Capacitive vs. eddy – current sensors, 2013.
[20] Geistler, A., Train location with eddy current sensors. *CompRail 2002*, pp. 12–14, 2002.
[21] Bhatt, D. et al., Expert systems with applications. A novel hybrid fusion algorithm to bridge the period of GPS outages using low-cost INS. *Elsevier/Expert Systems With Applications*, **41**(5), pp. 2166–2173, 2014.
[22] Bertran, E. & Delgado-Penín, J.A., On the use of GPS receivers in railway environments. *IEEE Transactions on Vehicular Technology*, **53**(5), pp. 1452–1460, 2004.
[23] Bajaj, R., Ranaweera, S.L. & Agrawal, D.P., GPS: Location tracking technology. *Computing in Science & Engineering*, pp. 92–94, 2002.
[24] Yao, Y. et al., A hybrid fusion algorithm for GPS/ NS integration during GPS outages. *Elsevier/Measurement*, **103**, pp. 42–51, 2017.
[25] Hu, Y., Duan, Z. & Zhou, D., Estimation fusion with general asynchronous multi-rate sensors. *IEEE Transactions on Aerospace and Electronic Systems*, **46**(4), pp. 2090–2102, 2010.

[26] Moreno, J. et al., A survey on future railway radio communications services: Challenges and opportunities. *IEEE Communications Magazine*, **53**(10), pp. 62–68, 2015.
[27] Albrecht, T., Lüddecke, K. & Zimmermann, J., A precise and reliable train positioning system and its use for automation of train operation. *IEEE ICIRT 2013 – Proceedings: IEEE International Conference on Intelligent Rail Transportation*, pp. 134–139, 2013.
[28] Manjula, T.R. & Raju, G., A comprehensive study of linear Kalman filter based tracking techniques under ionosphere scintillation. *Procedia Technology*, **25**, pp. 427–434, 2016.
[29] Groves, P.D., *Principles of GNSS, Inertial, and Multisensor Integrated, Technology*, 2008.
[30] Welch, G. & Bishop, G., An introduction to the Kalman filter. *SIGGRAPH*, pp. 1–81, 2001.
[31] Jiang, L., Bai Gen, C. & Tao Jian, W., A CKF based GNSS/INS train integrated positioning method. *2010 IEEE International Conference on Mechatronics and Automation, ICMA 2010*, IEEE, pp. 1686–1689, 2010.

TOWARDS MOBILE AND INTELLIGENT RAILWAY TRANSPORT: A REVIEW OF RECENT ERTMS RELATED RESEARCH

VAHID RANJBAR & NILS O. E. OLSSON
KTH Royal Institute of Technology, Department of Transport Science, Sweden
Norwegian University of Science and Technology, Department of Production and Quality Engineering, Norway

ABSTRACT

The development of the European Railway Traffic Management System (ERTMS) throughout Europe over the last two decades to remove obstacles to providing an interoperable railway network has been facing several challenges. This study briefly highlights several of these challenges by analysing the current architecture of ERTMS and main concerns such as its implementation, the interoperability of communication, formal methods, hybrid ERMS (Level 3), safety, and human factors. This study has taken a systematic approach to data collection and analysis through a review of the current literature. The studies examined illustrate that it is necessary to reduce the number of components in order to pivot the system toward mobile equipment, autonomous trains, and decentralised communication. In the area of formalisation, every supplier and responsible party conducting modelling and testing of the system uses a different tool. Here, we conclude it would be appropriate to develop a robust and reliable tool for modelling, formalisation, testing and the validation of critical safety systems in the railway industry according to its particular specifications and functionalities, e.g. for ERTMS L3, including moving block, virtual coupling, and Automatic Train Operation (ATO). To bridge gaps between human operation and new technologies, reduce human error, and increase performance in the railway industry, a stronger emphasis on the development of employees' skills with new technologies is highly recommended. In order to build upon previous investments in ERTMS and to guarantee backwards and forwards interoperability, it is necessary to exploit hybrid solutions and to gradually update the current systems.

Keywords: implementation of ERTMS, GNSS-based train positioning, adaptable communication system, interoperability, train integrity monitoring system, virtual coupling.

1 INTRODUCTION

This study aims to identify the current challenges facing the European Rail Traffic Management System (ERTMS) to achieve a mobile and intelligent railway transport system by carrying out a literature review. This literature review is based on a series of database searches using keywords such as "ERTMS OPERATION" and "ERTMS ISSUES". This study considered only research with a high number of citations, and a recent date of publication. Accordingly, this review draws on studies published since 2015. The studies reviewed are summarised and categorised according to their discussions of critical issues.

After three decades of operation, GSM-R cannot satisfy all the needs of European Train Control System (ETCS) due to issues related to its low capacity for data and voice transmission, latency, interferences, and a further lack of safety requirements for communication system in next generation railways [1].

There are some candidates to replace GSM-R, which is based on the 2nd Generation (2G) telecommunication system. One candidate is Long Term Evolution–Railway (LTE-R), which is based on 4G protocol; however, there is reason to hold off on implementing a new 4G system due to the imminent availability of better 5G networks. As a consequence, there is greater support from the International Union of Railways (UIC) to use a flexible 5G system, relying as well on other technologies [2].

WIT Transactions on The Built Environment, Vol 199, © 2020 WIT Press
www.witpress.com, ISSN 1743-3509 (on-line)
doi:10.2495/CR200061

The safe movement of trains during their operation along the railway network requires a safe and reliable positioning system. ERTMS Level 1 and Level 2 both rely on track circuits and axle counters for positioning information. In ERTMS HL3 (Hybrid Level 3), the positioning system relies on data provided by Trackside Train Detection (TTD) using either fixed block technology equipment (track circuit and axle counter) or Positive Train Detection (PTD). The train detection and positioning are the responsibility of an on-board subsystem, depending on the level of operation. For example, in Level 3, the position of the train should be provided continually to chasing and chased trains to fulfil the requirements of moving block technology [3]. This is illustrated in Table 1, which shows trackside components in different levels of ERTMS-ETCS.

Table 1: Trackside components in different levels of ERTMS-ETCS.

Level	Trackside subsystem							
	Track-side light	Track-circuit/ Axle-counter	Balise	LEU	Euroloop	RIU	RBC	GSM-R
NTC	✓	✓	✓	✓	-	-	-	-
1	✓	✓	✓	✓	✓	-	-	-
1+RIU	✓	✓	✓	✓	✓	✓	-	✓
2	-	✓	✓	✓	-	-	✓	✓
HL3	-	-/✓	✓	✓	-	-	✓	✓

Due to railway systems' increasing connectivity and complexity, the railway industry is relying increasingly on modelling, validation and testing to formally evaluate the functionality and safety of railway systems. A variety of different formal tools and languages have been used to accomplish this; however, no formal method dedicated specifically to the needs of the railway industry exists [4]. ERTMS is a multifaceted system and consists of different applications, interfaces, formal tools and languages. To integrate and design multi-tool systems like ERTMS using domain concepts, model-driven engineering is a useful and appropriate method [5].

Global navigation and satellite system (GNSS) is one of the proposed solutions for improved train positioning, but some important questions about safety and architectural arrangements to fulfil ERTMS requirements remain [6]. Some issues related to jamming and spoofing attacks require addressing with robust solutions in order to increase its safety and reliability [7].

Railway systems are known as Cyber Physical Systems (CPS), and they rely on interactions between distributed, connected, computational and physical components. A significant disadvantage of this kind of system is their vulnerability to cyber-attacks and safety issues due to their multimodal complexity. Consequently, cyber security and safety are crucial concerns when seeking to provide secure and reliable railway operation [8], [9].

The implementation of ERTMS to upgrade to a mobile and intelligent railway system thus requires step-by-step and continuous improvement to extant infrastructure and technologies, in combination with or replacement by new technologies (hybrid solutions e.g. HL3) in order to enjoy such benefits as reduced installation, operation and maintenance costs [10].

Table 1 shows which components are included in the trackside subsystem of following ERTMS-ECTS levels: National Train Control System (NTC), Level 1, level 1+ Radio Infill Unit, Level 2 and ERTMS Hybrid Level 3 (HL3). Trackside components include: Line

Electronic Unit (LEU), Radio Infill Unit (RIU), and Radio Block Centre (RBC). On board subsystems (Table 2) include: Balise Interface Unit (BIU), Train Interface Unite (TIU), Specific Transmission Module (STM), Driver Machine Interface (DMI), European Vital Computer (EVC), Juridical Recording Unit (JUR), and Train Integrity Monitoring System (TIMS).

Table 2: Interfaces and components of ERTMS-ETCS On-board subsystem.

Level	On-board subsystem								
	BIU	TIU	STM	DMI	EVC	Odometry	JUR	TIMS	GSM-R
NTC	✓	✓	✓	✓	✓	✓	✓	-	-
1	✓	✓	✓	✓	✓	✓	✓	✓	-
1+RIU	✓	✓	✓	✓	✓	✓	✓	-	✓
2	✓	✓	✓	✓	✓	✓	✓	-	✓
HL3	✓	✓	✓	✓	✓	✓	✓	-/✓	✓

This study provides an overview of the main topics relevant to ERTMS operation and related issues discussed in recent literature (since 2015). The next section explains the method of study and some preliminary statistics. The third section provides discussion about the study's findings and is followed by a conclusion.

2 METHODOLOGY

This literature review is based on a series of database searches using keywords such as "ERTMS OPERATION" and "ERTMS ISSUES" (as shown in Table 3). This literature review analysed 82 papers published since 2015, and categorises them as main themes, such as: formal methods, adaptable communication systems, GNSS-based train positioning, implementation of ERTMS, cyber security, safety, digitalisation, risk assessment ERTMS, virtual Coupling, human factors, interoperability and Hybrid ERTMS/ETCS Level 3. For each category, the most cited paper was chosen for closer reading and analysis.

Table 4 summarises the methodological guidelines designated for determining applicable papers to include in the review, such as period of publication, characteristics of the paper relevant to this study (including title, and electronic database where found), year, document type (conference, journal, book chapter), and institutional location (country). The results have been presented here as graphs, charts and tables.

Table 3: Search history.

Search string	Search number	Scopus, Springer, Elsevier
ERTMS ISSUES	1	38
ERTMS OPERATION	1	44

Table 4: Methodological guidelines.

Electronic database	Scopus, Springer, Elsevier
Time period	2015–2019
Collected data	Title, publication channel, document type
Additional coding	Country, main theme
Output format	Statistical summaries: graphs, charts, tables

3 RESULTS

This section presents some results from the literature review in the form of graphs, table and chats.

3.1 Preliminary statistics on 82 papers

Fig. 1 shows the proportional contribution of countries in the production of literature related to ERTMS. The majority of publications come from Italy, the UK, and France.

Fig. 2 shows the main themes of the publications, and number of publications which conform to each theme. Overall, the studies indicate that formal method is a prioritised issue in terms of the volume of publications dedicated to this topic. There is a high number of academic contributions regarding the formulation modelling and validation of railway and ERTMS subsystems and components.

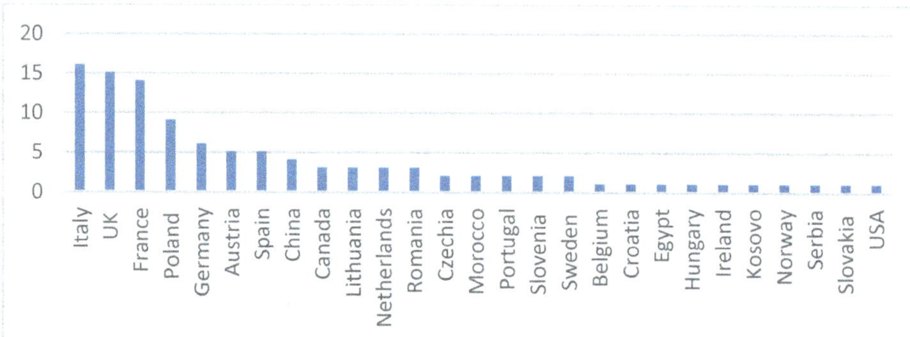

Figure 1: Number of publications by country.

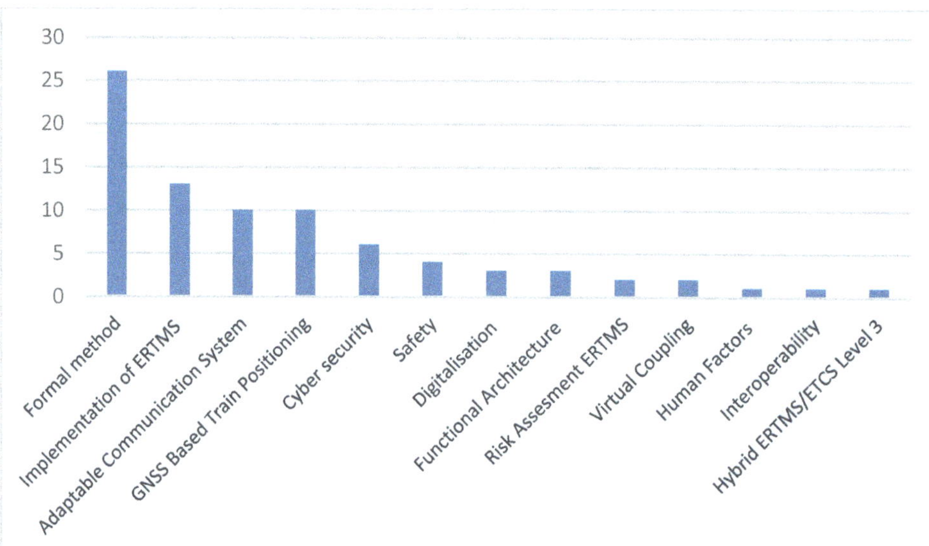

Figure 2: Main themes and their number of publications.

The following three themes were the next most-represented: the implementation of ERTMS, adaptable communication systems, and GNSS-based train positioning. Regarding adaptable communication systems, there are discussions about replacing the current communication system (GSM-R) with new backward and forward interoperable technologies that provide higher reliability and a greater capacity of data communication.

Table 5 shows the studies' publication years, the number of revised papers. Most of the reviewed papers were published in 2018, and represent a major increase in publication volumes since 2015. Table 6 shows the types of publication channel, the most literature has been published in conferences.

The reviewed studies have been categorised into 12 main themes; for each category, the most-cited or recently-published paper has been studied in depth. These main themes represent key issues to be addressed in future developments of ETRMS, particularly the move towards mobile and intelligent railway transport. Table 7 shows key references for each of the 12 main themes.

Table 5: Year of publication of reviewed papers.

Year	Number of publications
2019	15
2018	40
2017	16
2016	9
2015	2
SUM	82

Table 6: Year of publication of reviewed papers.

Type of publication	Number of publications
Journal article	11
Book chapter	3
Conference paper	68
SUM	82

Table 7: Main themes in the reviewed papers and key publications.

Main theme	Key references
Formal methods	Basile et al. [4]
Adaptable communication system	Mazzenga et al. [2]
GNSS-based train positioning	Beugin et al. [6]
Implementation of ERTMS	Kornaszewski et al. [10]
Cyber security	Armengaud et al. [11]
Safety	Gario et al. [12]
Digitalisation	Tokody et al. [8]
Risk assessment of ERTMS	Bloomfield et al. [13]
Virtual coupling	Flammini et al. [14]
Human factors	Naghiyev et al. [15]
Interoperability	Rajabalinejad et al. [16]
ERTMS-ETCS Hybrid Level 3	Hoang et al. [3]

Formal methods for modelling, testing, and validation play major role. Formal methods can also include semiformal models. Finding a new adaptable communication system is one solution to replace the current communication system, with the aim of enhancing data transmission and the level of service (LOS). The implementation of ERTMS involved several challenges. For example, ERTMS is cyber-physical system, and there is a potential risk of cyber-attacks and other safety issues due to the combined physical and digital nature of this infrastructure. Preparedness and capacity to face issues of cyber security and safety will play an important role in the development of railway communication systems going forward.

Table 8 presents the main themes and related critical issues of the reviewed literature, and which are discussed in greater depth in the next section, along with some proposed solutions.

Table 8: Main themes and related critical issues.

Main theme	Critical issues
Formal methods	The lack of a unified formal language and method dedicated to just modelling, verification and validation of railway systems necessitates the development of a formal tool specific to railways and moving blocks, with appropriate attributes [4].
Adaptable communication system	Increasing demand in railway networks requires the replacement of GSM-R with FRMCS that provide a higher QOS and capacity of data transmission. It must be able to guarantee safety requirements and forward, as well as backwards interoperability [1], [2].
GNSS-based train positioning	Calculation of potential hazardous situations caused by GNSS error or hardware/software failure and connection disruption, safety, and architectural criteria to fulfil ERTMS requirements [6].
Implementation of ERTMS	Deployment of Intelligent Transport Systems [10].
Cyber security	Difficulty in assuring the dependability hybrid cyber-physical systems (CPS) [11].
Safety	Estimate interactions between different data flows and software layers in order to estimate risk [12].
Digitalisation	Increasing the safety of digitalised systems, e.g. in level crossings and communication between rail convoys and road vehicles [8].
Risk assessment ERTMS	Higher connectivity and higher hazardous situation-related electronic attacks [13].
Virtual coupling	Train-to-train, train-to-wayside subsystem communication and the replacement of GSM-R with a new communication system [14].
Human factors	Driving models, unfamiliarity with new signalling schemes [15].
Interoperability	The diversity of goals and operating criteria between national and international levels for IMs and RUs reduces interoperability and increases complexity [16].
Hybrid ERTMS/ETCS Level 3	Operation of different trains with and without TIMS, on a Virtual Sub-Section (VSS) and 12 possible transition states, loss of connectivity of train communication [3].

4 DISCUSSION

The development of the European Railway Traffic Management System (ERTMS) throughout Europe over the last two decades to remove the obstacles for providing an interoperable railway network has faced several challenges. This study has aimed to indicate briefly some of the main challenges facing ERTMS by analysing its current architecture and operation. Table 9 is a summary of proposed solutions, as found in the reviewed literature. Significant challenges include the diversity of formal methods, tools, and languages being used to model, formalise, test and validate ERTMS products [4]. An adaptable communication system is needed to replace GSM-R. LTE/4G and 5G have been proposed; with whatever strategy this proceeds, a hybrid system ensuring interoperability with previous and future solutions is critical [1], [2]. GNSS-based train positioning for the calculation of potential hazardous situations, taking into account the principals of safety criteria of railway systems, hybridised architecture, and the calculation of potential hazardous situations are similarly important concerns [6]. Overall, the implementation of ERTMS requires a step-by-step upgrading of the railway system [10].

Table 9: Main themes and proposed solutions.

Main theme	Proposed solutions
Formal method	Developing a universal formal method for the railway industry [4].
Adaptable communication system	Replacing GSM-R with new solutions, including LTE/4G, 5G or a hybrid system, to ensure interoperability with previous and future solutions [1], [2].
GNSS-based train positioning	For the calculation of potential hazardous situations, taking into account the principals of safety criteria of railway systems, hybridised architecture, and the calculation of potential hazardous situations [6].
Implementation of ERTMS	Step-by-step upgrading of the railway system [10].
Cyber security	Designing proper tools to guarantee the safety and dependability of Cyber-Physical Systems (CPS) [11].
Safety	Employing clear graphical presentation of safety data and predicted interactions between different data flows and software layers to effectively estimate risk [12].
Digitalisation	Equip trains and road vehicles with automatic protection systems [8].
Risk assessment ERTMS	Higher connectivity and a higher number of hazardous situations caused by electronic attacks require the improvement of security in connected systems [13].
Virtual coupling	Replacing GSM-R with new communication systems (5G) [14].
Human factors	Training train drivers to address knowledge gaps between drivers and new train driving models [15].
Interoperability	Stabilise technical specification by a unique designer entity [16].
Hybrid ERTMS/ ETCS Level 3	Guaranteeing operation without collision by hazard mitigation of the system [3].

In the field of cyber security, the assurance of dependability of Cyber-Physical Systems (CPS) relies on the design of proper tools [11]. These must be capable of estimating the interactions between different data flows and layers of software in order to effectively estimate risk and present this in a clear, graphical form [12]. Increasing the safety of digitalised systems, e.g. in level crossings and communication between rail convoys and road vehicles, likewise needs to be addressed [8]. Risk assessment studies show that higher connectivity and an increasing incidence of hazardous situations due to electronic attacks requires the improvement of security in connected systems [13]. In the area of virtual coupling, train-to-train and train-to-wayside subsystem communication systems are at risk of failure because of the current GSM-R; these must be replaced with a new communication system, perhaps one based on 5G technology [14]. The unfamiliarity of train drivers with new driving models and new signalling schemes can be addressed through increased professional training [15]. The diversity of goals between nations and at the international level for IMs and RUs reduces interoperability and increases complexity. Rajabalinejad et al. propose to stabilise technical specification through the creation of a unique designer entity [16].

Some challenges related to Hybrid ERTMS/ETCS Level 3 include the operation of different trains with and without TIMS on a VSS, different possible transition states, loss of connectivity of train communication, and ghost and shadow trains. Addressing these challenges will require guaranteeing operation without collision by moderating risk within the system [3].

5 CONCLUSION

This study has presented an overview of the main issues discussed in recent research. The studies reviewed illustrate that in order to simplify the interaction of components and subsystems of ERTMS, to reduce the cost of implementation and maintenance, and to increase interoperability, it is necessary to reduce the number of wayside components and pivot the system as a whole toward mobile equipment, autonomous trains, and decentralised communication. In the area of formalisation, it is necessary to develop a robust and reliable tool for more standardised modelling, formalisation, testing and validation of critical safety systems specific to the needs of the railway industry and to be used by all involved parties. To address knowledge gaps between human operators and new technologies, reduce human error, and increase performance, an emphasis on developing the skills of railway employees and operators with new technologies is highly recommended. In order to save and build upon previous investments in ERTMS, it is necessary to pursue hybrid solutions and gradually update current systems in order to guarantee backwards and forwards interoperability.

REFERENCES

[1] Canonico, R., Marrone, S., Nardone, R. & Vittorini, V., A framework to evaluate 5G networks for smart and fail-safe communications in ERTMS/ETCS. *Reliability, Safety, and Security of Railway Systems. Modelling, Analysis, Verification, and Certification*, eds. A. Fantechi, T. Lecomte & A. Romanovsky, Springer: Cham, 2017.

[2] Mazzenga, F. et al., The adoption of public telecom services for the evolution of the ERTMS-ETCS train control systems: Challenges and opportunities. *Communication Technologies for Vehicles*, eds. M. Kassab, M. Berbineau, A. Vinel, M. Jonsson, F. Garcia & J. Soler, Springer: Cham, 2015.

[3] Hoang, T.S., Butler, M. & Reichl, K., The Hybrid ERTMS/ETCS Level 3 case study. *Abstract State Machines, Alloy, B, TLA, VDM, and Z. ABZ 2018*, eds. M. Butler, A. Raschke, T. Hoang & K. Reichl, Springer: Cham, 2018.

[4] Basile D. et al., On the industrial uptake of formal methods in the railway domain. *Integrated Formal Methods. IFM 2018*, eds. C. Furia & K. Winter, Springer: Cham, 2018.
[5] Bernardi, S. et al., Towards a model-driven engineering approach for the assessment of non-functional properties using multi-formalism. *Software and Systems Modeling*, **18**, pp. 2241–2264, 2019. https://doi.org/10.1007/s10270-018-0663-8.
[6] Beugin, J. Legrand, C., Marais, J., Berbineau, M. & El-Koursi, E., Safety appraisal of GNSS-based localization systems used in train spacing control. *IEEE Access*, **6**, pp. 9898–9916, 2018.
[7] Neri, A. et al., An anti-jamming and anti-spoofing digital beamforming platform for the GNSS-based ERTMS train control system. *Proceedings of the 30th International Technical Meeting of the Satellite Division of The Institute of Navigation (ION GNSS+ 2017)*, Portland, Oregon, pp. 3538–3556, 2017.
[8] Tokody, D., Mezei, I.J. & Schuster, G., An overview of autonomous intelligent vehicle systems. *Vehicle and Automotive Engineering. Lecture Notes in Mechanical Engineering*, eds. K. Jármai & B. Bolló, Springer: Cham, 2017.
[9] Schmittner, C. et al., Threat modeling in the railway domain. *Reliability, Safety, and Security of Railway Systems. Modelling, Analysis, Verification, and Certification*, eds. S. Collart-Dutilleul, T. Lecomte & A. Romanovsky, Springer: Cham, 2019.
[10] Kornaszewski, M., Chrzan, M. & Olczykowski, Z., Implementation of new solutions of intelligent transport systems in railway transport in Poland. Smart solutions in today's transport. TST 2017. *Communications in Computer and Information Science*, ed. J. Mikulski, Springer: Cham, 2017.
[11] Armengaud, E. et al., DEIS: Dependability Engineering Innovation for Industrial CPS. *Advanced Microsystems for Automotive Applications 2017*, eds. C. Zachäus, B. Müller & G. Meyer, Springer: Cham, 2018.
[12] Gario, A., Andrews, A. & Hagerman, S., Fail-safe testing of safety-critical systems: a case study and efficiency analysis. *Software Quality Journal*, **26**, pp. 3–48, 2018.
[13] Bloomfield, R., Bendele, M., Bishop, P., Stroud, R. & Tonks, S., The risk assessment of ERTMS-based railway systems from a cyber security perspective: Methodology and lessons learned. *Reliability, Safety, and Security of Railway Systems. Modelling, Analysis, Verification, and Certification. RSSRail 2016*, eds. T. Lecomte, R. Pinger & A. Romanovsky, Springer: Cham, 2016.
[14] Flammini, F., Marrone, S., Nardone, R., Petrillo, A., Santini S. & Vittorini, V., Towards railway virtual coupling. *2018 IEEE International Conference on Electrical Systems for Aircraft, Railway, Ship Propulsion and Road Vehicles & International Transportation Electrification Conference (ESARS-ITEC)*, pp. 1–6, 2018.
[15] Naghiyev, A., Sharples, S., Ryan, B., Coplestone, A. & Carey, M., Expert knowledge elicitation to generate human factors guidance for future European Rail Traffic Management System (ERTMS) train driving models. *Proceedings of the Institution of Mechanical Engineers, Part F: Journal of Rail and Rapid Transit*, **231**(10), pp. 1141–1149, 2017. https://doi.org/10.1177/0954409717695902.
[16] Rajabalinejad, M. & Schuitemaker, K., ERTMS Challenges for a safe and interoperable European railway system. *PESARO 2017. The Seventh International Conference on Performance, Safety and Robustness in Complex Systems and Applications*, pp. 17–22, 2017.

RESEARCH DIRECTIONS REGARDING THE ADOPTION OF FORMAL METHODS IN THE RAILWAY SIGNALING SECTOR: DETERMINANTS AND NEXT STEPS FOR FUTURE-PROOF RAILWAYS

DIMITRIOS RIZOPOULOS[1], NILS O. E. OLSSON[2], ANDERS LINDAHL[1] & OLOV LINDFELDT[3]
[1]KTH Royal Institute of Technology, Sweden
[2]NTNU Norwegian University of Science and Technology, Norway
[3]MTR Pendeltågen, Sweden

ABSTRACT

The terms Formal Methods (FMs) refer to a set of techniques and software toolkits that, based on mathematical rigor, can enhance safety, security, and the efficient operation of a wide range of systems. Considering that several innovative FMs applications have been performed over the last few decades, the utility of toolkits that are based on FMs has already been showcased in several industrial settings, such as the avionics and automotive industries, medical devices, computer software, and hardware systems, and finally, the railways and the railway-signalling sector. The current article focuses on the last of the aforementioned sectors, that of railway signalling, and aims to analyse research directions that regard the adoption of FMs in signalling. Despite the benefits and the availability of reports on the topic, the implementation of the adoption of FMs can be considered yet to be successful in most organizations that develop related systems. The authors have observed that this implementation lag in the adoption of FMs may stem from the absence of systematic approaches to the study of adoption. In that regard, a set of five Research Questions (RQs) is introduced in this article for the adoption of FMs in railway signalling systems. Furthermore, answers regarding two of those questions are given within the current article regarding: (i) studies on the adoption of FMs, and (ii) successful applications and benefits of FMs. The remaining three RQs are intended to point out issues for future research. By providing answers to all five RQs, it is expected that a map for the demand of FMs applications in railway signalling is introduced in the following years, as well as the role of the potential stakeholders in those applications.

Keywords: formal methods, railway signalling, technology adoption.

1 INTRODUCTION

Various meanings have been considered for the term *innovation.* In the study by Edison et al. [1], several definitions have been analyzed for coverage against aspects of the meaning of *innovation*, as identified in the scientific literature. The results showed that the most suitable was given in the article by Crossan and Apaydin in 2009 [2], who state that: "Innovation is: production or adoption, assimilation, and exploitation of a value-added novelty in economic and social spheres; renewal and enlargement of products, services, and markets; development of new methods of production; and establishment of new management systems. It is both a process and an outcome". With adoption being one of the key aspects of innovation, once an innovative method is identified, its implementation within organizations can take place, with its diffusion taking place from an individual or a group who may be called innovators, to other groups that can adopt the innovation and can possibly benefit from its utilization. While the aforementioned diffusion and adoption process of different innovations can take place across many organizations in several stages [3], the attempt to incorporate innovations to the way the companies operate can be proven to be a trivial task.

The current article discusses the adoption of FMs or semi-Formal Methods (semi-FMs) tools and toolkits, as an innovative technology, to the sector of railway signaling. Regarding

WIT Transactions on The Built Environment, Vol 199, © 2020 WIT Press
www.witpress.com, ISSN 1743-3509 (on-line)
doi:10.2495/CR200071

FMs, the terms include all the mathematical based methods that allow the rigorous specification of cyber-physical systems, their formal development, and verification through mathematical proof. In more detail, and concerning the railway-signaling sector and the traffic control systems, some of the benefits that FMs enable are:

- The enhancement of safety and security in railways thought the formal specification, modeling, development, and verification of signaling systems, in fully operational or degraded operation scenarios;
- Automated development of verified signaling systems and sub-systems that comply with international standards;
- Improvement of the specification, design, and architecture of a system, primarily through FMs or semi-FMs application in the early stages of a project life cycle;
- Re-usability of artifacts from a single product development process into multiple projects;
- Rigorous specification of requirements in procurement tasks;
- The standardization of protocols, the reduction or improvement of interfaces between signaling systems and sub-systems, and the efficient management of increasing complexity and automation;
- The cost reduction and schedule compression of projects' through the application of FMs or semi-FMs on multiple levels of the project.

Secondary long-term improvements in the signaling systems themselves can also affect the train services, with services' punctuality [4] or track capacity [5] being directly connected to signaling systems in previous studies. Despite the potential benefits, the adoption of FMs in railway signaling can be considered not to be complete yet as organizations hesitate to invest or utilize FMs tools in their projects. On a higher level, this lag in the adoption can be partially attributed to the general misconceptions that regard FMs regardless of the application sector. They are related to the perceived ease and benefits of use or even concerns regarding the support that FMs tools usually get from their developers. Those generic, as we may call them, misconceptions that experienced professionals in the field of FMs are already aware of, have been available in the scientific literature since 1990 in reviews [6], [7].

In order to extract the benefits of the utilization of FMs, product development routines, as followed by railway signaling companies, would need to include FMs tools or toolkits at some of their stages. The FMs toolkits, do not necessarily have to be the scientific methods themselves but can have a lower or higher level of abstraction. Although there are several mature studies [8]–[11] that review FMs tools and applications, there are only a few reports [12]–[17] that address the adoption of FMs, or include references to adoption and implementation.

The *purpose* of this article is to identify appropriate Research Questions (RQs) for further research regarding FMs adoption in railway signaling. Based on observations from both the academic and industrial settings, a set of five questions is given throughout this article, with *two* being answered within this article. Those first two RQs are:

1. What is the current state of adoption of FMs in railway signaling?
2. What are the potential benefits of FMs applications in railway signaling?

The latter three questions, and proposing ways to address them, are included in the conclusion. All of the five (first two and latter three) questions that are proposed in this article are defined on a general level so that future research is not constrained by the scope of those directions, as they are presented in this paper, but is instead guided within the goals that need to be attained. Our proposal is that those questions and answers are provided as a Layman's report in the field of FMs application to railway signaling.

Our method for the derivation of insights and results presented within this article is a literature review that sourced articles, reports, and books from several platforms such as ScienceDirect, Google Scholar, ResearchGate, and Web of Science. In that regard, Section 2 of this paper discusses the current state of adoption of FMs in railway signaling systems development. Section 3 gives an overview of applications that exist within signaling, where potential benefits are showcased from mainly the academic but also the industrial environment. Finally, in the last concluding section, we provide an overview of what has been discussed and what are some further research directions.

2 THE CURRENT STATE OF ADOPTION OF FORMAL METHODS IN SIGNALING

Regarding the adoption of FMs in railway signaling, it is a phenomenon that has not yet been studied in great depth from the academic perspective, and only very few articles can be found that address or even discuss this problem. All of the works tend to share empirical knowledge that has been gained through projects. Thus, the authors have observed that a systematic study of adoption is a research gap in the literature.

Let us first present the existing work. The first article by Bacherini et al. [12], several insights are presented regarding adoption that stem from the attempt to disseminate FMs tools in the railway signaling division of General Electric Transportation Systems (GETS) in collaboration with academic experts, when the problem of certifying software components within the CENELEC guidelines appeared. While several tools were considered, the main aim was to establish and disseminate knowledge about the semi-formal Stateflow tool of Matlab within GETS. As mentioned in the paper, the compatibility of Stateflow with other Matlab and Simulink tools, and the already existing experience in Matlab usage within the company, has been proven to be beneficial factors. At the time of the publication of the article, the authors discuss that while the adoption is not full yet, it is implemented on a project by project base. They point out that several factors benefited the process of adoption, such as: (i) collaboration with academia; (ii) prolonged time-to-market phase, and thus, more time for experimentation; (iii) the pressure as perceived by the organization and professionals to comply with CENELEC; (iv) the general market evolution; and (v) the pressure, as exercised by clients to include rigorous and formal specifications.

Another article was published in 2013 by Ferrari et al. [16], who share the challenges and learning lessons that GETS faced in adopting model-based development regarding the railway signaling systems developed at the company, in contrast to code-based processes that have been traditionally followed. One of the concluding remarks about the adoption of FMs and semi-FMs in GETS is that it should happen in steps, with the early stages focusing on less formal tasks and later move on to use FMs for more difficult tasks. Interestingly, here the authors point out the importance of a multi-stage approach regarding the technical aspect of adoption, so that industrial practitioners can assimilate the new knowledge on a task by task basis. We would like to state that this opinion resembles our thoughts, and further augments the already well-received multi-stage paradigm of analyzing technology adoption [3].

In the same year, Ferrari et al. [18] have published an article on a case study on the development of the driving support Automatic Train Protection (ATP) system Metrô Rio according to a formal model-based process, as well as another report [17] to discuss adoption regarding the project described in [18] alongside another two projects dealing with FMs application to ATP systems.

Additionally, in work by Huhn and Milius [13], insights are given regarding adoption, and the impression of the industrial partner to the case study presented for a Level Crossing (LC, LCs) controller system. The authors highlight that human expertise and perceived effort are two of the most critical barriers to adoption.

Moving on to the second article by Basile et al. [14], where the authors present the results of questionnaires that have been conducted within the frame of ASTRail project that concern the industrial uptake of FMs in railways. More specifically, the criteria for this questionnaire were (i) understanding adoption (i.e., uptake) of FMs and (ii) discover the features that the railways professional considered as more vital (i.e., relevant) for the application of FMs. The questionnaire was addressed to the participants of RSSRail17, and the results show that the B method is one of the most well-established tools and that the maturity of the tools, the learnability, and the documentation were among the most essential features. In this survey, the cost of the tools was not considered to be highly connected to the adoption of FMs since only 13.6% of the respondents consider “low-cost” as an attributable determinant for the use of the corresponding FMs toolkits.

Next, we will refer to the work by ter Beek et al. [15], where a summary of the results of three questionnaires that regard adoption are given that were conducted within ASTRail and X2Rail-2. Two of the questionnaires were conducted within X2Rail-2 and another one in ASTRail project (same with Basile et al. [14]). As given in the conclusion of the study [15], the results were harmonized and can be summarized in the following key points: (i) Most FMs applications are focused on interlocking systems within organizations; (ii) FMs are mainly used for formal specification and verification; (iii) The B-family tools are the most well-established with SCADE and Simulink being also already well-received; (iv) Essential features of tools are considered formal verification, formal modeling, traceability, simulation, test-generation and, finally, code generation; (v) regarding the adoption of tools, the most important determinants were documented to be tool maturity, usability, and learnability of the tools. An interesting insight also provided in the same article is that the industrial perception of FMs tends to fluctuate and is not solid, with FMs may sometimes be confused with semi-FMs. This insight confirms in a way our previous statement in this article that, possibly, similar misconceptions [6], [7] that exist for FMs on a general level, also, exist within railway signaling.

Overall, in this section all reports and articles have been included which discuss the adoption of FMs in railway signaling, and that can be found on online bibliographic databases at the time of publication of this article. One can notice a cohesion in the results of the articles in the literature of FMs adoption in railway signaling. From those observations, it could be claimed that it is only now that professionals in the field are starting to accumulate and formalize the knowledge that has been acquired through research in both the academic and the industrial settings regarding how FMs toolkits can be disseminated into organizations.

3 STUDIES AND APPLICATIONS THAT UTILIZE FORMAL METHODS IN RAILWAY SIGNALING

One of the driving forces for utilizing FMs in railway signaling the fail-safe operation principle. In other terms, signaling equipment must be guaranteed to operate the trains safely, even when some of the sub-systems have failed, so that human lives are not at stake. This is one of the main reasons for which signaling systems are categorized as safety-critical systems and are subject to several safety guidelines and imposed standards for their development. The most prominent related example is EN50128 guidelines by the European Committee for Electrotechnical Standardization (CENELEC). Although some professionals claim that the demand for FMs applications stems only from legislation, it is now obvious more than ever that there are other use-cases outside legislative requirements, for which FMs can be proven beneficial. In the next few Subsections 3.1 to 3.8, a review of articles and reports on studies and applications that have showcased beneficial results for the systems is given. The value-added result of the following review is not to be exhaustive in terms of articles within each application field, but its goal is instead to assemble all applications and use cases, for which articles and reports can be found within the scientific literature.

3.1 Level crossings (LC, LCs)

Although the terms railway level crossing could refer to the physical location where the railway line crosses the road or pedestrian transport network, in this subsection of the article we will refer to studies that report on successful applications of FMs approach to the safety-critical cyber-physical systems and sub-systems of the LCs that ensure the proper function of the equipment. Regarding LCs, one of the main aims of applying FMs is the assessment of safety, especially in the case of failed components locally or across the network. Of great importance is to reassure that the LC can work as an independent system and guarantees fail-safe operation as an automatic system since it is considered to be a black spot for railway transportation systems.

An approach to LC verification of safety that is worth mentioning is the one by Mekki et al. [19] that came in 2012. In their article, they focus on two hazardous situations that are connected to several LC accidents, and though the specification by formal notation and the verification by model-checking, they validate a new architecture for LC that can deal with those two specific situations. Another set of studies on LC is included by Mohamed Ghazel [20], [21]. A Petri net model-based simulation approach to the risk assessment of various dynamics (i.e., collision cases) within the LC environment is introduced in 2009 in the first article [20], where the model occurs progressively by integrating other simpler parts of the LC and their behavior, whether it regards how road or rail traffic affects the overall LC. Finally, we refer to the last study by Ghazel [21], where the author proposes a new LC control/command architecture in the context of ERTMS L2/3 operation. The article provides some beneficial insights regarding FMs capabilities, which on the one hand, can be the elimination of on-site testing, given that enough practical constraints are accommodated, or, on the other hand, the validation of new architectures that can be more efficient (fewer sensing devices).

3.2 Systems related to moving block

The moving block mode of operation for a railway network or line is the case when the signaling systems enable the fragmentation of routes of the network into *moving* blocks, in real-time, with the support of computer systems that consider parameters such as train positions, speeds, acceleration, and braking distances. While the operation of moving block lines is already applied at several transportation systems (London Underground and New York City Subway), it can be considered that further specification work could be useful for its global acceptance and application to networks.

FMs have been used for the performability evaluation of the ERTMS/ECTS Level 3 system in work by Biagi et al. [22]. The authors introduce an approach to evaluate emergency stops of a system under ERTMS/ECTC Level 3 operation that are caused by a combination of communication failures that could be burst noise in communication links and connection losses due to failure of devices or cell handovers when reaching borders of GSM-R cells. In the work by Hansen et al. [23], a full executable application based on Hybrid ERTMS/ECTS Level 3 (HBL3) is presented. HBL3 operational scheme does fully utilizing moving block functionality but instead divides the sections into virtual subsections which are used as fixed blocks, but provide finer granularity, as compared to total fixed block operation (such as in the case of ERTMS/ECTS Level 2). Insights are provided about FMs tools support as well as for the HBL3 as a system, too. A Virtual Block Function (VBF) is included in the approach, which behaves as an interlocking that transmits virtual signals to the RBC. In contrast, the implementation of the whole approach, by B Method family tools, can be considered as an executable reference specification for the demonstration and understanding of HBL3 and its core principles.

In the article by Qian et al. [24], a SCADE based development process is showcased to model and verify the Zone Controller of CBTC that enables the moving block function by calculating movement authority of trains. The authors of this article suggest a way to deal with the state explosion problem that occurs when many real systems are modeled and then verified according to FMs approaches. In general, the state explosion problem is considered to be a bottleneck for FMs applications to real systems; thus, solutions regarding it are beneficial within railway signaling applications. In more detail, they propose slicing criteria that can simplify the complexity of the SCADE models and can verify the correctness of movement authority calculation for a given train network. Such slicing criteria, and similar decomposition approaches, are deemed to be necessary for the efficient execution of calculations that are based on complicated models of vast size in other adjacent computer science applications fields [25]. Similarly, studies like the one by Qian et al. [24] could be necessary to be expanded in order to enable the construction and verification of models of signaling systems of large sizes.

3.3 Automatic Train Protection/Control (ATP/ATC) functions/systems

Those systems are generally considered to be protection systems that include speed and position control mechanisms that function based on inputs from external sources such as track-side equipment. They are used to guarantee safe speeds, braking distances, and breaking system triggering of the vehicles concerning the *braking curve* of the trains. They can be implemented as driver support systems, or they can be systems that coordinate the driverless operation of trains. ATP/ATC can be used to refer to the overall function of those systems, including both track-side (i.e., wayside) and on-board equipment. FMs can be used here to assure the safe operation of trains in fully operational scenarios or degraded scenario when some of the components of the system do not work as expected.

Successful applications can be found since 1993 [26], but other modern approaches can be found in the literature, such as the work by Wu et al. [27], where a mathematical problem is built-in ANSYS's SCADE Suite to calculate the dynamic curves of the ATP system. The curves are modeled and verified through several techniques such as a simulation test, a model test coverage (according to test cases), and formal verification (through SCADE's integrated tool). Another approach to FMs application on ATP systems has been presented by Ferrari et al. [18], wherein their case study Simulink and Stateflow have been used to produce a formal model of on-board ATP equipment of the Metro Rio Automatic Train Protection system. The wayside objects are modeled as an encoder and a transponder that provide the processed data to the on-board system, which is of the main interest in this study. Among the benefits that are reported in this study is the reduction of the bugs found during development by 71% and the average time of bug detection and correction from almost an hour to less than 30 minutes per bug.

3.4 Automatic Train Operation (ATO) function

Another notable instance of a successful application is the application of FMs to the ATO system for the Paris Metro line 14 (under the Meteor project/automatic train control system). A report by Behm et al. [28] presents processes followed for the development of the safety-critical software of this line, as well as insights on the organization and roles of teams within the project. Behm et al. [28] state in their conclusions that the application of FMs in the Meteor project has been considered cost-effective, with the increase in quality being noticeable even before systems' maintenance costs are considered. An even further successful application of the B method is reported in the paper by Essamé et al. [29] for the New York City Canarsie Line, whose critical safety software development was proven to be more demanding than that of the Meteor system. The study by Essamé et al. [29], which can be considered a continuation study when compared to the one by Behm et al. [28], states that B method tools are incorporated in the development of safety-critical software within the organization.

3.5 Communications-Based Train Control (CBTC)

Under the terms of CBTC, all signaling systems are included that are used for the communications between the trains running on a track of the network and the track-side equipment. CBTC systems are comprised of several software and hardware sub-systems that are adapted to each application, with both on-board and track-side equipment. In some cases, this adaptation happens through configuration data, which are considered to describe the arrangements of the equipment and their functions alongside the rail infrastructure. In contrast to dynamic data, static configuration data (e.g., geographical arrangement of the equipment) never change, and their safe operation and compliance with safety and functional requirements must be guaranteed in the initial life cycles of a project. Successful applications of FMs on configuration data have been conducted in the past [30]. FMs applications on CBTC systems can be found for other mechanisms/parts of the CBTC system, such as track circuits or emergency cancellation functions, as reported in the article by Comptier et al. [31]. In this article, it is described how Event-B and Atelier B were utilized to perform rigorous safety analysis of the multi-supplier interoperable *Octys* (Open Control of Train Interchangeable and Integrated system) CBTC system which has been deployed on Paris metro lines 3, 5 and 9 starting 2010 as reported in the article.

3.6 Railway interlocking systems

A railway interlocking system, or as it usually called interlocking(s), is one of the systems responsible for the safe guidance of trains through the network and is assigned with Safety Integrity Level 4 (SIL4) within the CENELEC EN50128 guidelines. As stated above, SSIL 4 is the highest level of safety integrity, as described in EN50128. Interlockings allow the movement of trains according to a set of rules that affect track-side elements. These movement authorities that are provided to train have to occur according to a specific way, or as it widely called according to interlocking logic. Interlocking logic has to be mathematically examined and proved to reassure the fail-safe operation of the network. Several successful applications have been showcased for interlocking, which can be considered a leading example in the field of FMs applications to railway signaling. It has been decided to not explain interlocking articles and reports in further detail, as instances of successful application can be easily found both within the academic environment [8], [32], but also from an industrial setting [33].

3.7 A short reference to other systems in railways

Some other instances of successful application can be found for the railways industry. One example is the work reported by Van Tang et al. [34], who report the application of FMs framework called the Vienna Development Method (VDM++) on a Train Fare Calculation and Adjustment System. These kinds of systems are the ones utilized to assist passengers at acquiring tickets at the right price for their trips within the transportation network. Because those systems tend to serve vast numbers of passengers on a daily basis, a small error in the system can lead to significant losses in short amounts of time for either the passengers or the operators. As in the case of other signaling systems mentioned above, the existence of errors once a system is deployed can lead to the need to fix those errors while the system is on-line and running, which can be a hard task for all stakeholders involved, such as software companies, services operators, infrastructure managers, and governments. In the case study, as presented [34], for a Train Fare Calculation and Adjustment System in Japan, the application of VDM++ has been viewed as very successful as it allowed the developers to come up with 32 erroneous issues in the informal specification document for the system that uses both newly developed modules but also legacy ones from an existing fare calculation system.

Platform Screen Doors (PSD) is another system that FMs have been applied to. PSD are the doors that are used to prevent passengers or users of rail or metro services from falling on to the track. Their use is of great importance for automatic/autonomous trains that do not have an on-board crew that manually checks whether all passengers are aboard. Also, their performance, correct operation, and relation to dwell times of services at stations, even in non-erroneous situations, has been linked to a decrease in capacity in some cases of study [35]. Thus, such a system is uniquely in need of certified proof of proper function to facilitate modern automatic/autonomous railway systems. A successful application of FMs on such a system has been implemented in the past [36], where the authors discuss an application where functional and safety requirements of the PSD and its controller software were verified. In the same article, it is reported that, at the time of publishing, the system successfully operated for eight months, controlling 96000 trains, without a single fault in its operation.

3.8 Security of railway signaling systems

Most of the articles and reports referenced in Section 3 of this article refer to the safety aspect of signaling systems and reassuring that the fail-safe operation of a railway line is never compromised, even in cases when some of the components fail. However, since railway signaling systems are mixed-criticality systems, their web resilience has to be evaluated and guaranteed in all cases. With the increase of phenomena like cyber-terrorism and malevolent attacks in general, FMs can be used to create fully secure architectures that allow the proper function of the railway network, while in the same time they enable the developers to make sure that security requirements are not violated across the engineering and development processes. Reports can be found on such cases, such as the one by Heinrich et al. [37], who propose several security requirements that architectures, such as the one proposed in this article, should cover.

Overall, the scope of Section 3 was to provide a comprehensive report, or a map, as we would like to call it, of FMs studies and applications in railway signaling (a few adjacent fields within railways). The goal for that section was to provide an answer to RQ 2 as provided in the introduction and provide a Layman report for what is possible within railway signaling by using FMs.

4 CONCLUDING REMARKS & FURTHER RESEARCH QUESTIONS

The current situation regarding FMs and their adoption to railway signaling applications can be considered to be in a stage where FMs are getting more widely acknowledged, distributed, and used over the years, but are not deeply integrated into systems development life-cycles yet at this point, and not utilized to their full potential. While there are enough tools and some successful applications, the understanding of the diffusion of their adoption remains a trivial task. Besides, from previous studies, it is noticeable that the use of a standard toolkit has not arisen yet, and different organizations use different approaches according to each team's or engineer's perspective and background. To gain a better understanding of FMs adoption, and from the literature study and several observations as presented in this article, the following five RQs arise:

1. What is the current state of adoption of FMs in railway signaling?
2. What are the potential benefits of FMs applications in railway signaling?
3. What is the current and future demand for FMs applications in railway signaling, and who are the stakeholders in the existing and potential projects?
4. What are the determinant factors of user acceptance of FMs as a new technology in organizations in railway signaling?
5. What are the benefits for organizations that invest in FMs tools and their adoption regarding cost reduction and schedule compression of signaling projects?

The first RQ question is answered by Section 2 of this article. From all of the available material that discusses adoption, it can be noticed that some potential crucial factors for adoption are: (i) adoption should be studied from both an academic and industrial setting; (ii) definition of the learning curve (considering ease of use, documentation and available support) for toolkits can be beneficial, especially on an individual user perspective; (iii) compatibility across tools of toolkits can be substantial; (iv) on a general level, formal specification, formal verification, and formal modeling are the most relevant features, as observed from questionnaire-based studies.

The second RQ is answered in this article, too, in Section 3. A comprehensive review is given, where articles and reports on studies and applications that showcase potential benefits have been presented. Overall, the main takeaway regarding this second question is that successful applications exist within the academic sector, but what would be more interesting is to see how those benefits that stem from academia translate into real-world practice and industrial applications. Reports on industrial projects can be beneficial for the dissemination of knowledge within railway signaling.

Regarding RQ three, it is a research question that is expected to be answered by an expert panel on FMs applications to signaling. The value-added result in the potential answer has to be supported by data on current and future projects. Discussion of stakeholders and their role in projects and applications could be highlighted.

Regarding RQ four, we propose the study of determinants of user acceptance of FMs within railway signaling. User acceptance by individual professionals within the sector is an important step, as also mentioned by Kamal [3]. In order to conduct such a systematic study of an adoption of a technology, several theoretical lenses can be used, such as the work by Venkatesh et al. [38] on the Unified Theory of Acceptance and Use of Technology (UTAUT). In UTAUT, the authors capture the most practical and essential aspects of eight previous models and consider four key constructs, and the relationship between them, as the most critical determinants for user intentions to utilize a new information system and their corresponding usage behavior. Namely, the constructs are (i) performance expectancy, (ii) effort expectancy, (iii) social influence, and (iv) facilitating conditions, while the research model also accounts for the role of key moderators such as gender, age, voluntariness, and experience. Regarding the application of such a framework, in the case of UTAUT, one could claim that is the most prominent framework used in modern literature with the original paper being cited 27529 times, several examples can be taken from previous studies that study technologies that range from course management software to internet banking. Given its widespread acceptance and cross-field compatibility, UTAUT can be proven to be suitable to the case of FMs application to railway signaling. By systematically studying the determinants of user acceptance in the case of signaling systems in railways, it is expected that a solid understanding of the structure of adoption will be gained in this direction of research. While the railways' sector is considered to be a conservative one, getting insights on which determinants will allow individuals to be more open to using new technologies can be beneficial not only in the case of FMs but in other cases of technology adoption in railways, such as machine learning and blockchain technologies.

Finally, there is RQ five, which concerns the economic aspect of FMs adoption and is a longer-term study to be conducted. It is proposed that this topic will be studied from a simplified perspective of individual signaling projects and how did they benefit in two aspects, the first being (i) cost reduction and the second being (ii) schedule compression. Although other aspects of projects could be measured too, we assume those two to be the main interests of stakeholders.

Moving forward and regarding the future, we consider that FMs will be applied in more and more projects over the years. Concerning prospective upgrade of the European railways' network to comply with higher standards of CBTC and ERTMS, and the interoperability of trains in the European Union, FMs is one of the essential scientific disciplines, if not the most important one, in the attempt to achieve this compliance with international standards and compliance in between networks of different countries. Applications will also vary to the level of abstraction of their application regarding the knowledge that is required to apply the tools. Some high-level tools will be necessary for most of the projects, but also low-level tools will be essential to guarantee the formal development and formal verification of properties of specific sub-systems.

REFERENCES

[1] Edison, H., bin Ali, N. & Torkar, R., Towards innovation measurement in the software industry. *Journal of Systems and Software*, **86**(5), pp. 1390–1407, 2013.

[2] Crossan, M.M. & Apaydin, M., A multi-dimensional framework of organizational innovation: A systematic review of the literature. *Journal of Management Studies*, **47**(6), pp. 1154–1191, 2010 Sep.

[3] Kamal, M.M., IT innovation adoption in the government sector: Identifying the critical success factors. *Journal of Enterprise Information Management*, **19**(2), pp. 192–222, 2006.

[4] Palmqvist, C.W., Olsson, N.O.E. & Hiselius, L.W., Some influencing factors for passenger train punctuality in Sweden. *International Journal of Prognostics and Health Management*, **8**(7), 2017.

[5] Goverde, R.M.P., Corman, F. & D'Ariano, A., Railway line capacity consumption of different railway signalling systems under scheduled and disturbed conditions. *Journal of Rail Transport Planning & Management*, **3**(3), pp. 78–94, 2013.

[6] Hall, A., Seven myths of formal methods. *IEEE Software*, **7**(5), pp. 11–19, 1990.

[7] Bowen, J.P. & Hinchey, M.G., Seven more myths of formal methods. *IEEE Software*, **12**(4), 1995.

[8] Haxthausen, A.E., An introduction to formal methods for the development of safety-critical applications, DTU Informatics Technical University of Denmark, 2010.

[9] Gruner, S., Kumar, A. & Maibaum, T., Towards a body of knowledge in formal methods for the railway domain: Identification of settled knowledge. *Communications in Computer and Information Science*, Springer, 2016.

[10] Fantechi, A., Flammini, F. & Gnesi, S., Formal methods for railway control systems. *International Journal on Software Tools for Technology Transfer*, **16**, pp. 643–646, 2014.

[11] ter Beek, M.H., Fantechi, A., Ferrari, A., Gnesi, S. & Scopigno, R., *Formal Methods for the Railway Sector*, ERCIM NEWS 112., 2018.

[12] Bacherini, S., Fantechi, A., Tempestini, M. & Zingoni, N., A story about formal methods adoption by a railway signaling manufacturer. *Lecture Notes in Computer Science (including subseries Lecture Notes in Artificial Intelligence and Lecture Notes in Bioinformatics)*, pp. 179–189, 2006.

[13] Huhn, M. & Milius, S., Observations on formal safety analysis in practice. *Science of Computer Programming*, **80**, pp. 150–168, 2014.

[14] Basile, D. et al., On the industrial uptake of formal methods in the railway domain. *Lecture Notes in Computer Science (including subseries Lecture Notes in Artificial Intelligence and Lecture Notes in Bioinformatics)*, pp. 20–29, 2018.

[15] ter Beek et al., Adopting formal methods in an industrial setting: The railways case. *International Synopsium on Formal Methods*, pp. 762–772, 2019.

[16] Ferrari., A., Fantechi, A., Gnesi, S. & Magnani, G., Model-based development and formal methods in the railway industry. *IEEE Software*, **30**(3), pp. 28–34, 2013.

[17] Ferrari, A., Fantechi, A. & Gnesi, S., Lessons learnt from the adoption of formal model-based development. *Lecture Notes in Computer Science (including subseries Lecture Notes in Artificial Intelligence and Lecture Notes in Bioinformatics)*, pp. 24–38, 2012.

[18] Ferrari, A., Fantechi, A., Magnani, G., Grasso, D. & Tempestini, M., The Metrô Rio case study. *Science of Computer Programming*, **78**(7), pp. 828–842, 2013.

[19] Mekki, A., Ghazel, M. & Toguyeni, A., Validation of a new functional design of automatic protection systems at level crossings with model-checking techniques. *IEEE Transactions on Intelligent Transportation Systems*, **13**(2), pp. 714–723, 2012.

[20] Ghazel, M., Using stochastic petri nets for level-crossing collision risk assessment. *IEEE Transactions on Intelligent Transportation Systems*, **10**(4), pp. 668–677, 2009.

[21] Ghazel, M., A control scheme for automatic level crossings under the ERTMS/ETCS Level 2/3 Operation. *IEEE Transactions on Intelligent Transportation Systems*, **18**(10), pp. 2667–2680, 2017.

[22] Biagi, M., Carnevali, L., Paolieri, M. & Vicario, E., Performability evaluation of the ERTMS/ETCS – level 3. *Transportation Research Part C: Emerging Technologies*, **82**, pp. 314–336, 2017.

[23] Hansen, D. et al., Validation and real-life demonstration of ETCS hybrid level 3 principles using a formal B model. *International Journal on Software Tools for Technology Transfer*, **22**, pp. 315–332, 2020.

[24] Qian, J., Liu, J., Chen, X. & Sun, J., Modeling and verification of zone controller: The SCADE experience in China's railway systems. *2015 IEEE/ACM 1st International Workshop on Complex Faults and Failures in Large Software Systems (COUFLESS)*, pp. 48–54, 2015.

[25] Saharidis, G.K.D. & Fragkogios, A., Open problems on benders decomposition algorithm, pp. 305–317, 2018.

[26] DaSilva, C., Dehbonei, B. & Mejia, F., Formal specification in the development of industrial applications: Subway speed control system. *Proceedings 5th IFIP Conference on Formal Description Techniques for Distributed Systems and Communication Protocols (FORTE 1992)*, Perros-Guirec, pp. 199–213, 1993.

[27] Wu, Q., Pan, T., Xue, X., Zhang, F. & Jia, J., Modeling and implementation for dynamic curves of onboard automatic train protection system based on SCADE. *Lecture Notes in Electrical Engineering*, pp. 93–103, 2016.

[28] Behm, P., Benoit, P., Faivre, A. & Meynadier, J.M., Météor: A successful application of b in a large project. *Lecture Notes in Computer Science*, pp. 369–387, 1999.

[29] Essamé, D. & Dollé, D., B in large-scale projects: The canarsie line CBTC experience. *Lecture Notes in Computer Science*, pp. 252–254, 2006.

[30] Abo, R. & Voisin, L., Formal implementation of data validation for railway safety-related systems with OVADO. *Lecture Notes in Computer Science*, pp. 221–36, 2014.

[31] Comptier, M., Deharbe, D., Perez, J.M., Mussat, L., Pierre, T. & Sabatier, D., Safety analysis of a CBTC system: A rigorous approach with Event-B. *Lecture Notes in Computer Science*, pp. 148–159, 2017.

[32] Vu, L.H., Formal development and verification of railway control systems: In the context of ERTMS/ETCS Level 2, Technical University of Denmark, 2015.

[33] Borälv, A., Interlocking design automation using prover trident. *Lecture Notes in Computer Science*, pp. 653–656, 2018.

[34] Van Tang, N., Souma, D., Hatayama, G. & Ohsaki, H., Modeling and validating the train fare calculation and adjustment system using VDM++. *Lecture Notes in Computer Science*, pp. 163–178, 2012.

[35] Lindfeldt, O., The impact of platform screen doors on rail capacity. *International Journal of Transport Development and Integration*, **1**(3), pp. 601–10, 2017.

[36] Lecomte, T., Servat, T. & Pouzancre, G., Formal methods in safety-critical railway systems. *Symposium on Formal Methods*, 2010.

[37] Heinrich, M. et al., Security requirements engineering in safety-critical railway signalling networks. *Security and Communication Networks*, pp. 1–14, 2019.

[38] Venkatesh, V., Morris, M., Davis, G. & Davis, F., User acceptance of information technology: Toward a unified view. *MIS Quarterly*, **27**(3), pp. 425–78, 2003.

VISIBLE LIGHT COMMUNICATIONS BASED TRAIN CONTROL

MEHMET ALI DAGLI
Aselsan Electronics Industry and Trade, Inc., Turkey

ABSTRACT

Within the rail environment, as well as in any part of life, communication is one of the most important topics. With the advanced features gained via the new communication technologies, train control systems are improving too, especially at the urban transport level with reduced distance between the running trains, more trips can be performed. Continuous communication is the most important role in that improvement. Within this paper a new communication utility is proposed for the urban transport signalling systems. Using the state of art visible light communication (VLC) technology, we show how to design the illuminations of the underground lines as the communication medium between the central processors and the trains. In addition to the signalling systems, thanks to the high capacity bandwidth of VLC, big data transfer used on the other systems like real time monitoring and managing of the trains can be supported. Another proposed design is for providing new features to the recent conventional signalling systems via upgrading the light bulbs with the VLC-enabled LED ones. With all of these aspects shown in the paper it is worth mentioning that existing conventional systems can be upgraded, new communication medium can be implemented or provide more redundancy and availability can be added to the present wireless communication systems.

Keywords: visible light communication, VLC, CBTC, signalling, railway.

1 INTRODUCTION

The ever-increasing population of people living in metropolitans brings with it increasing capacity needs in public transportation systems. As in many areas of life, this increase need in this area is tried to be solved by automation. At the same time, system risks that need to be taken into account with high frequency transportation increase. In other words, in the railway transportation system, in which it is included many sub-systems like infrastructure, energy, rolling stock, operation management, maintenance other than train control [1]. Each system receives, produces or transmits the data it needs. Looking at the entire ecosystem, daily data produced can be at the level of terabytes. This brings need for the high capacity transmission of these data.

2 CBTC TECHNOLOGY

With an effectively established infrastructure, railway metro systems can bear the biggest load in urban public transportation. The importance of operating at a high frequency has increased even more these days due to the density of passengers in the train can cause dissatisfaction. Communications based train control (CBTC) systems are used in many parts of the world with the need to operate at high frequency.

A CBTC system is a continuous, automatic train control system utilizing high-resolution train location determination, independent of track circuits; continuous, high-capacity, bidirectional train-to-wayside data communications; and train-borne and wayside processors capable of implementing automatic train protection functions, as well as optional automatic train operation and automatic train supervision functions [2]. With this system, operation can be performed in 90 seconds frequency [3].

Being able to operate this frequency requires the fastest communication and action of trains and all components in the railway system. The status of the trains and the equipment

www.witpress.com, ISSN 1743-3509 (on-line)
doi:10.2495/CR200081

on the track is monitored instantaneously; all the components are automatically monitored and controlled by real-time commands by evaluating the incoming data.

Increased levels of automation, as mentioned in the introduction, can also increase system risks and require the use of other supportive systems to ensure safety. This makes it important to transmit and store big data. For instance, in the driverless subway systems, many parameters of the trains and in-car camera images shall be transferred to the control centre. With this supporting system, hundreds of cameras may need to stream instantly.

3 VLC TECHNOLOGY

Visible light communication is an increasingly popular technology in the field of communication. It has become a technology that is in use worldwide with the widespread use of light-emitting diode (LED) technologies and the standardization studies of the related joint committees.

In optical wireless communications, data is transmitted by intensity modulating optical sources, such as LEDs and laser diodes, faster than the persistence of the human eye. Optical wireless communications merges lighting and data communications in applications such as area lighting, signboards, streetlights, vehicles, traffic signals, status indicators, displays, LED panel, and digital signage [4].

There are various advantages to use the visible light communication. One of these is visible light is very safe for human. And, the data can be transmitted by the visible light communication even through a high voltage of home/office lighting. Another one is the lightings are set everywhere. Then, the data wireless transmission system can be easily established through the visible light communication device attached to the lightings. On the other hand, the radio wireless communications have several problems although they are widely in use of cell phones and wireless LAN. One of these is the electric transmission power cannot be increased because of the bad effect to human body. Another one is due to the radio wave restriction, there is no room to use more radio frequency [5].

On the other hand, visible light frequencies vary from 430 THz to 770 THz, which is 10,000 times larger than the entire radio frequency spectrum [6]. Thanks to the high frequency of light waves, VLC searches have already obtained impressive results, reaching speeds of 100 Gbps [7].

There are various VLC structures consist of transmitter, receiver or transceivers. A sample diagram of a VLC transceiver is seen is Fig. 1 where OFDM is rate-adaptive orthogonal frequency division multiplexing and TIA is trans impedance amplifier [8].

3.1 VLC use cases

There are many use cases about the VLC. Main application areas are indoor internet or entertainment streaming applications, indoor positioning applications, underwater communication applications, transport and vehicular systems applications. Vehicle communication is most attractive area because of intense usage of LED light on the vehicles and the transport environments.

The use cases from the road traffic have similarities with the railway transportation. Both has regulations via the signalling systems. One of the use case from Japan focused on the implementation of the VLC technology by changing the infrared sensors with LED headlight and photodiodes.

As shown in Fig. 2, because of the distortion coming from the daylight they narrow down the field of view by using lens in front of the photo diode. As seen in Fig. 3, they performed

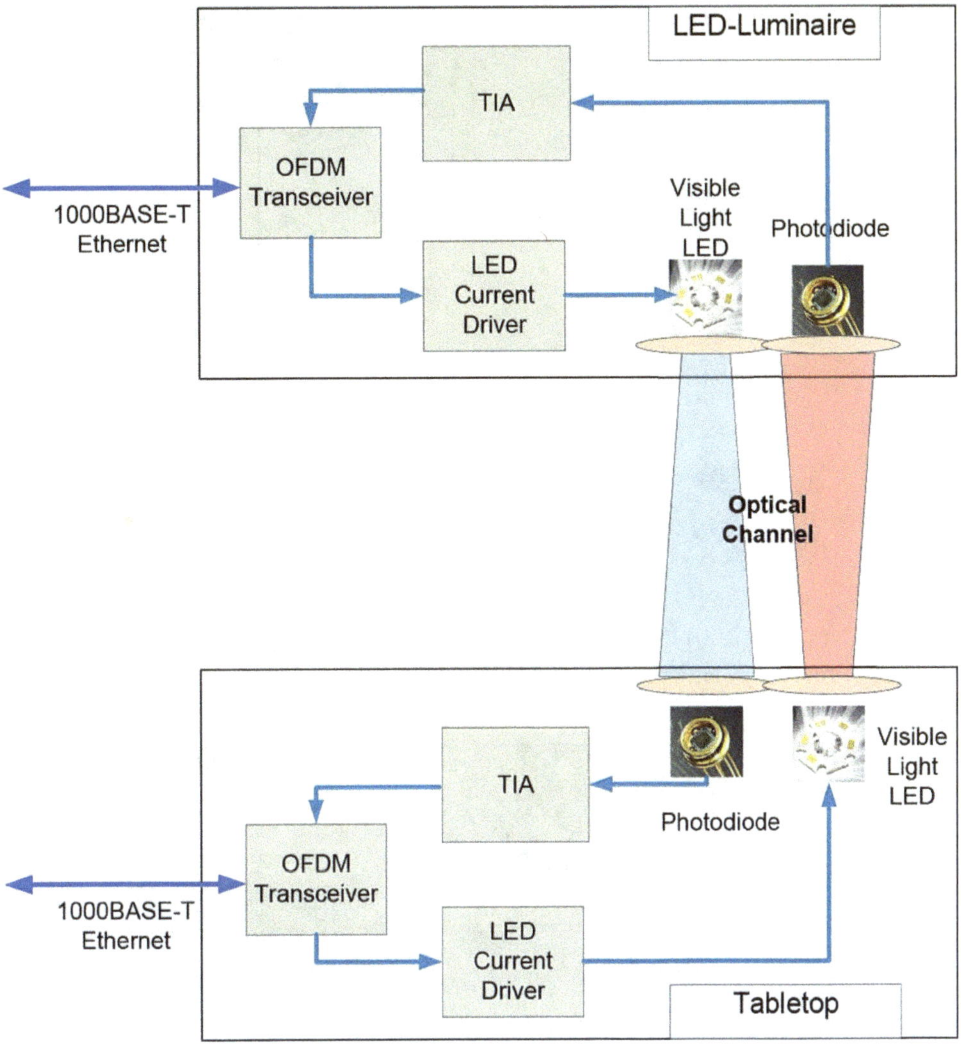

Figure 1: The overall scheme of a bidirectional real-time LOS visible light communication [8].

several measurements and they find the 25 m distance from the car headlight and 5 m height from the ground suitable for the receiving enough illumination values.

After experiments at a sunny day, with a car moving at 20 km/h speed they reached 3.1 Mbps throughput. They maintained higher throughput but packet success rate decreased [12].

There are other experimental VLC studies about the vehicular communication which used several measures to prevent the negative effect of other light sources like using colour filters [13]. Because of the environmental effects and the lack of off the shelf successful products, data rates are far from the theoretic rates that VLC can present.

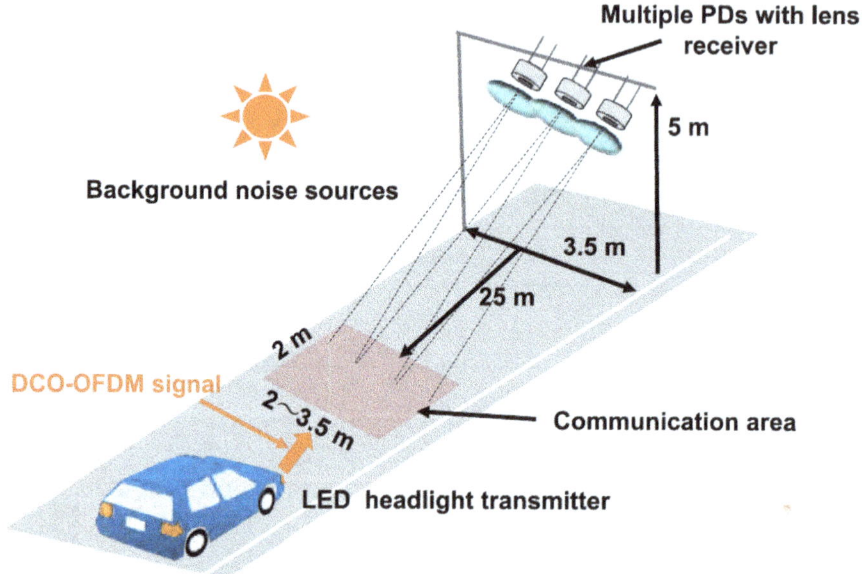

Figure 2: The uplink VLC beacon system. The transmitter is an LED headlight and the receiver is multiple PDs, each with a lens [12].

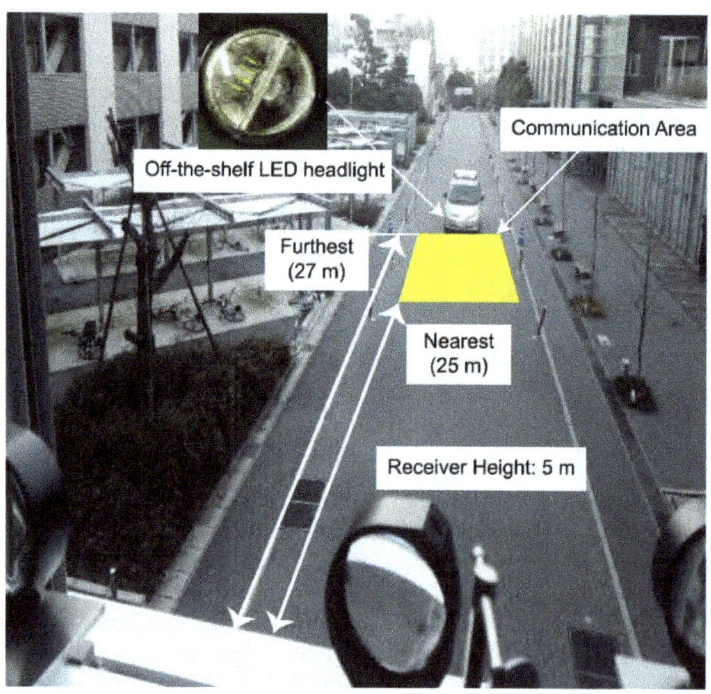

Figure 3: Experimental setups for SNR measurement at the farthest (27 m) and nearest (25 m) communications area [12].

3.2 VLC for CBTC

One of the important features that distinguishes the CBTC system from traditional fix block signal systems is that it reduces equipment maintenance costs in the field. Sensitive and frequent train position information can be obtained by dividing the line into small pieces with the moving block principle [9]. To provide the same frequency of information with rail circuits or axle counters, a large number of equipment will need to be placed on the site.

Although CBTC has provided a reduction in the number of equipment, it is a considerable number in the other equipment it brings with it. With this system, depending on the geometry of the tunnel, it is assumed it may be necessary to place a wi-fi access point with an average of 200–300 m [3].

There is a need for a large number of access points to meet the need to monitor vehicle information and in-car camera images centrally in CBTC operations, which are in unassigned train operation infrastructure. In this monitoring system, although efforts are made to use the common wi-fi infrastructure with the signalling system, it is still preferable to set up a separate network for security reasons. Also cyber security issues may be brought to the agenda in communications over wi-fi and extra measures are needed.

In the use of another communication equipment, balise or RFID transponder equipment on the track and reader units are installed in each vehicle. There may be damage to the equipment, various maintenance in the field or misplacement in renewal situations.

Visible light communication is a technology, which is suitable for use as a communications infrastructure in CBTC and the supporting systems, which ensures the maintainability of the operator company. Environments where lighting is provided, VLC can be used as a communication tool. This infrastructure can be provided by replacing the tunnel lighting on the subway lines with VLC technology LED lighting or using these lighting products directly on the new lines. Opposing these lamps, which will work as data transceivers, there should be transceivers on the vehicles. In this way, bi-directional communication environment can be provided. With these lighting products, which are frequently placed in the tunnel, continuous communication between the train and the centre will be provided. Transceivers can be installed more than once in the upper part of the vehicle, as demonstrated in Fig. 4, and the data accuracy can be increased and redundancy can also be provided. Since the lighting units in the tunnel will be at more frequent intervals than wi-fi access points, signal performance will also be high).

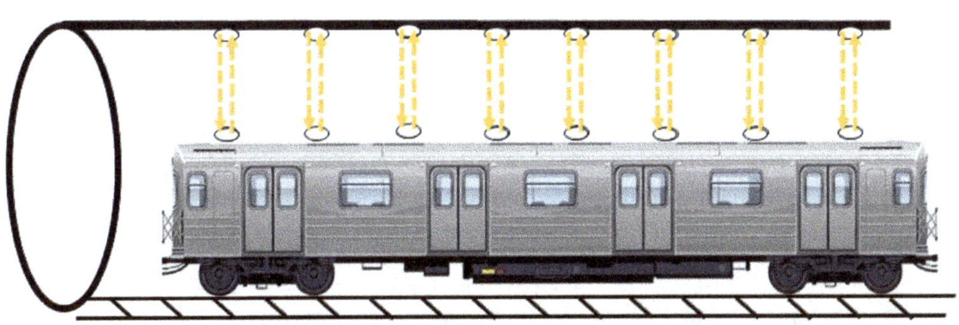

Figure 4: Tunnel lights for the bidirectional visible light communication.

3.3 VLC for main lines

The blocks separated in fix block signal systems are separated by signal lights, and the train drivers get the information that the region they are about to enter is occupied or not. In some cases, the vehicle can be stopped automatically with the balise connected to the signal at the wayside and a sensor on the vehicle. These signal indications, which are a visual warning for the train driver, can also be detected by VLC transceiver, a virtual eye to be placed in front of the vehicle as shown in Fig. 5. The information to be transmitted may be information pertaining to the stop or departure perceived by the train driver, and the information that is in use, such as automatic stop information obtained from the balise, and the speed limit information to be obeyed in the relevant region, can be transmitted with this new medium [10].

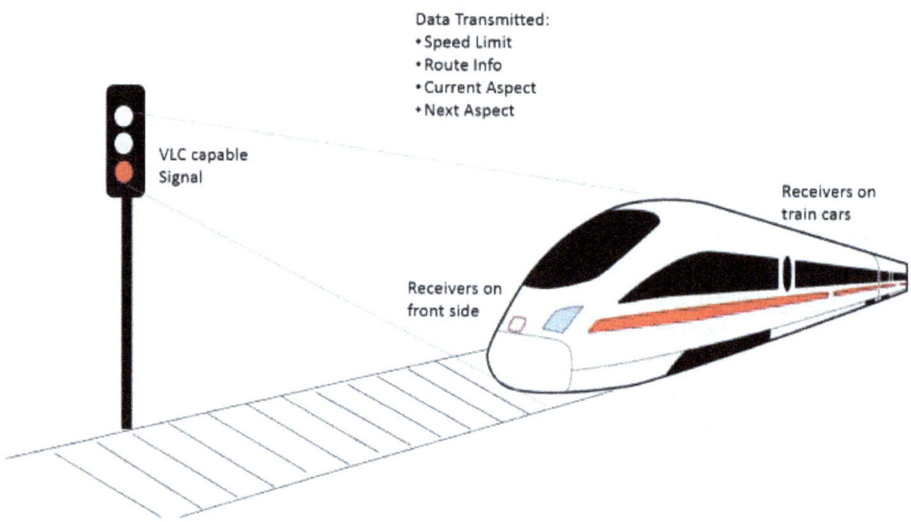

Figure 5: Signal lights for the communication transmission to the train [10].

In addition to these, information which comes with CBTC system technologies such as speed limit and movement authority information, information about arrival to the next station, information about the train in front of it, information about a few blocks ahead can be transmitted from the signal light to the train. Similarly, vehicle information such as direction, speed, fault status information can be transmitted from the vehicle to the centre. Signals at the station entrances and exits can also be enhanced with different information such as station stop/departure detail information or door status information. With new information to be added such as these, the information circulating and processed in traditional signal systems can be increased and features close to CBTC systems can be gained.

4 CHALLENGES

Light is spreading on the air by reflection and it cannot pass through the obstacles like walls. So the limit is the line of sight. Not only the walls also weather conditions like rain, fog, snow have effects on the line of sight, so to the light too. In subway tunnel applications, these environmental effects would be less but for the outdoor applications like main lines it will be a big issue.

In a wi-fi network, devices that transmit on the same frequency can interfere with each other. When light is used as a communication medium, natural light becomes a source of interference in communication, degrading VLC, especially in the case of outdoor applications. In addition to natural light, artificial lights also interfere with communication, and may even saturate the receiver.

Another challenge is the lack of required products. Especially at signalling system there are various standards or regulations to be ensure the safety and reliability of the system. The hardware chosen for these systems are railway specific products. For now there is lack of railway compliant proven VLC products to use.

A useful communication system built with VLC must allow uplink and downlink. LED light bulbs can be used for both, as a VLC transmitter and as a light source. In the receiver, a simple photodiode can be used to receive modulated light, which will be further decoded. In this sense, the downlink, i.e., the transmission from a LED light to devices, is straightforward.

However, sending data from devices to a LED light bulb is more challenging. A number of research papers consider the use of Visible Light Communication for both downlink and uplink. To remove the effect of signal reflection interference, researchers propose the use of techniques such as time-division duplex (TDD). TDD decreases the data rate since slots of specific duration are allocated both for downlink and uplink. However, high-level modulation mechanisms, such as OFDM can achieve better performance [11].

5 CONCLUSION

The importance of high speed big data management is increasing in the CBTC system and other supporting systems in underground train operations that need high frequency trips. This brings together the need for continuous improvement in communication systems. VLC has been increasing day by day in the developments in communication technologies. In many areas, efforts are being made to implement communication over light rather than RF communication. Although it is a technology that has not yet completed its development, thanks to its advantages, its usage in the railway business is promising as it is in many areas where the studies continue. With the production and the use of equipment suitable for the railway environment, the usage of VLC in CBTC systems will become widespread.

REFERENCES

[1] European Union Agency For Railways, Big data in railways: Common occurrence reporting programme, ERA-PRG-004-TD-003. www.era.europa.eu/sites/default/files/activities/docs/cor_big_data_en.pdf. Accessed on: 23 Jan. 2020.

[2] IEEE Standard 1474.1, Communications-Based Train Control (CBTC) Performance and Functional Requirements, 2004.

[3] Farooq, J. & Soler, J., Radio communication for communications-based train control (CBTC): A tutorial and survey. *IEEE Communications Surveys and Tutorials*, **19**(3), pp. 1377–1402, 2017.

[4] IEEE Standard 802.15.7, Local and Metropolitan Area Networks, Part 15.7: Short-Range Optical Wireless Communications, 2018.

[5] Visible Light Communication Consortium, About visible light communication. www.vlcc.net/modules/xpage0/. Accessed on: 29 Jan. 2020.

[6] Haas, H., High-speed wireless networking using visible light. *Proceedings SPIE Newsroom*, pp. 1–3, 2013.

[7] Gomez, A. et al., Beyond 100-Gb/s indoor wide field-of-view optical wireless communications. *IEEE Photon. Technol.*, **27**(4), pp. 367–370, 2015.

[8] The International Society for Optics and Photonics, Rate-adaptive visible light communication at 500Mb/s arrives at plug and play. https://spie.org/news/5196-rate-adaptive-visible-light-communication-at-500mb/s-arrives-at-plug-and-play?SSO=1. Accessed on: 13 Feb. 2020.

[9] Song, H. & Schneider, E., Availability and performance analysis of train-to-train data communication system. *IEEE Transactions on Intelligent Transportation Systems*, **20**(7), 2019.

[10] Ahamed, S., Visible light communication in railways. *Proceedings of the International Conference on Railway Engineering (ICRE)*, 2016.

[11] Song, H. & Schneider, E., Visible light communication: Concepts, applications and challenges. *IEEE Communications Surveys and Tutorials*, **21**(4), 2019.

[12] Yamazato, T. et al., The uplink visible light communication beacon system for universal traffic management. *IEEE Access*, **5**, pp. 22282–22290, 2017.

[13] Yoo, J.-H. et al., Demonstration of vehicular visible light communication based on LED headlamp. *Int. J. Autom. Technol.*, **17**(2), pp. 347–352, 2016.

SECTION 3
TIMETABLING AND RESCHEDULING

GENERATION OF DISPATCHING SOLUTIONS IN A ROLLING TIME HORIZON FRAMEWORK WITH THE AID OF TABU SEARCH

ULLRICH MARTIN, MARKUS TIDEMAN & WEITING ZHAO
Institute of Railway and Transportation Engineering (IEV), University of Stuttgart, Germany

ABSTRACT

To increase the robustness of railway operation, at the Institute of Railway and Transportation Engineering (IEV) of the University of Stuttgart, a proactive dispatching algorithm has been developed. At the beginning, the block sections of a railway network are classified automatically depending on their specific operational risk index. Therefore, numerous timetables are modified based on random disturbances generated in a Monte Carlo scheme, then the modified timetables are simulated in RailSys® software and, finally, the resulting waiting times in the studied network are calculated to obtain an operational risk map. Considering this operational risk map, near-optimal dispatching solutions are generated within a rolling time horizon framework. To this, the investigation time span is divided into so-called prediction horizons (PH), which, on their part, are divided into so-called dispatching intervals (DI). At the start of each DI the train runs are artificially prolonged according to the risk map. Then, it is checked for the duration of the current PH whether conflicts (overlap of blocking times of at least two different trains in a block section) occur or not. If the detected conflicts are slight, they are solved by retiming of the affected train runs. However, in case of significant conflicts, the train runs have to be reordered. This is achieved with the aid of Tabu Search. In this paper, the functionality of the proposed algorithm is not only described but also demonstrated according to a realistic reference example that is standardly used at the IEV for algorithm development. In this context, special emphasis is placed on several important dispatching-related parameters to determine their general influence on the dispatching process. In the light of the reference example, it is shown that the proposed dispatching algorithm has a sustainable impact on the operation quality, particularly in comparison to other dispatching principles.

Keywords: dispatching, rescheduling, rolling time horizon framework, Tabu Search, risk mapping, operational risk analysis.

1 INTRODUCTION

During railway operation, the occurrence of random operational disturbances like dwell time extensions, running time extensions, departure time extensions and entry delays can lead to a significantly lower operational quality, e.g. in the form of resulting delays. Hence, besides buffer times to avoid or reduce transmission of delays between following trains, recovery times are added already during timetabling to reduce the impacts of disturbances of a train itself on railway operation (e.g. Kroon et al. [1] and Lindfeldt [2]). Admittedly, those recovery times should not become too large because of two aspects: Firstly, in congested areas there is no or only a small margin for additional time reserves and secondly, the operating capacity would deteriorate. Furthermore, longer scheduled travel times would result [3]–[5].

For this reason, railway operators commonly use conventional dispatching algorithms to soften the remaining effect of the disturbances during railway operation (e.g. D'Ariano [6]). In contrast to conventional dispatching algorithms, the Institute of Railway and Transportation Engineering (IEV) at the University of Stuttgart developed a dispatching algorithm, which considers further potential disturbances [7]. Therefore, the proposed dispatching approach consists out of two main processes. The first one is the operational risk

WIT Transactions on The Built Environment, Vol 199, © 2020 WIT Press
www.witpress.com, ISSN 1743-3509 (on-line)
doi:10.2495/CR200091

analysis (Section 2) that provides the basis for the second process, which is the generation of dispatching solutions based on the Tabu Search (Section 3). Subsequently, the functionality of the proposed algorithm is demonstrated according to a realistic reference example (Section 4). At this one, special emphasis is placed on several important dispatching-related parameters to determine their general influence on the dispatching process. The present paper closes with an outlook for future research activities (Section 5).

To demonstrate the mode of action of the proposed dispatching principle, RailSys® software is used in different parts. This software not only allows to create a detailed infrastructure model and corresponding timetables but also to simulate the consequential railway operation [8]. As a main result of such a simulation, the software generates a log file which provides, inter alia, information about resulting delays of the train runs at any operating point of the investigated network.

2 OPERATIONAL RISK ANALYSIS

As this paper focusses especially on the generation of dispatching solution and the influence of dispatching-related parameters within the scope of a case study, the operational risk analysis is only described roughly in the following to support a basic understanding. For a detailed description of the operational risk analysis that is performed in offline mode reference is made to Zhao [9] and Zhao et al. [10].

At the beginning of this process, an investigated railway network is divided into appropriate block sections e.g. as shown in Li and Martin [11]. According to a fitting probability distribution of random disturbances a set of disturbed timetables is generated automatically in a Monte Carlo scheme for every block section separately. By imposing these so-called disturbance scenarios one by one on the railway operation on the corresponding block section and by simulating this in RailSys® software, the respective total weighted waiting time of each disturbance scenario can be obtained. Accordingly, the average total weighted waiting time can be calculated specifically for every block section. This average total weighted waiting time represents the operational risk of a specific block section. In general it can be said that the higher the average total weighted waiting is, the worse is the negative influence of disturbances occurring on the corresponding block section for the whole railway network.

To resort on the findings of the operational risk analysis in the context of the second process, which is the generation of dispatching solutions, the block sections are ranked according to their operational risk. This classification remains valid as long as the basic timetable and the infrastructure do not change significantly.

3 GENERATION OF DISPATCHING SOLUTIONS

Whereas the operational risk analysis is performed in offline mode, the generation of dispatching solutions is executed in online mode, viz. during railway operation. As shown in Fig. 1, this process consists out of four tasks [3], [9], [12].

Inter alia inspired by Zhan et al. [13], in the first task, the investigated time span is divided into so-called prediction horizons (PH) and dispatching intervals (DI) to enable a rolling time horizon framework that ensures a sustainable impact on the operational quality. Even though the respective duration has to be set by the user, it must be ensured that the investigated time span is an integer multiple of the length of the prediction horizons and that a prediction horizon is an integer multiple of the length of the dispatching intervals in turn.

At the beginning of every new dispatching interval, a conflict detection is performed (second task) for the length of the corresponding prediction horizon. To this goal, the blocking times of the train runs are artificially prolonged according to the results of the operational risk analysis. By doing so, potential disturbances occurring on the corresponding block sections are implicitly considered, whereby the robustness of the dispatching solution is increased. Then it is checked whether the current timetable contains conflicts in the shape of two or more overlapping blocking time stairways. In case of no detected conflicts, the current timetable will be maintained for the railway operation or rather its simulation (fourth task) until the beginning of the next dispatching interval. In the opposite case, a dispatching solution is generated for the current dispatching interval (third task). Depending whether a user-defined limit value for the target function, e.g. total weighted waiting time, is exceeded or not, the conflicts are treated as slight conflicts or significant conflicts.

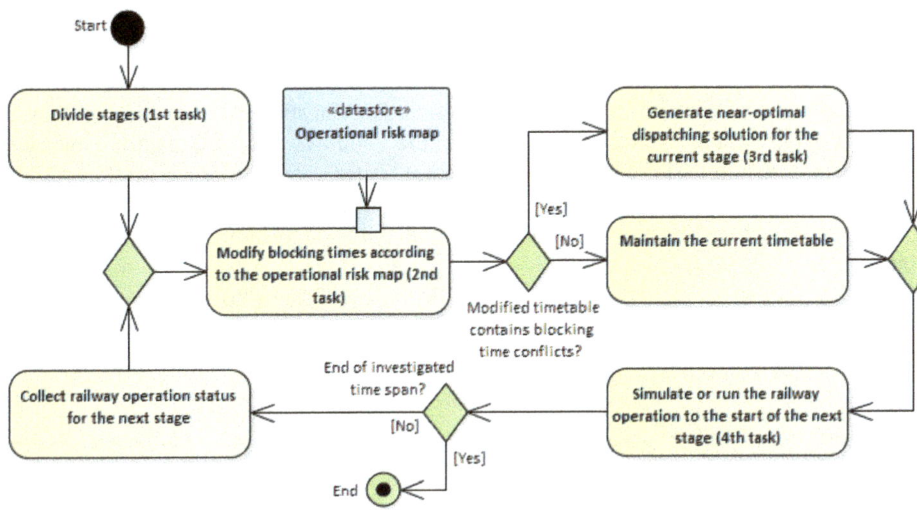

Figure 1: General workflow of the generation of dispatching solutions. *(Source: Martin et al. [12].).*

3.1 Generation of dispatching solutions for slight conflicts

Conflicts are considered as slight conflicts when the target indicator falls below a user-defined limit value. In order to save computation times and to minimize the deviation between the dispatched timetable and the original one, slight conflicts are solved only by retiming affected train runs. This means in particular that some train runs will be postponed, but the order of the train runs will not be changed [9].

3.2 Generation of dispatching solutions for significant conflicts based on Tabu Search

Whenever the value of the target indicator exceeds a user-defined limit value, the conflicts are solved by reordering the train runs. This means, that a better train sequence is gained. As a timetable usually consists out of a greater amount of trains, it would be very time consuming to calculate the value of the target function for every possible train order. This is why the proposed dispatching uses Tabu Search for reordering the train runs, which ensures a well-balanced compromise between calculation time and operational quality. Tabu Search method is a widely used in the field of transportation (e.g. Sterzik and Kopfer [14] and Corman et al. [15]). As depicted in Fig. 2, the user has to define aspiration and termination criteria, e.g. depending on the extension of the considered area or the expected number of trains to be scheduled. For example, as an aspiration criterion a value of the target function that is sufficiently good and as a termination criterion a maximum number of iterations could be set. The workflow of the reordering process starts with the current train order and an empty Tabu list. Then, all neighborhood train orders of the current train order are generated and saved as neighborhood timetables. This means just to switch the order of two trains in the train sequence. Afterwards, the value of the objective function for all these neighborhood timetables is calculated by simulating the railway operation in RailSys® software. Accordingly, the train order with the best objective function value is considered as the current train order and put on top of the Tabu list. As long as the termination criteria are not met, for this new current train order the neighborhood train orders are generated and their values of the objective function are calculated. If the best train order is not included in the Tabu list, the best train order is chosen as the new current train order and put on the top of the Tabu list. In case that the train order is included in the Tabu list, the sub-best train order is chosen as the new current train order and put on the top of the Tabu list, provided that the aspiration criteria are not fulfilled. Whenever the aspiration criteria are satisfied, the workflow interrupts and the current train order is treated as the dispatching solution. By using the Tabu list in the described manner, it can be avoided that the algorithm is stuck in a local minimum [9].

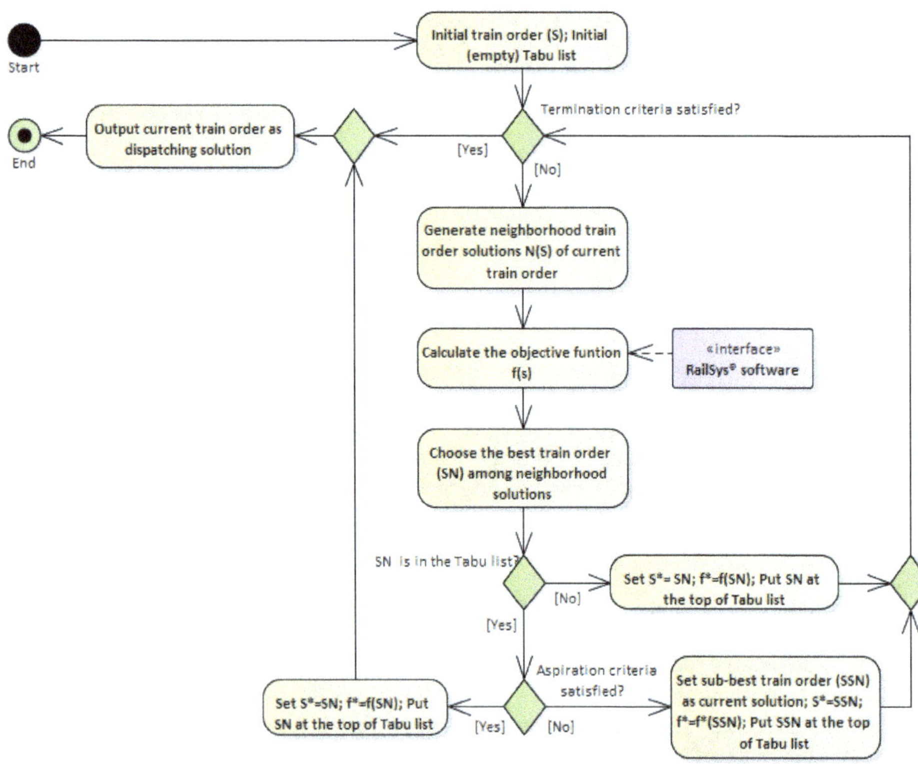

Figure 2: Workflow of the generation of dispatching solutions with the aid of Tabu Search.

4 CASE STUDY

To demonstrate the impact of the proposed dispatching approach on operational quality, a realistic reference example is applied, which is standardly used at the Institute of Railway and Transportation Engineering (IEV) of the University of Stuttgart for algorithm development. This reference railway network is depicted in Fig. 3 and consists out of 43 kilometre track length with up to 72 train runs in six hours. "Technically, the operational risk analysis (first process) as well as the dispatching algorithm (second process) is developed in Microsoft Visual Studio 2015 environment with C#. This code runs the simulations in RailSys® software automatically. Due to the large amount of disturbance scenarios that has to be imposed on every block section separately, the operational risk analysis has to be executed in an offline environment. For the reference scenario this process takes at least four hours (Fujitsu computer, Intel Core i5-4670 CPU @ 3.40 GHz, 8 GB RAM). After achieving the operational risk classification once, the classification will remain its validity in the future as long as there will not occur [significant] changes in infrastructure layout or operating program. The code for the dispatching algorithm, which is also written with C#, has to be started manually by the user. With the objective of using this second process in online mode, the calculation of both conflict detection and conflict resolution takes about [only] 20 seconds for each stage" [3]. In the case study, the runs of regional passenger, freight transport and long-distance trains are artificially prolonged according to Section 2 (operational risk analysis) and to the conflict detection within Section 3 (generation of dispatching solutions) with the aid of additional recovery times representing entry delays, dwell time extensions, departure time extensions and/or running time extensions. By using the proposed approach instead of first come first serve-principle significant improvements regarding the resulting total weighted waiting times can be achieved within every DI as Table 1 indicates. Regarding the sums of the total weighted waiting times, the proposed dispatching approach ensures almost an improvement of a third compared to first come first serve-principle.

In this paper, special emphasis is placed on several important dispatching-related parameters to determine their general influence on the dispatching process. In this regard, a sensitivity analysis concerning, inter alia, the length of prediction horizon (PH) and the length of dispatching intervals (DI) with respect to the investigated time span (TE) is conducted for the reference railway network described above. More precisely, the influence of different combinations of DI and PH by applying the dispatching solutions generated by the proposed approach on railway operation is explored in terms of the resulting total weighted waiting time, the average absolute retiming and the number of relative reordering.

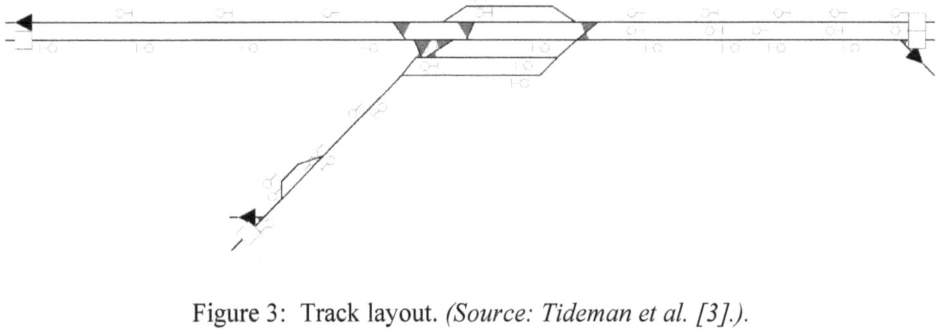

Figure 3: Track layout. *(Source: Tideman et al. [3].)*

Table 1: Percentage improvement of the total weighted waiting time by using the proposed algorithm instead of first come first serve-principle within the case study.

Dispatching interval (DI)	Time span	Improvement
DI-0	0:00–0:30	12%
DI-1	0:30–1:00	16%
DI-2	1:00–1:30	12%
DI-3	1:30–2:00	26%
DI-4	2:00–2:30	28%
DI-5	2:30–3:00	21%
DI-6	3:00–3:30	38%
DI-7	3:30–4:00	56%
DI-8	4:00–4:30	30%
DI-9	4:30–5:00	42%
DI-10	5:00–5:30	63%
DI-11	5:30–6:00	67%

In general, it is evident that the operational quality in the shape of total weighted waiting times gets better, the shorter DI and the longer PH is. This is due to two facts: First, the current railway operation status can be considered more effective by updating the dispatching solution very quickly (shorter DI). Second, dispatching solutions being generated by considering a greater time span (longer PH) enable a long-lasting effect. However, it would be far too time consuming to run the dispatching algorithm in online mode for such combinations of PH and DI. Hence, the total weighted waiting time (TotalwWT) is calculated for the reference example for several combinations and the corresponding results are shown in Fig. 4. It should be noted that due to limited time budget during research work not any combination of DI and PH have been investigated, which is why the horizontal axis are non-linear scale in Figs 4–6.

Based on Fig. 4, one could tend to set PH as 50–60% of TE and DI as 10% of PH with regard to computational burden as this combination guarantees almost the same results as the best combination (PH 100%, DI 5%). In fact, to determine the most appropriate combination of PH and DI not only the total weighted waiting time but also the average absolute retiming (AAR) and the number of relative reordering (NRR) have to be considered. Analogous to Fig. 4, in Fig. 5 the resulting average absolute retiming and in Fig. 6 the number of relative reordering for different combinations of PH and DI are depicted, in each case averaged over all time stages for every combination. Both indicators are of great relevance in the context of railway dispatching as a dispatching solution should not only be good regarding the theoretical operational quality. Instead, the practical feasibility is also important so that for instance dispatchers and train drivers will not be overextended by rapidly changing timetables. While the averaged average absolute retiming does not variate too much, the number of relative reordering increases especially for prediction horizons being longer than approx. 40% of the investigated time span. On the contrary, the length of the dispatching interval has almost no influence on this indicator. Concluding the findings of this reference scenario, the length of PH should amount within 30% to 40% of TE and the length of DI should amount within 20% to 25% of PH.

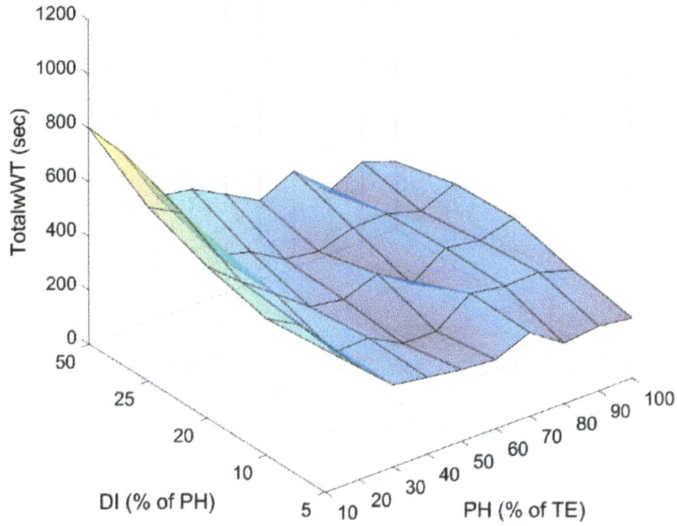

Figure 4: Variation of the total weighted waiting (TotalwWT) time against DI and PH for the reference example.

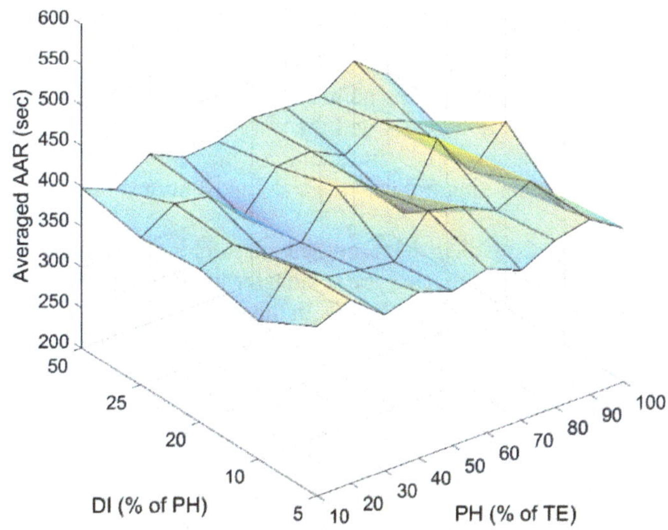

Figure 5: Variation of the averaged average absolute retiming (Averaged AAR) against DI and PH for the reference example.

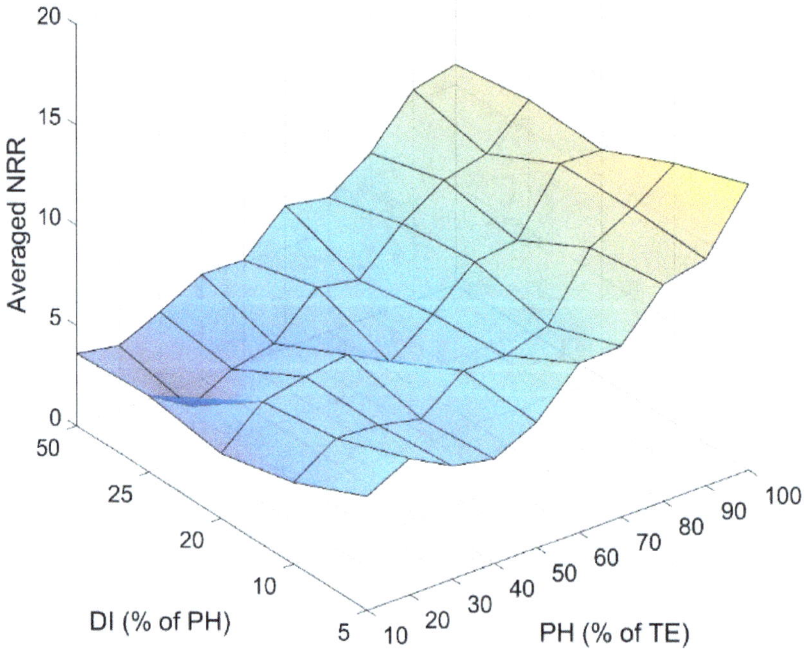

Figure 6: Variation of the number of relative reordering (Averaged NRR) against DI and PH for the reference example.

5 CONCLUSION AND OUTLOOK

The proposed dispatching algorithm offers numerous advantages, which are, for instance, its basic usability for every railway network and the improved operational quality compared with first come first serve-principle. In addition, the approach ensures practicable calculation times while generating robust dispatching solutions under consideration of future disturbances.

Based on the promising results, the proposed dispatching algorithm will be further developed. Suitable ideas and approaches are to transfer its operating principle to use cases belonging to other mode of transports [16], [17]. In this context, several adjustments have to be made. For instance, as in other transport modes no block sections exist a substitute is necessary to perform the operational risk analysis. In addition, the indicator representing operational risk might be replaced in cases where waiting times are not relevant. Furthermore, there are several use cases conceivable, in which the main objective of dispatching is not to better operational quality, but to reduce resource input or to use resources evenly.

Another promising intention is to implement methods of machine learning in the workflow of the operational risk analysis as well as the generation of dispatching solutions. In this field, the Institute of Railway and Transportation Engineering (IEV) at the University of Stuttgart has useful prior knowledge, especially originating from Tritschler et al. [18] and Bahamon et al. [19].

ACKNOWLEDGEMENT

We would like to thank the Deutsche Forschungsgemeinschaft (DFG) for funding that the research project Risk Oriented Dispatching of Railway Operation under the Consideration of Random Disturbances in Dynamic Circumstances (DICORD), on which the findings presented in this paper are based, made possible.

REFERENCES

[1] Kroon, L., Huisman, D. & Maróti, G., Railway timetabling from an operations research perspective. *Econometric Institute Report EI2007-22*, pp. 1–25, 2007.

[2] Lindfeldt, A., Railway capacity analysis. Methods for simulation and evaluation of timetables, delays and infrastructure. Doctoral thesis, KTH Royal Institute of Technology: Stockholm, 2008.

[3] Tideman, M., Martin, U. & Zhao, W., Proactive dispatching of railway operation. *Proceedings of the 8th International Conference on Railway Operations Modelling and Analysis (ICROMA) – RailNorrköping*, pp. 1605–1614, 2019.

[4] Cheng, Y., Hybrid simulation for resolving resource conflicts in train traffic rescheduling. *Computers in Industry*, **35**(3), pp. 233–246, 1998.

[5] Espinosa-Aranda, J.L. & García-Ródenas, R., A discrete event-based simulation model for real-time traffic management in railways. *Journal of Intelligent Transportation Systems*, **16**(2), pp. 94–107, 2012.

[6] D'Ariano, A., Improving real-time train dispatching: models, algorithms and applications. Dissertation, TU Delft, 2008.

[7] Martin, U., Tideman, M. & Zhao, W., Risk Oriented Dispatching of Railway Operation under the Consideration of Random Disturbances in Dynamic Circumstances (DICORD). DFG-project (MA 2326/22-1), in progress.

[8] Rail Management Consultants International GmbH. https://www.rmcon-int.de/railsys-en/railsys-suite/. Accessed on: 24 Apr. 2020.

[9] Zhao, W., *Hybrid Model for Proactive Dispatching of Railway Operation under the Consideration of Random Disturbances in Dynamic Circumstances*, Books on Demand: Norderstedt, 2017.

[10] Zhao, W., Martin, U., Cui, Y. & Liang, J., Operational risk analysis of block sections in the railway network. *Journal of Rail Transport Planning & Management*, **7**(4), pp. 245–262, 2017. https://doi.org/10.1016/j.jrtpm.2017.09.003

[11] Li, X. & Martin, U., Ursachenbezogene Engpassbewertung in der Eisenbahnbetriebssimulation – DFG-Forschungsprojekt EPSUR. *Proceedings of 25. Verkehrswissenschaftliche Tage Dresden*, 2016.

[12] Martin, U., Tideman, M. & Zhao, W., Generating robust dispatching solutions taking into account block sections' operational risk. *Proceedings of the Fifth International Conference on Railway Technology: Research, Development and Maintenance (RAILWAYS 2020)*, submitted for publication.

[13] Zhan, S., Kroon, L., Zhao, J. & Peng, J., A rolling horizon approach to the high speed train rescheduling problem in case of a partial segment blockage. *Transportation Research Part E: Logistics and Transportation Review*, **95**, pp. 32–61, 2016. https://doi.org/10.1016/j.tre.2016.07.015

[14] Sterzik, S. & Kopfer, H., A Tabu Search heuristic for the inland container transportation problem. *Computers & Operations Research*, **40**(4), pp. 953–962, 2013.

[15] Corman, F., D'Ariano, A., Pacciarelli, D. & Pranzo, M., A Tabu Search algorithm for rerouting trains during rail operations. *Transportation Research Part B: Methodological*, **44**(1), pp. 175–192,2010.

[16] Tideman, M. & Martin, U., Proaktive Disposition luftverkehrlicher Prozesse. *Internationales Verkehrswesen*, **70**(4), pp. 60–63, 2018.
[17] Tideman, M., Martin, U. & Zhao, W., Disposition von verkehrlichen Prozessen unter Einbeziehung von zufallsbedingten Unsicherheiten. *Eisenbahntechnische Rundschau*, **67**(10), pp. 22–25, 2018.
[18] Tritschler, S., Cui, Y. & Dobeschinsky, H., Störfallmanagement im ÖPNV. *Der Nahverkehr*, **23**(10), pp. 14–18, 2005.
[19] Bahamon, B., Rapp, S., Rupp. C., Liu, J. & Martin, U., Recognition of track defects through measured acceleration. *Proceedings of the 7th International Conference on Euro Asia Civil Engineering Forum (EACEF)*, pp. 995–1010, 2019.

TRANSIT ASSIGNMENT MODEL FOR EXPRESS TRAINS BASED ON INVERSE OPTIMIZATION

KYUNG MIN KIM
Industrial Management and Engineering Department, Myongji University, South Korea

ABSTRACT

A loading balance between express and local trains is a crucial issue on the train operation plan of urban railway, which provides two services on the same line. This is because passenger's preference is affected by the train operation plan, such as headway and halting patterns. Recently, Seoul Metro Line 9, in Korea, has experienced an imbalance, as the preference for express train over local train services is significantly higher, so the load factor for the express train was twice as much as for the local train. As such, to analyse load balance, a schedule-based transit assignment to determine the ridership by section and by train is needed when a timetable for trains and passengers' trips is given. This paper develops the cost function considering the transfer between express and local trains as well as the train load factor in the time-expanded network via inverse optimization. We suggest a convex optimization model to ensure the path that the passenger has selected will cost less than the alternative path, and from this model, the parameters of the deterministic cost function were estimated. We employed the Frank-Wolfe-based transit assignment algorithm to estimate the equilibrium flow using our cost function. The algorithm was tested to Seoul Metro Line 9 in order to verify its accuracy. The results show that the difference in the actual ridership by train was less than 10%. Finally, we simulated reducing the load difference between the express and the local train through re-scheduling to increase the service frequency of express trains.

Keywords: inverse optimization, transit assignment, express trains.

1 INTRODUCTION

Train operation planning for an urban railway, which provides both express and local train service simultaneously, is necessary to maintain load balance between the two services [1]. This is because passenger preference is affected by the train operation plan, such as headway and halting patterns. Recently, Seoul Metro Line 9 in Korea has experienced an imbalance, as the preference for express train over local train service is significantly higher, so the load factor for express trains was twice as much as for local trains. As such, to analyse load balance, schedule-based transit assignment to determine the ridership by section and by train is needed when the timetable for trains and passengers' trips is given.

Schedule-based transit assignment in a time-expanded network allows for the expression of the events of arrival and departure train and the complete passenger's flow – boarding, alighting, and transfer. In a time-expanded network, we can analyse individual passenger choices. Although GPS or a Smart Card system can be used to collect massive and reliable trip data, difficulties in handling a large size network and obtaining massive data reveal that preferences at the microscopic level may have hampered the implementation of schedule-based transit assignment in real cases. However, Hong et al. [2] recently developed a precise algorithm that detects an actual train-path for metro passengers based on an observation that tag-out times of passengers from Smart Card data were clustered.

In this paper, we propose a deterministic path cost function with three attributes – in-vehicle time, wait time, and transfer time between express and local trains – to develop a schedule-based transit assignment model. Congestion effect measured by step function of relative weight on in-vehicle time increasing to predetermined crowding level. Stochastic assignment models which are based on random utility maximization (RUM) theory such as

WIT Transactions on The Built Environment, Vol 199, © 2020 WIT Press
www.witpress.com, ISSN 1743-3509 (on-line)
doi:10.2495/CR200101

logit estimate the coefficient of the cost function to maximize likelihood. However, this paper estimates parameters using inverse optimization due to assuming the deterministic path cost.

An optimization (forward optimization) addresses the problem of finding a point, an optimal solution, in the region delineated by a set of functions, or constraints, which minimizes a cost function called an objective. An *inverse optimization*, on the other hand, is the problem of calibrating a set of parameters in terms of objective and/or constraints, so that a prescribed solution becomes optimal [3]. The parameters should be what best explains user-optimality, underlying the path choice of passengers. This implies that an inverse optimization has relevance in doing so. We will show that the inference problem can be posed as a convex quadratic optimization that allows an efficient solution method.

Tested on a daily set of trains, with parameters calibrated by the inverse optimization, from Seoul Metro line 9, the results showed a mean absolute percent error of deviation between actual and predicted passenger load of trains within 10%. Our model seems precise enough for the prediction of passenger behaviour required by the city of Seoul.

This paper is organized as follows. In Section 2, we describe the time-expanded network and cost function in the proposed transit assignment model. Section 3 describes the methodology of inverse optimization to infer the coefficient of parameters of the cost function. Section 4 evaluates the path choice predictability of our model in comparison with actual choices. Finally, Section 5 provides some concluding remarks, and suggests some directions for further research.

2 TIME-EXPANDED NETWORK AND COST FUNCTION

Time-expanded network $G = (N, E)$ is defined as Set N of place-time-defined nodes and Set E of edges, which is defined as Subset of $N \times N$ [4]. As shown in Fig. 1, each node $i \in N$ has place, i.e., station and time value as attribute, and each edge represents an event in which a train departs from each station or arrives at the station.

On the time-expanded network, path P can precisely represent the movement of passengers who use local and express trains. As shown in Fig. 1, for example, a passenger who moves to Station E from Station A where local both trains and express trains stop can have *P1, P2* and *P3* as an *alternative path.* That is, the *alternative paths set* of this passenger is {*P1, P2, P3*}. *P1* means a path in which a passenger transfers to an express train at Station C after using a local train, while *P2* and *P3* are the paths only for express train and local train, respectively.

Deterministic path cost function with three attributes – wait time, in-vehicle time and transfer time between express and local trains for schedule-based transit assignment model. The cost of path P on the time-expanded network is assumed to be a linear function, as shown in eqn (1). In-vehicle time is divided into 3 categories depending on the degree of in-vehicle crowding

$$c(P) = \alpha WT_P + \beta_1 IVT_P^L + \beta_2 IVT_P^N + \beta_3 IVT_P^H + \gamma TT_P, \tag{1}$$

- WT_P : Waiting time of Path P
- IVT_P^L : In-vehicle time given less than 100% congestion-rate on Path P (low-level ridership)
- IVT_P^N : In-vehicle time given 100%~150% congestion-rate on Path P (normal-level ridership)
- IVT_P^H: In-vehicle time given more than 150% congestion-rate on Path P (high-level ridership)
- TT_P : Transfer time between local and express trains on Path P

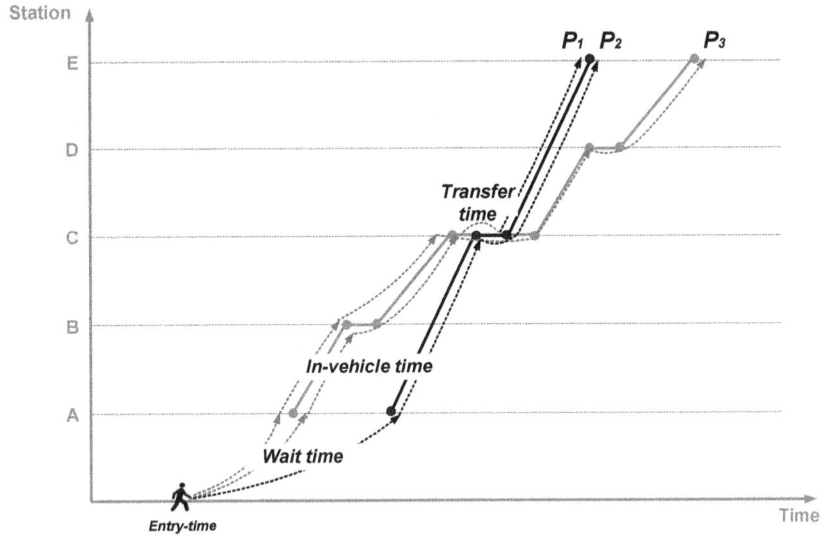

Figure 1: Time-expanded network and alternative paths set.

3 ESTIMATION OF COST FUNCTION VIA INVERSE OPTIMIZATION

In the stochastic model based on the Random Utility Maximization (RUM) theory including the logit model, the coefficient of cost function is estimated to be Maximum Likelihood Estimation (MLE). However, if the path cost is deterministic, as it is in this study, MLE is not available. As such, this study uses the inverse optimization model to estimate the coefficient of cost function.

Given that the path selection of passengers is selfish routing [5], $c(P) \leq c(Q)$ shall be fulfilled in the event that there are Path P selected by a passenger in equilibrium conditions, and alternative path Q. However, there are passengers who do not meet $c(P) \leq c(Q)$ due to an assumption of cost factor and function, which are not taken into account. As such, the convex quadratic optimization model was developed to minimize error related to those passengers

$$\min \quad \frac{1}{2}\sum_i e_i^2, \tag{2}$$

$$\text{s.t} \quad c(P^i) \leq c(Q^i) + e_i, \quad ,\forall i \in I, \tag{3}$$

$$\beta_1 = 1, \tag{4}$$

$$e_i \geq 0, \quad \forall i \in I, \tag{5}$$

$$\alpha, \beta_2, \beta_3, \gamma \geq 0. \tag{6}$$

Eqn (2) is an objective function to minimize error related to all passengers. Eqn (3) is a restriction in which the cost of the path that a passenger selects shall be less than that of an alternative path in equilibrium conditions. Eqn (4) means a restriction such that one of coefficients shall be fixed to ensure a feasible solution that makes all the coefficients 0 does not occur.

4 EXPERIMENT RESULTS

4.1 Data set and estimation result of inverse optimization

This section describes the implementation of the proposed model for the Seoul Metro 9. Seoul Metro 9 has operated both the local and express train since it launched commercial service in 2009. As shown in Fig. 2, there are 12 express stations and 18 express-local stations. Express and local trains are being serviced alternately every 3 min during the morning peak hour.

The RP-data of the real train choice of passengers over the interval was acquired through the path recovering method of Hong et al. [2] that was applied to two-day sets of Smart Card data: one on Monday, May 11 and another on Monday, May 18, 2015. The first data set is used for coefficient estimation, and the second data set for validation.

Table 1 shows coefficients of cost function per path estimated for 156,257 passengers during the peak hours (07:00 ~ 10:00) on Monday, May 11, 2015. The number of passengers with e_i more than 0 is 21, 148, which accounts for 13.5%. Coefficient for waiting time is 1.27, which appeared to be higher than in-vehicle time given less than 100% congestion. If the congestion-rate is 100%~150%, in-vehicle coefficient is 1.13, and if the congestion-rate is 150% or more, in-vehicle coefficient is 1.34. As such, the cost that the passengers feel is increased, even given the same in-vehicle time, if ridership increases. Finally, the coefficient of transfer between the local and express train is 4.86, and resistance to transfer appeared to be very high.

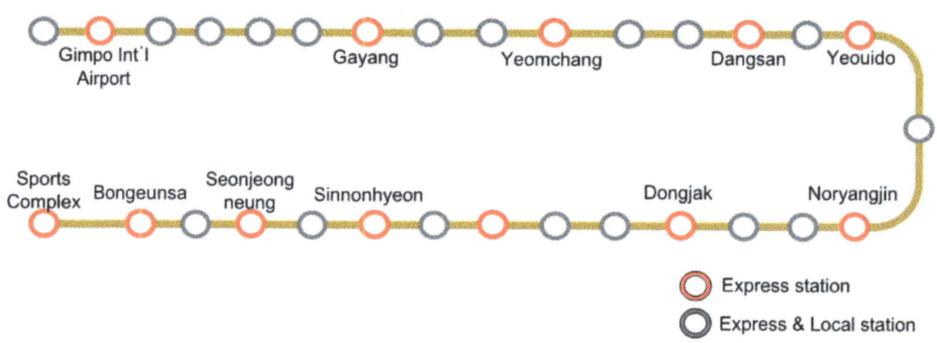

Figure 2: Map of Seoul metro line 9.

Table 1: Estimated parameters.

Parameter	α	β_1	β_2	β_3	γ
Coefficient	1.27	1	1.13	1.34	4.86

4.2 Performance evaluation

We conducted transit assignment using the cost function estimated in this study. The MSA-based Frank-Wolfe algorithm was applied, which is commonly used. Given that ridership of the observed train is f and flow of the estimated train is $\hat{f}$, the deviation of $\hat{f}$ from f is measured in both mean squared error (MSE) and mean absolute error (MAE), as in eqns (7) and (8)

$$\text{MSE} = \frac{1}{|E|}\sum_{e \in E}(\hat{f}_e - f_e)^2, \tag{7}$$

$$\text{MAE} = \frac{1}{|E|}\sum_{e \in E}\left|\frac{\hat{f}_e - f_e}{\hat{f}_e}\right|. \tag{8}$$

Table 2 shows the results of MSE and MAE calculated from the Estimation data set and Validation data set. For MSE, the difference in actual number of passengers is approximately 70 on average, and MAE is 0.09. Therefore, error in the transit assignment model presented in this study is approximately 10% or less.

Our transit assignment model implemented to Demand Responsive Metro Train Operation Planning Software (Dr.METRO) developed by KRRI [6]. Fig. 3 illustrates transit assignment results of Seoul metro line 9 using Dr.METRO. Dr.METRO display the ridership of each section and train by the color. Red, orange, yellow and green train-path indicate over 200%, 150%~200%, 150%~100% and less 100% congestion-rate, respectively. In time-space window 1 (from 07:20 to 08:40 and from Gayang station to Dongjak), we found almost every express and local train's congestion-rate are over 200%. Congestion-rate in express train is continued in time-space window 2, however, that of local train is decreased dramatically. So that, passengers prefer express trains.

Table 2: MSE and MAE of estimated link flows.

	MSE	MAE
Estimation set	66.2	0.08
Validation set	70.9	0.09

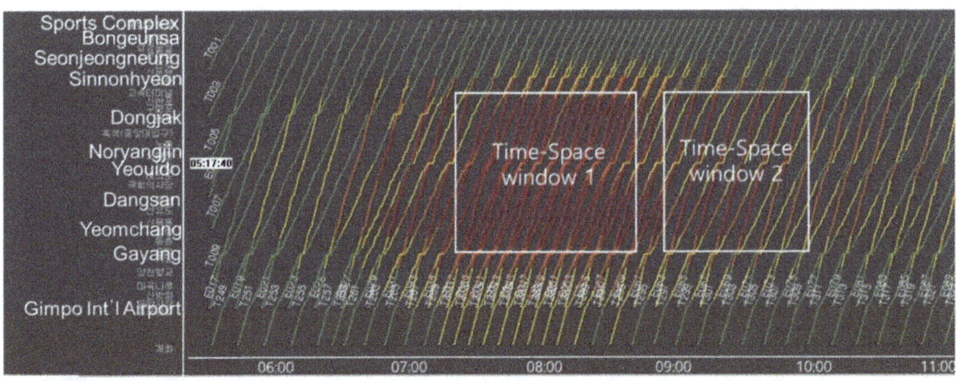

Figure 3: Transit assignment result of Seoul metro line 9 in a view of train-path diagram.

5 CONCLUSION

Express and local train service in urban railway provides an opportunity for passengers to meet the varying needs of passengers. After successfully launched on Seoul Metro Line 9, now the Korean government is considering an express-local train service on existing lines.

This study presents a methodology that estimates the coefficient of deterministic path cost for transit assignment by applying inverse optimization in the express-local train operation conditions. More specifically, our convex quadratic optimization model has estimated the coefficient of cost function to minimize passengers who violate the selfish routing condition. In addition, we validated the prediction of the model by applying the estimated cost function to the Seoul Metro 9. Through validation, it was shown that difference between actual and estimated ridership per train appeared to be less than 10%. Therefore, our model can be used to adjust train operation plan to reduce load imbalance between the local and express trains.

For future work, it would be interesting to develop the transit assignment model whose performance is better than the current ones and investigate the methodology that integrate the inverse optimization and stochastic model.

ACKNOWLEDGEMENT

This research was supported by Basic Science Research Program (2019R1G1A1085191) of the National Research Foundation of Korea (NRF)

REFERENCES

[1] Transit capacity and quality of service manual (3rd ed.), Technical Report TCRP Report 24 165, *Transportation Research Board of the National Academies*, 2013.

[2] Hong, S.-P., Min, Y.-H., Park, M.-J., Kim, K.M. & Oh, S.M., Precise estimation of connections of metro passengers from Smart Card data. *Transportation*, **43**(5), pp. 749–769, 2016.

[3] Ahuja, R.K. & Orlin, J.B., Inverse optimization. *Operations Research*, **49**(5), pp. 771–783, 2001.

[4] Hamdouch, Y. & Lawphongpanich, S., Schedule-based transit assignment model with travel strategies and capacity constraints. *Transportation Research Part B*, **42**, pp. 663–684. 2008.

[5] Correa, J.R., Schulz, A.S. & Stier-Moses, N.E., Selfish routing in capacitated networks. *Mathematics of Operations Research*, **29**(4), pp. 961–976, 2004.

[6] Oh, S.M. & Min, J.H., Dr. METRO: A demand-responsive metro-train operation planning program. *WIT Transactions on The Built Environment*, vol. 135, WIT Press: Southampton and Boston, pp. 595–601, 2014.

SECTION 4
SAFETY AND SECURITY

CERTIFICATION OF EGNOS SAFETY-OF-LIFE SERVICE FOR ERTMS ACCORDING TO IEC 61508 AND EN 50129

ALEŠ FILIP
Faculty of Electrical Engineering and Informatics, University of Pardubice, Czech Republic

ABSTRACT

A number of research projects are currently focused on the replacement of track balises (used for safe train position determination) with Virtual Balises (VBs) detected by GNSS to increase the efficiency of the European Rail Traffic Management System (ERTMS). The exploitation of the EGNOS Safety-of-Life (SoL) service, which is the European GNSS wide-area augmentation developed for aviation, for Virtual Balise Reader (VBR), should further strengthen the interoperability of ERTMS. In order to meet the high Safety Integrity Level required for VB detection (SIL 4), EGNOS-based train position determination in combination with other on-board sensors must be used. The adoption of EGNOS, which was not designed according to CENELEC railway safety standards, shall pass through a railway certification process to demonstrate that all demands on safety, dependability and interoperability have been met. This article proposes certification of the EGNOS SoL service for ERTMS via a so-called "pre-existing" item in the sense of the IEC 61508 and EN 50129 standards. The pre-existing item means an item that already exists and that was not specifically developed for the current project. The proposed solution starts with the use of Common Safety Method for Risk evaluation and Assessment (CSM-RA) that must be applied according to the EU Regulation No. 402/2013, because the safety change in ERTMS is significant. Since EGNOS alone is not able to meet the safety integrity requirement for VB detection, then EGNOS shall be correctly integrated within the VBR constituent. This article suggests that all essential information on EGNOS required for its integration into the VBR subsystem would be contained in the EGNOS Safety Manual. The Safety Manual, together with its assessment report, would be then referred in the section entitled "Related Safety Cases" of the overall Safety Case according to EN 50129.The presented results contribute to the introduction of EGNOS into ERTMS.

Keywords: certification, EGNOS SoL service, ERTMS, pre-existing item, safety case, virtual balise.

1 INTRODUCTION

The European Train Control System (ETCS), which is a part of ERTMS, employs track balises with known position for safe train position determination. These physical balises are detected on board of train by means of a so-called Balise Transmission Module (BTM). Besides Europe, ERTMS/ETCS has also become successful in Asia – especially in China, Turkey, Taiwan and South Korea. However, additional installations of ETCS with classical track balises on heavy haul and other lines may not be in some countries efficient (Australia, Russia, China, etc.) – mainly due to high investment and maintenance cost of ETCS balises.

The European Commission is aware of these ETCS limitations and therefore gradually initiated different R&D projects oriented on the use of Global Navigation Satellite System (GNSS) for signalling and ERTMS, such as 3InSat (2012–2015), ERSAT EAV (2015–2017), RHINOS (2016–2018), STARS (2016–2018), ERSAT GGC (2018–2020), STEMS (2018–2020), HELMET (2020–2022), etc. The exploitation of the European Geostationary Navigation Overlay Service (EGNOS), which is the European GNSS wide-area augmentation developed for aviation, for Virtual Balise Reader (VBR) within Virtual Balise Transmission Module (VBTM) is critical in this concept. The common objective of the above projects is to integrate the two large pan-European infrastructures (ERTMS and EGNOS) and, in addition to greater operational efficiency, to further enhance the interoperability of ERTMS.

WIT Transactions on The Built Environment, Vol 199, © 2020 WIT Press
www.witpress.com, ISSN 1743-3509 (on-line)
doi:10.2495/CR200111

The interoperability is a key feature of ERTMS. It ensures the correct technical interaction among on-board and track-side subsystems shared between different stakeholders, mainly Infrastructure Managers and Railway Undertakings. High safety and dependability requirements (i.e. RAMS: Reliability, Availability, Maintainability and Safety) for ERTMS must be met; also in cases when track balises are replaced with virtual balises and detected by GNSS – see e.g. ERSAT GGC project [1]. Therefore, it is necessary to pass the certification and approval process that guarantees all requirements for ERTMS (i.e. safety standards CENELEC EN 5012x [2]–[5], EN 50159 [6], Technical Specifications for Interoperability (TSIs), EU regulations, directives, etc.) are met. The EU Directive 2016/797 [7] extends authorization process of Control Command Systems (CCS) to the entire railway system – it supports concept of "cross-acceptance" as a steppingstone to the interoperability within the Trans European Network.

Railway stakeholders have to safely manage changes of the European railway system – including GNSS/EGNOS integration with ERTMS. Except Verification and Validation (VaV) and Safety Case development, system compliance with Control Command and Signalling Technical Specifications for Interoperability (CCS TSIs) should be checked – see [7]–[9]. VaV reports and Safety Case including assessment reports are important inputs for the certification process. The European railway sector utilises Common Safety Method for Risk evaluation and Assessment (CSM-RA) [10], [11] for harmonisation of risk assessment. CSM-RA harmonises in fact the whole Risk Management Process. It also creates a framework for certification and safety approval process of railway technical systems. It is evident that EGNOS shall also pass the certification process to ensure that all requirements for ERTMS safety and interoperability are met.

The safe exploitation of the aeronautical EGNOS SoL service in railway signalling primarily means that the SoL service must be employed according to CENELEC railway safety standards (EN 50126 [2], EN 50128 [4], EN 50129 [5], etc.). However, this has been a major obstacle to the EGNOS cross-acceptance for ERTMS for many years, since EGNOS was developed according to aviation safety standards and not according to railway CENELEC safety standards.

This article proposes that the EGNOS SoL service could be cross-accepted and certified for ERTMS via a so-called "pre-existing" item in the sense of the IEC 61508 [12] and EN 50129 [5] safety standards. It is further suggested that all essential information regarding EGNOS required for its integration into the ERTMS VBR subsystem would be contained in the EGNOS Safety Manual that would be then referred in the overall Generic Safety Case.

Section 2 describes basic requirements for safety assurance of railway Control Command and Signalling (CCS) systems. The cross-acceptance process according to CENELEC standards used in railway signalling is outlined in Section 3. The EGNOS cross-acceptance for ERTMS according to EN 50129 via pre-existing item is suggested in Section 4. Finally, the achieved results are summarised in Section 5.

2 SAFETY ASSURANCE OF RAILWAY SIGNALLING SYSTEMS

This section outlines a methodology to apply the current safety assurance procedures used in the railway sector to the exploitation of EGNOS and its related safety services for the application of the ERTMS Virtual Balise concept as explained in this article. It suggests that there should be an interface established between the current application of EGNOS in the aviation sector and its application in the railway sector. The approved EGNOS service provider then has the same relationship with the railway sector as with any aviation services. The adoption of EGNOS SoL service for ERTMS is then narrowed to the Safety and Risk Management processes of the EGNOS interface. Therefore, there will be an interface document setting out the services to be provided to the railway sector that will effectively be the same for all users of the service.

An important part of the interface document shall be a Safety Manual, which is a user guide for designers and systems integrators enabling safe EGNOS integration with ERTMS without compromising safety. The EGNOS Safety Manual is the fundamental document enabling cross-acceptance of EGNOS in the railway sector. The idea of the safety manual describing EGNOS interface and enabling safe EGNOS integration with ERTMS represents the main topic in this paper.

The safety manual shall be developed and approved in line with the railway Safety Management Process according to railway CENELEC standard EN 20129 [5] and also used within Risk Management Process according to CSM-RA, as it is required by the EU Regulation 402/2013 [10], [11]. The differences between the Safety and Risk Management processes are outlined in sections below.

2.1 Safety management and CENELEC standards

2.1.1 EN 50129 [5]

The standard EN 50129 is applied in conjunction with EN50126:2017 (part 1 and part 2; [2], [3]) which addresses safety-related requirements applied to the whole railway system. EN 50129 addresses the measures required to provide confidence in the safety of safety-related electronic system to be used in railway signalling systems. The interface between the two standards consist of two subjects; the overarching RAMS context, and the development of the safety case and the life cycle. Intrinsic to both these subjects is risk assessment, which is the subject of the CSM-RA regulation [10]. EN 50129 defines the conditions that shall be satisfied in order that a safety-related electronic railway system/sub-system/equipment can be accepted as adequately safe for its intended application. The conditions for safety acceptance are the following:

- Evidence of quality management;
- Evidence of safety management;
- Evidence of functional and technical safety.

All of these conditions shall be satisfied, at equipment, sub-system and system levels, before the safety-related system can be accepted as adequately safe. The documentary evidence that these conditions have been satisfied shall be included in a structured safety justification document, known as the Safety Case. The Safety Case forms part of the overall documentary evidence to be submitted to the relevant safety authority in order to obtain safety approval for a generic product, a class of application or a specific application.

2.1.2 Purpose of safety management process

Safety management process is that part of the RAMS management process which deals specifically with safety aspects – see EN50126-1 [2]. The above-mentioned evidence of safety management represents one of three basic conditions for safety acceptance of the railway safety-related system. The use of this safety management process is mandatory for Safety Integrity Levels (SILs) 1 to 4 inclusive. However, the depth of the evidence presented, and the extent of the supporting documentation should be appropriate to the SIL of the system/sub-system/equipment under scrutiny.

The documentary evidence to demonstrate compliance with all elements of the safety management process throughout the life cycle shall be provided in the Safety Management Report, which forms Part 3 of the Safety Case [5]. The purpose of this process is to further reduce the incidence of safety-related human errors throughout the life cycle, and thus minimise the residual risk of safety-related systematic faults.

2.2 Risk management process according to CSM-RA

Single implementation of Safety Management process in accordance with CENELEC safety standards is not enough for application of an EGNOS service in railway signalling and train control. Since the introduction of an EGNOS service into ERTMS/ETCS represents a significant change within the European railway network, then a so-called Common Safety Method for Risk evaluation and Assessment (CSM-RA) according EU legislation must be applied [10], [11].

CSM-RA (Regulation (EU) 402/2013 [10]) covers the following activities: 1) Risk assessment process and demonstration of compliance with the safety requirements, 2) Hazard Management and 3) Independent Assessment by CSM Assessment Body (AsBo).

CSM risk assessment is compliant with the CENELEC life cycle. However, the safety monitoring during real system operations is not covered by the harmonised risk assessment within CSM-RA. In order to achieve full compliance of CSM-RA with the CENELEC life cycle, CSM-RA requires a separate Safety Management System to be implemented and provided within activities of the proposer of the significant change. The proposer can be i.e. a railway undertaking, an infrastructure manager, an entity in charge of maintenance, etc.

CSM-RA enables mutual recognition of results including harmonization of risk acceptance and safety requirements in EU Member States. Harmonization and mutual recognition of safety requirements is performed via Risk Acceptance Principles (RAP) and Risk Acceptance Criteria (RAC). CSM-RA is in fact the iterative process. The iterative risk assessment process is completed when it is demonstrated that all safety requirements are fulfilled, and no additional reasonably foreseeable hazards have to be considered.

Widely acceptable Codes of Practice (e.g. CCS TSIs, CENELEC standards, etc.) such as Risk Acceptance Principles enables one to harmonise risk and thus also safety requirements across Europe. Both Codes of Practice and similar Reference Systems as RAP can be considered at the same time also as RAC.

In case of GNSS/EGNOS application for railway signalling including ERTMS, Codes of Practice (mainly ERTMS/ETCS subsets and CENELEC standards) have been utilized for derivation of safety requirements for Virtual Balise detection.

3 CROSS-ACCEPTANCE IN RAILWAY SAFETY CONCEPT

The aim of European railway authorities and European railway industry is to develop compatible railway systems based on common standards. Therefore cross-acceptance of Safety Approvals for sub-systems and equipment by the different National Railway Safety Authorities (NSA) represents basic approach in the system development and approval process. The issue of cross-acceptance also becomes critical in the area of exploitation of the aviation EGNOS SoL service within ERTMS.

3.1 Safety cases according to EN 50129

The cross-acceptance in the field of railway safety-related systems can be achieved and demonstrated via a so-called Generic Product and Generic Application. For a generic product, i.e. independent of application, and for a generic application, i.e. class of applications, it should be possible for safety approval granted by one safety authority to be accepted by other safety authorities (i.e. cross-acceptance or mutual recognition).

The CENELEC EN 50129 standard identifies three different kinds of safety cases:

- Generic Product Safety Case (GPSC) – independent of railway safety application;
- Generic Application Safety Case (GASC) – for a class of (ERTMS) applications;
- Specific Application Safety Case (SASC) – for a specific application.

The cross-acceptance can be achieved by means of the GPSC and GASC. The SASC is required for a specific application and will depend on mutually recognized GPSA and GASC. The Safety Case for a system may depend on the Safety Cases of other sub-systems or components. In such circumstances, safety acceptance of the main system is not possible without previous safety acceptance of the related sub-systems/components. If an Independent Safety Assessment/Assessor (ISA) report has been obtained for a generic product, or for a generic application, a reference may be made to this for safety acceptance of a specific application; it is not necessary to repeat the generic ISA process for each specific application. It is mandatory according to CENELEC that verification and validation (VaV) activities and development of the safety case for a product or application shall start at the early phases of the system life cycle (EN 50126 [2]). It is also required for GPSC and GASC.

Note: If a CSM-RA Assessment Body (AsBo) fulfils also the competence requirements of a CENELEC ISA, then the independent safety assessment carried out by such an AsBo can include also all necessary independent safety assessment activities that should be fulfilled by the CENELEC ISA. The objective is to avoid unnecessary duplication of independent safety assessments by different conformity assessment bodies and unnecessary duplication of inherent costs.

3.2 Role of generic safety case in certification and proposed solution

The interoperability means correct interaction between different interoperable constituents as defined in point (7) of Article 2 of Directive (EU) 2016/797 [7]. In order to guarantee the interoperable, safe and dependable operations, it is necessary to provide certification of individual constituents. A safety case is an important part of the certification process because it simplifies the process by providing a modular approach. It is assumed that the railway cross-acceptance approach based on generic safety cases (GPSC and GASC) would be utilised for demonstration of the essential safety requirements related to a Virtual Balise Reader (VBR) based on EGNOS. It can significantly simplify the certification of Interoperability Constituents implementing the VBR function.

As it is explained above, it is not considered that GPSC would be developed for EGNOS according to CENELEC and utilised for the cross-acceptance process since it is not practically possible. In addition, it seems it is not necessary to focus the attention on a generic safety case development, since as it is mentioned in Section 1 not always a safety case is required according to IEC 61508/EN50129:2018, especially for systems or services which already exist, such as EGNOS. Instead the EGNOS SoL service would be cross-accepted through mutual recognition of EGNOS SoL Service provider for railway usage based on the ICAO SARPs (International Civil Aviation Organization Standards and Recommended Practices) and GNSS MOPS (Minimum Operational Performance Standards) interpretation in terms of railway RAMS and elaboration of the Railway EGNOS SoL Service Definition Document (by EGNOS SoL service provider supported by rail industry). This document would also describe how EGNOS shall be used by designers and system integrators within the ERTMS VB concept in order to achieve safe integration of EGNOS with ERTMS. It represents a fundamental step in the EGNOS adoption for ERTMS. This would be then followed by the development of subsequent generic and also specific safety cases. The exact structure and contents of a safety case for railway safety-related systems is defined in EN 50129 [5]. The same structure is also required for generic safety cases.

The certified EGNOS SoL service for aviation already exists (since 2011) and it is compliant with ICAO SARPs. The EGNOS safety case has been already developed according to aviation safety standards and not according to CENELEC. Therefore, the major problem related to the EGNOS SoL service cross-acceptance for ERTMS is that EGNOS cannot be considered as a Generic Product from the CENELEC standards point of view. EGNOS as it is without additional supporting safety evidence and guarantees must not be used for safety applications in the rail sector. It is because the essential condition regarding VaV, safety case development and other activities to be performed during CENELEC safety life cycle was not fulfilled.

Aviation and railway safety regulations (i.e. ICAO SARPs, GNSS MOPS, CENELEC, CCS TSIs, etc.) are different and therefore it would be unrealistic to develop the EGNOS safety case according to CENELEC and use it as a basis for certification of EGNOS based ERTMS solutions. Therefore, another approach for the EGNOS SoL Service cross-acceptance should be utilised. It seems that the EGNOS SoL service could be adopted and certified by means a so-called "pre-existing" item that was originally defined in the standard IEC 61508 (2nd ed., 2010) [12].

This is the fundamental idea in this article that could open the door to the EGNOS cross-acceptance for ERTMS. The idea is further developed in Sections below.

4 EGNOS CROSS-ACCEPTANCE FOR ERTMS VIA PRE-EXISTING ITEM

This section deals with a possible EGNOS SoL service cross-acceptance for ERTMS through its certification as a pre-existing item, which is specified in detail in IEC 61508 [12] and CENELEC EN 50129 [5] standards.

Firstly, a common practice regarding certification of supporting services which are considered as critical elements of safety-related systems is described. Then the concept of a supporting service as a pre-existing item which was originally introduced in the standard IEC 61508 and subsequently adopted by the railway specific standard EN 50129:2018 is outlined. Finally, the possibility of EGNOS SoL service cross-acceptance for ERTMS via the IEC 61508/EN 50129 pre-existing item is discussed.

The analysis performed below is fully in line with the Return Experience (REX) [13] of the European Union Agency for Railways (ERA) with the use of CSM-RA (402/2013 and 2015/1136). The REX says that apart from widely used CENELEC EN 5012x standards within the CSM risk management process, the risk acceptance is also managed according to the standard IEC 61508 and Tolerable Hazard Rates (THRs), especially regarding generic components design for the Control Command Signalling (CCS) sub-system.

4.1 Certification of generic supporting services and utilities in industry

Generally, in the control systems market, manufactures' products are often certified for functional safety applications against IEC 61508. This is mostly done by a Certification Body (CB) which is accredited to do this work by an Accreditation Body (AB). Important element of the certification process is a safety case, which provides a systematic and complete way to show compliance to one or more functional safety standards. The IEC 61508 safety case demonstrates the fulfilment of the functional management requirements of the IEC 61508-1 to 3. This also applies to supporting services/utilities such as electricity, gas, pressure steam, etc. It is usually required to develop a safety case for such supporting services in the process industry to show that the supporting services that are needed to implement any measure defined in the safety case shall have suitable reliability and availability. To meet this criterion, the safety case relating to the safety services shall according to the Safety Case Assessment Guide [14] show:

- that the role and significance of the supporting services has been considered in design, construction, operation and maintenance to ensure that these supporting services and facilities will be available in the required quality when it is needed;
- the effect of the loss of key supporting services has been considered as part of a hazard identification and risk analysis process. This shall ensure that safety systems fail to a safe state and that the consequence of supporting service failure does not act as a major accident initiator;
- the reliability of services for safe shutdown and emergency response have been specified and independent back-up supplies provided where necessary, and
- those supporting services that are essential for operation of key safety systems and its back-up system.

However, it is not always possible to develop a safety case for a system or item according to required standards, e.g. IEC 61508 [12], CENELEC EN 50129 [5], etc. This also applies to the EGNOS system whose safety case was developed according to aviation standards because it was primarily designed for safety operations in aviation.

In order to utilise EGNOS SoL service for the ERTMS virtual balise concept, it should be adopted and certified according to railway specific standards and regulations. In sections below there are discussed possibilities to adopt and certify EGNOS for ERTMS as a supporting service via an IEC 61508/EN 50129 pre-existing item.

4.2 IEC 61508: Supporting service within safety-related system

The standard IEC 61508 [12] applies to safety-related systems that incorporate Electrical and/or Electronic and/or Programmable Electronic (E/E/PE) devices. It specifically covers hazards that occur when safety functions fail. The main goal of the safety standard is to reduce the risk of failure to a tolerable level. IEC 61508 is built on two fundamental pillars:

- the Safety Life Cycle intended to reduce or eliminate failures due to systematic causes; and
- the Safety Integrity Levels (SILs) to address dangerous random HW failures.

The safety life cycle is defined as a process that includes all necessary steps to achieve the required functional safety. It is also called Functional Safety Management. A Safety Integrity Level is one of four levels (i.e. SIL 1, 2, 3 and 4), each corresponding to a range of target average dangerous failure frequencies of a safety function. SIL is a measure of performance of a safety function, which is designed as a safety guard (safety provision) against the specific hazard. SIL 4 is used to protect users against the highest risks. SIL, according to IEC 61508 [12], is determined by the average frequency of dangerous failure per 1 hour (PFH) for systems working in continuous or high demand mode of operation – i.e. also computer-based railway signalling systems. Note that a SIL is a property of a safety function rather than of a system or its part.

IEC 61508 defines that a safety-related system is a designated system that both:

- implements the required safety functions necessary to achieve or maintain a safe state for the Equipment Under Control, and

- is intended to achieve, on its own or with other safety-related systems and other risk reduction measures, the necessary safety integrity for the required safety functions.

Note 6 of IEC 61508-4 [12] says that a safety-related system includes all the hardware (HW), software (SW) and supporting services (e.g. power supplies) necessary to carry out the specified safety function (sensors, other input devices, final elements, etc. are therefore included in the safety-related system). Therefore, IEC 61508 supporting services and related IEC 61508/EN 50129 pre-existing items could also represent a way to the EGNOS SoL service acceptance for the ERTMS Virtual Balise concept.

4.3 Pre-existing item according to IEC 61508 and EN 50129

The second edition of IEC 61508, [12] introduced a notion "pre-existing item" (for both HW or SW elements), which means an item or element that already exists (e.g. EGNOS) and that was not developed specifically for the current project (e.g. ERTMS). It is required by IEC 61508 that the compliant pre-existing item, which does not have a safety-case, should have an IEC 61508 certificate together with a certificate report and a safety manual. The safety manual should be developed by the system manufacture or service provider. The concept of pre-existing item was adopted later by CENELEC in EN 50129:2018 [5].

EN 50129 admits that safety of the system under consideration (e.g. ERTMS based on EGNOS) can rely on the specific properties of elements that have their own life cycle independent of the system safety life cycle. In this sense the EN 50129 pre-existing item is considered as an item which has been already developed (e.g. EGNOS) and cannot be changed for the system under consideration (e.g. ERTMS). EN 50129 further assumes that pre-existing items

- are complex (in their sense that a complete and deterministic functional model of the item is not possible or not completely available), and
- have not been developed in accordance with this document (EN 50129) or its previous versions.

It is further stated in EN 50129:2018 how pre-existing items shall be used within the railway safety-related systems, what are requirements for use of complete pre-existing systems (Section 6.2.2 of EN 50129:2018) and requirements for use of pre-existing elements, that performs only part of a safety-related function (Section 6.2.3 EN 50129:2018).

4.4 Role of IEC 61508 safety manual in safety acceptance process

The IEC 61508 [12] safety manual, which is not explicitly mentioned in the EN 50129 [5] standard, represents an important source of information regarding the pre-existing items for designers and system integrators. The manual includes all information about the compliant item, which is required for correct integration of the compliant item in the safety-related system or subsystem in order to meet the system or subsystems requirements without compromising safety. Development of the safety manual for the pre-existing item and its certification represents the required evidence which shall be contained in the overall Safety Case.

4.5 Contents of IEC 61508 safety manual

The content of the safety manual for the pre-existing is specified in detail in IEC 61508 [12]. For every compliant item, the safety manual shall contain (IEC 61508-2, Annex D, 2010):

- a description of the functions capable of being performed;
- identification of the hardware and/or software configuration of the compliant item to enable configuration management of the E/E/PE safety-related system;
- limitations regarding the use of the compliant item and/or assumptions on which analysis of the behaviour or failure rates of the item are based.

For every function, the safety manual shall contain according to [12] information on:

- the failure modes of the compliant item (in terms of the behaviour of its outputs) that:
    - are not detected by diagnostics internal to the compliant item;
    - are detected by diagnostics internal to the compliant item;
    - result in a failure of the diagnostics to detect failures of the function;
- estimated failure rates for the identified random/hardware failure modes including failure modes of the diagnostics internal to the compliant item;
- specification of the diagnostic test interval for every failure mode in that is detected by diagnostics internal to the compliant item requirements on any periodic proof test and/or maintenance;
- specification of outputs of the compliant item initiated by the internal diagnostics;
- requirements for any periodic proof test and/or maintenance;
- detailed information on the external diagnosis capable to detect identified failure modes of the compliant item;
- the hardware fault tolerance, etc.

For every function in terms of protection against systematic failure, the manual shall contain:

- the systematic capability of the compliant item that provides the function;
- any requirements or constraints relating to the application of the compliant item, relevant to the function, that should be observed in order to prevent systematic failures of the compliant item.

For additional requirements relating to software compliant items see 7.4.2.12 and Annex D of IEC 61508-3 [12].

4.6 Contents of safety evidence for pre-existing elements according to EN 50129

The standard EN 50129 [5] does not explicitly require that a “Safety Manual” for supporting elements having potential impact on functional safety shall be developed. Nevertheless, a very similar/same safety evidence of their contribution shall be documented in the overall Safety Case. EN 50129 states that pre-existing equipment performing only part of a safety-related function (like EGNOS), shall be correctly integrated in the technical and procedural safety management. The supporting safety evidence, which should be referred in the overall Safety Case, shall according to EN 50129 (Section 6.2.3) mainly include:

- a minimum set of information shall be available: described functionality; interfaces; hardware and/or software constraints; failure rate of the equipment; environmental conditions and other conditions of use;
- the pre-existing item shall be identified, included in the overall system definition and put under configuration control;

- the hazardous functional failure modes shall be identified. These failure modes should be identified by an interface hazard analysis. Note: This is particularly important for a complex system such as EGNOS;
- for each hazardous functional failure mode identified at the interface of the equipment, that would require a safety-related function with a safety integrity level from 1 to 4, the failure mode shall be externally (independently) negated and a safe state shall be enforced within the required time to fulfil the required safety target at the system level;
- failure modes observed at the boundary of the equipment are not always traceable to their actual causes. Therefore, for any equipment performing a SIL function, when it is not possible to identify the components of the equipment having a failure mode contributing to the hazard, the full failure rate of the equipment (e.g. EGNOS service) is used. This failure rate shall be demonstrated to be compatible with the TFFR (Tolerable Functional Failure Rate) required for the complete function;
- at the system level, when the pre-existing equipment is integrated in the system under consideration, all design, safety verification, safety validation and assessment activities shall be performed encompassing the pre-existing equipment as a black-box;
- for all SILs, a strategy shall be defined to manage the possible effects of product changes (e.g. disable automatic software updates).

EN 50129 does not explicitly give a specific name to the safety evidence file for the pre-existing item, for which it is impracticable to develop a safety case according to CENELEC. Nevertheless, it is obvious that the requirements regarding the contents of the IEC 61508 Safety Manual and the EN 50129 supporting safety evidence for the pre-existing item are practically the same. So why not name the supporting safety evidence required by EN 50129 as the Safety Manual. This Safety Manual as a safety evidence related to the pre-existing item shall be referred in the overall Safety Case (EN 50129).

4.7 Cross-acceptance of EGNOS for ERTMS

If the existing aviation EGNOS SoL service would be considered as a supporting service/utility and pre-existing item according to IEC 61508/EN 50129 for ERTMS Virtual Balise detection (generic application), then development of an additional EGNOS safety evidence would not be required. In this specific case only a supporting evidence (EN 50129)/Safety Manual (IEC 61508) describing all EGNOS SoL service features and characteristics, which are important for safe integration of the item into the ERTMS would be elaborated. Such guidance for designers and systems integrators is exactly what manufactures need.

It is assumed that the EGNOS Service Provider (currently it is ESSP) would develop in collaboration with the European Space Agency (ESA), European GNSS Agency (GSA), ERA and industry EGNOS SoL Service Definition Document (SDD) for Rail in compliance with the interpreted ICAO SARPs for Rail. The safety evidence/manual (EN 50129/IEC 61508) for EGNOS considered as the pre-existing item would be then important part of the EGNOS SDD for Rail. The EGNOS Safety Manual would be elaborated using EGNOS Usage Railway Requirements Document. The existing EGNOS safety case developed for safety operations in aviation would be used in this cross-acceptance process as well. Important fact is that all requirements for Safety Manual, that also represent a guidance for the safety evidence elaboration, are known. These are specified in detail in EN 50129:2018,

Section 6 and Annexes D of IEC 61508-2 and IEC 61508-3, and should be also used as a guidance for development of the EGNOS Safety Manual to be fully in line with CENELEC.

CSM-RA must be used for the EGNOS cross-acceptance process in any case since it represents a significant change in ERTMS. The applicability of CENELEC is within CSM-RA is optional. Nevertheless, the use of CENELEC standards within CSM Risk Assessment Process is generally highly recommended, as it also results from the ERA REX study [13]. The EGNOS Safety Manual including assessment report (i.e. supporting safety evidence in terms of EN 50129) for EGNOS as the pre-existing item shall be then referred in the overall Safety Case, Section named "Related Safety Cases", as it is outlined in Fig. 1.

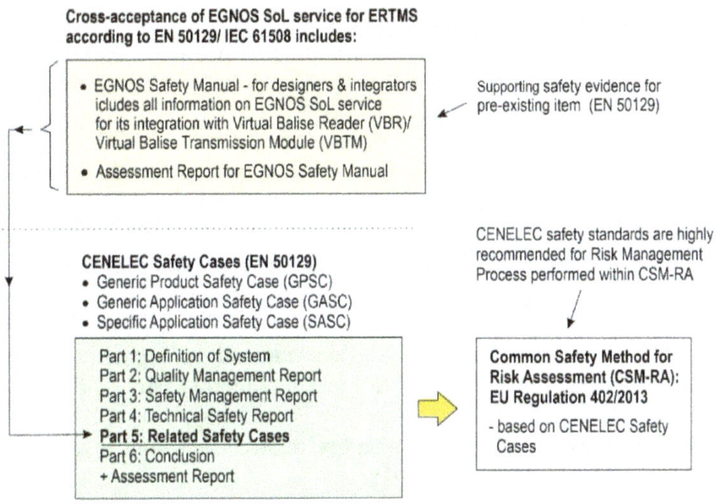

Figure 1: Cross-acceptance of EGNOS for ERTMS via CENELEC pre-existing item.

5 CONCLUSIONS

This article presents the original idea of the EGNOS Safety-of-Life (SoL) service adoption and certification for the ERTMS/ETCS Virtual Balise concept. The idea consists in fact that the EGNOS SoL service would be adopted and certified via a so-called "pre-existing" item in terms of the standards EN 50129/IEC 61508. Important is that EGNOS can be fully cross-accepted according to CENELEC railway safety standards and so this significant (safety) change in ERTMS can be fully controlled using CSM Risk Management Process in compliance with the CENELEC life cycle. It means that there is no need to develop a new EGNOS safety case specifically for rail safety applications according to CENELEC, which would not even be possible.

The idea of EGNOS certification for rail signalling by way of a pre-existing item is very fresh. This article was written with the intention to 1) show that a possibility of EGNOS SoL service certification according to CENELEC exists, and 2) initiate a discussion on the proposal within the GNSS and railway community. The European railway and GNSS researchers have been looking for a suitable solution to adopt EGNOS for railway safety systems for more than one decade. The imaginary closed loop consisting of the EGNOS safety case developed according to aviation safety standards on one hand and strict CENELEC standards for rail on the other hand did not offer any solution for EGNOS-based signalling. Therefore, it was necessary to untie the loop. In this paper it has been demonstrated that the suggested idea to certify the EGNOS SoL service by way of the pre-existing item offers a suitable solution. It is believed this proposal opens the door to the EGNOS certification for ERTMS/ETCS in Europe. In addition, the idea can be also utilised for GNSS SoL services adoption and certification in other land GNSS-based safety applications, such as self-driving cars, machine control, mobile robots, etc.

ACKNOWLEDGEMENTS

The work was supported: 1) from the ERDF/ESF grant "Cooperation in Applied Research between the University of Pardubice and companies, in the Field of Positioning, Detection and Simulation Technology for Transport Systems – PosiTrans (2018–2022)," No. CZ. 02.1.01/0.0/0.0/17_049/0008394, 2) by the ESA STEMS project (2019–2020), and 3) from the European research programme H2020 within the HELMET project (2020–2021).

REFERENCES

[1] H2020 ERSAT GGC project, 2019. www.ersat-ggc.eu/.

[2] EN50126-1, Railway applications – The specification and demonstration of Reliability, Availability, Maintainability and Safety (RAMS) – Part 1: Basic requirements and generic process. *European Standard*, CENELEC, 2017.

[3] EN50126-2, Railway applications – The specification and demonstration of Reliability, Availability, Maintainability and Safety (RAMS) – Part 2: Guide to the application of EN 50126-1 for safety. *European Standard*, CENELEC, 2017.

[4] EN50128, Railway applications – Communication, signalling and processing systems – Software for railway control and protection systems. *European Standard*, CENELEC, 2011.

[5] EN 50129, Railway Applications – Safety related electronic systems for signalling. *European Standard*, CENELEC, 2018.

[6] EN 50159, Railway applications – Communication, signalling and processing systems – Safety-related communication in transmission systems. *European Standard*, CENELEC, 2010.

[7] Directive (EU) 2016/797 of the European Parliament and of the Council of 11 May 2016 on the interoperability of the rail system within the European Union, 2016.

[8] Regulation (EU) 2016/919 of 27 May 2016 on the technical specification for interoperability relating to the "control-command and signalling" subsystems of the rail system in the European Union, 2016.

[9] Regulation (EU) 2019/776 of 16 May 2019 amending Commission Regulations (EU) No 321/2013, (EU) No 1299/2014, (EU) No 1301/2014, (EU) No 1302/2014, (EU) No 1303/2014 and (EU) 2016/919, 2019.

[10] Regulation (EU) No 402/2013 of 30 April 2013 on the common safety method for risk evaluation and assessment and repealing Regulation (EC) No 352/2009, 2013.

[11] Regulation (EU) 2015/1136 of 13 July 2015 amending Implementing Regulation (EU) No 402/2013 on the common safety method for risk evaluation and assessment, 2015.

[12] IEC 61508 (1-7), Functional safety of electrical/electronic/programmable electronic safety-related systems, *European Standard*, 2010.

[13] Jovicic, D., Report on the return experience (REX) with the use of the CSM for risk assessment (Regulations 402/2013 and 2015/1136). *ERA Report, Ref.: 004MRA1100*, 18/12/2018, page 46.

[14] Safety Case Assessment Guide, jointly developed by Ministry of Manpower, National Environment Agency, SCDF and SCIC (Singapore Chemical Industry Council), pp. 40–41, 2017.

DEVELOPING A GENERALISED ASSESSMENT FRAMEWORK FOR RAILWAY INTERVENTIONS

JOHN ARMSTRONG[1], GEORGIOS REMPELOS[1], JINGFU WEI[2], JOHN PRESTON[1], SIMON BLAINEY[1], JOHN EASTON[2] & CLIVE ROBERTS[2]
[1]Transportation Research Group, University of Southampton, UK
[2]Birmingham Centre for Railway Research and Engineering, University of Birmingham, UK

ABSTRACT

Work undertaken on the socio-economic assessment of the installation of under-sleeper pads (USPs) on a busy commuter railway route in Britain, as part of the Track to the Future (T2F) project, was presented at COMPRAIL 2018. This was followed by an expanded, comparative analysis of the same route, comparing the noise-reduction and other costs and benefits of USPs, rail dampers and noise barriers, published in the *International Journal of Transport Development and Integration.* Such analyses provide useful information and decision-making assistance to infrastructure managers, funders and other stakeholders, but typically require the repeated development of bespoke models of study locations, interventions and effects. The "one-off" nature of these models increases the time and costs required for such studies, and makes it more difficult to compare similar schemes in different locations, or different types of intervention in a single location. There is, therefore, a need for an improved, more generalised and standardised, transferable, replicable and comparison-enabling approach to the socio-economic assessment of such interventions. Ongoing work on the T2F project entails the development and use of standardised templates to describe the input costs and expected impacts of different infrastructure, rolling stock or operational interventions. These form the inputs to, and part of the development for, T2F of an improved, generalised, ontology-based assessment framework, which is being developed using the West Coastway railway line between Portsmouth and Brighton on England's south coast as an initial case study. This paper describes the context and rationale for the development of the assessment framework, its components and the methodology being adopted.

Keywords: railways, maintenance, renewals, enhancements, interventions, environmental impact, economic assessment, assessment template, ontologies.

1 INTRODUCTION

Recent decades have seen significant growth in railway traffic in Britain, as elsewhere in Europe and beyond [1], [2]. Recent changes in annual train kilometres in Britain are shown in Fig. 1, based on data published by the Office of Rail and Road (ORR) [3], [4], with growth in passenger train kilometres partly offset by recent declines in freight train kilometres. Such traffic growth is welcome for the railway industry, following a long period of decline in the mid-20th century, but can also present it with capacity challenges. Similarly, while modal shift to rail (especially from road and air travel) is generally desirable on environmental grounds, it does have its own environmental impacts, notably including the effects of noise and vibration on non-users living or working close to railway lines, which increase with traffic growth and may need to be reduced at source and/or attenuated.

In addition to such enhancements, the growth in traffic levels results in an increased need for infrastructure maintenance and renewals, and less opportunity to carry out any of these necessary works. There is therefore also a need for improved planning and implementation of maintenance and renewals (M&R) activities [5], and, more generally, for interventions to increase infrastructure working life. In order to achieve the desired outcomes in the most cost-effective manner, there is also a need for methods and tools to assess the relative costs and benefits of alternative approaches to such interventions. Previous work undertaken in the

WIT Transactions on The Built Environment, Vol 199, © 2020 WIT Press
www.witpress.com, ISSN 1743-3509 (on-line)
doi:10.2495/CR200121

Figure 1: Annual train kilometres in Britain, 2010–2011 to 2018–2019 [3], [4].

Track21 [6] and Track to the Future (T2F) [7] research programmes, whose aims are to improve the "engineering, economic and environmental performance of railway track" in Britain, developed bespoke cost-benefit analysis (CBA) tools for specific infrastructure interventions, as described in Ortega et al. [8] and Armstrong et al. [9].

While the outputs of these exercises provided useful information, and Armstrong et al. [9] included comparative analysis of alternative interventions, rather than assessing a single "do something" option against a "do minimum" base, the approaches used were still somewhat location- and situation-specific. It thus became apparent that, rather than conducting repeated, one-off assessments of possible interventions, there are significant potential benefits to be gained from the development of a standardised, flexible assessment framework and methodology, capable of assessing different types of intervention on different parts and extents of the national (or another) railway system. This paper sets out the approach being adopted to deliver these benefits.

Following this introduction, the background to and context of the work are first set out, including the rationale for the proposed new approach. The two main elements of the updated approach are then described. This is followed by a description of the case study being used for the initial implementation of the approach. Ongoing and further work are then described, followed by some conclusions and a list of references.

2 BACKGROUND AND CONTEXT

As indicated in the Introduction, considerable work has been done in the Track21 and T2F research programmes to develop methods and tools to model the whole-life costs and benefits of a range of infrastructure interventions. Much of the focus was on the reduction of environmental impacts (chiefly noise and vibration) on non-rail users (for example through the installation of under-sleeper pads (USPs), rail dampers or noise barriers), but the wider impacts on track quality, whole-life costs and M&R requirements were also considered, as were the effects on ride quality for rail users. There is considerable debate surrounding the economic valuations of noise and vibration reductions, and approximations are required to estimate the numbers of people affected by interventions, and the extent to which they benefit from or are disadvantaged by the resulting changes. These and other uncertainties (including valuations of ride quality changes) have been addressed by using Monte Carlo simulations and sensitivity analyses to identify a likely range of potential impacts.

Again, this approach has been applied to the assessment of a range of interventions in different locations, but, even for the comparative studies of alternative interventions in a single location, repeated and time-consuming, one-off analyses were required to obtain the required outputs, reducing the potential replicability of the approach for different interventions and locations, and the direct comparability of the results obtained. One of the wider objectives of the T2F research programme, therefore, is the development of a standardised framework for the assessment of the economic (and, ultimately, carbon) impacts of different infrastructure (and other, e.g. rolling stock) interventions in different locations and operating circumstances. Work to date to meet this wider objective has focussed on the use of standardised input "templates" for different interventions, and on moving towards an ontological approach to the handling and processing of the intervention data; these two aspects of the work are considered in the following sections of this paper.

3 STANDARDISED INPUT TEMPLATES

In a large research programme like T2F (and often more generally, in the assessment of large infrastructure and transport interventions and investments), the analysis of economic and carbon impacts tends to be undertaken by specialists in those fields, relying upon expert advice and input on the nature, costs and impacts of individual interventions. It is helpful to both parties if the data requirements for input to the economic and carbon assessments are clearly stated and specified, and the resulting inputs are of a standardised (or as much as is possible in the context of different interventions) layout and format.

A standardised input template was produced in an iterative process, by means of development of initial proposals, discussion with and feedback from colleagues, and subsequent refinement; the current version of the template is included as an Appendix, and its main features are summarised below.

Following a description of the type of intervention under consideration and the name of the investigator, the template consists of six main, generally qualitative, questions (questions 3–6 have associated sub-questions, seeking associated quantitative information). Additional comments are also elicited. The six main questions and associated sub-questions are:

1. Why will this intervention improve rail track systems?
2. In assessing the engineering impacts of this intervention what are the key input and output variables?
3. What are the main one-off (capital expenditure) and recurrent (operating expenditure) financial costs of the intervention?

   Sub-question: What is your judgement of the magnitude of these costs and can they be quantified? If so, how?
4. What are the main operator benefits? Operator refers to both infrastructure (Network Rail and its suppliers) and train services (train operating companies and their suppliers).

   Sub-question: What is your judgement of the magnitude of these benefits and can they be quantified? If so, how?

5. What are the main user benefits? User refers to customers of passenger and freight train services.

   Sub-question: What is your judgement of the magnitude of these benefits and can they be quantified? If so, how?

6. What are the main non-user benefits? Non-user refers to users if rival transport systems (air, road), residents or the wider community.

   Sub-question: What is your judgement of the magnitude of these benefits and can they be quantified? If so, how?

The template thus seeks a set of standardised qualitative inputs to the assessment of a proposed intervention, supported where possible by the associated quantitative inputs needed for detailed economic and/or carbon analysis. It is not generally anticipated that the subject matter expert for a given intervention will be able to directly provide all the data inputs needed for detailed analysis, but rather that (s)he will be able to "point the analysts towards" the sources of the required data.

4 AN ONTOLOGICAL APPROACH

4.1 The need for a new approach

Having developed a standardised template (informed by logic mapping, as in *The Magenta Book* [10]) for intervention-related inputs, outputs and outcomes/impacts, attention turned to the best means of standardising the analysis process, and developing a transferable, flexible framework to generalise the modelling processes. These processes should consider and include the whole life socio-economic (all relevant costs and benefits including carbon) and financial (all relevant cash flows, including Schedule 4 and 8 compensation payments for, respectively, planned and unplanned disruptions) impacts of track/railway system interventions, i.e. changes to infrastructure and/or the traffic using it.

It was envisaged that the result should take the form of a "whole-life cost dashboard", ideally interfacing with existing tools such as the Rail Safety and Standards Board's (RSSB's) whole life carbon model, and potentially contributing to a "digital twin" of the railway system. This would provide improved understanding, and enable easier and more effective management and handling, of industry datasets, including, for example, the identification, tagging and handling of missing data. Such a generalised lifecycle cost model would hold and process data from different sources and provenances, and, where necessary, of variable quality and criticality (i.e. the precision and accuracy of some datasets are more important than those of others).

In order to meet these needs, it was decided that an ontological approach should be adopted, as described and advocated in Morris and Easton [11] as a suitable means of integrating and processing disparate sources of railway data to generate useful information and knowledge. Ontologies (in the field of computer science, rather than philosophy) are used to study the existence of entities within a specific domain. Perhaps most easily described as the "knowledge model" component of what has previously been referred to as an Expert System, ontologies display some similarities with well-designed databases in terms of their ability to manage facts relating to the same entity, but with added functionality for the capture of semantics (i.e. data meaning, source, quality, shortcomings, etc.). Semantics and rules describing the relationships between concepts (e.g. different measures of temperature and the conversions linking them) are stored in models with the corresponding data, and transformations and translations then be undertaken within models, enabling updates or translations to be made once, within the scope of the model, rather than in multiple end-user software packages. This is of particular potential value in the context of Britain's railway industry, which uses and relies upon multiple poorly- or non-integrated legacy data systems; each system can request the data in the form it needs, and the model can be relied upon to provide either the raw data as stored (if it is already in the correct form), or to apply the appropriate translation to the data it does hold. From an implementational perspective, ontologies are typically represented using data "triples", a node–link–node structure comprising Observation–Value–Data, or Subject–Predicate–Object (or, in object-oriented programming terms, entity/instance–property–value): for example, "South West Main Line–Electrification Type–Third Rail DC". Graph-based databases known as triplestores are then used to hold the complete set of facts of interest, and to manipulate them efficiently.

In the context of the T2F and successor projects, an ontological approach will enable the combination of multiple and varied data sources into a more useful whole; this approach is particularly useful in broad analytical contexts, as in the case under consideration, i.e. the linking of infrastructure and/or rolling stock data with economic and/or carbon assessment.

4.2 Initial ontology development

As a first step in the movement towards the development and application of an ontological approach to the problem domain, an intervention modelling framework was developed, with six "top level" areas of primary interest:

- Impact categories (general);
- Inputs;
- Processes;
- Outputs;
- Outcomes;
- Impacts.

These areas were disaggregated and sub-categorised, as set out below, including their sub-division into "Real World" and "Scientific/Experimental" sub-categories where appropriate.

- Impact Categories (general)
  - Benefits – quantitative (e.g. track life increase, noise reduction) and qualitative
  - Disbenefits – quantitative (e.g. whole-life carbon footprint increase, maintenance cost increase) and qualitative
  - Additional comments (qualitative)
- Inputs
  - "Real World"
    - Time
    - Money
    - Resources/Materials
    - Railway Industry Data
      - Timetables (base cases)
      - Unit Cost data – e.g. USPs, fibre-reinforced ballast, M&R activities
      - Primary process-specific data related to the manufacturing of different track components

        - Secondary manufacturing and process-specific data
        - Detailed qualitative and quantitative data on the installation, maintenance, inspection and renewal procedures adopted by Network Rail (the infrastructure manager of Britain's railways)
        - Detailed information related to the end-of-life pathways of different track components (qualitative and quantitative), including locations of landfill sites, energy recuperation facilities, etc. from which detailed data may be obtained
        - Other Network Rail financial data: track access charges, schedule 4 and 8 payments, impact on Regulatory Asset Base etc.
        - Operational data – train movements (e.g. equivalent million gross tonnes per annum (EMGTPA))
        - Operator data: traffic levels (passenger kilometres, tonne kilometres, revenue)
  - Scientific/Experimental (exogenous to the intervention)
    - Demographics
    - Land use
    - Railway topography (e.g. embankment vs. cutting)
    - Asset condition and characteristics (i.e. what type of asset is it?)
        - Ambient air- and ground-borne noise levels
        - Carbon emissions by mode
        - Aspects of other modes – e.g. road congestion
        - Values of time, comfort, etc. – from e.g. WebTAG [12], although some values may be experimental and thus endogenous to an intervention
  - Scientific/Experimental (endogenous to the intervention)
    - Computational experiments
    - Laboratory tests
    - Field trials
    - Continuous measurements

- Processes
  - "Real World"
    - Use
    - Degradation
    - Maintenance and Renewals
    - Enhancements
  - Scientific/Experimental
    - Operational models
    - Economic models
    - Environmental models (primarily lifecycle analysis, based on ISO14000 series of environmental standards – in all cases, capturing and representing the results of detailed analyses of micro-level behaviour in macro-scale models is a challenge)
    - GIS modelling – data processing and integration

- Outputs
  - "Real World"

        - Engineering interventions – e.g. volume of USP installations, random fibre reinforcement etc.
    - Scientific/Experimental
        - Calculated costs – one-off (capital) and recurring (operating)
        - Calculated operator benefits (to Network Rail, operators and other industry suppliers)
        - Calculated user benefits (to passengers and freight customers)
        - Calculated non-user benefits (to local residents/workers, users of other modes, wider community)
        - Results of financial cost–benefit analysis
        - Results of social cost–benefit analysis
        - Results of carbon cost–benefit analysis, possibly including other, related impact factors

- Outcomes
    - "Real World"
        - Longer track life
        - Reduced maintenance
        - Reduced noise
        - Better track geometry
    - Scientific/Experimental
        - Extended and improved knowledge and understanding

- Impacts
    - "Real World"
        - Reduced maintenance costs, offsetting increased capital costs, and improved benefit:cost ratios (as indicated by the scientific/experimental outputs)
        - Improved user and non-user satisfaction
    - Scientific/Experimental
        - Application of research findings

The modelling framework and the standardised input template were then compared to identify and remedy any "gaps" between the two, and used to identify the available and potential data sources to provide the necessary inputs to the modelling process.

The impact categories, inputs, processes, outputs, outcomes and impacts are being structured in an ontology, building upon the existing Railway Core Ontologies (RaCoOn) [13] and ensuring compatibility with standard terminology used in life cycle costing standards and in WebTAG [12]. Created as a domain-specific representation of the railways, RaCoOn was originally developed to provide a model of the railway track and signalling, which could be used to automatically infer the set of routes available through a junction based on the status of the physical assets present. The model soon expanded to serve a broader range of railway use cases, with a "core" of generic railway concepts that are supported as needed by models for specific subdomains, such as infrastructure, signalling, and rolling stock [13]. The main strength of an ontology-based approach to modelling lies in its use of concepts to structure data. By providing a concept of a "track" that is then further refined by the data used to describe it, for example, RaCoOn is able to accommodate a wide range of viewpoints around the correct terminology and granularity of representation of that asset. This makes the ontology an ideal basis for the integration of data across different IT systems, with the fact-based, semantic representation of the data used by RaCoOn able to be translated to and from a wide range of other contexts without the introduction of ambiguity. In the work described in this paper, RaCoOn is being used to integrate GIS, ambient noise, and wheel-rail interaction data from a variety of systems, to enable the workflow shown in Fig. 2.

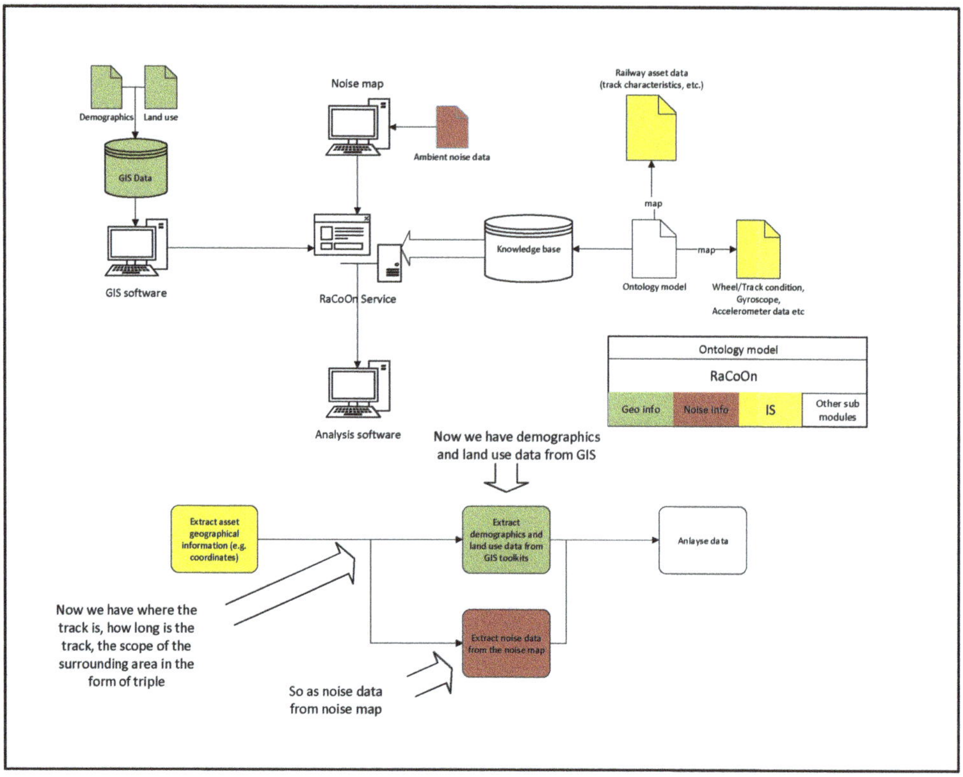

Figure 2: Workflow diagram.

The flow starts from the capture of railway asset data, wheel/track condition, etc. from various sources. A plugin is used to export relevant geographical data shapefiles, which converts geographical data to triples with reference to concepts defined in RaCoOn and stores them in the triplestore. Mapped data can then be handled by RaCoOn services.

RaCoOn services is a set of tools that helps to store converted data and subsequently retrieve it from storage. Its main purpose is to help users to retrieve data from storage and export the unified data with a selected format for further analysis. The key idea of this tool is to reproduce manual data consolidation and integration processes, and to help users to interact with extended RaCoOn, providing a solution to integrate rail-related data and shorten the data purification and consolidation processes using ontologies. It aims to provide a project-independent solution to help researchers and developers exploit the benefits brought by ontologies to the rail domain.

5 CASE STUDY

The work and findings described in Ortega et al. [8] and Armstrong et al. [9] are based primarily on two case study locations in Britain, the first on the East Coast Main Line between Newcastle and Edinburgh, and the second on the South West Main and Portsmouth Direct Lines between London Waterloo, Woking and Portsmouth Harbour stations. While the new, template- and ontology-based approach aims to replicate and generalise the processes used in Ortega et al. [8] and Armstrong et al. [9], there seemed little point in simply reproducing results already obtained (although checks for consistency will obviously be undertaken), and a new case study site was therefore identified and selected for the ongoing work. The chosen route is the "West Coastway" line along England's south coast between Portsmouth Harbour and Brighton stations, as shown in schematic form in Fig. 3.

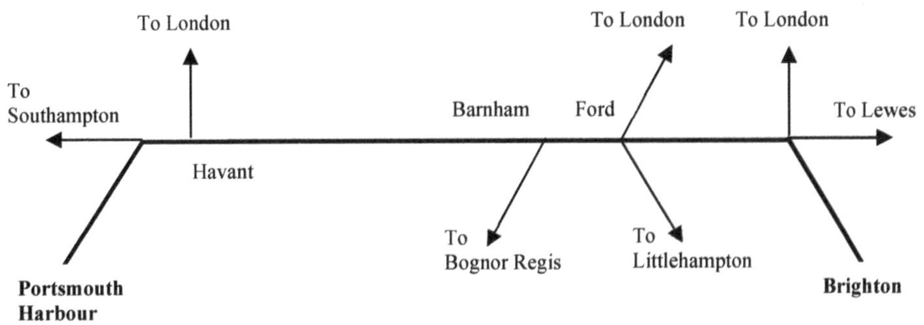

Figure 3: West Coastway schematic line diagram.

The length of the West Coastway is approximately 45.5 miles (73.2 km), and it is predominantly a two-track railway, with a few short, three-track sections. Line speeds along the route vary between 40 mph and 85 mph (approximately 64 km/h to 137 km/h), with a 20 mph (approximately 32 km/h) limit on the approaches to Brighton station. The route was chosen for a range of reasons, mainly because it is the subject of experimental and analytical activities in other research work being undertaken by members of the T2F team (see for example Young et al. [14]), and thus provides some potential synergies between activities.

6 ONGOING AND FURTHER WORK

Work is ongoing to develop and implement the ontology, and to replicate and extend the analytical processes described in Ortega et al. [8] and Armstrong et al. [9].

Looking beyond T2F, the project team aspires to develop and implement a more generalised version of the modelling and analytical framework, ideally including other transport modes and the interdependencies both between modes and with other infrastructure systems, in a similar manner to the models developed by the ITRC [15]. Another possibility is to link the framework with DAFNI (Data and Analytics Facility for National Infrastructure) [16], allowing direct access to and integration with the growing range of infrastructure datasets and models stored on this facility.

7 CONCLUSIONS

As traffic levels on legacy railway networks increase, there is a growing need for interventions to maintain and renew the infrastructure (and ideally to increase the time intervals between interventions), and to reduce the environmental impacts of increasing traffic levels on users and, particularly, non-users of the railway. At the same time, the increasing traffic levels reduce the opportunities for access to the infrastructure, and so judicious selection from the available range of interventions is increasingly important if overall benefits are to be maximised.

Previous work in this area has demonstrated the feasibility and merits of applying CBA to individual interventions and to the comparison of alternative options, but the tools and methods used were comparatively situation-specific and inflexible. An improved, standardised and flexible assessment framework would therefore be very useful for enabling the rapid and consistent assessment of alternative interventions and/or interventions in different locations and operating circumstances, thus also facilitating comparisons between alternatives.

This paper describes the work undertaken to date in developing such a framework for future application, the case study chosen for its initial application and testing, the ongoing work in this area, and anticipated extensions to other projects and research domains in which the authors are engaged.

ACKNOWLEDGEMENTS

This research has been carried out by the authors as part of the EPSRC-funded TRACK21 (EP/H044949/1) and Track to the Future projects (EP/M025276/1). The authors are grateful to Network Rail for the provision of industry data, information and expertise.

APPENDIX

TRACK TO THE FUTURE INTEGRATED MODELLING INFORMATION TEMPLATE

Type of intervention:
……………………………………………………………………………………………..

Name of investigator:
……………………………………………………………………………………………..

Q1. Why will this intervention improve rail track systems?

Q2. In assessing the engineering impacts of this intervention what are the key input and output variables?

Q3. What are the main one-off (capital expenditure) and recurrent (operating expenditure) financial costs of the intervention? *What is your judgement of the magnitude of these costs and can they be quantified? If so, how?*

Q4. What are the main operator benefits? *Operator refers to both infrastructure (Network Rail and its suppliers) and train services (TOCS and their suppliers).* *What is your judgement of the magnitude of these benefits and can they be quantified? If so, how?*

Q5. What are the main user benefits? *User refers to customers of passenger and freight train services* *What is your judgement of the magnitude of these benefits and can they be quantified? If so, how?*

Q6. What are the main non-user benefits? *Non-user refers to users of rival transport systems (air, road), residents or the wider community* *What is your judgement of the magnitude of these benefits and can they be quantified? If so, how?*

Any additional comments?

REFERENCES

[1] Bešinović, N. & Goverde, R.M.P., Capacity assessment in railway networks. *Handbook of Optimization in the Railway Industry*, eds R. Borndörfer, T. Klug, L. Lamorgese, C. Mannino, M. Reuther & T. Schlechte, Springer: Champaigne, pp. 25–45, 2018.

[2] Cacchiani, V. & Toth, P., Robust train timetabling. *Handbook of Optimization in the Railway Industry*, eds R. Borndörfer, T. Klug, L. Lamorgese, C. Mannino, M. Reuther & T. Schlechte, Springer: Champaigne, pp. 93–115, 2018.

[3] ORR, Passenger train kilometres by operator: Table 12.13, 2020. https://dataportal.orr.gov.uk/media/1486/passenger-train-kilometres-by-operator-table-1213.xlsx. Accessed on: 19 Mar. 2020.

[4] ORR, Freight train kilometres by operator: Table 13.25, 2020. https://dataportal.orr.gov.uk/media/1459/freight-train-kilometres-by-operator-table-1325.xlsx. Accessed on: 19 Mar. 2020.

[5] Armstrong, J. & Preston, J., Balancing railway network availability and engineering access. *Proceedings of the Institution of Civil Engineers – Transport*, 2019. DOI: 10.1680/jtran.19.00045.

[6] Track21, Railway track for the 21st century. http://track21.org.uk/. Accessed on: 25 Feb. 2020.

[7] Track to the Future, https://t2f.org.uk/. Accessed on: 25 Feb. 2020.

[8] Ortega, A., Blainey, S., Preston, J., Thompson, D., Squicciarini, G., Ntotsios, E. & Armstrong, J., Noise reduction for ballasted track: a socio-economic assessment. *WIT Transactions on Ecology and the Environment*, vol. 181, WIT Press: Southampton and Boston, pp. 461–472, 2018. DOI: 10.2495/CR180411.

[9] Armstrong, J., Ortega, A., Blainey, S., Preston, J., Thompson, D., Squicciarini, G. & Ntotsios, E., Noise reduction for ballasted track: A comparative socio-economic assessment. *International Journal of Transport Development and Integration*, **3**(1), pp. 15–29, 2019. DOI: 10.2495/TDI-V3-N1-15-29.

[10] HM Treasury, *The Magenta Book.* www.gov.uk/government/publications/the-magenta-book. Accessed on: 25 Feb. 2020.

[11] Morris, C. & Easton, J., Use of ontology for data integration in a degraded mode signalling system. *WIT Transactions on Ecology and the Environment*, vol. 181, WIT Press: Southampton and Boston, pp. 215–223, 2018. DOI: 10.2495/CR180191.

[12] Department for Transport, WebTAG, 2020. www.gov.uk/guidance/transport-analysis-guidance-webtag. Accessed on: 18 Feb. 2020.

[13] Tutcher, J., Easton, J. & Roberts, C., Enabling data integration in the rail industry using RDF and OWL: The RaCoOn ontology. *ASCE-ASME Journal of Risk and Uncertainty in Engineering Systems, Part A: Civil Engineering*, **3**(2), pp. F4015001-1-F4015001-12, 2017. DOI: 10.1061/AJRUA6.0000859.

[14] Young, M., Rempelos, G., Ntotsios, E., Blainey, S., Thompson, D. & Preston, J., A transferable method for estimating the economic impacts of track interventions: application to ground-borne noise reduction measures for whole sections of route. *Journal of Rail and Rapid Transit* (submitted).

[15] Lovric, M., Blainey, S. & Preston, J., A conceptual design for a national transport model with cross-sectoral interdependencies. *Transportation Research Procedia*, **27**, pp. 720–727, 2018.

[16] DAFNI, 2020. www.dafni.ac.uk/. Accessed on: 19 Feb. 2020.

INVESTIGATION OF FACTORS INFLUENCING THE EFFICIENCY OF RAILWAYS IN TERMS OF SAFETY AT LEVEL CROSSINGS

INARA WATSON, AMER ALI & ALI BAYYATI*
School of the Built Environment and Architecture, London South Bank University, United Kingdom

ABSTRACT

Road traffic accidents are in the top ten of all deaths worldwide. Regarding railways, in the EU there is one person killed or seriously injured every day in serious accidents involving level crossings (LCs). The number of accidents is almost unchanged in the last ten years despite the huge efforts that railways have made. There is a need to look at the factors that contributed to the occurrence of accidents at LCs beyond the drivers and pedestrian behavior or technical issues. The reported research investigates the efficiency of railways in terms of accident risk at LCs in 24 countries of the EU. It evaluates the efficiency of railways by applying the Data Envelopment Analysis (DEA) method and identifying the factors that influence the efficiency score of railways in terms of accident risk using the IBM SPSS. The results show that GDP per capita and density of population in the selected countries have a strong influence on the efficiency of railways. The expected outcome of this research may contribute to better understanding of the factors that influence the efficiency score of railways in term of accident risk at LCs and develop preventative measures.

Keywords: level crossings, accidents, efficiency score and influencing factors.

1 INTRODUCTION

A Level Crossing (LC) is a point where road and rail infrastructure intersect. As stated by United Nations Economic Commission for Europe (UNECE) in 1968, trains have priority and road users and pedestrians must give way to trains. There are many different types of LCs, but all of them must ensure the safety for road and rail uses. LCs are responsible for the second highest number of accidents on railways and there is a need to look for new solutions to reduce the number of accidents that would be economically acceptable. Apart from safety issues, LCs affect the speed of the trains and this reduces the capacity of lines and increases travel time [1]. They also cause disruption for road users and generates congestion [2]. In the future, the number of trains and cars will only increase. The best solution would be to remove LCs. This will improve the safety for road and railway users, reduce noise and air pollution from stopping and starting cars. Very often it is economically inappropriate. Railway authorities around the world are looking for cost-effective technological solutions that may improve the safety at LCs. To do so, there is a need to fully investigate factors that can influence the safety at LCs. Some of these did not get enough attention in the past, such as the population density, number of cars per 1000 inhabitants, GDP per capita, median age and the number of road accidents in selected countries. Whilst technology can help improve safety, but such safety can also be improved by educating the public and by means of law enforcement.

2 METHODOLOGY

The combination of several methods and different approaches may identify the major factors that influence the number of accidents at LCs. The approach is mainly qualitative and descriptive, supported by Data Envelopment Analysis (DEA) and IBM SPSS analyses of

**ORCID: http://orcid.org/0000-0002-4976-3776*

WIT Transactions on The Built Environment, Vol 199, © 2020 WIT Press
www.witpress.com, ISSN 1743-3509 (on-line)
doi:10.2495/CR200131

existing empirical evidence. In this research, aggregate statistics, which include numerical and non-numerical data, are collected and summarised to help having a clearer picture of safety at LCs. The DEA is used to evaluate the efficiency of performance of selected railways. For this, railways in 24 countries in Europe are studied and five years of data is analysed. Inefficient railways are made efficient through the proportional reduction of inputs. In this research, inputs are the number of accidents at LCs.

DEA calculates the lowest range of accidents for the given volume of passenger-km and thousands of tons of goods transported by railways in selected countries during the period 2012–2016. The model finds targets for the reduction of the number of accidents at LCs that need to be achieved by every country to be classified as an efficient railway. The outputs, number of passenger-km and tonnes-km proportions are kept constant. The selection of data is influenced by the data availability and difficulties in obtaining data. All data is derived from the UIC website, UNICE, Eurostat and annual reports and scientific publications.

3 TYPES OF LEVEL CROSSINGS

The type of LC depends on factors such as traffic, volume of trains and vehicles, types of use, area of location etc. LCs can be passive or active. Most LCs around the world are passive. In Australia it is 67% of the total number of LCs and 75% in the USA [3]. In 2011, in the UK, 75% of all LCs were passive but in Belgium only 15% [4]. In 2010, in the EU there were around 123,000 level crossings, 47% of them were passive and 53% active but only 28% provided barrier protection for road users. In Europe there are approximately five level crossings for every 10 km of line [5].

The passive LC is equipped with only warning signs such as “STOP” and a “ST ANDREW’s CROSS”, and road markings. It is up to the user to decide when it is safe to cross the line. There are 2,814 of this type of LC in the UK [6]. Regarding the active LC, in addition to warning signs and road markings, the LC is equipped with warning systems that warn road uses about approaching trains. The warning systems consist of flashing lights and sounds and some have a barrier between the train and road users that close when the train approaches. The effectiveness of any automatic warning system depends on a good visibility of the warning signals. The safest type of crossings is the grade separation LC that separates the rail and road traffic by building an overpass or underpass. This type of crossing is the most expensive option, but it is not always economically viable.

It is clear that, the higher the number of level crossings the higher possibility of accidents. The UK has 41 LCs per 100 km and that is slightly less than the average in the EU which stands at 50 LCs per 100 km of railway line [7]. The highest density of LCs is approximately 75 LCs per 100 km of railway line which belongs to Sweden, Austria, the Czech Republic, but the Netherlands, Bulgaria and Spain have the lowest density of LCs per km of line and it is less than 25 LCs per 100 km [8].

4 LC ACCIDENTS AND TYPES OF USERS INVOLVED

Approximately 25,700 people died in 2014 on the roads in the EU [9] whilst the railways had 2213 serious accidents and there were a total of 1928 people that had fatal or serious injuries. Moving rolling stock caused 1196 accidents and 638 accidents happened at level crossings [10]. The accidents and fatalities on railways cost the society in Europe more than €1.4 billion every year [11]. This cost is negatively affecting the economic sustainability of railways. There are two types of LC accidents, namely, collision of road vehicles with trains, and collision of road vehicles with LC devices. The latter is a huge problem for railways around the world. There was an increase in the number of drivers and pedestrians in UK that had near-miss accidents with trains from 140 drivers in 2009 to 161 in 2010 whilst the number of pedestrians for the same period rose from 270 to 297. These figures show that people ignore the warning signals at LCs [12]. Local drivers in rural areas are often overconfident and do not recognise that the risk of rural crossings is high and that may lead to accidents. Drivers do not look carefully for trains before crossing. There are hundreds of near-miss accidents every year [4].

On average, one person was killed or seriously injured every day at LCs in Europe. It was found that approximately 70% of all LC accidents occur approximately within 10 km of a car driver's home [14]. Fig. 1 shows the types of users involved in accidents at LCs in the EU.

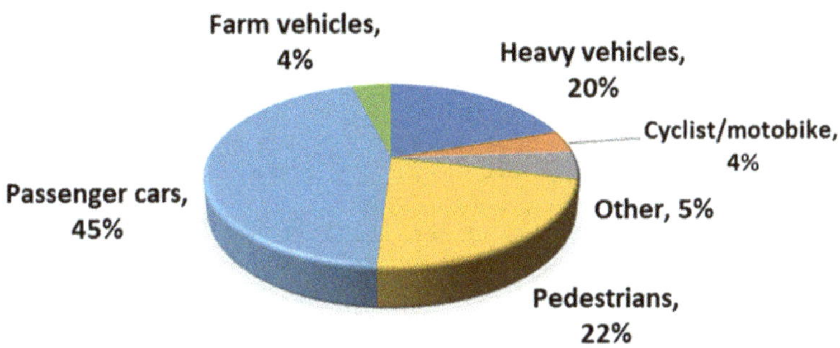

Figure 1: Types of the LC users involved in accidents in the EU. *(Source: ERA, 2012b.)*

Increased speeds of trains and vehicles on roads require greater sight distance for deciding whether to enter a LC or not. One of the main causes of crashes at LCs where there is a lack of barriers, is the behaviour of drivers who fail to judge correctly the speed and distance of approaching trains [3]. In 1996, the Transport Research Laboratory (UK) identified two categories of drivers that are most likely to be involved in accidents at LCs: drivers who believe that they have enough time to cross before train arrives, and do not stop or cannot stop because they are too close to the "stop line" when the amber light starts to flash, and drivers who are distracted or careless of the signals. Introduction of barriers at LCs can stop these types of drivers from being involved in accidents. Drivers are more likely to see the barrier as it is in front of their field of vision and will stop [6]. Drivers sometimes are trying to weave around the close half-barriers at LCs. These types of accidents are the second most common after over speeding when approaching the LC [15], [16]. To prevent this type of accident, two half-barriers can be replaced by two full barriers or four half-barriers. In the EU in 2016, 25,500 people died in road accidents, 296 of them at LCs. It is approximately 30% of all railway fatalities and 1.2% of all road fatalities [11]. Fig. 2 shows the number of accidents and fatalities at LC in EU 2009–2011.

The accidents and fatalities at LCs not only damage the safety and reputation of the rail industry, but also has a huge cost to society. In 2010 in Europe, this cost was estimated at €350 million [5]. Fig. 2 also shows the decreasing number of accidents and fatalities at LCs in EU between 2009 and 2011.

Approximately, three-quarters of all accidents at LCs involved pedestrians and fatality rate is higher for pedestrians than for car occupants. For example, in Australia 66% of all fatalities (excluding suicides) were pedestrians [17]. They are at greater risk than other road users. It was found that road users behave differently with respect to different warning systems at LCs. The LC with a passive warning system has higher accident rates, but a LC with barriers and flashing light has a lower accident rate [4], [18]–[20].

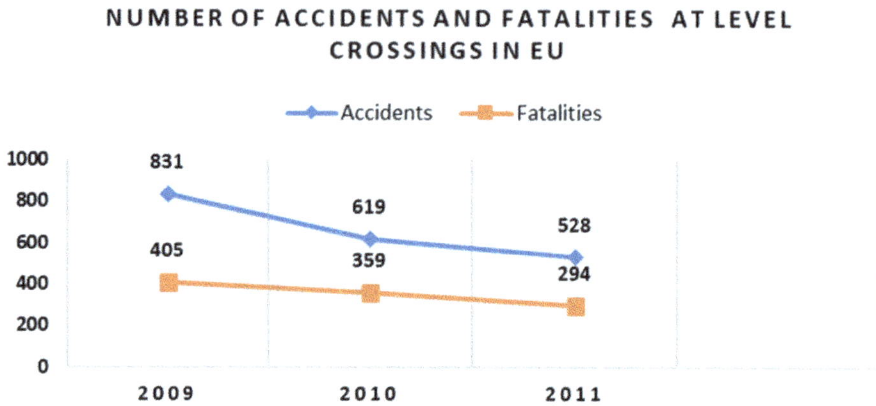

Figure 2: Number of accidents and fatalities at LCs in the EU. *(Source: EC, 2017.)*

The major reasons of accidents at LCs are distraction of car drivers or pedestrians, impatience, and the high speed of vehicles approaching LCs [15]. These users' behaviour is the reason for accidents in 95% of cases [7]. Weather conditions, using headphones, mobile phones, vehicle music systems and many more disruptions can affect the ability to hear approaching trains. Also, modern trains have a substantially reduced noise level. From time to time people cross the LC straight after the train passes whilst the alarms are still operating, but there can be another train that is coming from the opposite direction. Referring to the European Railway Agency, in 2013, UK's fatality risk at LCs was the lowest in the EU. The highest fatality risk at level crossings was in Greece which was more than 28 times higher than that of the UK and was the highest fatality rate in the EU [21].

Despite all efforts, the number of accidents at LCs in the EU in the last ten years remained stable [4]. Moreover, it has been found that the number of drivers and pedestrians who disobey the LC signals are increasing [12]. The LC represents a big operational risk and the most logical thing would be for all of them to close, but it can restrict mobility in some places whilst building grade separation crossings is not always possible in built up areas in countries such as the Netherlands or Japan. Building bridges or underpasses is very expensive, approximately €5–10 million in Europe. For example, replacing the LC at Elsenham, (UK) with a bridge it costed around £4 million [22].

5 INPUT ORIENTED DEA APPROACH

The DEA analysis has been acknowledged before as benchmarking the performance of Decision-Making Units (DMU) and found the best practice. The efficiency of DMUs depends on their distance to the frontier. This methodology uses the ratios between outputs and inputs and compares all units and their relative efficiency with respect to the best performing unit.

One advantage of applying the DEA is that it can operate with multiple inputs and outputs and is not needed to clarify their importance. Other advantages are that it is suitable for small samples and has a small run time [23]. The DEA compares each railway with all other railways and identifies railways that are operating inefficiently and finds the target values of output and input for inefficient railways. The DEA technical efficiency and service effectiveness study for railways was carried out by [24].

The CCR model is named after its developers Charnes, Cooper and Rhodes (1978) and BCC model is named after Banker, Charnes and Cooper (1984). The CCR model is based on an assumption of constant returns-to-scale (CRS). For this research, the input oriented CCR model has been applied. Input oriented model means that outputs will be kept constant, but inputs must be reduced to reach the frontier line. For this research one input was selected that is the number of accidents at LC and two outputs were selected, that are the thousand tonnes of goods transported by railway and millions of passenger-km. The efficiency is defined as the ratio of the output to input, and this ratio must be equal or less than 1. If the ratio is equal to 1, it points to the most efficient DMUs. Table 1 shows the efficiency of CRS input oriented CMI DEA for selected countries in the period 2012–2016.

Table 1: Efficiency of CRS input oriented CMI DEA for selected countries 2012–2016.

Country	2012	2013	2014	2015	2016	Mean
AT	23.97	10.88	19.79	4.6	8.66	13.58
BG	5.41	4.46	3.81	3.36	4.31	4.27
HR	4.23	2.05	5.3	1.08	4.2	3.37
CZ	14.4	8.88	10	5.02	7.22	9.1
DK	19.71	16.66	15.1	9.89	6.47	13.57
EE	85.39	39.12	37.01	14.47	14.95	38.19
FI	28.06	56.9	56.61	5.75	11.84	31.94
FR	38.38	32.79	23.05	6.69	20.58	24.3
DE	43.15	46.7	31.02	9.48	16.74	29.42
GR	2.17	2.68	1.48	0.29	12.18	3.76
HU	8.78	6.41	8.84	2.81	5.27	6.42
IT	38.41	49.83	44.76	11.91	40.65	37.11
LV	72.32	100	87.25	19.16	46.99	65.14
LT	94.28	34.41	33.31	33.09	20.07	43.03
NL	23.95	15.47	25.7	5.75	37.05	21.58
PL	22.72	11.18	20.5	4.93	7.54	13.37
PT	7.12	4.86	7.69	3.28	5.45	5.68
RO	5.27	2.84	3.41	3.26	3.53	3.66
SK	11.3	10.22	10.76	3.76	8.76	8.96
SI	15.11	5.12	11.01	2.83	6.09	8.03
ES	50.15	42.01	31.05	9.77	24.65	31.53
SE	36.94	20.19	27.75	12.21	28.42	25.1
UK	100	100	100	100	86.87	97.37
CH	82.2	85.7	100	34.13	100	80.41
Mean	34.73	29.56	29.82	12.81	22.02	25.79

The technical efficiency of railways shows that the efficiency of railways in terms of accident risk per million passenger-km and thousand tonnes of goods transported by railways in selected countries. The highest efficiency scores for the selected period 2012–2016 was UK and CH, 97.37% and 80.41% respectively. The railway in the UK was efficient 4 out 5 of years. It was efficient between 2012–2015, but in 2016 efficiency score dropped to 86.87%. This decrease happened despite the reduced number of LC in UK from 6,617 in 2012 to 6117 in 2016. Decreased tonnes of good transported by railways in UK from 115,225 in 2012 to 78,549 in 2016 affected the efficiency scores. The decrease in the selected period was 31.83%. This decrease in transporting goods can be partly explained by congestion of railway lines in the UK that was caused by the increases in passenger-km from 60,783 million passenger-km in 2012 to 68,010 million passenger-km in 2016. The increase in million passenger-km was 11.89%.

The average efficiency score of selected countries for the period 2012–2016 was 25.79%. It is widely believed that the reduction of the number of LCs will reduce the accidents at LCs. Despite the substantial reduction in number of LC in Europe in the last ten years, the number of accidents is still high. In 2017, there was 209 people injured at LCs around Europe [25]. The lowest number of LCs per 100 km of railway line in 2016 was NL, BG and SE, but the efficiency scores for these countries were 37.05%, 4.31% and 24.65% respectively. NL and SE had efficiency scores slightly higher than the average of 22.02%. The lowest average efficiency score in 2016, which was less than 5%, was BG, HR, GR and RO. The highest number of LCs per 100 km in 2016 was SE, AT and CZ. The efficiency scores in these countries were 28.42%, 8.66% and 7.22% respectively. AT and CZ had efficiency scores substantially lower than the average, which was 22.02%. It appears that the low number of LC per 100 km of railway line does not always have the huge impact on efficiency of railways in terms of accident risk. There is a need to look for other factors that can influence the efficiency scores. For this reason, there was selected seven factors that can also influence the efficiency scores.

6 IBM SPSS ANALYSES

To find the relationship between selected variables and efficiency of railways in term of accident risk the IBM Statistical Program for the Social Sciences (SPSS) analytic software was applied. Table 2 shows the descriptive statistics selected variables.

For variable “Number of cars per 1000 inhabitants”, skewness value is negative that indicates a clustering of scores at the high end. Positive kurtosis values indicated that the values clustered in the centre with long thin tails. Kurtosis values below 0 indicated a distribution is relatively flat [26]. To analyse the relationship between selected variables it is needed to assess the normality of the distribution of scores. To find outliers, cases with values well above or well below most of the other cases [26] the boxplot was built.

To explore the relationship between “Efficiency of railways in term of accident risk” and selected variables the scatterplot was build. The scatterplot gives an indication that the variables have a linear or curvilinear relationship. Only linear relationship is suitable for correlation analysis [26].

The correlation analyses are performed in order to find the relationship between variables. Pearson correlation coefficient take values between –1 and +1. The sign in front of the values indicates whether there is a positive or negative correlation. Negative sign indicated that as one variable increases the other decreases and positive sign indicated that if one variable increases the other increases too. The size of the absolute value provides an indication of the strength of the relationship. A perfect correlation of +1 or –1 means that the exact value of one variable can be found by knowing the value of the other variable. A correlation of 0 means that there is no relationship between selected variables [26].

Testing the Pearson correlation coefficient has been following a guideline [26] which states that the strength of the relationship is:

Small $r = .10$ to $.29$; Medium $r = .30$ to $.49$; Large $r = .50$ to 1.

Positive skewness values indicated that scores clustered to the left at the low values [26].

Table 3 shows the relationship between efficiency of railways in term of accident risk with selected variables. The coefficient of determination shows how much variance is shared by two selected variables or how much overlap there is between two variables [26].

Table 2: Descriptive statistics. *(Source: Extract from SPSS simulation.)*

Variables	No	Min. stat	Max. stat	Mean stat	Standard deviation	Skewness	Standard error	Kurtosis	SE
Efficiency of railways	24	3.53	100.00	22.0204	25.21366	2.144	0.472	4.388	0.918
Millions of passenger-km	24	280.00	95465.00	19083.5000	28136.36483	1.921	0.472	2.704	0.918
Goods transported by rail	24	1094.00	363512.00	67629.7083	78406.30685	2.790	0.472	8.992	0.918
Number of LC by country	24	322.00	16678.00	4455.7917	4575.60072	1.464	0.472	1.498	0.918
Population density	24	18.10	498.10	125.4583	103.02057	2.261	0.472	6.801	0.918
Number of LC accidents	24	1.00	87.00	19.7500	21.98863	1.764	0.472	2.805	0.918
Total length of railway lines	24	1161.00	38623.00	9052.4583	9448.45209	1.785	0.472	3.276	0.918
No. of cars per 1,000 inhabits	24	261.00	625.00	474.5000	87.33643	–0.592	0.472	0.345	0.918
GDP per capita	24	6050.00	58200.00	23982.5000	14118.16968	0.756	0.472	–0.281	0.918
Median age	24	40.50	47.10	42.7375	1.59587	0.894	0.472	1.156	0.918
Road accidents	24	2.6	9.90	5.4000	2.12009	0.567	0.472	–0.251	0.918

Table 3: Summary of Pearson correlation test. *(Source: Author's creation.)*

	Variables	Pearson correlation coefficient	Interpretation of relationship	Coefficient of determination (%)
Efficiency of railways in term of accident risk at LC	Number of LC by country	–0.29	weak	8.41
	Population density	0.427	moderate	18.23
	Road accidents	–0.352	moderate	12.39
	GDP per capita	0.538	strong	28.94

7 DISCUSSION AND ANALYSES

The key factors affecting the efficiency of railways in term of accident risk are:

7.1 Factor 1 – Number of LCs in a country

The number of LCs in a country has a negative weak relationship with efficiency of railways in term of accident risk at LC. This means that increasing the number of LCs decreases the efficiency of railways in terms of accident risk at LCs. To improve the efficiency of railways in terms of accident risk at LCs, there is a need to work on a reduction of the number of LCs where it is economically viable. Countries such as France (16678), Germany (14054) and Poland (13109) have the highest number of LCs, but the efficiency of railways in term of accident risk at LC is quite low as France has 20.58%, Germany 16.74% and Poland 7.54%.

7.2 Factor 2 – Population density

Population density has a positive moderate relationship with efficiency of railways in term of accident risk at a LC. The increases of population increase the efficiency of railways in terms of accident risk at LC. In countries with a higher density of population, there is an observed higher efficiency of railways in term of accident risk at LC. Regarding railways in countries with a higher density of population, improving safety at LCs requires more radical measures. From five countries (NL, IT, GB, CH and DE) with the highest level of density of population in Europe, four countries (NL, IT, GB and CH) scored the highest efficiency of railways in term of accident risk at LCs score. The score of efficiency of railways in term of accident risk at LC for 2016 was 37.05%, 40.65%, 86.87% and 100% respectively. Only DE has lower score of 16.74%.

7.3 Factor 3 – Number of deaths per 100,000 people in road accidents in 2016

The number of deaths has a negative moderate relationship with efficiency of railways in terms of accident risk at a LC. The increasing number of deaths in road accidents decreases the efficiency of railways.

7.4 Factor 4 – GDP per capita

GDP per capita has strong positive relationship with efficiency of railways in term of accident risk at LC. This means that, countries with a higher GDP per capita observed higher efficiency of railways in term of accident risk at LC score. This means that railways authorities in countries with a stronger economy invest more money to eliminate accident risk at LCs. Countries with the highest GDP per capita in 2016 of the selected ones in this study are CH, DK, SE, NL, AT, FI, DE, FR and GB. Five from them have the highest score of efficiency of railways in term of accident risk at LC for 2016. They are CH, GB, IT, NL and SE.

8 CONCLUSIONS

Railways around the world have benefited from the continuing improvement in the control and communication systems, but LCs remain a substantial safety risk. The LC is regarded as the largest single risk element of train accidents. The consequences of each accident are not limited to the loss of life but also the damage to the infrastructure and rolling stock, traffic disruption and damage to the reputation of the rail travel as a safe mode of transportation. To reduce the number of accidents at LCs, there is a need to close them where it is necessary to do so and to improve the related risk mitigation measures. Closing LCs, replacing them with bridges or underpasses have an economic benefit for both railways and for local communities. Uninterrupted road traffic flow can reduce the congestion, delays, air and noise pollution and overall make the transport system more sustainable. Railways can benefit from reducing LC maintenance and operational costs. It reduces the number employees, improves the speed and reliability of trains and increases the line capacity. Eliminating a LC can be cost-effective but upgrading the LC very often can be expensive and not proportional to the risk reduction that was achieved.

It was found that many factors can influence the efficiency of railways in term of accident risk. In this research, a number of factors were selected to investigate their relationship with efficiency scores, and it was found that the strongest relationship with efficiency scores have a number of LCs in a country, population density and GDP per capita.

From the results of this investigation, conclusions can be made that when evaluating the safety at LCs specific railways there is a need to take into consideration the level of economic development and density of population in the country. For some railways, improving the safety level at LCs is easier than for others where stagnation in economy and low density of population negatively affect performance of railways. There is an exception for this, the Baltic State countries [27]. GDP per capita in Euro in 2016 in LV was €11030, LT was €12,040 and in EE was €13650. Nevertheless, LV has 46.99% efficiency score, the third highest score of efficiency of railways in terms of accident risk at LC after CH with 100% and GB with 86.87%. This can be explained by the low number of passenger cars on the roads and by low railway traffic. In 2016 LV had one of the lowest numbers of passenger cars per 1000 inhabitants 341 cars. Romania had 261 and Hungary had 338 cars per 1000 inhabitant. Also, LV had low passenger traffic by railway and goods transported by rail which was only 47819 million tonnes which was lower than the average which was 67629 million tonnes.

To reduce the number of accidents, it is a need to better understand the local and human factors. It is crucial to increase the number of awareness campaigns and improves the cooperation between different road and railway institutions and stakeholders. Although, technology is fundamental to improving the safety at LCs, but safety can also be improved by educating the public and by enforcing related laws and regulations.

REFERENCES

[1] Djordjević, B., Krmac, E. & Mlinarić, T., Non-radial DEA model: A new approach to evaluation of safety at railway level crossings. *Safety Science*, **103**, pp. 234–246, 2018.

[2] Nguyen-Phuoc, D., Currie, G., De Gruyter, C. & Young, W., New method to estimate local and system-wide effects of level rail crossings on network traffic flow. *Journal of Transport Geography*, **60**, pp. 89–97, 2017.

[3] Larue, G., Filtness, A., Wood, J., Demmel, S., Watling, C., Naweed, A. & Rakotonirainy, A., Is it safe to cross? Identification of trains and their approach speed at level crossings. *Safety Science*, **103**, pp. 33–42, 2018.

[4] Laapotti, S., Comparison of fatal motor vehicle accidents at passive and active railway level crossings in Finland. *IATSS Research*, **40**(1), pp. 1–6, 2016.

[5] ERA, Level crossings safety in the European Union 2012a. www.era.europa.eu/Document-Register/Documents/Level_crossing_safety_EU_2012.pdf. Accessed on: 9 Jan. 2018.

[6] Raib, Fatal accident at Halkirk level crossing – GOV.UK 2009. www.raib.gov.uk/publications/investigation_reports/reports_2010/report162010.cfm. Accessed on: 9 Jan. 2018.

[7] Pacts, Safety at level crossings: PACTS submission to the Transport Select Committee 2013. www.pacts.org.uk/wp-content/uploads/sites/2/Safety-at-Level-Crossings-PACTS-submitted.pdf. Accessed on: 9 Jan. 2018.

[8] ERA, Railway safety performance in the European Union, 2016. www.era.europa.eu/Document-Register/Documents/Railway%20Safety%20Performance%202016%20final%20E.pdf. Accessed on: 9 Jan. 2018.

[9] European Commission (EC), Road safety in European Union, trends, statistics and main challenges, 2015. https://ec.europa.eu/transport/sites/transport/files/road_safety/pdf/vademecum_2015.pdf. Accessed on: 9 Jan. 2018.

[10] EC, File: Number of persons killed and injured by type of accident and category of persons in EU-28, 2014 final.png – Statistics Explained, 2017. http://ec.europa.eu/eurostat/statistics-explained/index.php/File:Number_of_persons_killed_and_injured_by_type_of_accident_and_category_of_persons_in_EU-28,_2014_final.png. Accessed on: 9 Jan. 2018.

[11] Arrivealive, Level crossing safety supplement, 2012. www.arrivealive.co.za/documents/Level-Crossing-Safety-Supplement-2012.pdf. Accessed on: 9 Jan. 2018.

[12] Theaa, Level crossing safety, AA, 2011. www.theaa.com/motoring_advice/news/level-crossing-safety-more-drivers-playing-russian-roulette.html. Accessed on: 9 Jan. 2018.

[13] ERA, Railway safety performance in the European Union, 2012. www.era.europa.eu/Document-Register/Documents/SafetyReport2012.pdf. Accessed on: 9 Jan. 2018.

[14] Global Railway Review, Level crossing safety in-depth focus, 2017. www.globalrailwayreview.com/article/33865/level-crossing-strategy-in-depth-focus-2017/. Accessed on: 9 Jan. 2018.

[15] Bahloul, K., Defossez, F., Ghazel, M. & Collart-Dutilleul, S., Adding technological solutions for safety improvement at level crossings: A functional specification. *Procedia – Social and Behavioral Sciences*, **48**, pp. 1375–1384, 2012.

[16] Liang, C., Ghazel, M., Cazier, O. & El-Koursi, E., A new insight on the risky behavior of motorists at railway level crossings: An observational field study. *Accident Analysis & Prevention*, **108**, pp. 181–188, 2017.

[17] Freeman, J. & Rakotonirainy, A., Mistakes or deliberate violations? A study into the origins of rule breaking at pedestrian train crossings. *Accident Analysis and Prevention*, **77**, pp. 45–50, 2015.

[18] Rudin-Brown, C., Lenné, M., Edquist, J. & Navarro, J., Effectiveness of traffic light vs. boom barrier controls at road–rail level crossings: A simulator study. *Accident Analysis and Prevention*, **45**, pp. 187–194, 2012.

[19] Beanland, V., Salmon, P., Filtness, A., Lenné, M. & Stanton, N., To stop or not to stop: Contrasting compliant and non-compliant driver behaviour at rural rail level crossings. *Accident Analysis & Prevention*, **108**, pp. 209–219, 2017.

[20] Tey, L., Wallis, G., Cloete, S. & Ferreira, L., Modelling driver behaviour towards innovative warning devices at railway level crossings. *Accident Analysis and Prevention*, **51**, pp. 104–111, 2013.

[21] Kelly, J., The level crossing wars 2014. *BBC News*. www.bbc.co.uk/news/magazine-27037582. Accessed on: 9 Jan. 2018.

[22] Lo, C., SafeRail level crossing study: putting ideas into action – Railway Technology 2015. www.railway-technology.com/features/featuresaferail-level-crossing-study-putting-ideas-into-action-4526758/. Accessed on: 10 Jan. 2018.

[23] Nataraja, N. & Johnson, A., Guidelines for using variable selection techniques in data envelopment analysis. *European Journal of Operational Research*, **215**(3), pp. 662–669, 2011.

[24] Yu, M., Assessing the technical efficiency, service effectiveness, and technical effectiveness of the world's railways through NDEA analysis. *Transportation Research Part A: Policy and Practice*, **42**(10), pp. 1283–1294, 2008.

[25] EC, Rail accident fatalities in the EU – Statistics Explained 2020. https://ec.europa.eu/eurostat/statistics-explained/index.php/Rail_accident_fatalities_in_the_EU. Accessed on: 4 Jan. 2020.

[26] Pallant, J., *SPSS Survival Mannual*, 1st ed., McGraw-Hill: London, pp. 59–77, 2013.

[27] Watson I., Ali A. & Bayyati A., Sustainability of HSR-comparative study. ICE *Innovation in Railway Civil Engineering Journal*, 2019. www.icevirtuallibrary.com/doi/pdf/10.1680/jtran.18.00131.

VALIDATION OF A BRAKE MONITORING SYSTEM USING A MULTI-AXLE ROLLER-RIG

NICOLA BOSSO*, MATTEO MAGELLI† & NICOLÒ ZAMPIERI‡
Department of Mechanical and Aerospace Engineering, Politecnico di Torino, Italy

ABSTRACT

A widespread diffusion of monitoring systems is the key to improving safety and reliability of railway vehicles and to reducing maintenance costs by increasing the lifetime of structural and mechanical components. Several on-board devices have been conceived in recent years to detect faults on critical parts, by monitoring the real-time conditions of wheelsets, bearings, brakes, bogie, carbody, etc. These systems rely on powered sensor nodes, mounted on different parts of the vehicle, which acquire signals of interest and use proper algorithms to detect component faults or train derailments. Roller-rigs represent a good solution to develop and validate new monitoring devices with good repeatability of test conditions. The research group from Politecnico di Torino realized an innovative 1:5 scaled roller-rig, consisting of four wheelsets running on the same pair of rollers, originally used for investigations on wheel–rail adhesion. Then, minor changes were made to simulate and test a braking system monitoring device. Each wheelset is provided with a braking system, consisting of two brake discs and two brake callipers, while the rollers are powered using one AC brushless motor. A freight train pneumatic system is reproduced thanks to three electropneumatic regulators, which simulate the pressure in brake pipe, auxiliary reservoir and brake cylinder. The test bench and the motor drive are managed by an industrial PC using the software LabVIEW, allowing to perform braking operations of a real vehicle. Sensors are installed to measure wheelset speed, calliper temperature and pressures in the braking system. These signals are sent to an electronic control unit, which could be provided with a fault detection logic. The paper deals with the experimental setup of the bench and the preliminary results obtained by laboratory tests, which highlighted that the apparatus can be a reliable tool to develop new monitoring algorithms for train braking systems.
Keywords: roller-rig, railway monitoring, braking system, fault detection, railway brake plant simulation.

1 INTRODUCTION

Safety and reliability of railway vehicles can be improved by the adoption of monitoring systems able to detect anomalies in vehicle running behaviour and faults of the critical components of the bogie. Usually the maintenance of the vehicle components is scheduled considering a fixed inspection period. This approach is simple and easy to schedule, but it does not allow to optimize costs and safety. In fact, failures and malfunctions could occur in advance of the subsequent inspection, while other inspections could take place when not necessary. The adoption of specific monitoring systems to check the condition of critical components represent a valid solution able to improve the predictive maintenance interventions based on data recorded during the vehicle operation.

Monitoring systems currently available can be divided into two main categories according to the installation site: on-track systems and on-board systems (installed on the vehicle). The first solution has several advantages such as easy installation and available power supply. However, this solution does not guarantee a continuous monitoring of the railway vehicle, since detections are performed only when the vehicle runs on track sections where the

* *ORCID: http://orcid.org/0000-0002-5433-6365*
† *ORCID: http://orcid.org/0000-0002-2962-7873*
‡ *ORCID: http://orcid.org/0000-0002-9197-1966*

WIT Transactions on The Built Environment, Vol 199, © 2020 WIT Press
www.witpress.com, ISSN 1743-3509 (on-line)
doi:10.2495/CR200141

monitoring system is installed. On the other hand, on-board systems assure a continuous monitoring, and are therefore capable to detect any faults and failures as soon as they occur. The main drawback of on-board systems is the more complex implementation and assembly, since the available space on the vehicle is extremely limited. Furthermore, the main obstacle to the spread of on-board systems is the difficulty in supplying electrical power to the sensors and control units. In fact, normally the freight wagons do not have an electrical power supply, but even where the electrical power is available, the wiring for all the devices distributed along the train is difficult to realize in compliance with current standards [1]. However, recent advances in the field of energy harvesting techniques [2], [3], which allow to obtain electricity from environmental energy, could be an interesting opportunity for the development of on-board monitoring systems. The state of the art proposed by Bernal et al. [4] shows that monitoring systems are composed by sensors for measuring quantities of interest (accelerometers, thermocouples, thermistors, etc.), devices for signal conditioning and acquisition, and control units running specific algorithms [5], [6] able to detect anomalies. To increase the safety level of railway vehicles, great interest is obviously directed to the development of monitoring systems for the brake system, such as the one proposed by Somà et al. [7], Zampieri et al. [8] and by Bosso et al. [9].

The validation of new monitoring systems and related algorithms for data analysis requires extensive experimental test campaigns, typically carried out on track, with high costs, line occupation and poor repeatability of operating conditions. A valid alternative to on-track tests is the use of roller-rigs [10], i.e., test devices, in which the track is replaced by pairs of rollers. This solution allows to reproduce with good accuracy the geometry and the kinematics of the wheel–rail contact, as well as the dynamics of the railway vehicle [11]. The aim of the paper is to demonstrate how the innovative 1:5 scale multi-axles roller-rig, designed by the research team of the Politecnico di Torino, can be adopted to test and validate monitoring systems of the braking system. The experimental system, proposed in this work, can in fact be considered a Hardware In the Loop (HIL) system of the braking system of a railway vehicle.

The braking manoeuvre is replicated on the roller-rig by means of a braking system, and it is managed by a control algorithm developed in LabVIEW. Sensors are installed on the bench in order to measure the quantities of interest and their signals are acquired by the monitoring unit. The acquired signals are processed by specific algorithms in order to evaluate the condition of the braking system. The architecture proposed in this work will allow us to develop and validate braking system monitoring algorithms without requiring on-track tests, at least during the early stages of development. The paper shows, both the mechanical design of the test bench and the pneumatic system designed to replicate the braking system of a freight vehicle.

2 DESCRIPTION OF THE MULTI-AXLE ROLLER-RIG

The multi-axle roller-rig, shown in Fig. 1, is an evolution of the single axis roller-rig, already designed by the authors which was previously designed and used for the study of degraded adhesion phenomena [12], [13] and wear [14]. More recently the attention of the research group focused on the adhesion recovery phenomenon [15]. At first, the research team tried to study this phenomenon adopting a traditional single axis roller-rig [16], which is composed by a single wheelset suspended on a pair of rollers. The results show that the inefficiencies of the system used to remove the contaminant made the experimental results difficult to interpret. Therefore, the innovative 1:5 scaled multi-axle roller-rig was designed and realized [17], [18]. The peculiarity of the testing device is to dispose of four wheelsets running on the same pair of rollers. This solution is in fact essential to correctly study and reproduce the

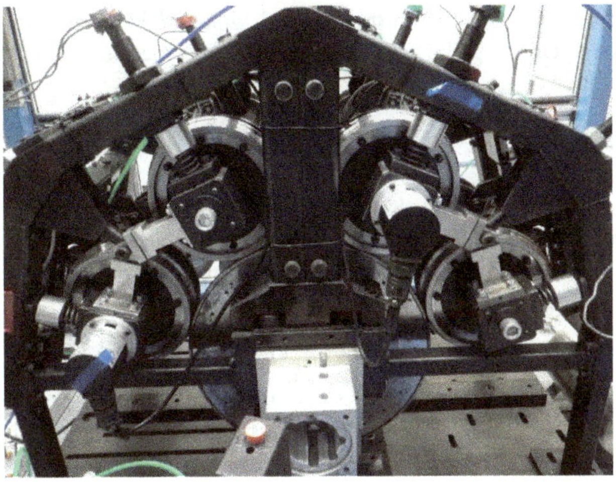

Figure 1: Lateral view of the multi-axle roller-rig.

different levels of adhesion of the wheelsets of a vehicle when running on a contaminated section of rail. The test bench has been designed according to the Jaschinski's similitude, which keeps a 1:1 scale for the accelerations.

2.1 Mechanical design of the roller-rig

Considering the mechanical arrangement, the roller-rig is composed by the frame, two rollers and four 1:5 scaled wheelsets. The two rollers, in this case secured each other thanks to a rigid joint, are driven by a single six poles brushless motor, controlled by means of a digital drive. The rollers consist of a disk to which a rim is fixed. The rim is machined to realize a 1:5 scaled UIC60 rail profile canted 1:20. Each wheelset is composed by an axle on which are fixed two brake discs and two 1:5 scaled wheels. The last ones consist of a disk and an external rim, which is machined to obtain the S1002 wheel profile. The nominal diameter of the wheels d_W and the rollers d_R are respectively 184 mm and 368 mm.

The two axle-boxes, installed at the extremities of each wheelset, are rigidly connected to the frame in the longitudinal and lateral direction while the flexibility in the vertical direction is guaranteed by two helical springs (one on each side) which act as primary suspension. The preload generated by each spring can be independently regulated thanks to a dedicated screw loading system. In this way, it is possible to simulate different axle-load values and compensate for the differences in the normal component of the weight force due to the inclination of the wheelsets with respect to the vertical direction.

In order to perform the braking tests and testing the monitoring system, the test bench is equipped with several sensors able to measure the quantities of greatest interest and an industrial PC (NI PXIe-8840) to control the braking manoeuvre. All the axle-boxes are equipped with load button cells, with a maximum load of 1000 lb (454 kg) and a resolution of 0.5 lb (0.227 kg), for regulating the wheel load. During the braking tests, however, only the signal generated by the load cell installed on one wheelset is used. In fact, the freight wagons are usually equipped with weighting cells, able to estimate the weight of the entire bogie from the load generated by one of the 8 springs composing the primary suspension of the bogie. However, since each axle-box of test bench is equipped with a single spring, the axle-load Q is estimated starting from the signal F_S generated by one of the two load cells. The eqn (1) is used to calculate the axle load and m is the wheelset mass (15.16 kg) and g is the gravity acceleration

$$Q = 2F_S + mg\cos 72°. \tag{1}$$

During the experimental tests shown in this paper only one wheelset is braked. The braking torque is measured by two S-shaped load cells, with a maximum load of 300 lb (136 kg) and resolution of 0.15 lb (0.07 kg). The load cells measure the reaction forces of the brackets supporting the two brake callipers. The breaking torque C_B acting on the wheelset can be estimated according to eqn (2).

$$C_B = \left(F_{B,1} + F_{B,2}\right) \cdot r_B. \tag{2}$$

In eqn (2), $F_{B,1}$ and $F_{B,2}$ are the braking force values measured by the load cells, while r_B is the equivalent braking radius, which has been estimated during the calibration procedure of the test bench and is equal to 79.24 mm. The wheelset is also equipped with an incremental encoder (resolution of 1024 ppr), which allows to obtain the angular position of the wheelset, from which the speed is calculated by numerical differentiation.

Three proportional pneumatic valves, which regulate the outlet pressure in the range 0–9 bar against a voltage signal 0–10 V, are used to simulate the braking system of the vehicle. One valve simulates the pressure in the brake pipe, another one regulates the pressure to the auxiliary reservoir and the last one allows to supply the desired pressure value to the brake cylinder. The three valves receive as input a reference voltage signal (set), and suitably regulate the outlet pressure, supplying the feedback pressure value as a voltage signal in the range 1–5 V.

The experimental setup includes two Pt1000 thermistors used to measure the calliper body temperature. The sensors perform temperature measurements in the range –50°C to 300°C and have a maximum resistance error of 1.2 Ω.

2.2 Design of the pneumatic braking system

In order to test new prototypes of monitoring systems for freight vehicle brakes, the multi-axles roller-rig is equipped with a braking system that replicates the one normally installed on freight trains [19], [20]. Fig. 2 shows the architecture of the typical braking system of a freight railway vehicle.

Under normal operating conditions the brake pipe is maintained at a pressure of 5 bar. The train driver can change the pressure of the brake pipe by acting on the brake valve (V), which receives compressed air from the main reservoir (MR), kept by the locomotive compressor at a pressure of about 8 bar. Each wagon is equipped with a triple valve (D), which is the decision-making element of the braking manoeuvre. This valve compares the pressure in the pipeline with the pressure in the control tank (CR). When the driver activates the braking operation the pressure in the brake pipe drops below 5 bar, and the triple valve connects the auxiliary reservoir (AR), at the pressure of 3.8 bar, and the brake cylinder chamber. Traditional freight trains are equipped with a single brake cylinder for bogie. The force generated by the cylinder is increased by means of a linkage which allows to press each pair of brake pads against the respective wheel. Emergency braking differs from service one since it generates a fast discharge of the brake pipe and it is not reversible, i.e., once the manoeuvre is activated, the pipeline is completely discharged. In contrast, at the end of a service braking, the release of the brake cylinder requires that the triple valve connects the brake cylinder

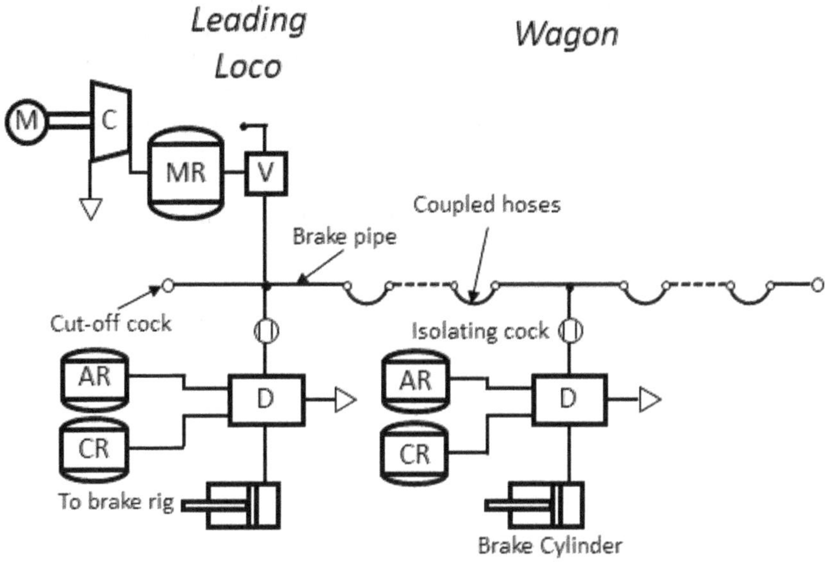

Figure 2: Typical architecture of the braking system of a railway freight vehicle.

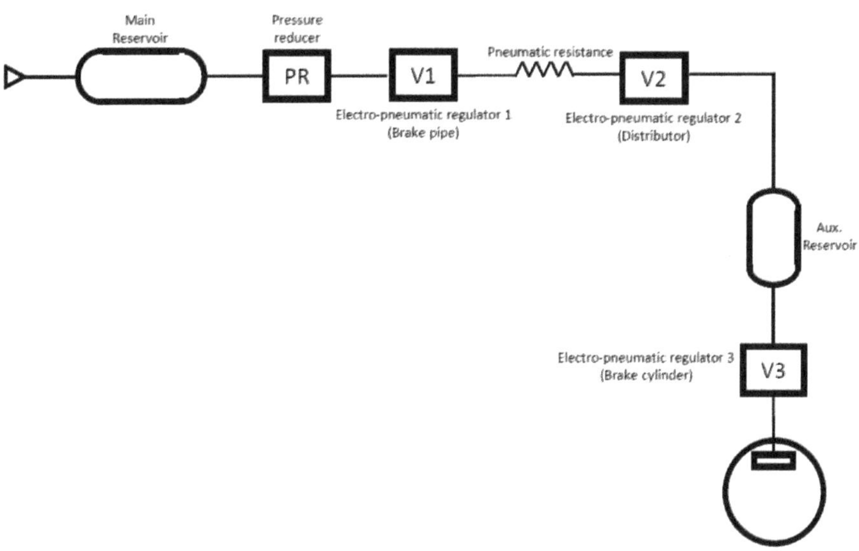

Figure 3: Architecture of the braking system adopted for the multi-axle roller-rig.

chamber and the atmosphere, while the auxiliary reservoir is refilled by the brake pipe at the pressure of 3.8 bar.

The architecture of the braking system built on the test bench to replicate the real one is shown in Fig. 3.

The compressed air, supplied by an industrial compressor, is stocked in the main reservoir (MR). Downstream the compressor there is a pressure regulator which maintains a pressure of 8 bar upstream of the first proportional valve (EP Reg. 1). The output of this valve reproduces the brake pipe, and in the absence of braking effort the pressure is kept at 5 bar. Downstream of the first proportional valve there is a long pipe, which simulates a distributed resistance and allows to physically reproduce the propagation delay of the pressure wave along the brake pipe. The output of the second proportional valve (E-P Reg. 2), set at 3.8 bar, is connected to a tank that simulates the auxiliary reservoir. This one supplies the third proportional valve (E-P Reg. 3), which regulates the pressure of the two brake callipers installed on the wheelset, replacing the distributor. The pressure downstream of the third proportional valve, which represents the pressure at the brake cylinder, is regulated as a function of the pressure drop detected in the brake pipe, according to the transfer function shown in Fig. 4.

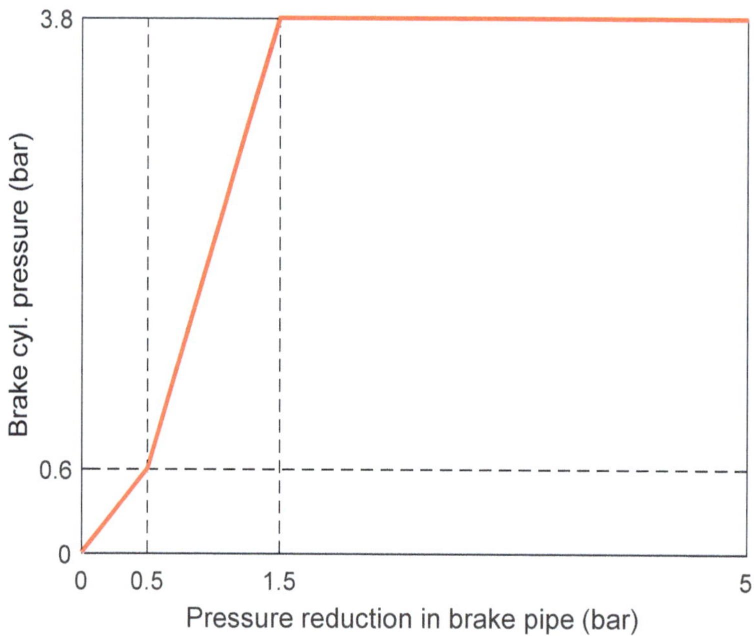

Figure 4: Transfer function of the brake cylinder pressure as a function of the pressure drop in the brake pipe.

The transfer function has been defined according to the following considerations. First, the indications of UIC 540 [20] standard have been followed, which suggests that the maximum pressure to the brake cylinder (3.8 bar) should occur in correspondence of a pressure drop of 1.5 bar of the brake pipe. Furthermore, experimental tests on the roller-rig show that the pads exert a braking effort on the disc for pressure value higher than 0.6 bar. In order to considering this effect a brake pressure reduction of 0.5 bar has been considered for the definition of the transfer function. In this way, when the braking manoeuvre starts, pressurized air flows to the brake cylinder, keeping the pads close to the disc as soon as the brake pipe pressure drops.

3 SIMULATION OF THE BRAKING MANEUVER

The simulation of realistic braking manoeuvres on the test bench is managed by a dedicated control software (LabVIEW VI), developed in LabVIEW 2016 and implemented in the industrial PC. The analogue output (AO) module NI-9263 (4 channels, 16 bits resolution and maximum sampling rate of 100 kS/s/ch) is used to adjust the outlet pressure of the 3 electro-pneumatic regulators, which reproduce the brake pipe, the distributor and the brake cylinder. When no braking effort is applied, the 3 set values are equal to 5, 3.8 and 0 bar respectively. The feedback values are measured by the electro-pneumatic regulators and acquired through an USB analogue input (AI) module NI-9239 (4 channels, 24 bits resolution and maximum sampling rate equal to 50 kS/s/ch). The tangential forces measured by the two S-beam load cells installed on the first wheelset and the normal force measured by the button load cell mounted on the axle box on the non-powered side of the same wheelset are acquired by the strain bridge input module NI PXIe-4330, with 8 channels, resolution equal to 24 bits and maximum sampling rate equal to 25.6 kS/s/ch. A servo mode control is performed on the brushless motor powering the rollers, with the reference speed being an analogue signal coming from the AO module PXIe-4322, which also generates three analogue signals corresponding to the feedback pressure at the outlet of the electro-pneumatic regulators and an analogue signal reproducing the weighing cell output.

These analogue output signals, as well as the signals from the encoder mounted on wheelset 1 and the measurements from the Pt1000 thermistors installed on the brake pads are acquired by the monitoring unit, which was used by the research team in the past to develop and test a monitoring system able to control the vehicle running behaviour and to diagnostic the condition of the main parts of the bogie [21]. The monitoring unit is managed by a 32-bit ARM7 processor, provided with a multichannel synchronous acquisition system with proper signal conditioning depending on the type of acquired signal (ICP accelerometers, encoders, thermocouples and thermistors). Data acquisition and transfer operations, which can be performed by means of the TCP/IP protocol or a CAN network, are supervised by an editable built-in resident code. The vehicle condition parameters and possible anomalies can be stored in a SD card included in the monitoring unit. Data operations are performed by an embedded PC which is connected to the monitoring unit TCP/IP network. A single embedded PC could be used to managed data from several monitoring units, however the typical arrangement comprises a monitoring unit mounted on each bogie and an embedded PC for each vehicle.

In order to replicate a braking manoeuvre on the multi-axle roller-rig, it is essential to supply compressed air to the third electro-pneumatic regulator, representing the brake cylinder, and to simulate the vehicle inertia on the roller shaft. Focusing on the longitudinal dynamics of a railway vehicle [22]–[24] and assuming that coupling systems are infinitely stiff so that the entire train moves as a single rigid body and slows down with a deceleration $\ddot{x}$, the equilibrium in the longitudinal direction is expressed through eqn (3)

$$\sum_{j=1}^{N_V}\sum_{i=1}^{N_{W,j}} F_{X,ji} = \sum_{j=1}^{N_V}\sum_{i=1}^{N_{W,j}} M_{,ji}\ddot{x} - \sum_{j=1}^{N_V} F_{res,j}\,. \quad (3)$$

In eqn (3), F_x is the longitudinal force acting on each wheelset-rail interface, M is the axle load (in kg), F_{res} is the total retarding force on each vehicle, subscript j refers to each of the N_V vehicles composing the train, while subscript i is related to the number of wheelset N_W of each vehicle. Fig. 5(a) shows the forces acting in the longitudinal direction on the jth vehicle in the train composition.

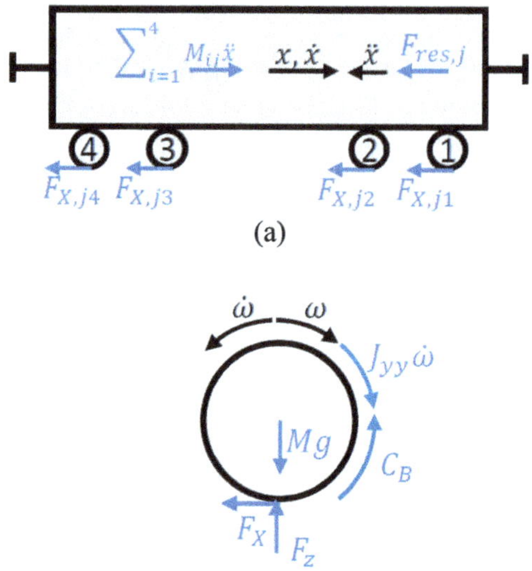

Figure 5: Force equilibriums. (a) Equilibrium in the longitudinal direction; (b) Rotational equilibrium of the wheelset.

On the hypothesis that the vehicle is running on a straight track with no grade and curves, that the force acting at each wheel–rail interface is the same for all wheelsets and neglecting the aerodynamic resistances, the longitudinal equilibrium can be simplified, as eqn (4) states

$$F_X = M\ddot{x}. \tag{4}$$

The total longitudinal force acting on each wheelset can be calculated from the rotational equilibrium of the wheelset, expressed by eqn (5) and shown in Fig. 5(b), where C_B is the total braking force, r_W is the wheel radius, $\dot{\omega}$ is the wheel deceleration and J_{yy} is the total inertia of the wheelset

$$F_X = \frac{C_B - J_{yy}\dot{\omega}}{r_W}. \tag{5}$$

If perfect adhesion conditions are assumed, the relationship between the angular deceleration of the wheelset and the vehicle longitudinal deceleration is expressed by eqn (6), and the longitudinal deceleration can be calculated from eqn (7)

$$\dot{\omega} = \frac{\ddot{x}}{r_W}, \tag{6}$$

$$\ddot{x} = \frac{C_B r_W}{M r_W^2 + J_{yy}}. \tag{7}$$

Since the bench is realized following Jaschinski's similitude model, which guarantees that the acceleration of the scaled system is equal to the acceleration of the full-scale system, equation (NN) is valid also using the scaled quantities. The braking torque is obtained from the two load cells installed on wheelset 1, while the wheelset mass is calculated from the button load cell installed on the axle-box of the non-powered side of wheelset 1. The total inertia of the wheelset is estimated from the CAD model of the wheelset and equal to 0.051 kg·m^2.

Therefore, at each time step of the simulation of the braking manoeuvre the LabVIEW VI can calculate a value for the vehicle deceleration, and an analogue signal is sent to the motor drive in order to set the proper motor speed.

4 EXPERIMENTAL SETUP AND RESULTS

At this first stage of the work, preliminary tests were performed to assess the capability of the new bench design to simulate realistic braking manoeuvres. Simple experimental tests were carried out by imposing a ramp of pressure reduction in the electro-pneumatic regulator reproducing the brake pipe, and correspondingly increasing the brake cylinder pressure according to the feedback value of the brake pipe pressure. The brake pipe was not totally released, but the pressure ramp decrease ramp was stopped at a limit value. Three tests were performed in order to investigate the influence of both the pressure ramp and the limit value. Table 1 summarizes the values of the main parameters of the 3 experimental braking tests, i.e., the maximum pressure reduction in the brake pipe, the corresponding maximum pressure in the brake cylinder and a ramp slope parameter. The latter is the time that the ramp signal would require to reach a pressure decrease equal to 1.5 bar in the brake pipe, if no previous saturation was superimposed. As shown in Table 1, the same ramp signal was used for tests 1 and 2, but a lower saturation level for the brake pipe pressure was superimposed in test 2, thus leading to a lower maximum pressure in the brake cylinder. The ramp used in tests 1 and 2 is such that a maximum pressure reduction of 1.5 bar in the brake pipe, corresponding to a maximum brake cylinder pressure equal to 3.8 bar, would take 18 s, which is the same as the time usually needed in G regime to reach the maximum braking pressure of 3.8 bar in the brake cylinder. On the other hand, test 3 was performed by superimposing the same limit value for the brake pipe reduction as test 1, thus leading to the same maximum brake cylinder pressure, but a faster ramp signal was generated. In all tests, the brake forces measured by the S-beam cells were acquired by the LabVIEW VI and used to estimate the deceleration of the vehicle, thus modifying accordingly the set value for the roller speed and reproducing the vehicle inertia.

Table 1: Main parameters for experimental braking tests.

Test	Limit value of pressure reduction in brake pipe (bar)	Max. brake cylinder pressure (bar)	Time for 1.5 bar pressure reduction in brake pipe (s)	
			Scaled	Real vehicle
1	0.71	1.16	8	18
2	0.66	1	8	18
3	0.71	1.16	6.5	14.5

Fig. 6 shows the roller peripheral speed as a function of time (up) and the braking curves (down), i.e., the speed vs position curve, obtained in the three tests. Please note that the plots refer to scaled values of time, position and speed. The initial roller speed in all three tests is set to about 260 rpm, corresponding to a peripheral speed of 5 m/s and an initial speed for a real vehicle of 40 km/h.

The braking manoeuvres starts as soon as a pressure reduction is imposed in the brake pipe, but in all three tests a few seconds elapse before the rollers start to slow down, since no braking effort is applied when the brake cylinder pressure is below a certain threshold, experimentally determined to be equal to about 0.6 bar. It can be noticed that this time interval

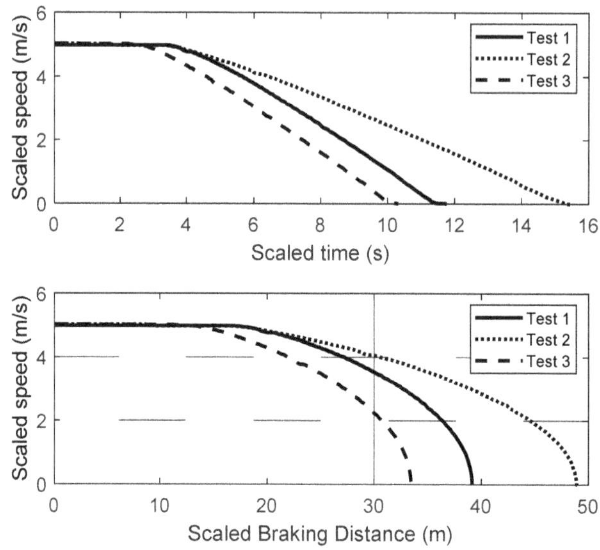

Figure 6: Scaled values of speed and time during the braking manoeuvres (up) and scaled braking curves in all three tests (down).

is approximately the same for tests 1 and 2, which are performed with a similar pressure reduction ramp, while a lower time is required in test 3, which is carried out with the fastest ramp. When the brake cylinder pressure reaches the limit value set for the test, a linear trend of speed reduction in time can be noticed in all three tests, i.e., the wheelset is braked with an approximately constant deceleration.

Moreover, Fig. 6 highlights that the minimum barking distance is obtained in test 3, which is carried out with the same limit value as test 1, but with a faster ramp. At the same time, the comparison of tests 2 and 3, which are performed with the same ramp but different limit values of brake cylinder pressure, shows that the braking distance and time are higher for test 2, which has a lower saturation value of brake cylinder pressure and is thus braked with reduced forces.

The preliminary tests shown in this section are not reproductions of a real braking manoeuvre for a railway freight vehicle. In fact, the influence of several parts of railway braking system was neglected in the experimental activity. Moreover, the pressure reduction in the main brake pipe is typically not linear in time, as registered in several experimental works. However, data obtained from this experimental campaign were in total agreement with the expectations, since higher ramps and higher values of maximum pressure in the brake cylinder led to reduced braking times and distances. Therefore, the preliminary experimental activity described in the paper can be considered as successful, since the new arrangement of the multi-axle roller-rig proved a suitable solution for both the reproduction of the braking manoeuvres of the railway vehicles and the acquisition of the signals usually required by braking system monitoring algorithms, which are acquired by the monitoring unit.

5 CONCLUSIONS

The paper describes an experimental activity aiming to demonstrate the feasibility of using an existing multi-axle roller-rig as a laboratory device for the development and testing of new railway braking plant monitoring systems. The roller-rig had been originally designed to investigate degraded adhesion conditions and adhesion recovery phenomena occurring during braking operations. Therefore, a few changes to the original arrangement were made in order to adapt the test bench to the new experimental activity. First, a pneumatic plant, reproducing the main features of the typical air brake system installed on freight trains, was specifically designed. Moreover, a new test bench control software was developed and implemented in LabVIEW. This software is provided with an algorithm able to reduce the roller speed according to the measured braking force, thus reproducing the vehicle inertia on the roller shaft. At the same time, the control software manages data acquisition and transmission to a monitoring unit, which acquires the same signals as required by typical concepts of railway braking plant monitoring systems.

Some preliminary tests, concerning the reproduction of simplified braking operations, were performed on the new arrangement of the bench and the results highlighted a good agreement with the expectations. Therefore, the goal of the activity can be considered as accomplished, since the new arrangement proved to be an effective solution to reproduce realistic railway braking operations by means of a laboratory device and to acquire with the monitoring unit the signals required by typical monitoring algorithms. In future work the current arrangement could thus be exploited to develop and test new air brake plant monitoring systems and algorithms.

The future developments of the experimental activity could aim to reduce the discrepancies between the current arrangement of the bench and the real freight vehicles air brake system. For instance, the current disc brakes could be replaced by a mechanical system with brake blocks and eventually a brake rig, while changes to the pneumatic system or to the control software could reproduce the influence of the relay valve. Finally, future tests could be performed using experimental data of pressures in the brake plant and vehicle speed, eventually investigating the behaviour of monitoring algorithms in simulated failure mode.

REFERENCES

[1] EN, 50343: Railway applications – Rolling stock – Rules for installation of cabling, 2014.

[2] De Pasquale, G., Somà, A. & Zampieri, N., Design, simulation, and testing of energy harvesters with magnetic suspensions for the generation of electricity from freight train vibrations. *Journal of Computational and Nonlinear Dynamics*, 7(4), 2012.

[3] Bosso, N., Magelli, M. & Zampieri, N., Application of low-power energy harvesting solutions in the railway field: a review. *Vehicle System Dynamics*, pp. 1–31, 2020.

[4] Bernal, E., Spiryagin, M. & Cole, C., Onboard condition monitoring sensors, systems and techniques for freight railway vehicles: A review. *IEEE Sensors Journal*, **19**(1), pp. 4–24, 2019.

[5] Li, C., Luo, S., Cole, C. & Spiryagin, M., An overview: modern techniques for railway vehicle on-board health monitoring systems. *Vehicle System Dynamics*, **55**(7), pp. 1045–1070, 2017.

[6] Bosso, N., Gugliotta, A. & Zampieri, N., Wheel flat detection algorithm for onboard diagnostic. *Measurement*, **123**, pp. 193–202, 2018.

[7] Somà, A., Aimar, M. & Zanardelli, A., Experimental data measured with an on board unit for condition monitoring of freight wagons. *Ingegneria Ferroviaria*, **72**(7–8), pp. 547–567, 2017.

[8] Zampieri, N., Bosso, N. & Gugliotta, A., Innovative monitoring systems for onboard vehicle diagnostics. *Civil-Comp Proceedings*, **110**, 2016.

[9] Bosso, N., Gugliotta, A. & Zampieri, N., Design and testing of an innovative monitoring system for railway vehicles. *Proceedings of the Institution of Mechanical Engineers, Part F: Journal of Rail and Rapid Transit*, **232**(2), pp. 445–460, 2018.
[10] Bosso, N., Allen, P.D. & Zampieri, N., Scale testing theory and approaches. *Handbook of Railway Vehicle Dynamics*, 2nd ed., CRC Press, pp. 825–867, 2019.
[11] Bosso, N., Gugliotta, A. & Somà, A., Dynamic behavior of a railway wheelset on a roller rig versus tangent track. *Shock and Vibration*, **11**(3–4), pp. 467–492, 2004.
[12] Bosso, N., Gugliotta, A. & Zampieri, N., Study of the wheel-rail adhesion under degraded conditions using a multi-axles roller-rig. *Dynamics of Vehicles on Roads and Tracks*, vol. 2, CRC Press, pp. 803–808, 2017.
[13] Bosso, N., Gugliotta, A., Magelli, M., Oresta, I.F. & Zampieri, N., Study of wheel-rail adhesion during braking maneuvers. *Procedia Structural Integrity*, **24**, pp. 680–691, 2019.
[14] Bosso, N. & Zampieri, N., Experimental and numerical simulation of wheel-rail adhesion and wear using a scaled roller rig and a real-time contact code. *Shock and Vibration*, 2014.
[15] Bosso, N., Magelli, M. & Zampieri, N., Investigation of adhesion recovery phenomenon using a scaled roller-rig. *Vehicle System Dynamics*, pp. 1–18, 2019.
[16] Bosso, N., Gugliotta, A. & Zampieri, N., Strategies to simulate wheel–rail adhesion in degraded conditions using a roller-rig. *Vehicle System Dynamics*, **53**(5), pp. 619–634, 2015.
[17] Bosso, N., Gugliotta, A. & Zampieri, N., A test rig for multi-wheelset adhesion experiments. *Civil-Comp Proceedings*, **110**, 2016.
[18] Bosso, N., Gugliotta, A., Magelli, M. & Zampieri, N., Experimental setup of an innovative multi-axle roller rig for the investigation of the adhesion recovery phenomenon. *Experimental Techniques*, **43**(6), pp. 695–706, 2019..
[19] UIC 541-1, Brakes – Regulations concerning the design of brake components, 2003.
[20] UIC 540, Brakes – Air brakes for freight trains and passenger trains, 2016.
[21] Bosso, N., Gugliotta, A., Magelli, M. & Zampieri, N., Monitoring of railway freight vehicles using onboard systems. *Procedia Structural Integrity*, **24**, pp. 692–705, 2019.
[22] Aboubakr, A.K., Shabana, A.A. & Bosso, N., Numerical study of the noninertial systems: Application to train coupler systems. *Nonlinear Dynamics*, **68**(1–2), pp. 215–233, 2012.
[23] Wu, Q. et al., International benchmarking of longitudinal train dynamics simulators: results. *Vehicle System Dynamics*, **56**(3), pp. 343–365, 2018.
[24] Bosso, N. & Zampieri, N., Long train simulation using a multibody code. *Vehicle System Dynamics*, **55**(4), pp. 552–570, 2017.

DEVELOPMENT AND VERIFICATION TEST OF ELECTRICAL INDICATOR FOR 1.5 KV DC FEEDER IN ORDER TO VISUALIZE ENERGIZED CONDITION

YOSHIHISA MITOMA & KEITA HASEGAWA
East Japan Railway Company, Japan

ABSTRACT

The overhead contact line system is essential to supply traction power to trains. 1.5 kV DC is mainly adopted in the metropolitan area. The overhead wires are congested near the traction substations and the large scale passenger stations. When on-site workers start maintenance or improvement work for the overhead contact line system, the overhead wires should be de-energized individually in order to avoid electrical shock. We have developed the electrical indicator for 1.5 kV DC feeder in order to visualize energized or de-energized condition. The electrical indicator consists of a cylindrical electrode, detection circuits, rechargeable batteries, charging circuits, solar cells and LEDs. The electrical indicator can visualize energized or de-energized condition on the overhead wires by the LEDs. The cylindrical electrode covers the overhead wire and detects the ripple voltage. The earthing wire has been successfully removed by focusing on the electric field, although the detection of 1.5 kV DC voltage basically needs the earthing wire to settle the reference potential. A fail-safe system which will activate the LED in event of a malfunction of the electrical indicator has been also devised, even during de-energized condition. Subsequently, the verification test in the laboratory, where an overhead wire on 1.5 kV DC is able to be energized, have been conducted. Threshold voltage level of the LED for the energized condition has been founded by measuring the ripple voltage using the oscilloscope. Moreover, the visibility of LEDs at a distance of 5 m in de-energized condition have been confirmed. In conclusion, we have developed the electrical indicator for 1.5 kV DC feeder and conducted verification test, enabling visualization of energized or de-energized condition. We will continue to make further improvements on electrical indicator and aims to introduce the device into the field for practical use in the near future.
Keywords: electrical indicator, 1.5 kV DC feeder, visualizing energized or de-energized condition.

1 INTRODUCTION

East Japan Railway Company manages and operates many electrical facilities such as power stations, substations, electric power transmission lines, power distribution lines and so on. The overhead contact line system is essential to supply traction power to trains. 1.5 kV DC is mainly adopted in the metropolitan area. The overhead wires are congested near the traction substations and the large scale passenger stations. When on-site workers start maintenance or improvement work for the overhead contact line system, the overhead wires should be de-energized individually in order to avoid electrical shock. Basically, on-site workers start maintenance or improvement work after checking de-energized condition and attaching the earthing wire to the overhead contact line system based on the work permit and procedure.

However, it is sometimes difficult for the congested overhead wires near the traction substation and the large scale passenger stations to be de-energized at the same time due to their operating schedule. This condition might cause the electrical shock especially when on-site workers misunderstand the de-energized area due to the complexity of the overhead wires or a safety lapse might happen during maintenance. We have developed the electrical indicator for 1.5 kV DC feeder in order to visualize energized or de-energized condition. It would be beneficial to avoid both electrical shock and mistakenly attaching the earthing wire

WIT Transactions on The Built Environment, Vol 199, © 2020 WIT Press
www.witpress.com, ISSN 1743-3509 (on-line)
doi:10.2495/CR200151

to the energized wire. General outline of East Japan Railway Company's electrical facilities is shown in Fig. 1. The image of the congested overhead wires near the traction substation is shown in Fig. 2.

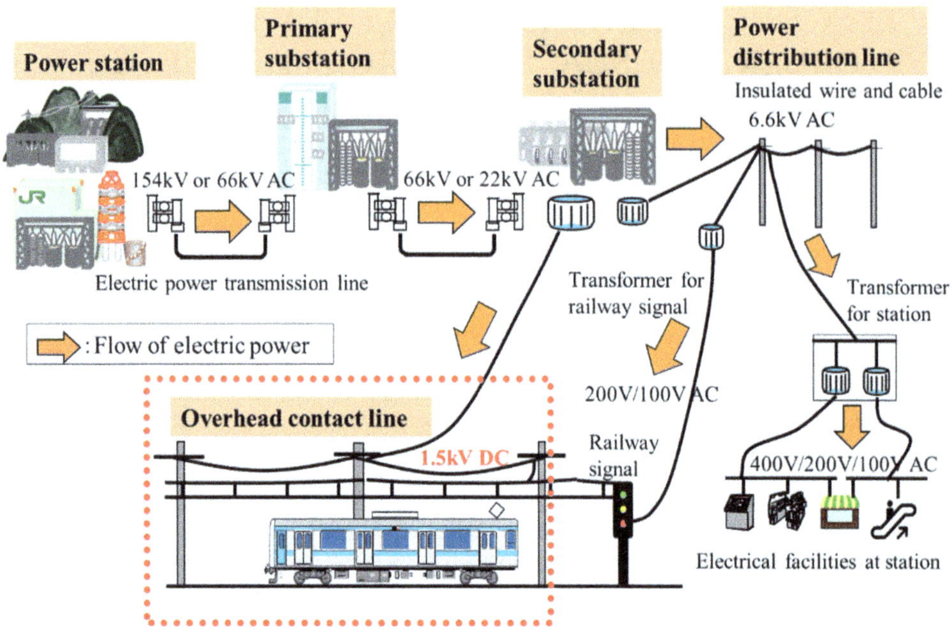

Figure 1: General outline of East Japan Railway Company's electrical facilities.

Figure 2: The image of the congested overhead wires near the traction substation.

2 THEORY AND FEATURE OF ELECTRICAL INDICATOR

We focus on 1.5 kV DC overhead contact line system. The theory and feature of electrical indicator are described as below.

2.1 Detection method

When the overhead wires are energized, they contain the ripple voltage. It is occurred when the rectifier converts AC voltage to DC voltage by three-phase full-wave rectification. The cylindrical electrode covers the overhead wire and detects the ripple voltage. The earthing wire has been successfully removed by focusing on the electric field, although the detection of 1.5 kV DC voltage basically needs the earthing wire to settle the reference potential.

2.2 Indication method

The energized or de-energized condition on the overhead wires is able to be visualized by attaching the LEDs on the electrical indicator. When the 1.5 kV overhead wires are energized, the electrical indicator is not activated and LEDs remains deactivated. On the other hand, when the 1.5 kV overhead wires are de-energized, the electrical indicator is activated and LEDs remains activated. A fail-safe system which will activate the LED in event of a malfunction of the electrical indicator has been also devised, even during de-energized condition. Small rechargeable batteries have been installed inside the electrical indicator and solar cells on it because the power source of lighting LEDs is lost and LEDs cannot be activated in the event of de-energized condition.

2.3 Attachment method

Basically, on-site workers start maintenance or improvement work after checking de-energized condition and attaching the earthing wire to the overhead contact line system based on the work permit and procedure. We decided to attach the electrical indicator directly to the overhead wire for the purpose of the subsidiary measures of checking de-energized condition and attaching the earthing wire to the overhead contact line system. The target overhead wire would be around 6.5 m from the ground. Taking into consideration the average human eye height of 1.5 m, LEDs which have a visibility range of 5 m have been selected.

3 CONFIGURATION OF ELECTRICAL INDICATOR

The electrical indicator consists of a cylindrical electrode, detection circuits, rechargeable batteries, charging circuits, solar cells and LEDs. The electrical indicator can visualize energized or de-energized condition on the overhead wires by the LEDs. The cylindrical electrode covers the overhead wire and detects the ripple voltage. Outline of DC electrical indicator is shown in Fig. 3. Images of the cylindrical electrode are shown in Fig. 4.

4 OUTLINE OF VERIFICATION TESTS

The verification test in the laboratory, where an overhead wire on 1.5 kV DC is able to be energized, have been conducted. The applied voltage to the overhead wire was measured by the voltage recorder, and the ripple voltage, which was incorporated in the electrical indicator, was measured by the oscilloscope. Threshold voltage level of the LED for the energized condition has been founded by measuring the ripple voltage using the oscilloscope. Moreover, the visibility of LEDs at a distance of 5 m in de-energized condition have been confirmed. Experimental circuit composition is shown in Fig. 5. Image of experimental circuit composition is shown in Fig. 6.

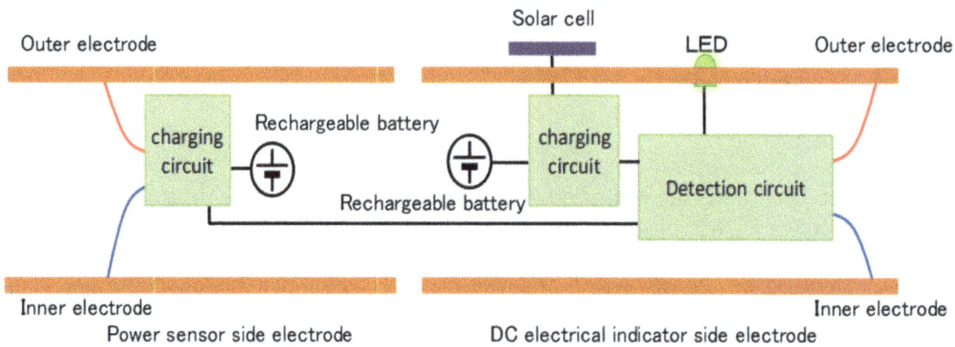

Figure 3: Outline of DC electrical indicator.

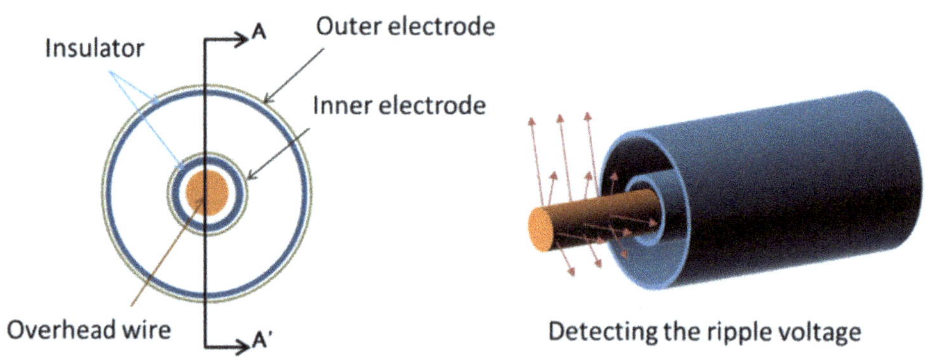

Figure 4: Images of the cylindrical electrode.

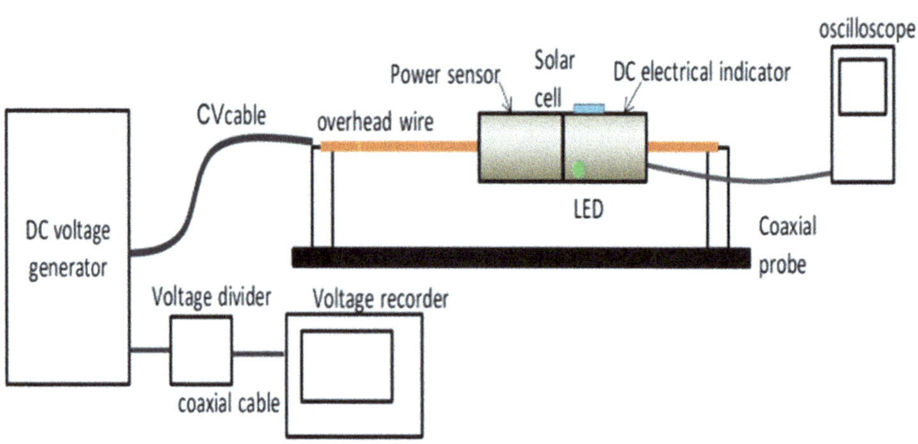

Figure 5: Experimental circuit composition.

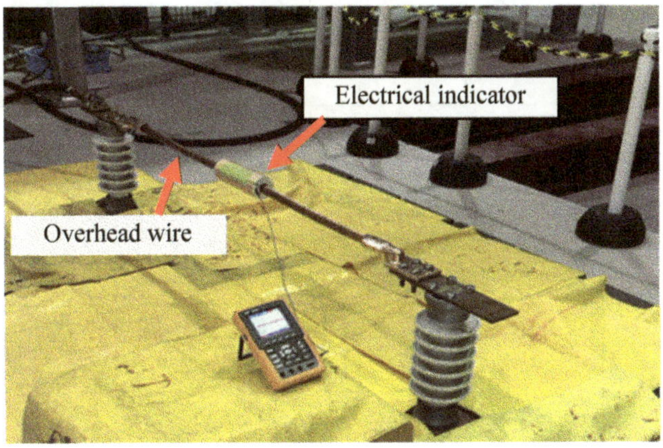

Figure 6: Image of experimental circuit composition.

5 RESULT OF VERIFICATION TESTS

The detection voltage level of the voltage detector which on-site workers usually use at site is more than 750 V DC referring to the voltage detector. We tried to calculate the threshold voltage level by measuring the applied voltage and the ripple voltage based on the several transformer setting voltage. Then, we have calculated the approximate equation as

$$y = 0.0111x + 24.375, \quad (1)$$

where x = the applied voltage (V); y = the ripple voltage (V).

When the overhead wire is applied in 750 V DC, the ripple voltage is 32.7 Vp-p and the threshold voltage level of the detection circuit of the electrical indicator is 620 mV. Therefore, we have decided the threshold voltage level of LEDs for the energized condition as 620 mV. Relation between applied voltage and ripple voltage is shown in Table 1 and Fig. 7.

We have selected the LEDs colour as green, which is adopted in the signal of the public transportation as safe. And green also has good visibility. The visibility of LEDs at a distance of 5 m in de-energized condition have been confirmed through the experiment. Images of LEDs of the electrical indicator are shown in Fig. 8.

Table 1: Relation between applied voltage and ripple voltage.

Transformer setting voltage [V]	(Voltage recorder) DC voltage [V]	(Voltage recorder) Ripple voltage [Vp-p]	Ripple to DC voltage Voltage Proportion[%]	(Oscilloscope) Detection voltage [mV]	Detection for ripple voltage Voltage ratio
1788	2130	48	2.25	880	0.018
1491	1750	44	2.51	820	0.019
1192	1400	40	2.86	780	0.020
893	1050	36	3.43	680	0.019
				Average value	0.019

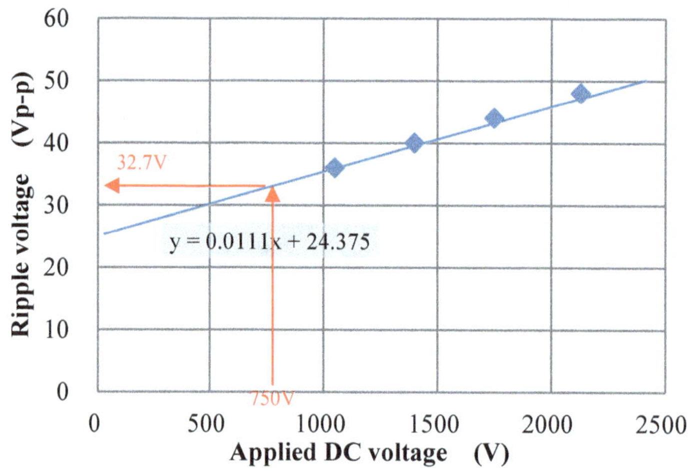

Figure 7: Relation between applied voltage and ripple voltage.

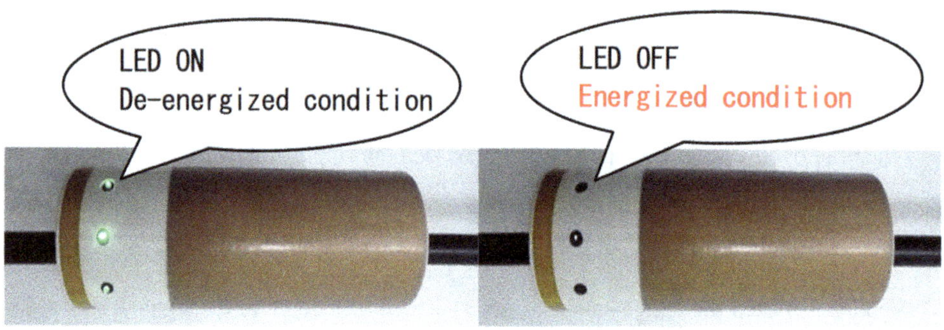

Figure 8: Images of LEDs of the electrical indicator.

6 CONCLUSION

We have developed the electrical indicator for 1.5 kV DC feeder and conducted verification test, enabling to visualize energized or de-energized condition. Threshold voltage level of the LED for the energized condition has been founded by measuring the ripple voltage using the oscilloscope. Moreover, the visibility of LEDs at a distance of 5 m in de-energized condition have been confirmed. We will continue to make further improvements on electrical indicator and aims to introduce the device into the field for practical use in the near future.

ACKNOWLEDGEMENTS

We would like to thank Masahiro Tsuchiya of National Institute of Information and Communications Technology, Kazuhisa Iwata of A.C.S. Inc. and Michiaki Kouda of East Japan Railway Company for supporting development and verification test of electrical indicator for 1.5 kV DC feeder.

REFERENCES

[1] Tsuchiya, M., Shiozawa, T. & Harakawa, S., Electric field sensing and imaging by noninvasive parallel-plate sensor. *IEICE Electronics Express*, **11**(18), pp. 1–7, 2014.

[2] Hasegawa, K., Mitoma, Y. & Kouda, M., Basic verification of DC electrical indicator. *IEE Japan*, pp. 383–384, 2019.

ACCIDENT ROOT CAUSES IDENTIFICATION USING A TAXONOMY

JEROME MAGOUET[1], MATTHIEU PERIN[2] & SIMON COLLART-DUTILLEUL[2]
[1]Vulcain Ingénierie, France
[2]Gustave Eiffel University, IFSTTAR/ESTAS, France

ABSTRACT

Railway accidents are rare, but for the sake of safety improvement they are all investigated by experts. Whereas this solution has proven its correctness to increase the overall safety in transportation, the lack of a methodological analysis formalization in case of similar accidents may lead to the non-detection of common root causes and to forfeit their removals. The aim of this work is to propose a formalized taxonomy based on the accident causes, classified by main categories such as "technical causes", "human causes" or "organizational causes". This classification is then used as a seed for safety analysis, like fault tree analysis, to detect and quantify possible common root causes for similar railway accidents. Along with this paper, the methodology is exemplified on a specific accident type: the derailments caused by over-speed in curves.

Keywords: taxonomy, railway, accident, root causes, fault tree analysis.

1 INTRODUCTION

Although railway accidents are subject to expert investigations, there is a lack of methodological analysis formalization in case of similar accidents, allowing an identification of the root causes by the mean of correlation. Actually, multiple occurrences of a given factor may lead to stronger conclusion and will allow the definition of corrective actions to avoid the considered class of accident.

The aim of this work is to propose a methodology to prepare taxonomy of the accidents causes, classified by main categories such as "technical causes", "human causes" or "organizational causes", and then use this result as a seed for safety analysis: root cause analysis and fault tree analysis (FTA).

First step of the method proposed is to build the taxonomy based on expert knowledge for main categories, and then refine in subcategories based on analysis of existing events. Using this taxonomy, a causal tree analysis is performed in order to track all the actions and conditions (or causal factors) that are necessary and sufficient for a given consequence to occur.

The second and last step is to setup safety analysis using our previous results. In this paper, a fault tree is built. Its main purpose is to help identifying potential causes of system failures. It should also be used to evaluate the probability of the top event using analytical or statistical methods. These calculations involve system quantitative reliability and maintainability information, such as failure probability, failure rate and repair rate.

Once completing the FTA, it is easier to focus the efforts on improving system safety and reliability on pinpointed core causal factors and their base event counterparts.

Along this paper, the methodology is exemplified on a specific accident: the derailments caused by over-speed in curve. For this case study, the available worldwide railway accident data are analysed in order to build taxonomy. The adequacy of the taxonomy is discussed with regards to the socio-technical context of accident reports. Indeed, fifty years old accident occurrences have to be analysed while keeping in mind the shift in scientific and technical knowledge. Cultural and legal evolution is something to be taken into account too when reviewing old accident reports.

WIT Transactions on The Built Environment, Vol 199, © 2020 WIT Press
www.witpress.com, ISSN 1743-3509 (on-line)
doi:10.2495/CR200161

As prospects, consulting the scientific state of the art in order to propose further formalization of the methodology is proposed.

2 BUILDING THE TAXONOMY

2.1 Approach for preparing the taxonomy

2.1.1 Definition of taxonomy

Taxonomy is a term issued from biology. It is the science of naming, describing and classifying organisms and includes all plants, animals and microorganisms of the world. Using morphological, behavioural, genetic and biochemical observations, taxonomists identify, describe and arrange species into classifications, including those that are new to science. Carl Linnaeus (1707–1778) is known as the father of taxonomy.

In the proposed methodology, this approach is transposed to perform the identification of root causes of train accidents.

As a minimum, train accident root causes can be classified into three main categories:

- Technical causes;
- Human causes;
- Organizational causes.

Each main category will be divided into sub-categories, as per the results of the analysis. Any new main category could be added if it is highlighted by the analysis.

2.1.2 Steps proposed to prepare the taxonomy

In order to prepare the taxonomy, the following main steps are performed:

- Definition of the set;
- Exhaustive analysis of the set;
- Definition and classification of the categories and subcategories.

Each activity will be detailed in the following sections.

2.1.3 Presentation of the example

We will illustrate how the methodology is applied by using a type of train accident: derailment in curve due to overspeed. This kind of accidents occurs more than 20 times in the last century.

2.2 Definition and analysis of the representative set of events

2.2.1 Definition of the set

A representative sample is a small quantity of something that accurately reflects the larger entity. An example is when a small number of people accurately reflects members of an entire population. In a classroom of 30 students, in which half the students are males and half are females, a representative sample might include six students: three males and three females.

When a sample is not representative, the result is known as a sampling error. Using the classroom example again, a sample that includes six students, all of whom are males, would not be a representative sample.

A representative sample parallels the key variables and characteristics under examination. Some examples include sex, age, education level, socio-economic status or marital status. Using a larger sample size increases the likelihood that the sample more accurately reflects what actually exists in the population. Any information collection with biased tendencies is unable to generate a representative sample.

A representative sample allows the collected results to be generalized to a larger population. For most marketing or psychology studies, it is impractical in terms of time, finances and effort to collect data on every person in the target population. This is especially impractical for large population such as an entire country or race.

The use of sample groups poses risks, as the sample may not accurately reflect the views of the general population. One of the largest risks is developing a sample that is not truly representative. This most likely occurs because the population group is too small. For example, when comparing data relating to gender, a representative sample must include individuals of different ages, economic status and geographical locations. Such information typically requires a diversification of information-collecting sites.

Random sampling involves choosing respondents from the target population at random, to minimize bias in a representative sample. While this method is more expensive and requires more upfront information, the information yielded is typically of higher quality. Purposive sampling is more widely used and occurs when the managers target individuals matching certain criteria for information extraction. Ideal interview candidates receive profiles. Although this leads to the potential of bias in the representative sample, the information is easier to collect, and the sampler has more control when creating the representative sample.

2.2.2 Application

Table 1 provides the list of train derailments in curve. This list has been prepared using existing accidents database such as the European Railway Agency accidents base.

For each event, it is assessed if the available input information allows an exhaustive analysis of the accident in order to go deeply into the root causes.

2.3 Definition of categories

2.3.1 Listing categories and subcategories

For each item of the representative sample, a review is performed by an expert of the topics to be analyzed. The expert will identify the categories that meet the targets of the review. He is free to use any available input.

Categories and subcategories should not be defined in advance, they will be identified throughout the review process without any limit.

When each item of the representative sample has been reviewed and categories have been defined, the taxonomy can be defined.

2.3.2 Application

As explained in Section 2.2.2, the events to be analysed are chosen based on the available inputs. Main inputs used are the accident reports (as far as available, official reports following investigations).

These reports are analysed by a railway safety expert whose target is to identify the root causes of the accidents. It has to be reminded that investigation techniques are the ones used when the reports were issued.

The main target of this review is to focus on the analysis on the facts provided in the reports. Conclusions of the reports are only checked at the end in order to compare the results with the ones from the railway safety expert.

Table 1: Set of events.

Main inputs used	Date	No. of fatalities	Comments	Place of accidents
[1]	01/07/1906	28	–	Salisbury, UK
[2], [3]	01/11/1918	98	–	New York, Malbone Street, USA
[4], [5]	1947	24	Not analyzed because of the lack of input information	Pennsylvania, USA
	1947	16	–	Camp Mountain, Australia
–	02/10/1947	184	–	Hachiko Line, Japan
[6]	1955	17	–	Sutton Coldfield, UK
[7], [8]	1969, 1984, 1994	6	Six fatalities in three separate accidents	Morpeth, UK
[9]	1972	6	–	Eltham Well Hall, UK
[10], [11]	1989, 1996	8	Eight fatalities in two separate accidents	Cajon Pass, USA
–	1997	8	Not analyzed because of the lack of input information	Piacenza, Italy
[12]	2000	9	–	Bruhl, Germany
[13], [14]	31/01/2003	7	–	Waterfall, Australia
[15]	15/11/2004	0	–	Cairns Tilt train, Australia
[16], [17]	25/04/2005	107	–	Amagasaki, Japan
[18]	2006	41	–	Valencia, Spain
–	06/07/2013	47	Not analyzed because of the lack of input information	Megantic Lake, Canada
[19], [20]	24/07/2013	79	–	Santiago de Compostela, Spain
[21]	01/12/2013	4	–	Spuyten Duyvil, NY, USA
–	09/11/2014	0	Not analyzed because of the lack of input information	Colebrook, Tasmania
[22]	12/05/2015	8	–	Philadelphia, USA
[23]–[25]	14/11/2015	11	–	Eckwersheim, France

Table 2 summarizes the root causes identified by the railway safety expert for three representative accidents. These accidents have been chosen among the 20 others because of the involvement of the human and the organisation in the main causes of the accidents.

As a result of the expert review, root causes have been sorted into three main causes' categories: Infrastructure/Hardware, Human, Society. This classification is the taxonomy.

Tables 3 and 4 provide for the taxonomy for human and society route causes.

Table 2: Results of the analysis.

Location of accidents	Root causes identified
Amagasaki, Japan	• Horizontal alignment – strong radius of the curve • Driver error (overspeed, no respect of signaling information) • Lack of driver's knowledge or training • Lack of driver's experience • Lack of conscience of risks • Problem of organization • Pressure of organization
Santiago de Compostela, Spain	• Horizontal alignment – strong radius of the curve • Lack of safety signaling system • Driver error (overspeed, no respect of signaling information) • Error in procedures • Lack of conscience of risks • External perturbations
Eckwersheim, France	• Horizontal alignment – strong radius of the curve • End of line, connection between two lines • Driver error (overspeed, no respect of signaling information) • Error in procedures • Lack of driver's knowledge or training • Lack of driver's experience • Lack of conscience of risks

Table 3: Taxonomy – Human causes.

Factors	Code
Driving error	*H01*
Driver error	H01.01
Error in procedures	H01.02
Knowledge/training	*H02*
Lack of driver's knowledge or training	H02.01
Lack of driver's experience	H02.02
Human behaviour	*H03*
Use of alcohol, drugs	H03.01
Lack of conscience of risks	H03.02
Driver's health problem/personal problem	H03.03
External perturbations	H03.04

Table 4: Taxonomy – Society causes.

Factors	Code
Organization issue	*S01*
New organization	S01.01
Problem of organization	S01.02
Pressure of organization	S01.03
Strike in the company	*S02*
Behavior of travellers	*S03*
Competition	*S04*

2.4 Finalization of the taxonomy

The first revision of the taxonomy is prepared based on the expert review of the representative set of events.

It could be decided to use the taxonomy as it is or to try to develop the first results to make it as exhaustive as possible. The possible solutions to do so are:

- To use dependability tools (such as causal tree or FTA) to identify an exhaustive list of scenario;
- To review and complete the taxonomy during a workshop organized with several experts with experience on similar or connected subjects (will be detailed in a future publication) [27];
- To extend the representative set of events (will be detailed in a future publication).

Section 3 provides details on the way to use and enhance the taxonomy.

3 USING THE TAXONOMY RESULTS

3.1 Statistical approach

Once quantitative data have been collected, statistical analysis can be carried out in order to make sense, and draw some inferences from the data.

There is a wide range of possible techniques that can be used [26].

The first thing to do with any data is to summarise it, which means to present it in a way that best tells the story.

One of the most common techniques used for summarising is using graphs, particularly bar charts, which show every data point in order, or histograms, which are bar charts grouped into broader categories.

In our analysis of 20 accidents cases, first conclusions are:

- More than 75% of the accidents are caused by a human error;
- Among this 75%:
    - 80% are caused by lack of training, knowledge or experience;
    - 66% are caused by under estimation of the possible risk;
    - 33% are caused by errors in procedures;
- Only 25% of the accidents are caused by rolling stock failures;

- 55% of the accidents are caused by issue with the organization, and among this 55% because of pressure on the driver.

As a first conclusion of this statistical analysis, if we want to decrease the number accidents, we have to focus on the driver itself in order to provide the adequate level of training and safety systems onboard to assist in the driving activities, especially during stressful conditions.

3.2 Causal factor tree analysis

3.2.1 Listing the categories and subcategories

Causal factor tree analysis is a root cause analysis technique used to record and display, in a logical, tree-structured hierarchy, all the actions and conditions (or causal factors) that were necessary and sufficient for a given consequence to have occurred.

This technique:

- Provides structure for the recording of evidence and display of what is known.
- Through application of logic checks, gaps in knowledge are exposed.
- Tree structure is familiar and easy to follow.
- Can easily be extended to handle multiple (potential) scenarios.
- Can incorporate results from the use of other tools.
- Works well as a master investigation/analysis technique.

In the contrary, this

- Cannot easily handle or display time dependence.
- Sequence dependencies can be treated, but difficulty increases significantly with added complexity.
- Shows where unknowns exist, but provides no means of resolving them.
- Stopping points can be somewhat arbitrary.

3.2.2 Application

All the categories and subcategories identified in the taxonomy have been integrated in the causal tree (Fig. 1) in order to show all possible scenario.

This figure allows to check the overall consistency of the model and the possible lack in the taxonomy [28].

3.3 Fault tree analysis

3.3.1 Listing the categories and subcategories

FTA is a top down, deductive failure analysis in which an undesired state of a system is analysed using Boolean logic to combine a series of lower-level events. This analysis method is mainly used in the fields of safety engineering and reliability engineering to understand how systems can fail, to identify the best ways to reduce risk or to determine (or get a feeling for) event rates of a safety accident or a particular system level (functional) failure.

FTA methodology is described in several industry and government standards, including NRC NUREG-0492 for the nuclear power industry, an aerospace-oriented revision to NUREG-0492 for use by NASA, SAE ARP4761 for civil aerospace, MIL-HDBK-338 for military systems, IEC standard IEC 61025 is intended for cross-industry use and has been adopted as European Norm EN 61025.

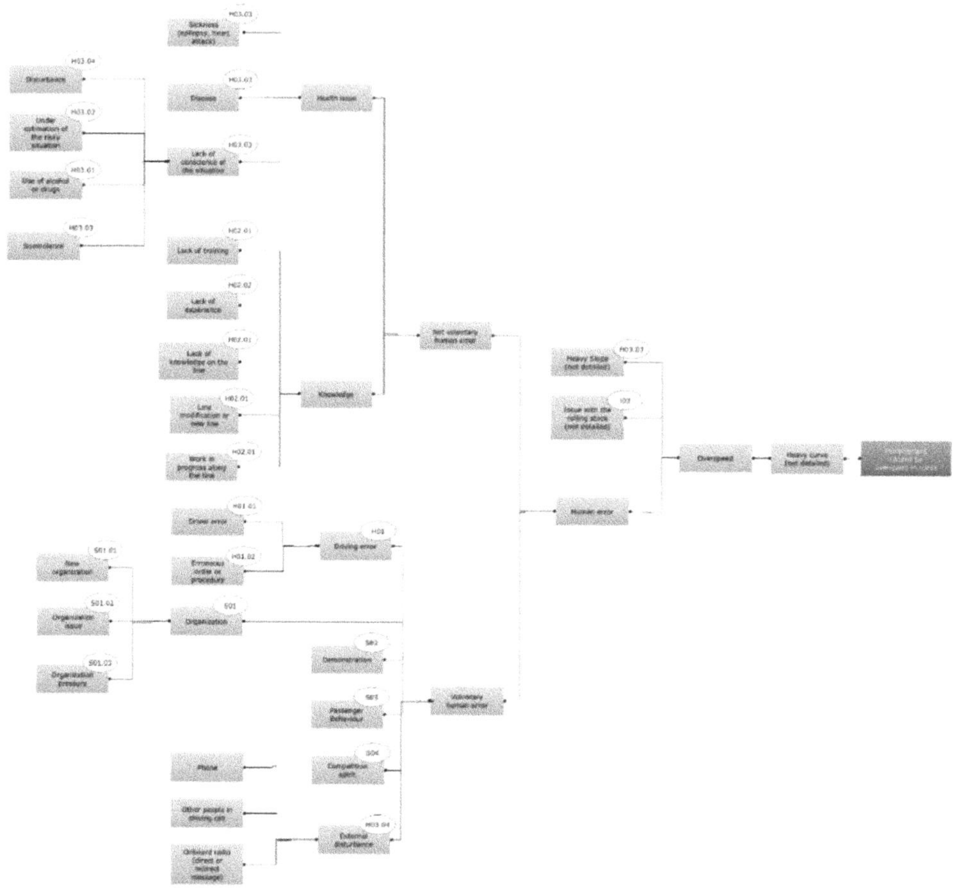

Figure 1: Extract of the causal tree for "Derailment in curve due to overspeed".

Any sufficiently complex system is subject to failure as a result of one or more subsystems failing. The likelihood of failure, however, can often be reduced through improved system design. FTA maps the relationship between faults, subsystems, and redundant safety design elements by creating a logic diagram of the overall system.

The undesired outcome is taken as the root ("top event") of a tree of logic. Working backward from this top event we might determine there are two ways this could happen: during normal operation or during maintenance operation. This condition is a logical OR. Considering the branch of occurring during normal operation perhaps we determine there are two ways this could happen: the press cycles and harms the operator or the press cycles and harms another person. This is another logical OR. We can make a design improvement by requiring the operator to press two buttons to cycle the machine – this is a safety feature in the form of a logical AND. The button may have an intrinsic failure rate – this becomes a fault stimulus we can analyze. When fault trees are labeled with actual numbers for failure probabilities, computer programs can calculate failure probabilities from fault trees. When a specific event is found to have more than one effect event, i.e. it has impact on several subsystems, it is called a common cause or common mode. Graphically speaking, it means this event will appear at several locations in the tree. Common causes introduce dependency relations between events. The probability computations of a tree which contains some common causes are much more complicated than regular trees where all events are considered as independent.

The tree is usually written out using conventional logic gate symbols. A Cut Set is a combination of events, typically component failures, causing the top event. If no event can be removed from a Cut Set without causing the top event, then the Cut Set is called a Minimal Cut Set.

3.3.2 Application

A fault tree has been prepared with the following top event “derailment in curve due to overspeed”. A sample of this tree is shown in Fig. 2.

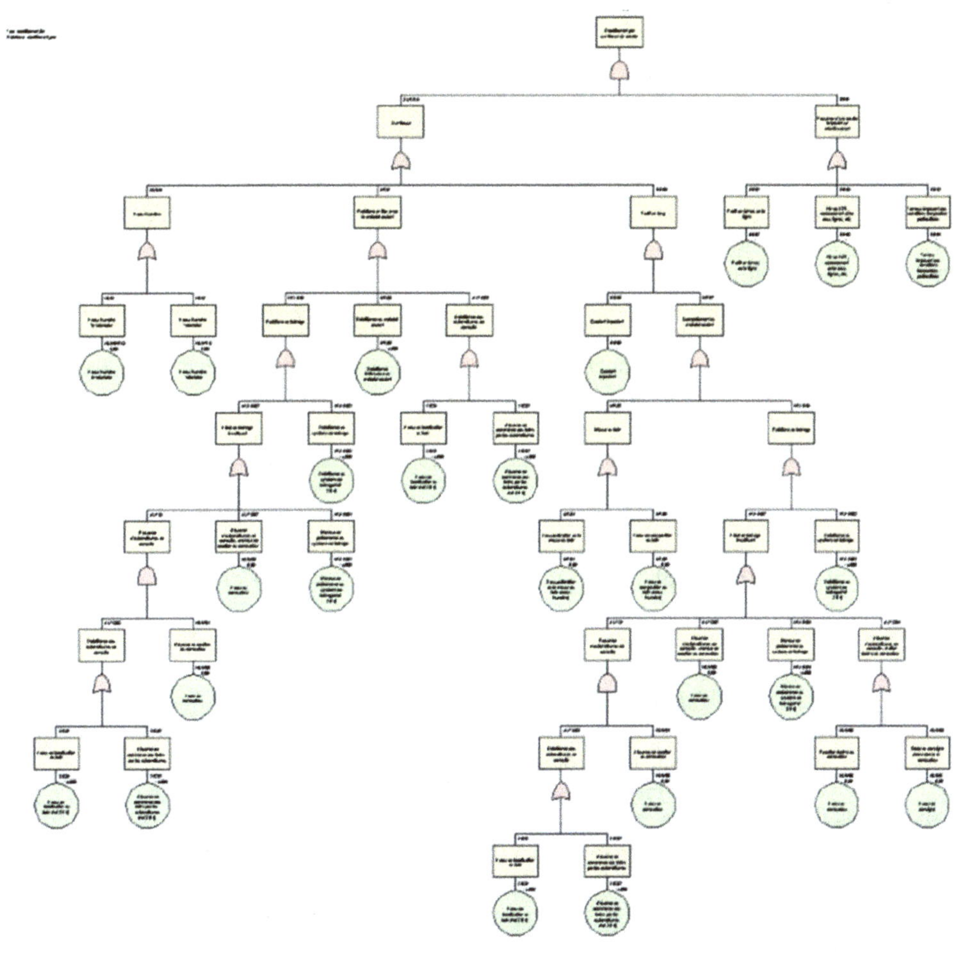

Figure 2: Fault tree for “Derailment in curve due to overspeed” (extract).

This tree could be used:

- To define the probability of the top event;
- To define the possible common cause failures;
- To define the minimal cut sets;
- To propose enhancement to reduce the overall probability of the top event.

4 CONCLUSIONS

This article provides the demonstration that it is possible to combine the taxonomy approach with traditional dependability tools to analyse and determine common root causes of a list of similar accidents.

The proposed short term development of this methodology could be:

- To split causes between direct and indirect causes… and to assess the ratio between each kind of causes;
- To use the analysis results to perform simulation of accidents;
- To apply the methodology on other type of accidents;
- To cross check the above results with the accidents taxonomy prepared by the European Railway Agency [29].

Considering the knowledge engineering perspectives, the taxonomy can be handled using an ontology-based framework like Protégé. The Ontorail (https://ontorail.org) initiative is clearly helping, but it only mainly defines infrastructures. Rolling Stocks automatisms and operating rules are still not considered.

Using conceptual computing will probably need strong analysis efforts before providing efficient results. Anyway, using GORO, the basic principle that can be found in accident may be exported to quite different scenarios [30]. It is to be explained that this conceptual modelling is the correct answer to the following situation: accidents are rare, so the time range of the analysis is wide, whereas the technological context is rapidly evolving. "Technologies changes, but concepts remain", conceptual modelling seems to be a good tool to apply this principle.

REFERENCES

[1] Pringle, J.W., Board of Trade, Accident returns: Extract for accident on Salisbury on 1st July 1906, 1906.

[2] www.nycsubway.org, Malbone Street wreck (New York Times, 1918). www.nycsubway.org/wiki/Malbone_Street_Wreck_(New_York_Times,_1918).

[3] Cudahy, B., *The Malbone Street Wreck*, Fordham University Press: New York, 1999.

[4] Queensland Railway's Interest Group, Camp Mountain disaster: Labour Day 1947. www.qrig.org/articles/camp-mountain-disaster-labour-day-1947.

[5] Queensland Parliamentary Debates, Legislative assembly, Thursday 20 Nov. 1947, pp. 1528–1530, 1947.

[6] UK Ministry of Transport, Report on the derailment which occurred on 23/01/1955 at Sutton Coldfield

[7] UK Ministry of Transport, Report on the derailment that occurred on 7 May 1969 at Morpeth.

[8] UK Ministry of Transport, Report on the derailment that occurred on 24 Jun. 1984 at Morpeth.

[9] UK Ministry of Transport, Report on the derailment that occurred on 11 Jun. 1972 at Eltham (Well Hall) Station.

[10] NSTB, Railroad accident report: Derailment of southern pacific transportation company freight train on main 12, 1989.
[11] National Geographic, Crash scene investigations: Runaway train (San Bernardino disaster), (Video).
[12] Eisenbahn-Bundesamt (EBA), Untersuchungsbericht: Entgleisung des D203 im Bahnhof Brûhm am 06.02.2000.
[13] Special Commission of Inquiry into the Waterfall Rail Accident, Final report, Volume 1, Jan. 2005.
[14] Special Commission of Inquiry into the Waterfall Rail Accident, Final report, Volume 2, Jan. 2005.
[15] Australian Transport Safety Bureau and Queensland Transport Rail Safety Investigation Report QT1472 into the derailment of Cairns tilt train VCQ5 north of Berajondo on 15 Nov. 2004.
[16] Aircraft and Railway Accidents Investigation Commission, Railway accident investigation report: Train derailment accident between Tsukaguchi and Amagasaki stations of the Fukuchiyama Line of the West Japan Railway Company.
[17] National Geographic, Seconds from disaster: Runaway train (Amagasaki disaster), (Video).
[18] Comision de investigacion del accidente de la linea 1 de metrovalencia de 3 de Julio de 2006.
[19] Informe final sobre el accidente grave ferroviaro, 0054/2013 ocurrido el dia 24.07.2013 en las proximidades de la estacion de Santiago de Compostela.
[20] Railvolution, Tragic derailment at Santiago de Compostela, Volume 13, No. 4.
[21] National Transportation Safety Board, Railroad accident brief: Metro-North railroad derailment.
[22] National Transportation Safety Board, Derailment of Amtrak passenger train 188, Philadelphia, Pennsylvania, 12 May 2015.
[23] SNCF, Direction Générale de l'Audit et des Risques: Rapport d'enquête immédiate no. 2015-AS-67 – Déraillement du Train d'essai no. 814521 sur le raccordement de Vendenheim de la LGV Est Européenne en construction le 14 Nov. 2015.
[24] BEATT, Note d'étape sur le déraillement de la rame d'essai de la LGV Est-Européenne survenu le 14 novembre 2015 à Eckwersheim.
[25] BEATT, Rapport d'enquête technique sur le déraillement d'une rame TGV lors d'une marche d'essai sur la LGV Est-Européenne le 14 novembre 2015 à Eckwersheim (67), 2017.
[26] Wilson, P.F., Dell, L.D., Anderson, G.F., *Root Cause Analysis: A Tool for Total Quality Management*, ASQ Quality Press: Milwaukee, WI, pp. 8–17, 1993.
[27] International Electrotechnical Commission, Fault Tree Analysis, Edition 2.0, EC 61025, 2006.
[28] Suryanto, H. & Compton, P., Learning classification taxonomies from a classification knowledge based system, University of Karlsruhe.
[29] European Railway Agency, Designing the common occurrences and taxonomy for Common Occurrence Reporting Programme ERA-PRG-004-TD-002 V 2.0.
[30] Negri, P.P., Souza, V.E.S., de Castro Leal, A.L., de Almeida Falbo, R. & Guizzardi, G., Towards an ontology of goal-oriented requirements. *CIbSE*, pp. 469–482, 2017.

TRAIN COLLISION AVOIDANCE WITH MOBILE DEVICES FOR SPARSELY USED RAILWAY LINES

THOMAS STRANG[1,2], ANDREAS LEHNER[1,2] & BENNO KÜHNL[2]
[1]German Aerospace Center (DLR), Institute of Communications and Navigation, Germany
[2]Intelligence on Wheels (IoW) GmbH, Germany

ABSTRACT

This paper shows the results of a communication range measurement campaign performed with mobile onboard units of a fully de-centralized train collision avoidance safety overlay system in typical environments of sparsely used railway lines. The most critical sections of the track with respect to a direct train-to-train communication channel were identified and selected for measurements in different antenna configurations. The measurement results justify the conclusion that the direct train-to-train communication channel along typical sparsely used railway lines fits the requirements of mobile onboard units of the train collision avoidance system.
Keywords: TrainCAS, TETRA, collision avoidance, safety, railway, trains, NGT.

1 INTRODUCTION

Although rail transport is extremely safe, collisions of railway vehicles happen occasionally. The safety overlay system described here adopts a very successful concept from aeronautics for avoiding the collision of trains [1]. It combines three core technologies: A direct train-to-train communication system, an accurate localization system, and a cooperative situation analysis and decision support system. Each equipped train determines its track-selective position on the line as well as other relevant parameters such as the current calculated braking distance, and broadcasts them per direct train-to-train communication to all other trains in communication range. Intelligent algorithms on board the receiving trains evaluate this information and raise an alarm in case of danger.

Acting as independent overlay system there is no need for a 100% penetration rate before being able to take advantage of the additional safety provided through the system. It does intentionally not interface with any other railway signalling technology which might or might not be in place or in operation along the line, to not inherit remaining failures or operational inadequacies. Thus it lifts the level of safety with every vehicle equipped with the technology. Despite being a technology to be used on rolling stock and relying on onboard technology only, it must be considered a fully de-centralized "virtual infrastructure", usually under the responsibility of the infrastructure management.

Whereas the best performance of the system can be achieved with fix mounted onboard technology, there are circumstances where operators prefer at least partially a mobile version of the technology which can be used only to temporarily equip vehicles on the track (see Fig. 1). This includes situations where there are additional guest or yellow fleet vehicles on the track, as well as situations when only a small varying set of vehicles out of a large vehicle fleet shall be used on a line.

One of the few restrictions of the mobile version compared to the fix mounted unit is the transmit power classes supported by the train-to-train transmitter hardware. Transmit power is a key factor determining the transmission range. Due to the large braking distances required for rolling stock once in movement, the transmission range defines the maximum speed at which the railway vehicles can be stopped early enough to avoid a potential collision.

WIT Transactions on The Built Environment, Vol 199, © 2020 WIT Press
www.witpress.com, ISSN 1743-3509 (on-line)
doi:10.2495/CR200171

Figure 1: Mobile TrainCAS onboard unit.

This paper elaborates on the transmission ranges achievable with mobile units in typical sparsely used railway line environments.

2 ANALYSIS

Each mobile unit periodically broadcasts a status update containing position and braking distance information, utilizing a TETRA radio module using the short data service (SDS) in direct mode operation (DMO). The frequency is in the protected band of 380–430 MHz using EN300392-2 power class 3L, i.e. with up to 1.8 W. Each single SDS received at a train's receiver is a successful communication between two equipped trains. The first successfully transmitted messages between approaching trains determine whether an upcoming critical situation can be avoided by raising an alarm timely enough.

The achievable communication range can be as large as almost 40 km with fix mounted onboard units in highspeed trains [2] using TETRA DMO-SDS without any base stations involved, well serving braking distances of several kilometres of trains running up to 300 km/h each. It is unlikely that mobile onboard units will reach this communication range due to several reasons. One reason is the limited transmit power already mentioned. Another reason is the strong influence of the wave propagation channel [3] along urban and sub-urban railway lines with a variety of curves, reflectors (e.g. walls) and scatterers (e.g. trees). Finally, from an operational perspective the railway operators prefer to use the mobile units without any antennas on the roof of the trains, i.e. just with an antenna inside the driver cabin.

To assess the communication range of the mobile units in the aforementioned setup, a series of measurements were performed at typical sparsely used regional railway lines. In particular the most critical sections of the railway lines with respect to direct train-to-train communication range were identified and selected for the measurements. This encompassed for instance curved sections in forests and challenging sections through rural villages.

The measurements were performed with one mobile device acting as transmitter placed inside a train driver's cabin in a regional train during regular operations, running a service back and forth on a sparsely used regional line with one train per direction and hour. The antenna configuration was changed between runs to see differences. The communication peer was another onboard unit acting as receiver located in a road vehicle, stationary placed next to the track at the selected most critical sections of the line. At the receiver side two different antenna setups have been measured: One on the roof and one inside the vehicle.

3 RESULTS

The measurement results – as expected – clearly show a difference whether a roof antenna was used at the receiver or not. For example, Fig. 2 (rural village) and Fig. 3 (timberline) show a minimum reception range of 2.7 km and 2.0 km respectively. Note that in all the figures the train with the transmitter was approaching the stationary receiver from one side, before passing and leaving it to the other side. Thus each figure illustrates the influence of the wave propagation channel on two sides, with the x-axis being normalized to the point in time of passing. Due to the strong influence of the channel it is very typical that the reception range of the two directions differ significantly, whereas the communication channel between the units is symmetric at any time, i.e. if receive and transmit functions are switched.

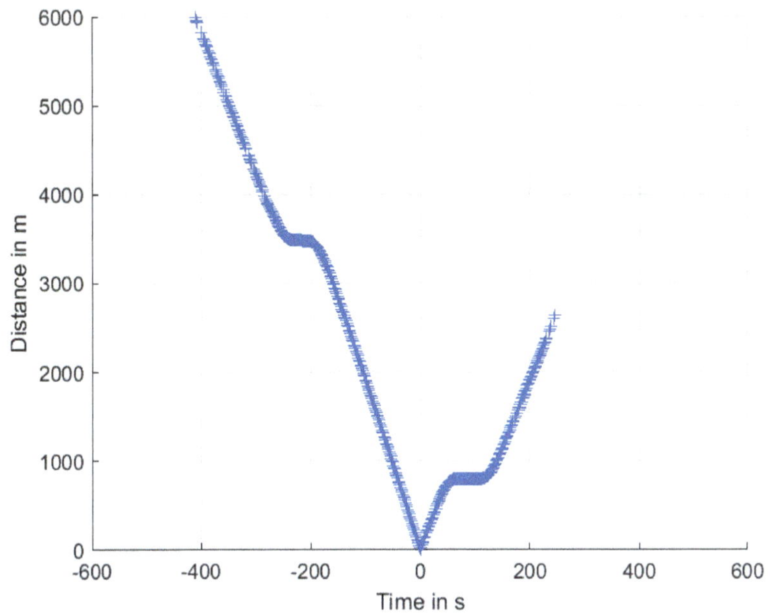

Figure 2: Reception ranges of train passing in rural village area (transmitter with internal antenna, receiver with roof antenna).

For the same scenarios (rural village, timberline), the results look different if no roof antenna is used at the receiver. Here the reception range degrades to 1.6 km (rural village) and 1.2 km (timberline) respectively as illustrated in Figs 4 and 5.

Despite the fact that the full potential of the use of an external antenna would be achieved if mounted on the roof of the vehicle, Figs 6 and 7 illustrate the potential of an external antenna used inside the driver's cabin of a train. Especially in the most demanding scenario (forest) with the reception range in Fig. 6 being slightly worse (0.9 km) compared to Fig. 5 (timberline), the use of an external antenna inside the driver's cabin can significantly improve the reception range to 2.4 km as seen in Fig. 7.

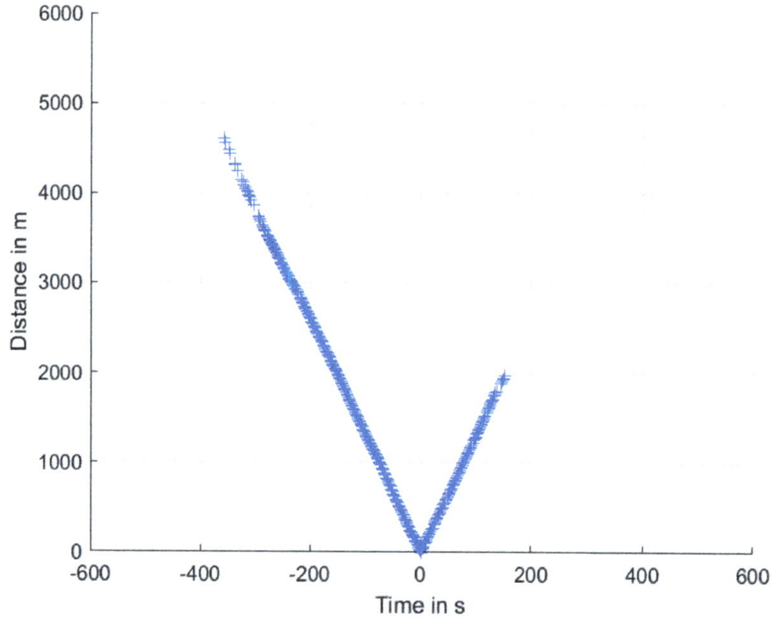

Figure 3: Reception ranges of train passing in timberline area (transmitter with internal antenna, receiver with roof antenna).

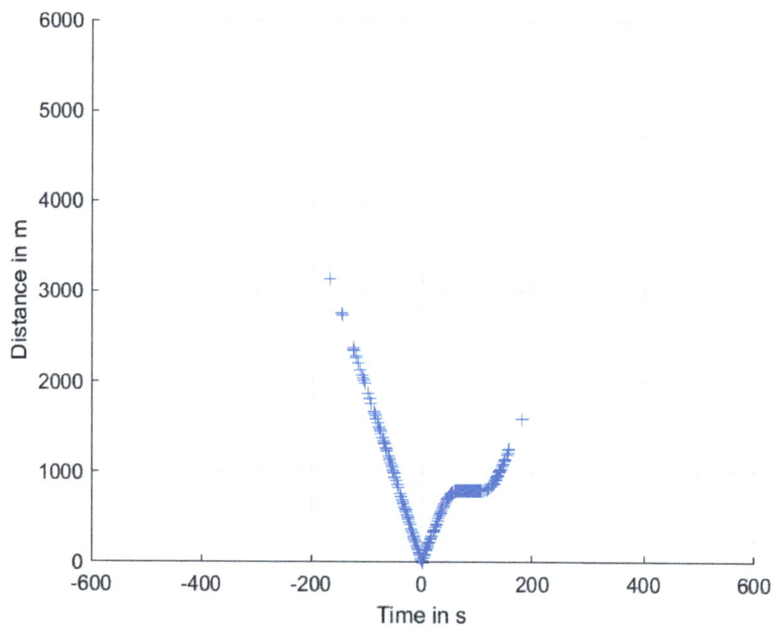

Figure 4: Reception ranges of train passing in rural village area (transmitter with internal antenna, receiver with internal antenna).

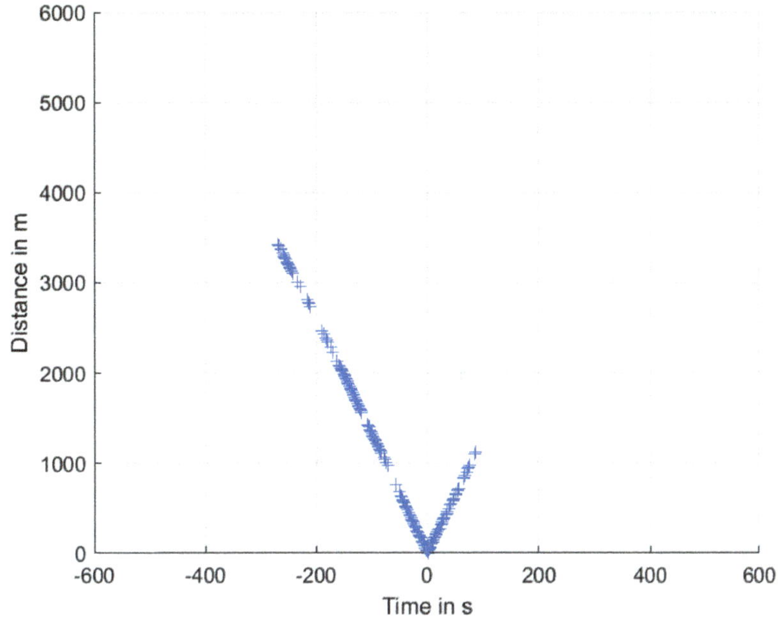

Figure 5: Reception ranges of train passing in timberline area (transmitter with internal antenna, receiver with internal antenna).

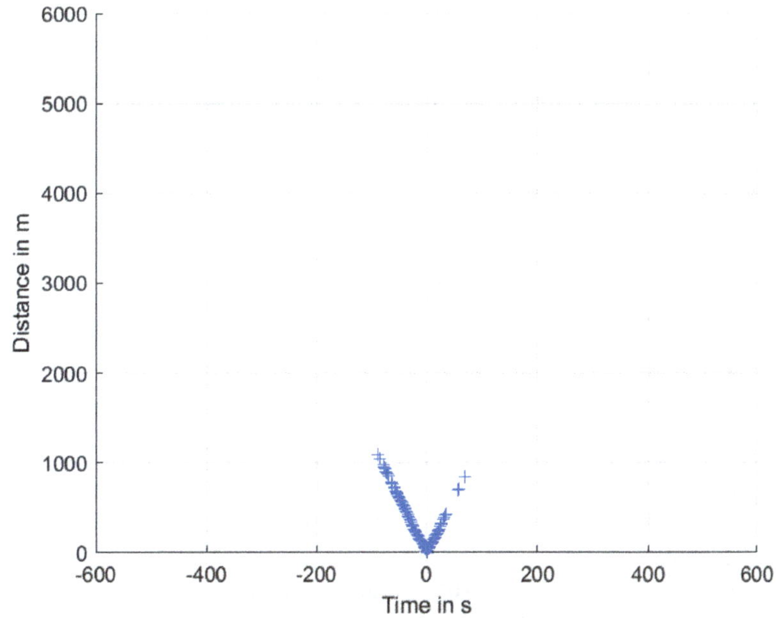

Figure 6: Reception ranges of train passing in forest area (transmitter with external antenna used inside driver's cabin, receiver with internal antenna).

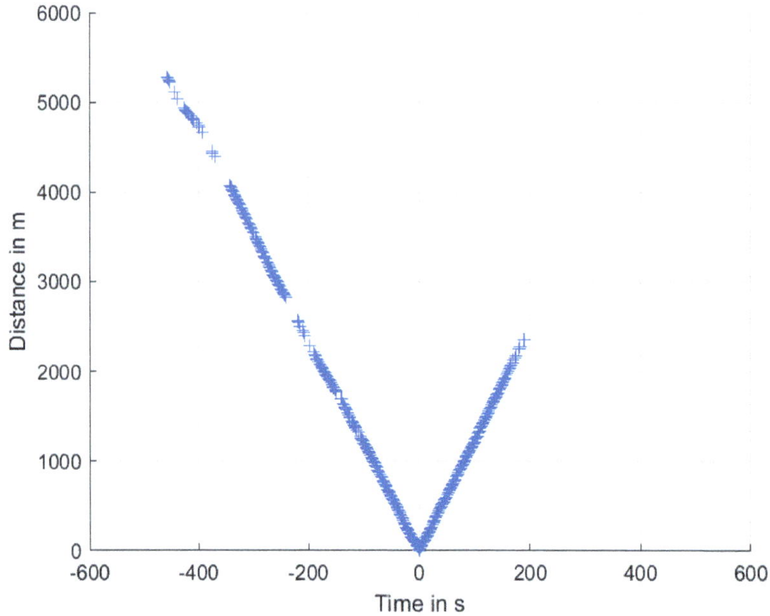

Figure 7: Reception ranges of train passing in forest area (transmitter with external antenna used inside driver's cabin, receiver with roof antenna).

4 CONCLUSIONS

Not surprisingly, the measurement results confirm the reduced communication ranges of TETRA DMO-SDS transmissions of mobile onboard units along typical sparsely used regional railway lines. Whereas a few kilometres of range can be achieved in less challenging parts of the railway network, the range can go down to about a kilometre or even less in challenging sections (environments), depending also on the used antennas and their position (inside driver cabin or on the roof).

For collision avoidance, the range has to be sufficiently large to be able to receive a message of an approaching train early enough to initiate a timely braking manoeuvre. This must include also any technical and human reaction times. The investigated railway line had a block size of 400 m, i.e. all trains have to guarantee to be able to come to a complete standstill within 400 m – if necessary through an emergency braking manoeuvre – which is mainly ensured through a maximum speed limit to be obeyed along the line and a minimum braking ability of the train. That means for two opposite trains under protection of the collision avoidance overlay system, any reception range above 800m would be sufficient. The collision avoidance system foresees an additional alert margin in the order of 15%, which takes even larger variances in technical and human reaction times into account. This appears more than sufficient for instance for the operational circumstances experienced at the lines of the measurements, where a speed limit of 50 km/h (equals 13.8 m/s) was to be obeyed by the train drivers. Thus there are 8s of extra margin on the total reaction time after successful message reception even in the most critical sections along the line.

It should be further noted that the speed limits along the track vary from the general speed limit. Permanent speed restrictions are usually assigned to sections of the line with steep curves or other potential of danger. As there is high degree of overlap of those areas with areas of challenging communication propagation conditions, there is even more reaction time available.

To finally conclude, mobile onboard units are not negatively affected by the communication conditions along typical sparsely used railway lines from the collision avoidance application perspective. Whereas the device's internal antennas provide a performance already above the design boundary, the use of external antennas even inside the train can further improve the performance.

ACKNOWLEDGEMENT

This work was partly supported by the Federal Ministry for Economic Affairs and Energy on the basis of a decision by the German Bundestag.

REFERENCES

[1] Haas, K.-G., Better railway collision prevention with TrainCAS, the autonomous train collision avoidance system. *Signal + Draht*, **110**, 2018.

[2] Lehner, A., Strang, T. & Unterhuber, P., Direct train-to-train communications at low UHF frequencies. *IET Microwaves, Antennas and Propagation*, **12**(4), pp. 486–491, 2017. DOI: 10.1049/iet-map.2017.0597.

[3] Rico Garcia, C., Lehner, A. & Strang, T., Channel model for train to train communication using the 400 MHz band. *IEEE 67th Vehicular Technology Conference*, Singapore, pp. 3082–3086, 2008.

SECTION 5
MODELLING AND SIMULATION

ON THE DISTRIBUTION OF THE RANDOM ARRIVAL TIME

BORIS DAVYDOV[1], VLADIMIR CHEBOTAREV[2] & KSENIYA KABLUKOVA[2]
[1]FESTU, Russia
[2]CC FEB RAS, Russia

ABSTRACT

This paper proposes the generalised model for formation the patterns of train arrival time distribution. The model provides that the arrival time has a random component. A feature of the model is the allowance for the stochastic dependence of travel interval on train departure time deviation. Within the framework of the model, a formula for the distribution function of the arrival time is obtained. The paper describes the method for checking the degree of stochastic relationship of the above random variables. Theoretical formulations are confirmed by analysis of statistical data reflected by the passenger train traffic on Russian railways.

Keywords: train traffic, stochastic model, operation time distribution, statistical dependency.

1 INTRODUCTION

Stochastic models are used for probabilistic forecasting of the traffic breakdowns when random disturbances occur (see, for example, [1] and [2]). They make it possible to identify intertrain conflicts and predict their further propagation over the railway network. The models proposed allow you to eliminate the delays or to reduce their duration. Distributions of the following random variables such as departure moment, travel time, headway, etc. can be considered as input data of such a model. Stochastic simulation results in the arrival delay probability distribution such as functional relation from the input random variables. In turn, knowledge of the delay distribution is the basis for predicting the delay duration as well as a set of measures to eliminate it.

It is usually assumed that the type of input distributions is standard when arrival delay function calculating. As a rule there are used an exponential distribution function or its modifications in a majority of papers. Such the model describes only particular cases and do not cover other possible variants of input distributions observed in reality. Our study shows that a more general approach is to represent the input random variables using the more comprehensive class of probability distribution.

This article proposes the stochastic model of the train run along some railway section which includes several stations. The input data are the distributions of departure moments and travel times. The result of simulation process is the set of distributions the arrival time deviation which is obtained for each station. The analytical method assumes that there are no restrictions on the form and the type of input distributions.

The statement of the problem analysis arose as a result of studying the real statistics of train traffic on the Russian railways. The research has shown the need to consider two cases of formation the output densities. The first case relies on the situation when the input random variables (departure time and travel time) are independent. In this case, the output distribution is a convolution of the input distributions. In the second case, a dependence of input random variables is observed (the travel time depends on the departure time). Based on the available statistics we create the output distribution only for those stations at which the input random variables are linearly dependent. The analysis is shown how well the output distribution of arrival times is consistent with the real statistics.

WIT Transactions on The Built Environment, Vol 199, © 2020 WIT Press
www.witpress.com, ISSN 1743-3509 (on-line)
doi:10.2495/CR200181

An ability to use arbitrary input distributions expands the possibilities of applying the model to various situations with different types of input random variables behave differently.

2 LITERATURE REVIEW

Modeling the process of random delays occurrence and their propagation along the train chain makes it possible to assign rational traffic adjustments and predict their likely consequences. The analytical method which uses convolutions of distributions of the initial (at the site entry) and newly arising deviations to obtain the delay density is proposed in Muhlhans [3]. Applications of this approach are limited since only the uniform distribution the secondary delays is used. In reality, this assumption is rarely confirmed. One effective approach which allows us to predict statistical characteristics of the train traffic is Situational-Heuristic Method (SEMN) [4]. The expected values of operating times are obtained from statistics of previous periods taking the current on-site situation into account. A similar approach is used in Karetnikov [5] which models the local traffic using average values of train processing time. It should be noted that there are attempts to derive some average train schedule based on the emerging situation. Such the attempts are fraught with conservation of technological problems that exist on the site.

One of the fundamental papers is devoted to the generic problem of traffic stochastic modeling under consideration [1]. The authors describe a model of delay formation in a train set using a probabilistic approach. Total run time of the train is considered as a sum of random running intervals along the section when analyzing the process of delay propagation. Buker and Seybold [6] show that calculation of the arrival time distribution at the terminal station is based on repeated use of convoluting the distributions of random variables characterizing the elementary operations.

Below mentioned papers use this approach to analyze the process of delay propagation along the train chain. Boucherie and Huisman [7] solve the problem of interacting the two trains that have different speeds and are influenced by random impacts. The result of analysis is used to model the train traffic in a dense heterogeneous train flow. This idea is developed further in the papers [2], [6], [8]. Description of the traffic process [2] occupies an intermediate position between the macro-models and models with very detailed description of the process, i.e. micro-models. Approximation method is proposed for exact representation of delay distributions. Cumulative distribution is calculated as the result of sequence the elementary activities; the activity chain is determined by a stochastic event-graph.

The problem of predicting the random delay appearance and propagation across the large railway network most fully is considered in Berger et al. [8]. Scattering of arrivals is treated as a random variable generated with joint accounting the departure times and the running intervals. The main subject of research is the process of adjusting the schedule at transfer stations in the presence of multiple delays. The model operates with discrete distributions of travel, dwell times and deviations from the schedule. The authors use a simplified description of input random variables so that they receive a small computing time.

The proposed approach was further developed in Buker and Seybold [6] which attempts to take the realistic distribution of operation time into account and choose the appropriate approximation. The framework proposed uses a probabilistic operational graph which considers passenger transfer operations and conflict situations. The authors argue that mesoscopic modelling of traffic is the purposeful approach to compute the delay propagation. The model considers only station and crossing as the operating points where conflicts can occur. The purpose of the study is to reduce computing time.

The general characteristic of the schedule is the ability to absorb small current deviations [9], [10]. The process of random deviations calculation is the basis of the methodology for determining corrective actions that reduce the time required to restore the normative schedule [11].

3 A GENERALIZED MODEL FOR THE FORMATION OF THE DISTRIBUTION OF THE ARRIVAL TIME

In this section we propose a model for the formation of the arrival times taking into account that the departure moments, travel times and possible delays can have the random nature. Random trajectories of train run are depicted in Fig. 1 as "beams" of segments emanating from the stations S_1 and S_2.

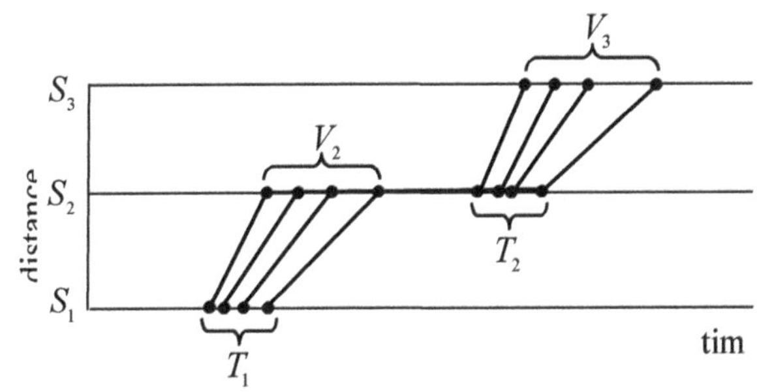

Figure 1: Scheme of the formation of arrival moments.

In Fig. 1 symbols T_i, $i = 1, 2$, denote the departure times from stations S_i respectively, and symbols V_k, $k = 2, 3$, the arrival times at stations S_k respectively. For each $i = 1, 2$, the values of T_i are scattered, and to emphasize this, some possible values of T_i are indicated by means of a brace. Similarly, for each $k = 2, 3$, some possible values of V_k are indicated by means of a brace too.

To describe the proposed model, we need the following notations:

d_i is the scheduled departure time from the station S_i, $i = \overline{1, N-1}$, where N is the total number of stations; $\delta_i = T_i - d_i$;

$\rho_{i,i+1}$ is the travel time from station S_i to station S_{i+1};

a_i is the scheduled arrival time at the station S_i, $i = \overline{2, N}$; $\xi_i = V_i - a_i$;

τ_i is the stop duration at station S_i, $i = \overline{2, N-1}$.

Distribution functions of random variables T_i, $\rho_{i,i+1}$ and V_i are denoted by $F_i(t)$, $L_{i,i+1}(t)$ and $G_i(t)$, respectively, and are defined by the following equalities: $F_i(t) = \mathrm{P}(T_i < t)$, $L_{i,i+1}(t) = \mathrm{P}(\rho_{i,i+1} < t)$ and $G_i(t) = \mathrm{P}(V_i < t)$.

Corresponding density functions, if they exist, will be denoted by the same letters as distribution functions, only lowercase ones: $f_i(t)$, $l_{i,i+1}(t)$ and $g_i(t)$. Fig. 2 shows a scheme of the formation of the following random variables: V_i, T_i and V_{i+1}.

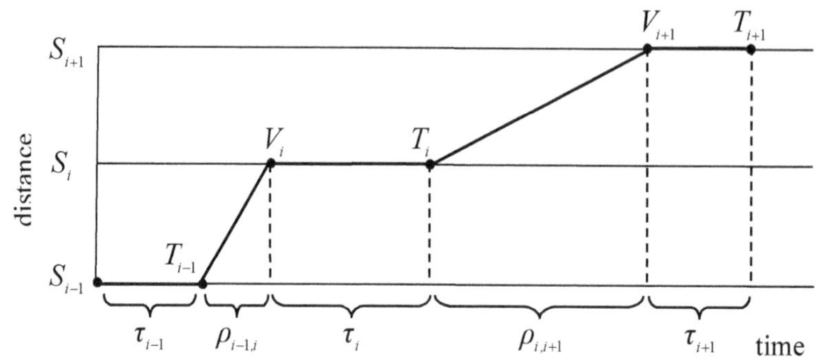

Figure 2: Scheme of the formation of the random variables: V_i , T_i and V_{i+1} .

According to this scheme, in particular, we have,

$$V_{i+1} = T_i + \rho_{i,i+1} . \tag{1}$$

Now we will show that the following equation is valid for the distribution function of V_{i+1}:

$$G_{i+1}(t) = \int_{-\infty}^{\infty} \mathrm{P}\left(\rho_{i,i+1} < t - x \mid T_i = x\right) dF_i(x). \tag{2}$$

We will use nonstrict reasoning for the derivation of (2) not to clutter up the paper.

First, suppose that for some constants $a < b$ the equality $\mathrm{P}(a \le T_i \le b) = 1$ holds. Divide the segment $[a,b]$ into n equal parts by points: $a = x_0 < x_1 < \ldots < x_n = b$. Using the well-known total probability equation we obtain from (1) that for each $i = \overline{1, N-1}$,

$$\begin{aligned} \mathrm{P}(V_{i+1} < t) &= \mathrm{P}(T_i + \rho_{i,i+1} < t) \\ &= \sum_{\{x_k\}} \mathrm{P}(x_k \le T_i < x_{k+1}) \mathrm{P}\left(\rho_{i,i+1} < t - T_i \mid x_k \le T_i < x_{k+1}\right), \end{aligned} \tag{3}$$

where $\mathrm{P}\left(\rho_{i,i+1} < t - T_i \mid x_k \le T_i < x_{k+1}\right)$ is the conditional probability of event $\{\rho_{i,i+1} < t - T_i\}$, provided that event $\{x_k \le T_i < x_{k+1}\}$ has occurred. Let us consider in eqn (3) the following term:

$$\mathrm{P}(x_k \le T_i < x_{k+1}) \mathrm{P}\left(\rho_{i,i+1} < t - T_i \mid x_k \le T_i < x_{k+1}\right).$$

At first, $\mathrm{P}(x_k \le T_i < x_{k+1}) = F_i(x_{k+1}) - F_i(x_k)$. In addition, assuming that the difference $x_{k+1} - x_k$ is small enough, we have the approximate equality $T_i \approx x_k$, since $x_k \le T_i < x_{k+1}$. Hence, rearranging the factors in (3), we obtain the equality

$$G_{i+1}(t) \approx \sum_{\{x_k\}} \mathrm{P}\left(\rho_{i,i+1} < t - x_k \mid x_k \le T_i < x_{k+1}\right)\left[F_i(x_{k+1}) - F_i(x_k)\right],$$

the right hand-side of which is the Stieltjes sum. When infinitely grinding the segment $[a,b]$ this sum converges to the Stieltjes integral

$$\int_a^b \mathrm{P}\left(\rho_{i,i+1} < t - x \mid T_i = x\right) dF_i(x).$$

Letting that a tends to $-\infty$ and b tends to $+\infty$, we obtain eqn (2).

If density function $f_i(x)$ exists, then

$$G_{i+1}(t) = \int_{-\infty}^{\infty} \mathrm{P}\left(\rho_{i,i+1} < t - x \mid T_i = x\right) f_i(x)\, dx.$$

In addition, if the conditional density $\frac{d}{dt}\mathrm{P}\left(\rho_{i,i+1} < t - x \mid T_i = x\right)$ exists, then density $g_{i+1}(t)$ exists and is equal to

$$g_{i+1}(t) = \int_{-\infty}^{\infty} \frac{d}{dt}\mathrm{P}\left(\rho_{i,i+1} < t - x \mid T_i = x\right) f_i(x)\, dx.$$

The conditional distribution function $\mathrm{P}\left(\rho_{i,i+1} < t - x \mid T_i = x\right)$ is sometimes written differently. Let $f_i(x,y)$ be the density of the joint distribution of T_i and $\rho_{i,i+1}$. Let x be such that $f_i(x) \ne 0$. For every such x the conditional distribution function of the random variable $\rho_{i,i+1}$ relative to the random variable T_i is defined as

$$L_{i,i+1}(y \mid x) := \mathrm{P}\left(\rho_{i,i+1} < y \mid T_i = x\right) = \frac{1}{f_i(x)} \int_{-\infty}^{y} f_i(x,u)\, du.$$

Thus,

$$G_{i+1}(t) = \int_{-\infty}^{\infty} L_{i,i+1}(t - x \mid x) f_i(x)\, dx = \int_{-\infty}^{\infty}\int_{-\infty}^{t-x} f_i(x,u)\; du\; dx. \tag{4}$$

Note that $l_{i,i+1}(u \mid x) := \dfrac{f_i(x,u)}{f_i(x)}$ is called the conditional density function of $\rho_{i,i+1}$ relatively to T_i.

Eqns (2) and (4) contain the conditional distribution that allows one to take into account the dependence of $\rho_{i,i+1}$ on T_i. These eqns are universal, but their application requires additional study of the joint distribution of random variables T_i and $\rho_{i,i+1}$.

If random values T_i and $\rho_{i,i+1}$ are independent, then (2) takes the form of convolution:

$$G_{i+1}(t) = \left(F_i * L_{i,i+1}\right)(t) \equiv \int_{-\infty}^{\infty} L_{i,i+1}(t - x)\, dF_i(x) = \int_{-\infty}^{\infty} L_{i,i+1}(t - x) f_i(x)\, dx. \tag{5}$$

To obtain equalities for density $g_{i+1}(t)$, it is necessary to replace the distribution functions in (5) with the corresponding densities. Note that the eqn for the moments of arrival/departure in the discrete case is presented in Berger et al. [8]. This eqn, although in other notation, is a special case of equality (2).

Remark: let us say a few words about question of the existence of the Stieltjes integral in (2). According to Fichenholtz [12, p. 92] this integral exists if the conditional distribution function $\mathrm{P}\left(\rho_{i,i+1} < t - x \mid T_i = x\right)$ is continuous with respect to the argument x.

4 A METHOD FOR CHECKING THE DEGREE OF STOCHASTIC DEPENDENCE BETWEEN INPUT VARIABLES

The stochastic model described in Section 3 operates with random variables T_i, $\rho_{i,i+1}$ and V_{i+1} for which equality (1) holds. Recall the following property of variances: two random variables are uncorrelated if and only if the variance of their sum is equal to the sum of the variances. Using this property and equality (1), one can check whether random variables T_i and $\rho_{i,i+1}$ are correlated or not.

Suppose that for each station S_i we have a three-dimensional sample of volume n, from the distribution of the three-dimensional random vector $\left(T_i, \rho_{i,i+1}, V_{i+1}\right)$. We will denote by symbol $\mathrm{D}^*\left[X\right]$ the sample variance of a random variable X (there is mean the asymptotically unbiased estimate):

$$\mathrm{D}^*\left[X\right] = \frac{1}{n}\sum_{i=1}^{n}\left(x_i - \overline{X}\right)^2,$$

where x_i are the sample values, $\overline{X}$ is the sample mean. For each fixed $1 \le i \le N$, we will check the equality

$$\mathrm{D}^*\left[V_{i+1}\right] = \mathrm{D}^*\left[T_i\right] + \mathrm{D}^*\left[\rho_{i,i+1}\right], \tag{6}$$

admitting an approximate equality in (6). In the case when equality (6) is fulfilled at least approximately, we will test the hypothesis that the correlation coefficient for the random variables T_i and $\rho_{i,i+1}$ is equal to zero, using Student's *t*-test. To this end, we calculate the value of the following statistic:

$$\mathrm{T}^* = \frac{r^*\sqrt{n-2}}{\sqrt{1-r^{*2}}}, \tag{7}$$

where r^* is the sample correlation coefficient, i.e.

$$r^* = \frac{\mathrm{cov}^*\left[T_i, \rho_{i,i+1}\right]}{\sqrt{\mathrm{D}^*\left[T_i\right]\mathrm{D}^*\left[\rho_{i,i+1}\right]}} = \frac{\sum_{j=1}^{n}\left(T_{ij} - \overline{T}_i\right)\left(\rho_{(i,i+1)j} - \overline{\rho}_{i,i+1}\right)}{\sqrt{\sum_{j=1}^{n}\left(T_{ij} - \overline{T}_i\right)^2 \sum_{j=1}^{n}\left(\rho_{(i,i+1)j} - \overline{\rho}_{i,i+1}\right)^2}}, \tag{8}$$

where $\mathrm{cov}^*[\cdot,\cdot]$ denotes the sample covariance, $\overline{T}_i$ and $\overline{\rho}_{i,i+1}$ are the sample means, T_{ij} denote sample values of T_i, and $\rho_{(i,i+1)j}$ sample values of $\rho_{(i,i+1)}$, $1 \le j \le n$; $1 \le i \le N$. Note that for Student's statistic we use the traditional notation T^*.

Let r be the Pearson correlation coefficient. The hypothesis $r = 0$ is considered as not contradict to experimental data with a significance level α if $|\mathrm{T}^*| < t_{\mathrm{crit}}(\alpha, n-2)$. The critical value $t_{\mathrm{crit}}(\alpha, n-2)$ is taken from the Student's distribution table.

If the hypothesis $r = 0$ is not rejected for the random variables T_i and $\rho_{i,i+1}$, then we can assume their independence (although it is well known that the non-correlation of random variables does not mean their independence). In this case, to find the distribution function (and, as a consequence, the density function) of the arrival times, one should use eqn (5) containing the convolution operation.

In the opposite case, if the hypothesis $r = 0$ is rejected, and the equality of the sample variances (6) is not satisfied, we make an assumption about the stochastic dependence of $\rho_{i,i+1}$ on T_i. If the modulus of the sample coefficient of correlation is close to 1, then we can consider that random variables $\rho_{i,i+1}$ and T_i are linearly dependent, i.e.

$$\rho_{i,i+1} = aT_i + b, \tag{9}$$

where a and b are the coefficients of linear regression and are calculated by the known eqns:

$$a = r^* \frac{\sqrt{\mathrm{D}^*[\rho_{i,i+1}]}}{\sqrt{\mathrm{D}^*[T_i]}} = \frac{\sum_{j=1}^{n}(T_{ij} - \overline{T}_i)(\rho_{(i,i+1)j} - \overline{\rho}_{i,i+1})}{\sum_{j=1}^{n}(T_{ij} - \overline{T}_i)^2}, \quad b = \overline{\rho}_{i,i+1} - a\overline{T}_i. \tag{10}$$

We proceed to check the dependence of the random variables $\rho_{i,i+1}$ and T_i on the example of a specific section of the railway.

5 CHECKING THE DEPENDENCE OF INPUT RANDOM VARIABLES BY USING STATISTICAL DATA

Consider the real section of the "Moscow-Tver" suburban line with intense train traffic. The line is limited by the terminals "Moscow" (point 1) and "Tver" (point 13) and includes 11 intermediate stations. The average travel time at the line is about three hours. Statistics are collected for departure, travel and arrival times for morning rush hours. Immediately, we note that all sample values are expressed in minutes. In addition, samples for departure and arrival times are shifted to the left by the minimum observed sample value of departure time from station 1. This is done in order to take the zero time point as the departure time from station 1.

Let us find out whether equality (6) holds for some stations of the considered railway. Table 1 contains the calculated values of the sample variances $\mathrm{D}^*[V_{i+1}]$, $\mathrm{D}^*[T_i]$ and

$D^*[\rho_{i,i+1}]$, the sum $D^*[T_i]+D^*[\rho_{i,i+1}]$, as well as the values of the sample correlation coefficient r^* for monitoring calculations. Using data provided in Table 1 one can see that the independence of the quantities $\rho_{i,i+1}$ and T_i is possible for $i = 3, 4, 12$.

Next, we test the hypothesis $r = 0$ by applying Student's *t*-test. To this end, we calculate the sample correlation coefficients (8) and statistic (7). The hypothesis is tested with two significance levels, $\alpha_1 = 0.05$ and $\alpha_2 = 0.1$. The calculation results are presented in Table 2.

Table 1: Verification of statistical dependence the input variables (eqn (6)).

i	$D^*[V_{i+1}]$	$D^*[T_i]$	$D^*[\rho_{i,i+1}]$	$D^*[T_i]+D^*[\rho_{i,i+1}]$	r^*
1	23.9097	25.0556	0.6875	25.7431	-0.2209
2	29.7431	23.9097	0.5556	24.4653	0.7241
3	31.0764	30.0556	0.2431	30.2986	0.1439
4	28.6667	28.0764	0.5764	28.6528	0.0017
5	17.5764	24.8056	0.8542	25.6597	-0.8780
6	35.6389	17.2500	4.7222	21.9722	0.7571
7	34.5208	35.6389	0.2431	35.8819	-0.2312
8	41.0000	33.7431	2.2431	35.9861	0.2882
9	24.7934	31.7025	0.6281	32.3306	-0.8445
10	0.5600	0.8100	0.2900	1.1000	-0.5571
11	0.2400	0.4900	0.2100	0.7000	-0.7170
12	1.6100	0.5600	1.2500	1.8100	-0.1195

Table 2: The results obtained by applying Student's *t*-test.

i	r^*	n	T^*	$t_{crit}(\alpha_1, n-2)$	Conclusion with $\alpha_1 = 0.05$	$t_{crit}(\alpha_2, n-2)$	Conclusion with $\alpha_2 = 0.1$
1	-0.2209	12	-0.7161	2.2281	1: accepted	1.8125	1
2	0.7241	12	3.3196	2.2281	0: rejected	1.8125	0
3	0.1439	12	0.4598	2.2281	1	1.8125	1
4	0.0017	12	0.0055	2.2281	1	1.8125	1
5	-0.8780	12	-5.8017	2.2281	0	1.8125	0
6	0.7571	12	3.6650	2.2281	0	1.8125	0
7	-0.2312	12	-0.7516	2.2281	1	1.8125	1
8	0.2882	12	0.9516	2.2281	1	1.8125	1
9	-0.8445	11	-4.7313	2.2622	0	1.8331	0
10	-0.5571	10	-1.8974	2.3060	1	1.8595	0
11	-0.7170	10	-2.9093	2.3060	0	1.8595	0
12	-0.1195	10	-0.3405	2.3060	1	1.8595	1

According to presented data, it can be assumed that at stations 4 and 12, the independence of random variables $\rho_{i,i+1}$ and T_i (i = 4, 12) takes place. To find the distribution density function $g_{i+1}(t)$, it is required to use (5).

We can see in Table 2 that for i = 5 and 9, the coefficient r^* is close to (-1). In these cases, we find out the linear dependence (9). Then the distribution function of the arrival time can be written as follows:

$$G_{i+1}(t) = \mathrm{P}(T_i + aT_i + b < t) = \mathrm{P}\left(T_i < \frac{t-b}{1+a}\right) = F_i\left(\frac{t-b}{1+a}\right). \tag{11}$$

In the next section we will obtain the density functions for the arrival times and check the degree of their consistency with real-world data.

6 COMPARISON OF THE MODEL WITH REAL STATISTICS EXAMPLES

Let us verify the consistency of the model with real data. We begin the verification with the case when $\rho_{i,i+1}$ and T_i are independent. In section 5, we found out that there is a reason that for $i = 4$, 12 random variables $\rho_{i,i+1}$ and T_i are independent. Therefore, for these stations we will check how well the result obtained after applying the equality, similar to (5), for density functions will be consistent with real data. The results of the verification are presented in Table 3, which contains histograms based on statistics for departure times T_4, T_{12} (row 1) and travel times $\rho_{4,5}$, $\rho_{12,13}$ (row 2). We approximate these histograms by the following density functions:

$$f_4(t) = I(t \geq 32)0.4e^{-0.4(t-32)},\ f_{12}(t) = I(t > 143)10\, e^{-2.9(t-143)}(t-143)^{1.4},$$

$$l_{4,5}(t) = I(t > 11.7)35.3\, e^{-3.8(t-11.7)}(t-11.7)^{2.75} \text{ and}$$

$$l_{12,13}(t) = I(27.5 \leq t \leq 32.5)0.3564e^{-0.36(t-29.5)^2}, \text{ respectively.}$$

Let us explain what $I(\cdot)$ means. The multiplier $I(\cdot)$ is an indicator, i.e.

$$I(x \in A) = \begin{cases} 1, & \text{if } x \in A, \\ 0, & \text{if } x \notin A, \end{cases}$$

where A is an arbitrary set. The functions $g_5(t)$ and $g_{13}(t)$ are determined by using an analog of eqn (5) for densities, i.e.

$$\begin{aligned} g_5(t) &= \int_{-\infty}^{\infty} l_{4,5}(t-x) f_4(x)\, dx \\ &= I(t > 43.7)14.12\int_{32}^{t-11.7} e^{-3.8(t-x-11.7)-0.4(x-32)}(t-x-11.7)^{2.75}\, dx, \end{aligned}$$

$$
\begin{aligned}
g_{13}(t) &= \int_{-\infty}^{\infty} l_{12,13}(t-x) f_{12}(x)\,dx \\
&= 3.564 \int_{143}^{\infty} I(27.5 \le t-x \le 32.5)\, e^{-0.36(t-x-29.5)^2 - 2.9(x-143)} (x-143)^{1.4}\, dx.
\end{aligned}
$$

Calculations using the latest eqns are performed by means the technical computing system Wolfram Mathematica.

It can be seen from row 3 of Table 3, the resulting densities visually are in good agreement with the statistics of arrival times.

Table 3: Histograms and approximating densities when $\rho_{i,i+1}$ and T_i are independent.

Station 4	Station 12
T_4 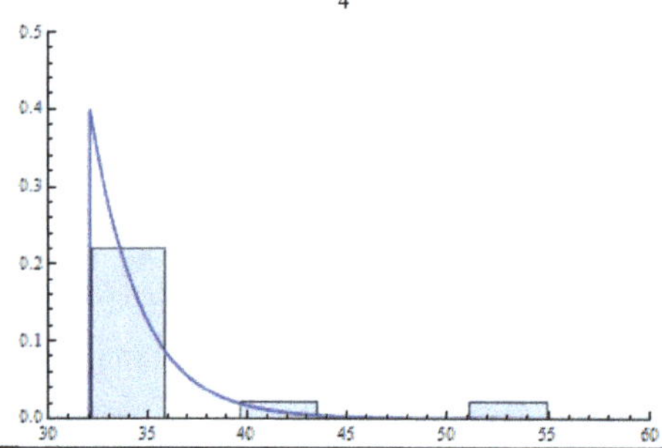	T_{12} 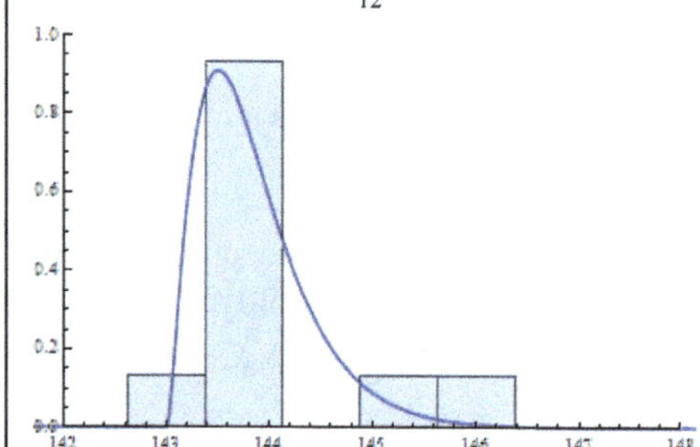
$\rho_{4,5}$ 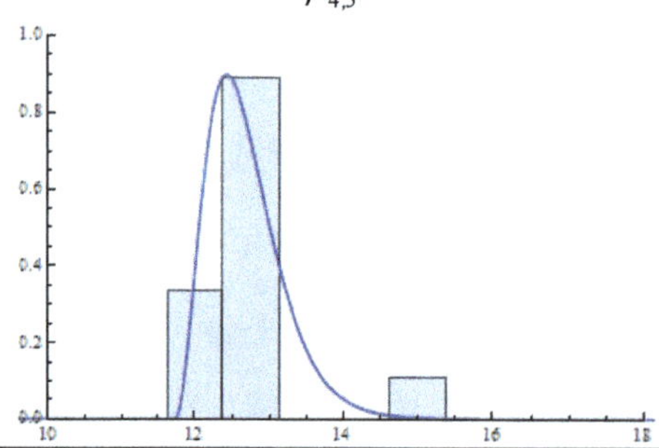	$\rho_{12,13}$ 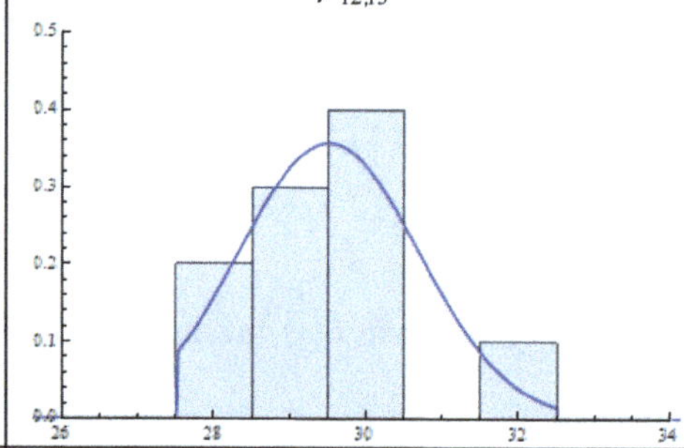
V_5 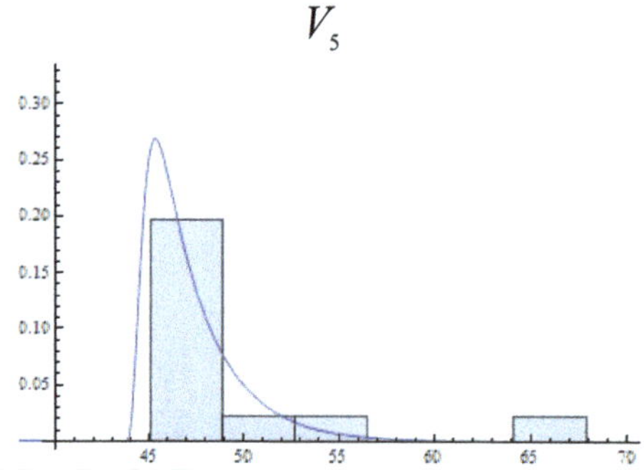	V_{13} 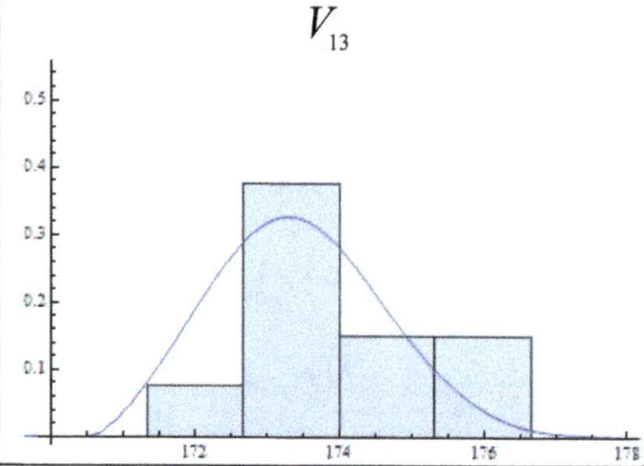

Now we check the consistency of the model with real data in the case of dependence $\rho_{i,i+1}$ and T_i. We restrict ourselves by the linear dependence of these values at stations 5 and 9, identified in Section 5 (see Table 1). The results of the verification are given in Table 4. Travel time (row 2) depends on the departure time (row 1) in accordance with (9). The histograms in row 1 are approximated by the following density functions:

$$f_5(t) = I(t \geq 47)0.23e^{-0.23(t-47)}, \quad f_9(t) = I(t \geq 104.5)0.21e^{-0.21(t-104.5)}.$$

Table 4: Histograms and approximating densities when $\rho_{i,i+1}$ and T_i are dependent.

Station 5	Station 9
T_5 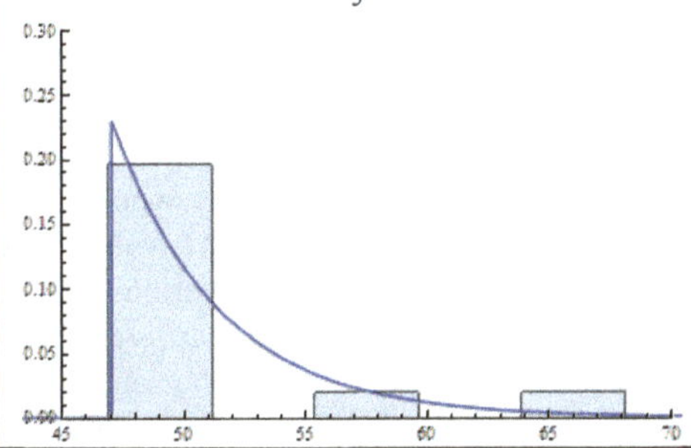	T_9 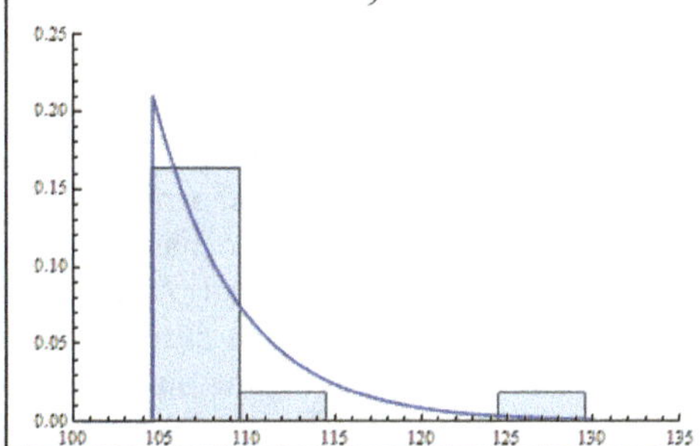
$\rho_{5,6}$ 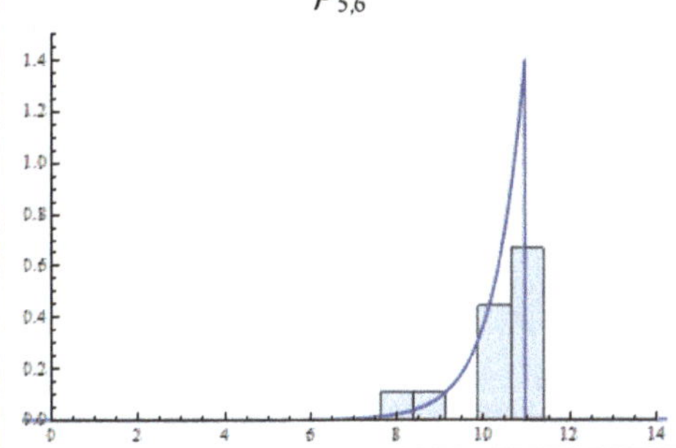	$\rho_{9,10}$ 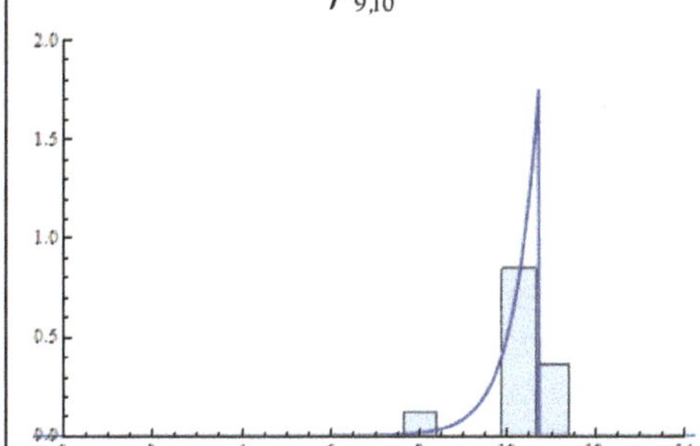
V_6 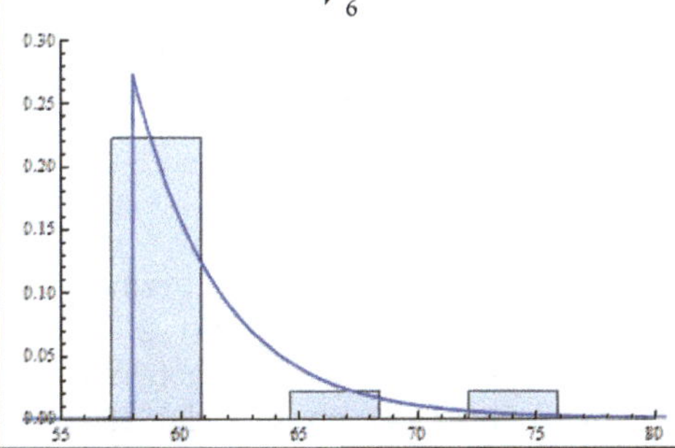	V_{10} 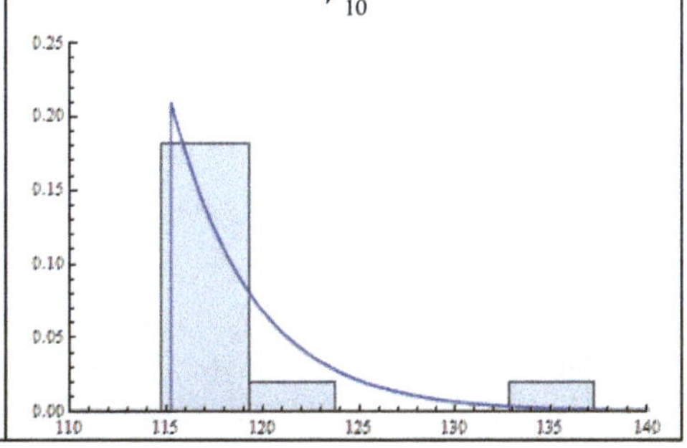

Density functions in row 2 of Table 4 are obtained from $f_5(t)$ and $f_9(t)$ as follows:

$$l_{5,6}(t) = I\left(\frac{t-b_1}{a_1} \geq 47\right)\frac{0.23}{|a_1|}e^{-0.23\left(\frac{t-b_1}{a_1}-47\right)}, \; l_{9,10}(t) = I\left(\frac{t-b_2}{a_2} \geq 104.5\right)\frac{0.21}{|a_2|}e^{-0.21\left(\frac{t-b_2}{a_2}-104.5\right)}.$$

The parameters a_1, b_1, a_2, b_2 are determined from the real data according to (10):

$$a_1 = -0.1629,\; b_1 = 18.5868,\; a_2 = -0.1189,\; b_2 = 23.1022.$$

The densities $g_6(t)$ and $g_{10}(t)$ are determined by using eqn (11), i.e.

$$g_6(t) = \frac{dF_5(y)}{dt} = \frac{1}{1+a}f_5(y),\quad g_{10}(t) = \frac{dF_9(y)}{dt} = \frac{1}{1+a}f_9(y),\quad y = \frac{t-b}{1+a}.$$

It can be observed from row 3 of Table 4, the output densities are in good agreement with the arrival time statistics.

7 CONCLUSIONS

The arrival time of a train can be described using two input random variables: departure time and travel time. The form of the arrival time distribution law is determined by whether the input random variables are stochastically dependent or independent. If the values are independent, then the output distribution is a convolution of two input distributions. In the opposite case, it is necessary to describe the relationship between input random variables: the dependence of the travel time on the departure time. And the output distribution will be like a distribution of the departure time converted by a certain way. In practice, on one section of the rail line, consisting of several stations, both dependent and independent input random variables are observed. It is shown in the work that in the case of input data independence, the density function obtained by the eqn containing convolution integral gives a good approximation to real data of arrival times.

In the present paper, only a linear dependence of the travel time on the arrival time is described. We specially selected examples when the modulus of the sample correlation coefficient is close to unity. In these examples, it is shown that the output density function obtained as a transformation of the departure time density function also gives a good approximation to real data of the arrival times.

It should be noted that cases when the modulus of the sample correlation coefficient takes intermediate values between 0 and 1 will be considered in a further study of the problem.

ACKNOWLEDGEMENT

This work is partially supported by the RF Ministry of Transport according to the research work plan of scientific and educational organizations for 2020, project "Improving the operations management of the freight train traffic".

REFERENCES

[1] Carey, M. & Kwiecinski, A., Stochastic approximation to the effects of headways on knock-on delays of trains. *Transportation Research Part B*, **28**(4), pp. 251–267, 1994.
[2] Meester, L.E. & Muns, S., Stochastic delay propagation in railway networks and phase-type distributions. *Transportation Research Part B*, **41**, pp. 218–230, 2007.
[3] Muhlhans, E., Berechnung der Verspatungsentwicklung bei Zugfahrten/ Eisenbahntechn. *Rundschau ETR*, **39**(7/8), pp. 465–468, 1990.

[4] Shapkin, I.N., Yusipov, R.A. & Kozhanov, E.M., Simulation of train functioning on the basis of multivariate regulation of technological operations. *Bulletin VNIIZhT*, **4**, pp. 30–36, 2006. (In Russian.)
[5] Karetnikov, A.D., *Schedule of Train Traffic*. Transport: Moscow, p. 301, 1979. (In Russian.)
[6] Buker, T. & Seybold, B., Stochastic modelling of delay propagation in large networks. *Journal of Rail Transport Planning and Management*, **2**(12), pp. 34–50, 2012.
[7] Boucherie, R.J. & Huisman, T., Running times on railway sections with heterogeneous train traffic. *Transportation Research Part B: Methodological*, **35**(3), pp. 271–292, 2001.
[8] Berger, A., Gebhardt, A., Muller-Hannemann, M. & Ostrowski, M., Stochastic delay prediction in large train networks. *Proceedings of the 11th Workshop on Algorithmic Approaches for Transportation Modelling, Optimization, and Systems (ATMOS'11)*, Saarbrucken, Germany, pp. 100–111, 2011.
[9] Salido, M.A., Robustness in railway transportation scheduling. *Proceedings of the 7th World Congress on Intelligent Control and Automation (WCICA 2008)*, Chongqing, China, pp. 2833–2837, 2008.
[10] Al-Ibrahim, A., Dynamic delay management at railways. A semi-Markovian decision approach, PhD thesis, Universiteit van Amsterdam, 2010.
[11] Goverde, R.M.P., Punctuality of railway operations and time-table stability analysis, PhD thesis, Technical University of Delft, 2005.
[12] Fichtenholtz, G.M., *The Course of Differential and Integral Calculus*. Vol. 3. Moscow, 2001.

OPTIMISING ACCELERATION AND BRAKING PERFORMANCE PROFILES OF RAILWAY VEHICLES FOR THE ROBUST OPERATION OF ULTRA-HIGH FREQUENCY TIMETABLES

RYO TAKAGI & MASAO WATANABE
Department of Electrical and Electronic Engineering, Kogakuin University, Japan

ABSTRACT

Ultra-high frequency train timetables, which will double the capacity of urban railway lines, are expected to eliminate the chronic congestion and minimise passengers' disbenefit. However, it is well known that the higher the train frequency, the lower the so-called "robustness" or "resilience" of the timetable. Also, it is generally accepted that the fundamental, and in many cases the only, method to improve robustness and/or resilience is to add larger buffer times to it. This will cause disbenefits for the operators and the passengers alike because this will make trains slower. Instead, we examined the method to replace the vehicles with the ones with considerably higher acceleration and braking performance. According to our findings, this method is effective in allowing larger buffer times without slowing down trains for most cases but, unfortunately, not always. Also, problems in the trackside power supply system, such as overloaded substations, excessive voltage drop or too many losses, are almost inevitable because the power consumed or regenerated by trains become too high. Through these investigations, we concluded that the optimisation of the acceleration and braking performance of trains was necessary. However, few literatures were found on this problem. Many previous works on the optimisation of running profiles of trains could be found. In these optimisation attempts, however, the performance of trains was treated as the "given" condition for them. In this paper, we discuss the impact of the acceleration/braking performance of the railway vehicles on the whole system performance, based on the results of both the literature survey and some calculations. We then discuss how train acceleration/braking performance optimisation can be performed.

Keywords: railway timetabling, robustness, resilience, buffer time, accelerating and braking performance of railway vehicles.

1 INTRODUCTION

Ultra-high frequency train timetables, such that will double the capacity of urban railway lines, are expected to eliminate the chronic congestion of urban railways and minimise passengers' disbenefit [1]. However, it is well known that the higher the frequency, the lower the so-called "robustness" or "resilience" of the timetable as defined by Goverde and Hansen [2]. This is mainly due to the fact that a small delay of a train will affect movements of larger number of trains as train frequency goes up. It is also generally accepted that the fundamental, and in many cases the only, method to improve robustness and/or resilience of a train timetable is to add larger buffer times to it. Increasing buffer times generally results in trains running slower, which of course causes disbenefit both for the operators and for the passengers.

Buffer times can be defined as the difference between the scheduled and the fastest possible run times of trains between the same pair of stations. Increasing buffer times can be realised by either extending the scheduled run times or cutting the fastest possible run times. Although extending the scheduled run times is clearly easier, it will result in slower timetables. If we can consider replacing the vehicles with those having considerably higher accelerating and braking performance, it may result in shorter fastest run times and hence larger buffer times, even when the scheduled run times are unchanged.

WIT Transactions on The Built Environment, Vol 199, © 2020 WIT Press
www.witpress.com, ISSN 1743-3509 (on-line)
doi:10.2495/CR200191

However, this replacement of vehicles may not yield good enough reduction in fastest run times. Also, this method may result in high power demands by the accelerating trains, causing overload at the traction substations. Therefore, assessment of the run time reductions and the impact on traction system loadings, energy consumption, etc., is required when considering such a replacement.

In this paper, we first explain we explain what the ultra-high frequency train operation is and how this can be achieved. Then we discuss the impact of ultra-high frequency operation on the timetable "resilience", and the need for increasing buffer times. Finally, we present a set of preliminary calculations to verify: (a) how buffer times can be increased by higher accelerating and braking performance of the railway vehicles, and (b) the impact on the traction power supply systems and energy consumption.

2 WHAT IS ULTRA-HIGH FREQUENCY TRAIN OPERATION AND HOW IT CAN BE REALISED

For urban railways using long trainsets (~ 200 m), the practical maximum train frequency is around 30 tphpd (trains per hour per direction). On many commuting railway lines in large cities like Tokyo, unacceptable level of congestion is observed even with this level of frequent train operation, and the increase in transport capacity is desperately needed.

In this paper, the word "ultra-high frequency train operation" refers to the operation that will realise the transport capacity that is nearly twice as much as the currently realised level. This will mean the frequency level close to 60 tphpd when trains of the same length on the same railway line are assumed, or even higher frequency if the train lengths are to be shortened.

Such frequency will be realised by the introduction of very complicated scheduling schemes. For example, assume there is a railway line where the simple "parallel and local only" scheduling scheme, i.e. the scheduling scheme in which all trains are local trains that will call at every station en route, is applied when planning its train timetable. If there is, on this line, an intermediate station with only one platform per direction where all trains should dwell for more than 60 seconds due to a very high demand, it would be very difficult to raise the train frequency higher than 30 tphpd.

In fact, such stations can be found on many existing commuting lines in Tokyo. Those stations form the bottleneck on these lines, where the minimum headway between trains is limited to around 130 seconds, which yields 27 tphpd.

If we can assume that half of the trains can pass this station, then it would be possible to think of doubling the frequency. Fig. 1 [3] shows that the headway between two trains stopping at a station, trains #1 and #3, can be maintained at 130 s while another train, train #2, can be inserted in between these two trains so that the capacity is doubled to 54 tphpd. The calculation shown in Fig. 1 is based on the following set of assumptions: (1) pure moving block signalling is introduced, (2) vehicles are designed capable of constant acceleration and deceleration of 3.6 km/(h·s) or 1 m/s^2 regardless of their speed, and (3) the passing train is allowed to make an out-of-station halt before entering the platform.

The resulting timetable for the entire line would be much more complicated than the current one based on the "parallel and local only" scheduling scheme, especially if the possibility is to be maintained for passengers to travel between any pair of stations without having to use trains running in the opposite direction. Therefore, the implementation of a sophisticated system of guidance for passengers, such as the "IPASS" concept as proposed by Takagi [4], would be crucial in the realisation of this kind of timetables. Even with these issues, the doubled transport capacity will be a clear benefit to all users of this railway.

It should also be noted that, even in the central area of large cities such as Tokyo, stations with such high demands are very limited. Investment in infrastructure, such as increasing platforms at these stations, in combination with the introduction of the complicated scheduling scheme and the passenger assistance measures such as IPASS, is expected to yield even better timetables.

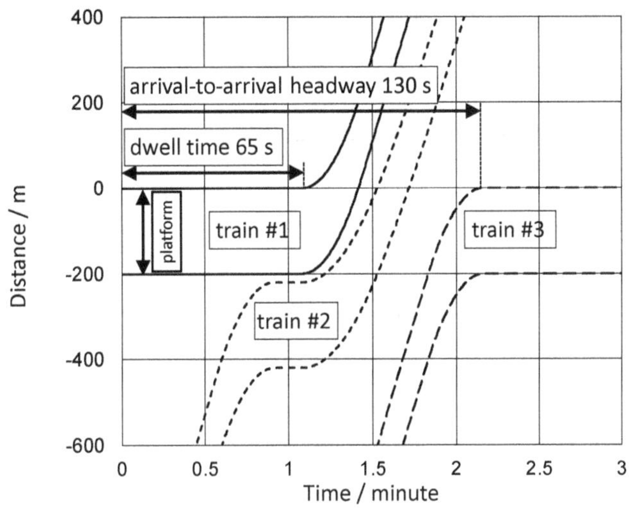

Figure 1: Train movement of an example ultra-high frequency operation at a large station with insufficient platforms.

3 THE TIMETABLE RESILIENCE OF ULTRA-HIGH FREQUENCY TRAIN OPERATION AND THE NEED TO INCREASE BUFFER TIMES

An ultra-high frequency train timetable is, however, believed by the train operators to be difficult to operate, especially because it is susceptible to disruptions. In such a timetable, the initial small disruptions that frequently occur, such as those caused by passengers or their belongings being caught in opening or closing doors, will easily grow into large disruptions as they create secondary delays to more trains than in the conventional timetables. These initial small disruptions are very difficult to prevent, and will be more frequent as congestion gets worse.

Therefore, for an ultra-high frequency timetable to be operable successfully, it must have enough buffer times. The buffer time for a scheduled run of a train from any one station to another is, as has been explained in Section 1, defined as the difference between the scheduled run time and the fastest possible run time of that train. The fastest possible run time is dependent on the track profile and the accelerating and braking performance of the vehicles to be used with the train, and is not easily shortened. It is much easier to extend the scheduled run times to realise larger buffer times; however, this will lead to slower timetables.

According to the existing timetables on the commuting lines in Tokyo with very heavy peak-hour traffic, trains are running significantly slower in those peak hours during which the train frequency is higher than in other hours of the day. Of course, the "high" frequency during peak hours on these lines is not "ultra-high" as we expect in this paper. We can therefore safely deduce from this observation that adding buffer times even further is crucially needed when realising the ultra-high frequency operation which has enough "robustness" or "resilience".

At the same time, it is undesirable to slow trains even further to increase buffer times from the point of view of the passengers' and operators' disbenefits. Although not an easy option, replacing the railway vehicles currently in use with higher accelerating and braking performance to shorten the fastest possible run times for realising the buffer time increase without slowing trains should be considered.

4 DISCUSSION ON HOW TO ENHANCE ACCELERATING AND BRAKING PERFORMANCE OF RAILWAY VEHICLES

Generally, the accelerating performance of the railway vehicles can be considered as the combination of three different modes, namely: (a) "constant-torque" at low speed region, where the torque of the traction motors, and hence the tractive effort of the vehicles, is constant regardless of their speed; (b) "constant-power" at middle speed region, where the output power of the vehicles at wheel rim is constant regardless of their speed; and (c) "free-run" at high speed region, where the output power of the vehicles is in inverse proportion to their speed. Out of these three, the speed region for (c) is sometimes considered as above the practical maximum speed of the vehicles. The tractive effort at region (a) is constrained by the limit of adhesion, while the power at region (b) is limited by the rated output of the traction equipment onboard the trains.

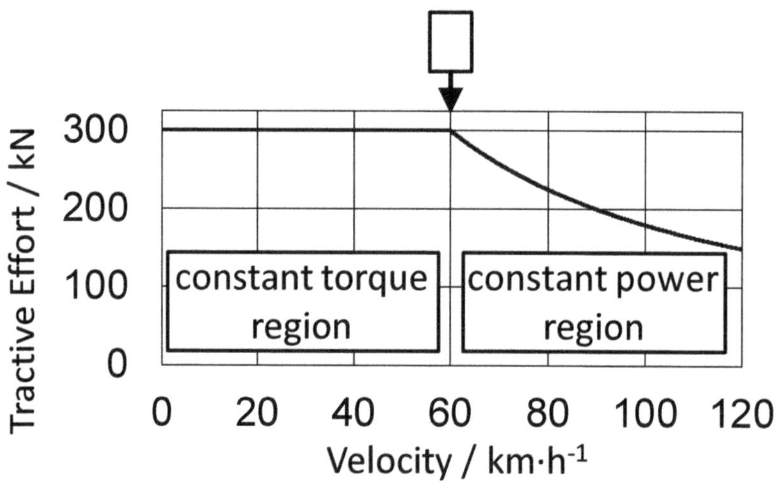

Figure 2: Train movement of an example ultra-high frequency operation at a large station with insufficient platforms.

Fig. 2 shows an example tractive effort-speed profile of a vehicle with only two regions, (a) and (b), under the assumption that the tractive effort at (a) and the output power at (b) are 300 kN and 5000 kW, respectively. The parameter v_1 in the figure shows the boundary speed between the two regions, which becomes 60 km/h under the above assumptions.

Let us discuss how the enhancement of accelerating performance of the vehicles can be done using the performance profile of Fig. 2. The possible methods are:

1. Increasing the tractive effort at region a;
2. Increasing the output power at region b; and
3. Doing both (1) and (2).

Method (1) requires better utilisation of adhesion. One possible way to achieve this is by increasing the ratio of the number of driving axles to the number of all axles in a train set. Because the output power at region b, and therefore the vehicle performance at the higher speed region, remain unchanged, v_1 in Fig. 2 becomes lower as the tractive effort increases. On the other hand, method (2) requires increasing the power output of the traction equipment. Because the tractive effort at region a, and therefore the vehicle performance at the lower speed region, remain unchanged, v1 in Fig. 2 becomes higher as the output power increases. Method (3) is the combination of two methods, and therefore enhancement of performance at all speed regions is realised.

Here, it must be noted that, to achieve meaningful increase in buffer times, reduction of the fastest run times between the adjacent stations by a couple of seconds at least should be realised. Also note that the discussion in this paper is mainly for urban railways, where the distance between any adjacent stations is short (e.g. 1 km). Therefore, the necessary enhancement of the performance will be a significant one. If the output power of the vehicles are to be significantly increased as methods (2) and/or (3) are to be implemented, very high overload may occur at the traction substations as the result. It is unlikely that method (1) causes the similar kind of problem, but the fact that the performance at higher speed region will not be enhanced may mean insufficient reduction of the fastest run times.

5 PRELIMINARY EVALUATION OF THE RUN TIME REDUCTION

We have performed a number of preliminary calculations to see how much the fastest inter-station run times of a train can be reduced by method (1) in the previous section.

Assumed is a 200-m-long, 313-t train set with the interior design which is typically seen in Japan, fully loaded with passengers, and with the traction equipment having the output power at wheel rim of 5 500 kW. Assumed acceleration at region a are: 2.5 km/(h·s) (= 0.69 m/s^2), 3.0 km/(h·s) (= 0.83 m/s^2), 3.5 km/(h·s) (= 0.97 m/s^2), 4.0 km/(h·s) (= 1.11 m/s^2), 4.5 km/(h·s) (= 1.25 m/s^2), 5.0 km/(h·s) (= 1.39 m/s^2) and 5.5 km/(h·s) (= 1.53 m/s^2). For deceleration, the same value as the acceleration in region a is assumed for the entire speed region. The track profile is assumed as straight and level all the way. The train is to start from halt at the origin station, accelerate, decelerate and finally come to a complete halt at the destination station. In the calculation, infinite jerk is allowed. When the train speed reaches 100 km/h during acceleration, the train is assumed to stop the acceleration and keep this speed of 100 km/h until it starts braking in order to stop at the appropriate position in the destination station.

Table 1 shows some of the calculation results. The difference between the run time for the acceleration of 2.5 km/(h·s) and the run time for one of the other acceleration values will be the reduction in the run times when raising the acceleration at region a from 2.5 to the selected value. As an example, the difference between the 2.5 km/(h·s) case and the 5.5 km/(h·s) case for each assumed distance has been calculated and added at the bottom row of Table 1.

It is generally expected that, as the inter-station distance becomes longer, this difference becomes larger. However, looking at the calculation results, the differences seem more or less independent of the inter-station distance; This is due to: (1) the acceleration performance at higher speed is unchanged, and (2) the maximum speed reaches the assumed maximum of 100 km/h in many calculated cases. Especially, because of (2), this difference when even longer inter-station distance is assumed is almost identical to the case with the assumed inter-station distance of 1400 m.

Table 1: Calculated fastest inter-station run times over the assumed distances.

Acceleration and deceleration	Inter-station distance				
	600 m	800 m	1000 m	1200 m	1400 m
2.5 km/(h·s)	59.02 s	68.47 s	76.97 s	84.79 s	92.10 s
3.0 km/(h·s)	54.46 s	63.49 s	71.64 s	79.16 s	86.37 s
3.5 km/(h·s)	51.24 s	59.99 s	67.91 s	75.25 s	82.45 s
4.0 km/(h·s)	48.88 s	57.42 s	65.18 s	72.43 s	79.63 s
4.5 km/(h·s)	47.07 s	55.47 s	63.10 s	70.31 s	77.51 s
5.0 km/(h·s)	45.66 s	53.93 s	61.47 s	68.67 s	75.86 s
5.5 km/(h·s)	44.51 s	52.70 s	60.15 s	67.35 s	74.55 s
Difference (5.5 km/(h·s) case − 2.5 km/(h·s) case)	14.51 s	15.77 s	16.82 s	17.44 s	17.55 s

6 DISCUSSIONS

The acceleration of 2.5 km/(h·s) or 3.0 km/(h·s) are not particularly high for the urban railway vehicles, but are the commonly seen values in many commuting lines in Tokyo. 5.5 km/(h·s) is the value typical for tram and light rail vehicles rather than the vehicles for heavy metro lines, and therefore there is no reason why this high acceleration cannot be adopted for heavy metro trains. Therefore, although at the preliminary stage, it seems we can conclude that raising the acceleration at region a, i.e. at low speed region, without increasing the maximum output power, is effective in reducing run times. The run time reduction, of course, can be used to increase buffer times by keeping the scheduled run times to the level achievable by the current vehicles with lower performance.

However, the following should be considered.

6.1 Proper consideration of track profiles are necessary

In the calculations in Section 5, track profiles are assumed as straight and level, and no speed restriction has been taken into account. This is, of course, very unrealistic; we have to take up a more realistic model to evaluate the possibility.

6.2 Constant deceleration is assumed for the entire speed region

Also, constant deceleration has been assumed for the entire speed region. This means that, for the cases with higher assumed acceleration, such as 5 km/(h·s), the maximum braking power will also become very high (for the case of 5 km/(h·s), it will become twice as much as the case for 2.5 km/(h·s)). Designing a traction system that can absorb this level of high regenerative power will be a very difficult work.

6.3 The nature of ultra-high frequency timetables: Trains are to call at fewer stations

In Section 2, we discussed that the ultra-high frequency timetables are complicated timetables in terms of the patterns of station calls made by trains. However, from another perspective, they can be called as the timetables in which trains make much fewer calls at stations. This means that, for at least some of the trains in these timetables, the "start-to-stop" run times between stations that are close by may not be the issue. We may have to consider cases in which trains which pass many stations are requested to control their inter-station run times, for which acceleration performance at the middle to high speed region may become more important.

As has been discussed in Section 4, this requires the adoption of methods (2) or (3). However, the required output power of the train becomes large, which have a very significant impact on the trackside traction power supply system. Already, we have performed some simulations on the idea to increase power of trains so that the robustness or resilience is improved [5]. However, it was quickly revealed that the high performance led to the overload of substations. A number of countermeasures, including the method to allow the use of high power only when the train is delayed and needs it for recovery, have been tried but without satisfactory results in suppressing substation loads to the safe level. We believe that the application of techniques proposed by Takagi and Sone [6] will help when it is implemented together with the introduction of very-high-power vehicles.

6.4 Optimisation of the vehicle performance should be attempted

From the discussions in 6.3, obviously there is a need for optimisation of the acceleration and braking performance of the railway vehicles.

There have been a number of attempts to optimise the operational pattern of a train for an inter-station run. One of the most impressive examples in which this kind of optimisation has been performed would be the one presented by Ko and Miyatake [7]. However, in this kind of research, the train performance is the assumption and not the parameters to be optimised.

We could find very few works have been found in which optimisation of the railway vehicle performance is performed in any way. In one example [8], simulations are repeated for railway vehicles with different specifications for the traction equipment on board. Among the conclusions therein is the finding that, from the point of view of energy saving, different optimal designs of the traction equipment exist for different operating conditions, especially the inter-station start-to-stop run times.

For the purpose of the ultra-high frequency train operation, we should perform optimisations taking into account not only the energy consumption but also the load limitations of the trackside power supply network.

6.5 Proper consideration of timetable resilience is required

Throughout this paper, we have not distinguished timetable "robustness" and timetable "resilience", although Goverde and Hansen [2] made a clear distinction between these two.

According to Goverde and Hansen [2], timetable robustness is the ability of a timetable to withstand design errors, parameter variations, and changing operational conditions, while timetable resilience is the flexibility of a timetable to prevent or reduce secondary delays using dispatching (re-timing, re-ordering, re-routing). Generally, adding buffer times are directly connected to timetable robustness. We understand that the improvement in timetable robustness will lead to the improvement in timetable resilience.

However, it must also be noted that the complicated nature of the ultra-high frequency timetable as outlined in Section 2 may make train control (dispatching) very difficult. Although we can generally state that more buffer times will open up wider possibility for good train control, the relationship with the buffer times and the timetable resilience should be seriously sought, together with the effective and easy-to-perform method of train control operations.

7 CONCLUSIONS

We have investigated the possibility of increasing buffer times of the ultra-high frequency train timetables by means of significantly enhancing the accelerating and braking performance of railway vehicles to be used with them. The preliminary calculation results give hope that significantly higher performance leads to effective reduction in the fastest run times, leading to more buffer times and hence more resilient operation.

Ultra-high frequency train operation, once realised, will give clear benefits to passengers, rail operators and the wider society. Although we are still at the preliminary stage and there are still so many challenges to be tackled as listed in Section 6, through the future works we believe we can move this idea closer to reality from the train control points of view.

REFERENCES

[1] Shimizu, T., Watanabe, M. & Takagi, R., Scheduling techniques to achieve ultra-high frequency operation over existing infrastructure for realization of ultra-convenient rail transport. *Proceedings of International Symposium on Speed-up and Service Technology for Railway and Maglev Systems (STECH 2015)*, Chiba, Japan, 3D24, 2015.

[2] Goverde, R.M.P. & Hansen, I.A., Performance indicators for railway timetables. *IEEE International Conference on Intelligent Rail Transportation (IEEE ICIRT 2013)*, Beijing, China, 2013.

[3] Takagi, R., *You Can Make a Railway Timetable*, Ohmsha Ltd.: Tokyo, pp. 169–175, 2019. (In Japanese.)

[4] Takagi, R., Reservation-compulsory commuting railways: Innovation that will be made possible by UCRT/IPASS. *WIT Transactions on the Built Environment*, vol. 181, WIT Press: Southampton and Boston, pp. 225–233, 2018.

[5] Watanabe, M., Shimizu, T. & Takagi, R., Measures against disruptions on ultra-high frequency operated railways: operation of high performance vehicles and substation peak load suppression. *IEEJ Joint Technical Workshop on Physical Sensors/ Transportation & Electric Railways*, TER-16-018/PHS-16-012, Nagoya, Japan, 2016. (In Japanese.)

[6] Takagi, R. & Sone, S., More intelligent DC railway electrical power systems with traction power control. *Electrical Engineering in Japan*, **114**(7), pp. 91–101, 1994.

[7] Ko, H. & Miyatake, M., A numerical algorithm for run-curve optimization of trains considering a DC feeding circuit. *WIT Transactions on the Built Environment*, vol. 88, WIT Press: Southampton and Boston, pp. 787–796, 2006.

[8] The energy economy of train operations and the specifications of traction equipment for the railway vehicles. *IEEJ Technical Report*, **112**(2), 1981. (In Japanese.)

INTEGRATED VIEW OF RAILROAD AND PORT TERMINALS: A STUDY ON THE PORT OF SANTOS, BRAZIL

ULISSES FERNANDES ALVES
Production Planning and Control, Rumo Logística, Brazil

ABSTRACT

The Development and Zoning Plan (PDZ) for the port complex of Santos estimates that the share of the railway system in cargo handling will reach about 40% by 2040. Thus, it is necessary to implement projects that improve productivity and increase the railway system's capacity; therefore, it may function with such higher demand. To decide and evaluate investments, we developed a simulation model on the Arena® software, considering the randomness and dynamism of the interaction between the railway and the terminals on the right bank of Santos port. To validate our model, we verified the adhesion results for the simulation for the realized values. From this analysis, we obtained 95% for the number of unloaded cars and 90% for the cycle time. Then, we evaluated the cycle time gain within two investment scenarios. The first scenario comprises a project that aims to increase the productivity of the discharge of grain hoppers from Terminal G; the second encompasses the construction of viaducts and walkways to remove the speed restrictions of the trains for circulation. In the first scenario, our simulation showed a ~4% decrease in the wagon cycle on the right bank, which corresponds to a saving of around 1.2 million dollars. For the second scenario, the simulation showed a 7% decrease in the right bank cycle, thus saving 2.1 million dollars. Therefore, the model presented in this study allows the evaluation of several scenarios and also provides a consistent technical basis for decision making from an integrated view of the railway and port terminals.
Keywords: rail yard simulation, discrete events, railway, productivity, Port of Santos.

1 INTRODUCTION

The Port of Santos, located in Brazil, is the largest port complex in Latin America and is connected by a wide network of highways, railroads and waterways with six states that concentrate more than 60% of the Brazilian GDP [1]. There has been a 39% growth in the millions of tons handled by this port over the last 10 years, as shown in Fig. 1 [2].

Currently, 33% of the cargo handled at the Port of Santos is transported by railways; this share continues to increase and should reach 40% by 2040 [3]. For that purpose, several investments must be made in the railway system. However, as the investments are high, it is crucial to carefully study the feasibility of the projects or search for cheaper alternatives to improve the efficiency of the process.

Besides, one of the main concerns that investors have is whether the railway system that currently serves the port can meet the expected demand [4]. Such observation is relevant, as the arrival of cargo at the terminal is crucial to gauge its productivity. If the cargo does not arrive at the port, idleness will increase and, consequently, investments in the terminal will decrease.

Therefore, it is necessary to build accurate models of the railway system to provide more in-depth information on this port's capacity, to examine the gains from the implementation of expansion projects, and to verify the overall effects of individual changes. Given the complexity of the rail processes involved, more appropriate mathematical and computational approaches are required. There are several possibilities and restrictions to the occupation of the railroad, besides the need for an adequate representation of the situations involving human

WIT Transactions on The Built Environment, Vol 199, © 2020 WIT Press
www.witpress.com, ISSN 1743-3509 (on-line)
doi:10.2495/CR200201

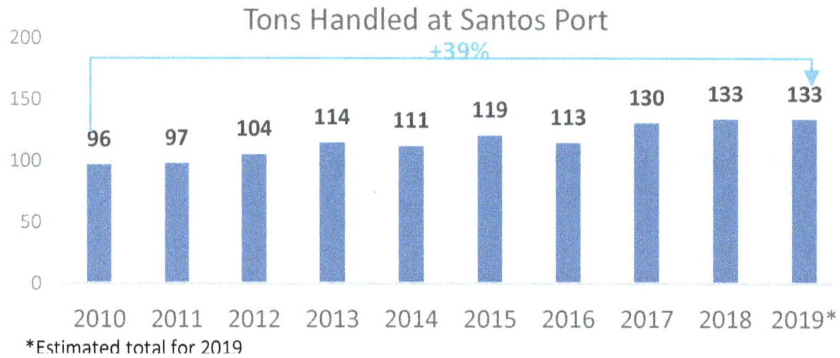

Figure 1: Movement in Santos Port in the period from 2010 to 2019.

decision to move trains. All these factors contribute to the complexity of the models that represent the railway transportation system.

Rumo Logística, a Brazilian logistics company [5], and the consulting firm Paragon [6], made a partnership to develop a model that allows assessing the benefits of projects for the railway systems. This development considered the integration between the railway system and the port terminals on the right bank of the Port of Santos. Importantly, all the needs, restrictions, and challenges mentioned above were also considered. The computational model was implemented using Arena® software developed by the company Rockwell Automation and which allows the simulation of discrete events with resources for statistical analysis, process modelling, animation and results analysis.

2 LITERATURE REVIEW

In the literature, railway systems are usually described as employing queuing theories and computational simulations [7]. Petersen [8], [9] used queuing theory to develop a model for yard operations; his model allowed for calculating the probability distribution of the wagon connection times for the various traffic levels. One of the disadvantages of using queuing theory to develop complex models is the need to formulate simplifying hypotheses to obtain an analytical solution – which, in some cases, can distance the results from reality [10].

For the purposes of this study, the computational simulation approach suited our goals more adequately. In this regard, Ramos [11] developed a computational model to analyze the railroad capacity of load processing at the Port of Santos. The author used the Arena® software and evaluated five alternative scenarios that considered both changes in railway operation and the increased demand for trains. According to the author, the limitations of this model are (a) not considering the wagon distribution process for each terminal, (b) disregarding the capacity for wagons at the yards, and (c) not accounting the processing capacity of each terminal separately. We solved these limitations by developing our model which considers the capacity restriction of all resources involved in the railroad operation. This includes the controlled use of each track for the displacement of trains between the yards, the number of locomotives available for maneuver, the management of the movements involved for the distribution of the wagons and the capacity to receive wagons both in yards and at the terminals.

Brito [12] analyzed the implementation proposal of a container regulator terminal in the Cubatão region, located near the Port of Santos, through a computer simulation technique on the Arena® software. The author considered the road, water, and rail transport systems to analyze the productivity on the terminal's operation and demand.

Rosa [13] developed a computer simulation model for a railway terminal; the terminal would be used for unloading wagons with agricultural products and loading them with fertilizer. The author generated three scenarios to assess the system behavior and concluded that the simulation model was efficient for carrying out strategic studies while being able to foresee possible difficulties and operational bottlenecks. As potential improvements to the model, the author suggested introducing unexpected interruptions in the operation of any railway line then analyzing their impact in the generation of queues.

Marinov and Viegas [7] proposed a methodology for the simulation, analysis, and evaluation of flat-shunted yard operations. The authors' goal was to divide the yard into areas, so it would be possible to describe and analyze the behavior of each area separately. Thus, it would be equivalent to a queuing network model. A queueing network model consists of interconnected queue systems that influence each other; since it captures the overall impact of railway operations, it is possible to study the processing capacity of the yard.

3 CASE STUDY

3.1 Processes and railway system at the Port of Santos

The company Portofer (a subsidiary of Rumo Logística) operates and maintains the railway system that serves the terminals at Port of Santos. The companies Rumo, MRS and VLI are three Brazilian railway operators that use Portofer's services to access around 100 km of railway tracks that extend along the right and left banks of the port. In the right bank under Portofer's control are the rail yards of Santa, Macuco, and Corredor Exportação, which are responsible for a large portion of the maneuvers and the formation and dismemberment of the train compositions. The operational management of the Valongo yard is shared between the companies Portofer and MRS so that there are regions in this yard where each company is responsible for railway control (see division of Valongo's yard in Fig. 2). In the left bank,

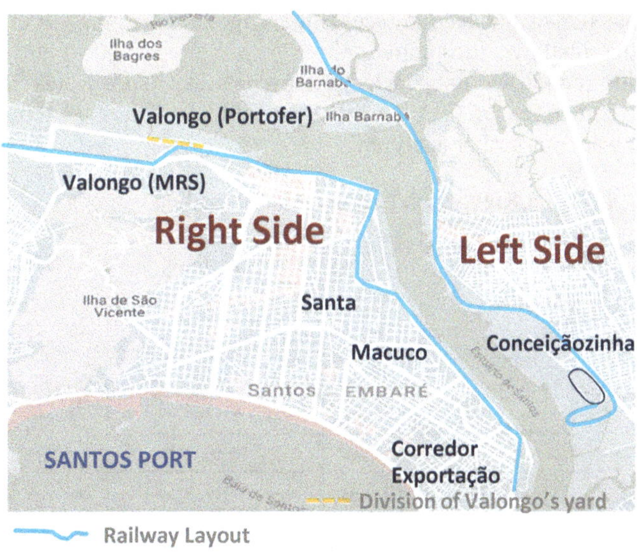

Figure 2: Railway system at the Port of Santos.

there is only the Conceiçãozinha yard that serves to railway operations. Fig. 2 shows the railway system near the port and the location of the yards on the right and left banks.

In this study, we will only consider the operations on the right bank of the port. The railway operators' compositions start from the MRS rail yard located in Valongo. From this point on, in some cases, train locomotives can be replaced with Portofer's locomotives to transport the wagons to the right parking site. In general, the train composition goes through the processes shown in Fig. 3.

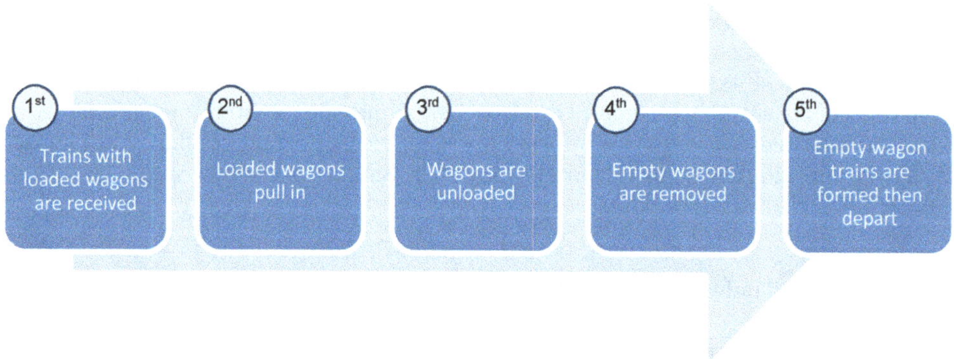

Figure 3: Main rail processes on the right bank.

In the first process, the train is licensed by Portofer's Operational Control Center (CCO) to travel from the Valongo yard to the yard of its destination terminal. In general, once they arrive at their destination yard, the wagons are separated into lots. The lots are moved to unload the wagons according to the availability of the terminal. If the terminal is busy or unavailable, the lots will wait on the parking lines in the yard. The second process starts as soon as the terminal signals the need to receive the wagons. Then, the lots are shunted and transported to the terminal by a locomotive.

In the simulation model, we only included terminals that unload bulk freight wagons (sugar, corn, soy, corn bran, or soy), as listed in Table 1. The other terminals are less representative and were not further analyzed.

Table 1: Right-bank terminals for unloading wagons.

Yard	Terminals
Santa	A
	B
	C
	D
Corredor Exportação	E
	F
	G

After the wagons are shunted, the terminals unload them; the empty wagons are then moved to specific tracks (third process). When there are enough empty wagons, the CCO orders the shunting locomotive to remove the ones close to the terminal and take them to the tracks dedicated to forming empty wagon trains (fourth process). Finally, the fifth process starts when sufficient wagons from the same concessionaire unload the same type of product. This procedure consists of forming the train that will transport these empty wagons. Thus, it is necessary to join the wagons, perform brake tests, and await Portofer CCO's license for the train to move to the Valongo yard. There, the train awaits a license from MRS concessionaire's CCO to depart from the right bank and proceed to a destination outside the Port of Santos.

3.2 A simulation model for the right bank rail processes

To build the model, we have considered the steps defined by the simulation methodology presented by Chwif and Medina [14], shown in Fig. 4.

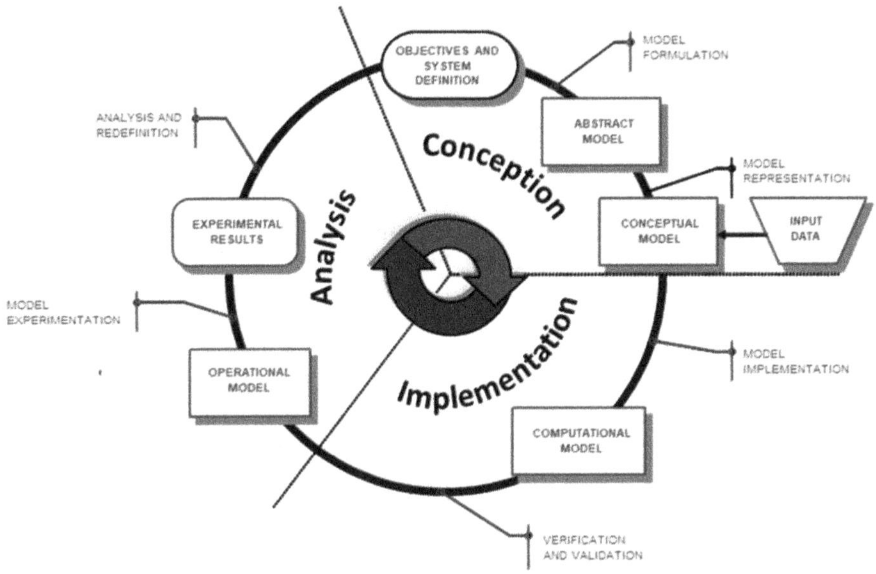

Figure 4: Simulation methodology [14].

The Conception stage consists of defining the objectives of the system and preparing the conceptual model, which consists of documenting tacit knowledge and making it explicit. Implementation activities include the development of the computational model and the correction of possible errors. In the last stage of the cycle, the model becomes operational, as it has passed verification and validation tests and is fit for analyzing the system in question.

During the implementation of the model, the wagons and locomotives were represented as individual entities within the system. The Arena® modeling features allow grouping wagons and locomotives into compositions and ungroup them whenever it is convenient; the individual information of each component is maintained, such as the type of wagon, weight, the concessionaire in charge of the asset, origin, and destination.

As in the real world, the model's rationale for managing circulation in the railway infrastructure is like the CCO's. This role is crucial to avoid train collisions or even simultaneous use of the same track beyond its capacity. For such purpose, we used a programming approach in Arena® that combines the use of resources and variables. This combination provides flexibility in detailing behaviors and situations present in the railway processes.

To move around the rail yard, one must make a reservation of the railways that will be used. The reservation avoids conflicts with other trains and deadlock situations in which there is a dispute between two entities to access the same route.

For this study, we built and ran the model in the Arena® software version 15.1 Professional. We used ten replications to generate the results, with each replication lasting one month of train operation and circulation on the right bank. The warming time was 24 hours (one day), and the total simulation time equivalent to 7,740 hours (310 days). We chose this configuration to ensure that the variables were able to reach the stationary state.

To validate the model, we evaluated the adhesion of the simulation results to the indicators' real values, according to the eqn (1)

$$Adhesion\ [\%] = \frac{Simulation\ value}{Real\ value}100. \tag{1}$$

The adhesion result indicates how close the simulation value is to the realized one. Thus, an adhesion close to 100% means that the simulation result is close to the realized value. The two indicators evaluated in this study are the number of wagons unloaded at the terminals and the cycle time, which comprises the entire period in which the wagon remained on the right bank. The cycle starts when the wagon enters the Valongo yard and ends when it returns to the entry point in a train with empty cars. In the case of the number of unloaded wagons, adhesion is expected to be equal or less than 100% because the arrival times of the trains are defined so that the real number of wagons that arrived at the port in a specific month is reached. That is not the case for the cycle time adhesion, as in this case it can assume values above 100% depending on the dynamic and random interaction of the system.

After validating the model, we assessed the indicators for different processes and productivity scenarios of the terminals. To this purpose, we compared the scenario simulation values with the model simulation results validated for the current processes, as shown in the eqn (2)

$$Ratio[\%] = \left(\frac{Simulation\ value\ in\ evaluated\ scenario}{Simulation\ value\ in\ current\ process\ (validated\ case)} - 1\right)100. \tag{2}$$

4 RESULTS AND DISCUSSION

To validate the simulation model, we compared the real values for July 2019 and the simulation results. Rumo Logística allowed us to publish this paper under the condition of not disclosing operational data. Thus, Table 2 shows only the percentage values of adhesion between the simulation results and the real values.

Table 2: Adhesion of the simulation to the values realized on the right bank in July 2019.

Indicator	Adhesion (%)
Total number of unloaded wagons	95
Cycle time	90

The results in Table 2 show good adhesion to the real number of wagons unloaded on the right bank. Fig. 5 shows the opening of this indicator for each evaluated terminal.

The values of the terminals did not vary significantly. Terminal D had the lowest adhesion and reached 94%, which is satisfactory. Similarly, the cycle time indicator was also opened for the evaluated terminals, as shown in Fig. 6.

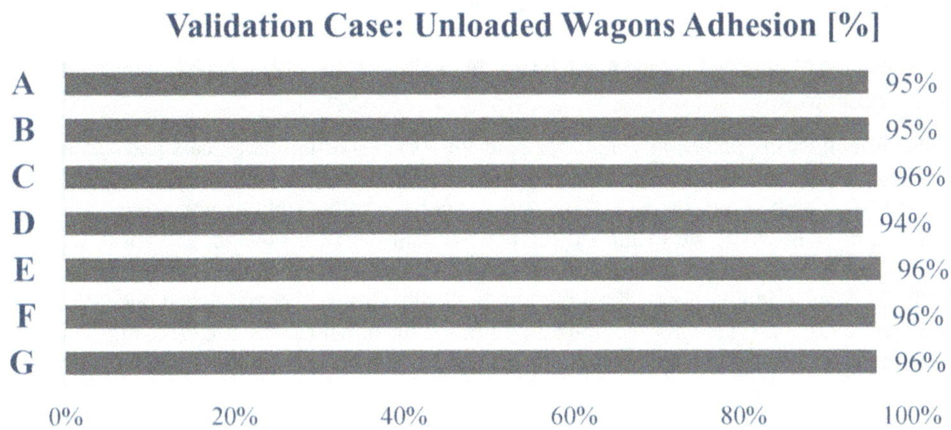

Figure 5: Adhesion percentage of the number of unloaded wagons per terminal.

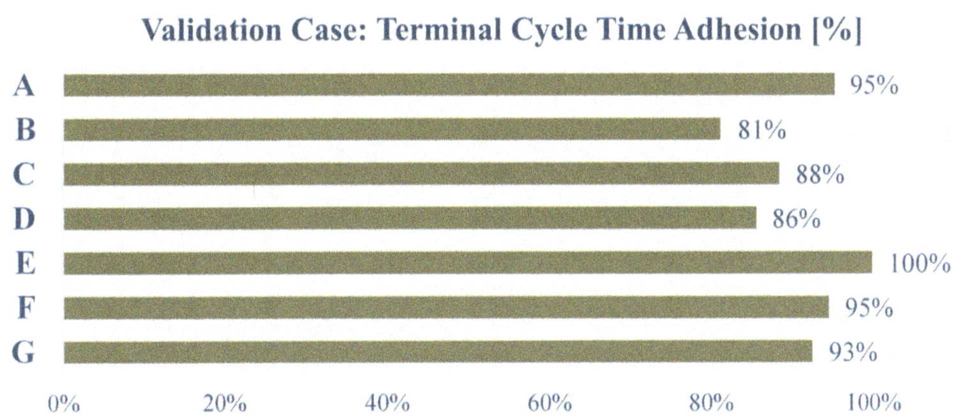

Figure 6: Adhesion percentage of the cycle time per terminal in the simulation of the validation case.

In this case, the adhesion to the cycle time varied more substantially. Low adhesion can be explained by the fact that there are several setbacks during the mentioned processes in Fig. 3, which are either poorly or not recorded at all. Among these setbacks, we highlight the following:

- Terminal maintenance times, when hoppers do not unload, even though wagons are available.

- Using a wagon type that employs a door for unloading, which is on average 45% less productive.
- Interference from rain and other elements. In this case, the terminals do not load ships and may not receive wagons for unloading if they are close to their static capacity.
- Interference from trucks during the movement of the wagons towards the hoppers.
- Failures both on the permanent track and on locomotives that delay the moving of wagons.
- Invoicing issues or even mechanical breakdowns in the wagons, which delays their removal.

Nevertheless, the adhesion values for both the number of unloaded cars and cycle time were satisfactory and enough to validate the model. Therefore, this model can be considered operational, which allows us to perform several analyses of the right bank system.

Thus, in Scenario 1, we evaluated the overall gain in cycle time for the entire railway process due to the investment to reduce the unloading time of the wagons at Terminal G by 36%. For comparison purposes, the departure table of trains in Scenario 1 remained the same as in the validation case; only the unloading time at Terminal G was changed. Fig. 7 shows the percentage ratios between the cycle time results obtained in Scenario 1 and the values of the validated simulation.

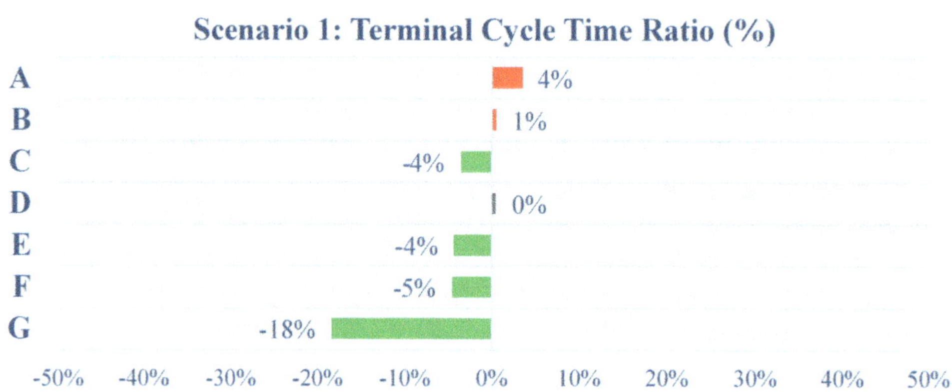

Figure 7: Percentage ratio for cycle time per terminal in the simulation of Scenario 1.

These results show an 18% decrease in the cycle time of Terminal G when compared with the validated case. There was no decrease for Terminal D; however, we observed a 1% increase in the cycle time of Terminal B and 4% for Terminal A. Thus, it is necessary to evaluate the cycle time stages in order to investigate the reason for such an increase. Fig. 8 shows the corresponding results.

There was an increase in Terminal A train entry and exit times, which suggests that the simulation model prioritizes the flow of trains from other terminals. We can explain such an increase by the fact that the number of cars for Terminal A corresponds to only 6% of the total unloaded bulk freight wagons on the right bank. Also, there is a restriction rule on the right bank, according to which there can only be one train for each terminal. Thus, the decrease in cycle time in Scenario 1 was due to more available time to unload the wagons, which made these terminals have priority. However, even with the occasional increase in

Terminal A, the overall cycle time of the right bank reached a 4% decrease. This value reduced around 17 wagons from the fleet needed to meet the grain flow and resulted in savings of around 1.2 million dollars.

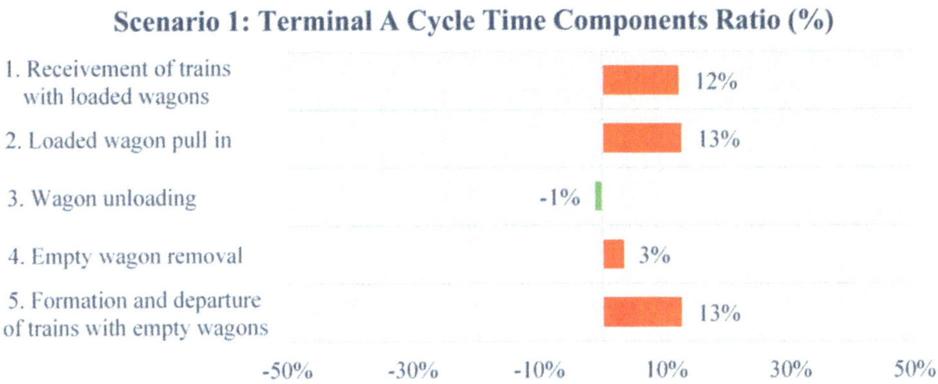

Figure 8: Percentage ratio for the cycle time components of Terminal A in Scenario 1.

Scenario 2 included the evaluation of investment to remove 10 level crossings (PNs) that interfere with train circulation along the right bank. The PNs implies crossing between the railway track and the road for the traffic of people, cars, and trucks. Therefore, for safety reasons, when passing through a PN, the train reduces its speed, meaning an increase of three to 12 minutes in its travel time. Thus, this project consists of establishing viaducts and walkways to eliminate these intersections. Similar to the method applied in Scenario 1, we applied the departure table of trains used in the validation case. Fig. 9 shows the percentage ratios between the cycle time results obtained in Scenario 2 and the validated simulation values.

Investment to remove PNs results in benefit for all evaluated terminals and generates a 7% decrease in the global cycle time of the right bank, allowing for a decrease of 30 cars in the fleet needed to meet the grain flow and yielding monthly savings of at least 2.1 million dollars.

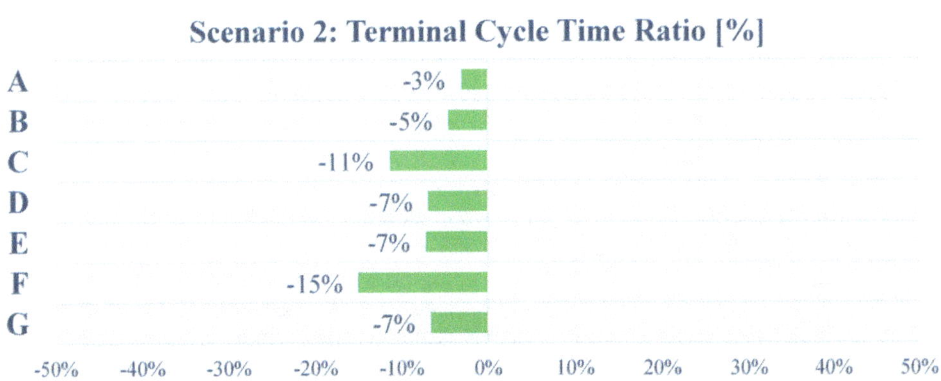

Figure 9: Percentage ratio for the cycle time per terminal in the simulation of Scenario 2.

5 CONCLUSIONS

This study presents the development and application of a model for the integrated simulation between the railway and port terminals located at the right bank of Santos port. This integration allowed us to quantify the effects that would unlikely be evaluated by queuing theory models or other statistical approaches. The validation of the simulation model reached 95% adhesion for the number of unloaded wagons on the right bank and 90% for the cycle time. These results are satisfactory, and the lower cycle time adhesion was due to restrictions on reliability or even absence of the times of some rail activities. One of the main challenges and difficulties in developing this simulation model was the definition of the operational rules that allow the entry, exit and circulation of trains in each yard. The solution to this problem came only after several months of interaction with the team responsible for daily operation, which made it possible to understand and record decisions that avoid deadlock or that make the railway operation more efficient.

The evaluations of the projects presented in Scenarios 1 and 2 are critical to direct investments; such assessments allow us to estimate the payback and provide essential information for building a strategic plan. In Scenario 1, the investment in Terminal G only had consequences for the system as a whole. There were benefits to decreasing the cycle time of Terminals C, E and F, but the cycle time of Terminal A increased. This increase occurred as a consequence of the smaller number of unloaded cars at Terminal A compared to other terminals. However, we met the entire monthly discharge volume planned for Terminal A. Thus, Scenario 1 benefits most of the rail system.

On the other hand, Scenario 2 benefited all the terminals in the system. These results can stimulate terminal representatives to support this project before the port authorities to accelerate its release. In this sense, the simulation model is beneficial not only from a technical point of view but also in the political sphere, as it provides an argument and logical basis for stimulating concrete actions by stakeholders.

Future studies are required to analyze the railway capacity on the right bank and evaluate other projects, such as the implementation of new railway tracks and the remodeling of the service tracks at the terminals of Corredor Exportação.

ACKNOWLEDGEMENT

The author thanks Rumo Logística for providing data and for granting the opportunity to develop and present this paper.

REFERENCES

[1] Porto de Santos, www.portodesantos.com.br/institucional/o-porto-de-santos. Accessed on: 22 Jan. 2020.
[2] Porto de Santos, Mensário Estatístico de Novembro de 2019. intranet.portodesantos.com.br/docs_codesp/doc_codesp_pdf_site.asp?id=128174. Accessed on: 22 Jan. 2020.
[3] Porto de Santos, Plano de Desenvolvimento e Zoneamento – PDZ. www.portodesantos.com.br/wp_porto/wp-content/uploads/2020/02/apresentacao-coletiva-pdz.pdf. Accessed on: 1 Mar. 2020.
[4] Agência Infra, www.agenciainfra.com/blog/preocupacao-com-acesso-ferroviario-e-destaque-em-audiencia-publica-no-porto-de-santos/. Accessed on: 22 Jan. 2020.
[5] Rumo Logística, www.rumolog.com. Accessed on: 1 Mar. 2020.
[6] Paragon Consulting, www.paragon.com.br. Accessed on: 1 Mar. 2020.

[7] Marinov, M. & Viegas, J., A simulation modelling methodology for evaluating flat-shunted yard operations. *Simulation Modelling Practice and Theory*, **17**(6), pp. 1106–1129, 2009.

[8] Petersen, E.R., Railyard modeling: Part I. Prediction of put-through time. *Transportation Science*, **11**(1), pp. 37–49, 1977.

[9] Petersen, E.R., Railyard modeling: Part II. The effect of yard facilities on congestion. *Transportation Science*, **11**(1), pp. 50–59, 1977.

[10] Freitas Filho, P.J., *Introdução à modelagem e simulação de sistemas: com aplicações em Arena®*, Visual Books: Florianópolis, p. 59, 2008.

[11] Ramos, G.M., Análise prospectiva da capacidade de processamento de cargas pela ferrovia no porto de Santos. Master's dissertation, University of São Paulo, 2003. DOI: 10.11606/D.18.2003.tde-11102003-153159.

[12] Brito, T.B., Modelagem e simulação de um terminal regulador de contêineres. Graduation work, Polytechnic School of the University of São Paulo, 2008.

[13] Rosa, R.A, *Uma proposta de um modelo matemático para a programação das atividades de um terminal ferroviário de carga.* XXI Congresso de Ensino e Pesquisa em Transportes (ANPET), Rio de Janeiro, 2007.

[14] Chwif, L. & Medina, A., *Modelagem e simulação de eventos discretos: Teoria e aplicações*, Elsevier Brasil: São Paulo, 2015.

NEW APPROACH FOR MISSING CONSUMPTION RECONSTRUCTION AND VALIDATION OF ENERGY METER DATA

LUIS MARIA ALONSO[1], AURELIEN WATARE[2], LAURENT-DIDIER ROUX[3],
LIONEL TAUNAY[4], CHRISTOPHE SAUDEMONT[1] & BENOÎT ROBYNS[1]
[1]University of Lille, Centrale Lille Institute of Technology Arts et Metiers, France
[2]Substation Control-Command RTE, France
[3]CNER/DP/TCC RTE, France
[4]Direction Stratégique Réseau SNCF RESEAU, France

ABSTRACT

The opening of the railway use all around Europe carried plenty of technical challenges never seen before in the railway field. An important challenge in energy billing filed, is to precisely do the billing of the consumed energy for each railway company in a reliable way. For that, in ground applications, RTE (the French national transmission system operator, who is in charge of the energy metering in railways by law in France) normally installs two energy meters to ensure the reliability of the measurements. Nevertheless, in agreement with the European law, the installation of only one energy meter is mandatory in the new/refurbished rolling stock since 2014. To ensure the correct billing of the railway companies, the data collected by the energy meters must be validated. In other cases no measure is available due to different technical problems, therefore a system that estimates the energy consumptions of those journeys is necessary. In this paper, a validated dataset is used to reconstruct the amount of energy consumed by trains whose energy meter data is missing or not valid. The validated data will be the start point for different estimators of the consumed energy by the trains. This dataset is composed of the information collected from the energy meters (date information and GPS position of the train) and train run data (transit times on fixed points of the line). Different types of estimators for the energy reconstruction are compared with real measurement data from a regional train circulation for a couple of years. Finally, a discussion of the performance of the different estimators will be shown.

Keywords: energy billing, consumption estimation, neural networks, measurement reliability.

1 INTRODUCTION

The creation of the Single European Railway Directive 2012 [1] started the process of the market opening of the railway use all around Europe. This opening carried plenty of technical challenges (interoperability with different electrification systems, different security systems even different track gauge) not seen for a long time in the railway field. Another important challenge is to precisely do the billing of the consumed energy for each railway company in a reliable way. For that, in ground applications, RTE normally installs two energy meters to ensure the reliability of the measurements. However, the European law only expects the installation of one energy meter (norm EN 50463) in the new/refurbished rolling stock since 2014. To ensure the correct billing of the railway companies the data collected by the energy meters must be validated. For that in Alonso [2] a method based on physical models is shown in the aim of validating the measurement done by the energy meters. This validated data-set (which represents around the 25% of all consumptions) is a great starting point to develop different types of estimators of total consumptions. Four different estimators will be shown and compared in this paper in the aim to find the most performant one.

2 AI TO RECONSTRUCT THE MISSING CONSUMPTIONS

When the energy meters give erroneous data or directly does not communicate, a method has to be developed to estimate the energy consumption. The idea of making a prediction of this

WIT Transactions on The Built Environment, Vol 199, © 2020 WIT Press
www.witpress.com, ISSN 1743-3509 (on-line)
doi:10.2495/CR200211

consumption based on the physical models presented in Alonso [2] may look as the best idea in a first time. However, the physical model is highly sensitive to different parameters (as the passenger number or the climatology of the journey) that are unknown when the reconstruction of the consumed energy is done. As can be seen in Tables 1 and 2, those parameters generate big variations in the total consumption. A strong relation between the train mass and the minimal consumption can be observed in Table 1. Also, under DC traction current, the determination of the factor of regenerative braking energy acceptance of the substations (the μ factor) is nearly impossible. As in DC the substations are not reversible in power, the calculation of this μ factor has to take into account other trains or other fixed loads (as stations, maintenance centres…) in the surroundings of the analysed train as well as his positions respect to these loads. Even more, as shown in Andre-Philippe et al. [3] (a study of consumptions based on real data made by SNCF), only the factor of the driving style of the train driver can make to vary the total consumption up to a 30% under DC traction current and near a 12% under AC traction current (whose substations are reversible in power naturally).

Table 1: Impact of the train mass in the minimal and maximal consumption.

Total mass	Minimal consumption	Maximal consumption
Minimal mass, 0 person/m^2 (–24%)	295.88 kWh (–20%)	834.02 kWh (–4.24%)
Standard mass, 4 person/m^2	356.54 kWh	869.44 kWh
Maximal mass, 6 person/m^2 (+10%)	390.70 kWh (+9.5%)	882.71 kWh (+1.5%)

Table 2: Impact of wind in the minimal and maximal consumption.

Wind	Minimal consumption	Maximal consumption
Heavy storm 60 km/h	423.10 kWh (+18.67%)	922.99 kWh (+6.15%)
Light storm 30 km/h	383.88 kWh (+7.67%)	894.51 kWh (+2.88%)
No wind 0 km/h	356.54 kWh	869.44 kWh
Light storm –30 km/h	338.57 kWh (–5.3%)	850.88 kWh (–2.22%)
Heavy storm –60 km/h	329.33 kWh (–8.26%)	838.81 kWh (–3.65%)

The addition of all the impacts in consumption from all the sources of incertitude shows clearly that an estimation based on physical models is far from being the answer to the problem of estimating the consumption in an accurate way. Therefore, other kind of estimators have to be developed; in this paper the performance of four statistical estimators (a mean, a multiple linear regression, a support-vector machine (SVM) and neural networks) will be compared with data from a regional line.

3 CASE STUDY: REGIONAL PASSENGER TRAIN

The studied regional train line is 58.17 km long and has 11 stops. This line has two speed limits, one of 160 km/h for the first 45 km and a second one of 100 km/h for the last 13 km. The line is nearly flat with one end (STA 1) at 7 MASL (metres above sea level) and the other end at 8 MASL (STA 2), the 23rd km being the highest point of the line with 60 MASL. The line is electrified at 1500 V DC with non-reversible substations. In Figs 1 and 2 the repartition of the total consumptions can be observed for years 2016 and 2018, even if the

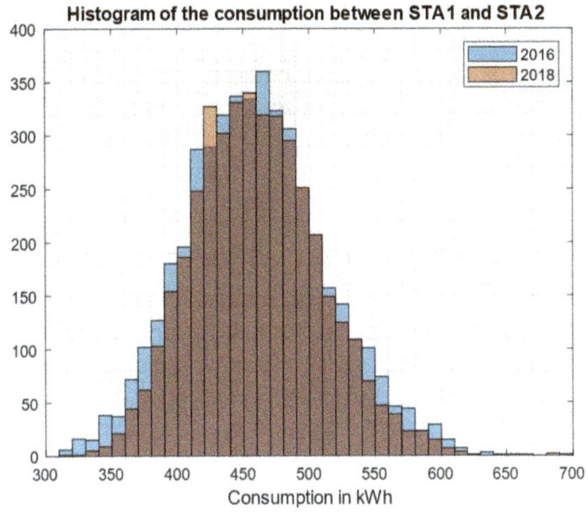

Figure 1: Consumption histogram for 2016 and 2018 for the journey between STA 1 and STA 2.

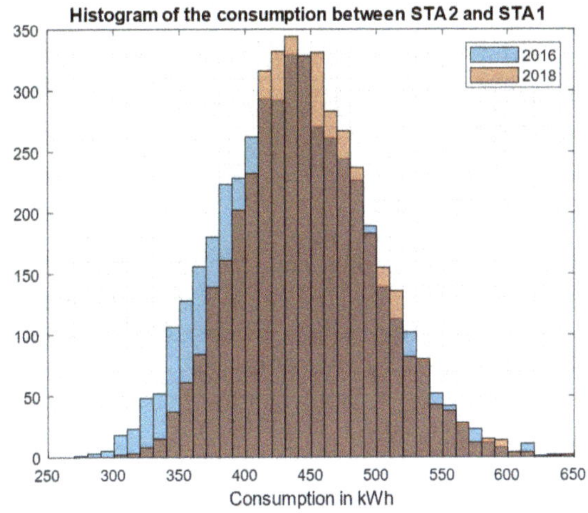

Figure 2: Consumption histogram for 2016 and 2018 for the journey between STA 2 and STA 1.

majority of the consumptions are between 400 kWh and 500 kWh a gap of a double can be found between the minimal consumption and the maximal consumption.

As mentioned before four different estimators will be compared. For all estimators the same procedure has been followed. A cross validation with a kfold (the number of groups into which data is divided) of 10 has been followed in this procedure. All the data is divided in 10 groups randomly, then nine groups will be used to determinate the estimator values and the performance will be tested over the group that has not been used for the estimator training. This procedure is repeated 10 times changing each time the group where the estimator is tested. Finally, an average performance is calculated with all the 10 performances already calculated.

For each type of estimator 100 tries are calculated, to evaluate the performance of each indicator the mean error of the 100 tries is calculated as well as the standard deviation of this error.

3.1 Mean estimator

To compare all the estimators a benchmark has to be established. For that the mean has been chosen as it is quite simple to calculate. Two type of mean estimator have been calculated. The first one takes the consumptions of all the year and applying the cross-validation method, this is the "year mean" indicator in Table 3. A second indicator has been developed with the same cross-validation method, but in this case a mean is established for each month of the year. This indicator is the "month mean" indicator in Table 3. In Tables 3–6 the percentage shown is the mean error (over a 100 tries) of the estimator with between parenthesis the standard deviation of this error.

Table 3: Mean estimator performances for the case study.

Estimator	STA 1–STA 2 2016	STA 1–STA 2 2018	STA 2–STA 1 2016	STA 2–STA 1 2018
Year mean	11.57% (0.003%)	10.34% (0.002%)	12.82% (0.002%)	10.97% (0.002%)
Month mean	10.13% (0.009%)	10.06% (0.011%)	11.71% (0.011%)	10.69% (0.013%)

3.2 Multiple linear regression estimator

As the main factors of impact in the consumption are unknown (the number of passengers to calculate the mass of the train, the train driver identification or the meteorological conditions) other sources of data have to be found to estimate the impact of those factors. In the following three estimators, the following listed data (calendar data and meteorological data basically) is used to estimate the total consumption of each circulation:

- Weekend or bank holiday (binary)
- Day of the week (1 to 7)
- Month (1 to 12)
- Day of the year (1 to 365)
- Departure hour (by truncation)
- Journey duration (in minutes)
- Temperature (in °C)

Parametric estimators are widely used in engineering to try to estimate a quantity from different input data with a linear relation. In Ahmad [4] an example is shown where from data related to a car (model, type of gearbox, engine size…), their fuel consumption wants to be reconstructed. The Multiple linear regression is the most basic option of parametric estimators. With input data listed before the following performances (Table 4) have been found.

Table 4: Multiple linear regression estimator performances for the case study.

Estimator	STA 1–STA 2 2016	STA 1–STA 2 2018	STA 2–STA 1 2016	STA 2–STA 1 2018
Multiple linear regression	10.29% (0.098%)	9.9% (0.105%)	12.37% (0.112%)	10.79% (0.108%)

3.3 Support-vector machine estimator

The SVM estimator is the first recommended type of AI to try when a model from data is wanted to be created. The SVM integrates non-linearities than can help to get a better result in comparison to a multiple linear regression. This estimator has been already used in other works [5], [6] to estimate the state of charge (SOC) of lithium batteries that have a high non-linear behaviour. In the case study, the SVM estimator gives the following results.

3.4 Neural network estimator

Neural networks are nowadays the most famous AI tool to build model from data. In the electrical domain they have been used widely for different purposes. In Rad et al. [7] neural networks are used to diagnose fault in electric machines, in Hannan et al. [8] they are used to estimate the SOC of batteries and in Khan et al. [9] they are used to forecast the load of the grid in London. All those studies show how with a good database, neural networks are able to create a model of any system. In the case of energy consumption, a good database can be constructed with all the validated measurements from the energy meters. However, the determination of the topology is not obvious. As the goal of the neural network is to build a model from data, the number of layers and the number of neurons per layer will depend on the complexity of the problem that is being modelled.

In the case study a series of tests have been done to try to find the optimal structure for the neural network. A first approach with only one layer where the number of neurons were varied between two and 40 leads to the following result.

In Fig. 3 the evolution of the mean standard error with respect to the number of neurons used. As it can be observed when the number of neurons is not high enough the mean standard error is high because of the underfitting. Meanwhile when the number of neurons is too high, the mean standard error is also high due to the overfitting in this case. An optimal number of neurons if found between 4 and 10 neurons, this result was also found when the data of 2016 was analysed.

A second test was performed to find if a two layers neural network would improve the performance. For that a series of test were done varying the number of neurons of the first and the second layer between one and 20. In Fig. 4 the results can be observed, the optimal configuration for a two layers neural network will be one neuron in the first layer and between 10 and 15 neurons in the second layer. Nevertheless, the obtained performance is nearly the same as the one found for the one-layer configuration neural network. Due to this reason the one-layer configuration has been retained as the best structure, as it gives the same performance and takes less time to train.

Using a one layer with eight neurons neural network with a Bayesian regularization backpropagation training algorithm, the following results were found in the case study.

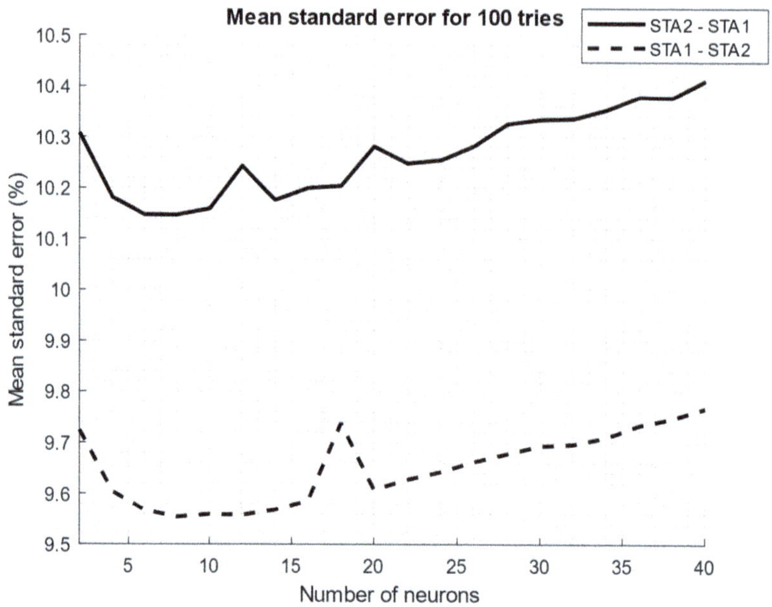

Figure 3: Mean standard error of different configurations of neural networks with data from 2018.

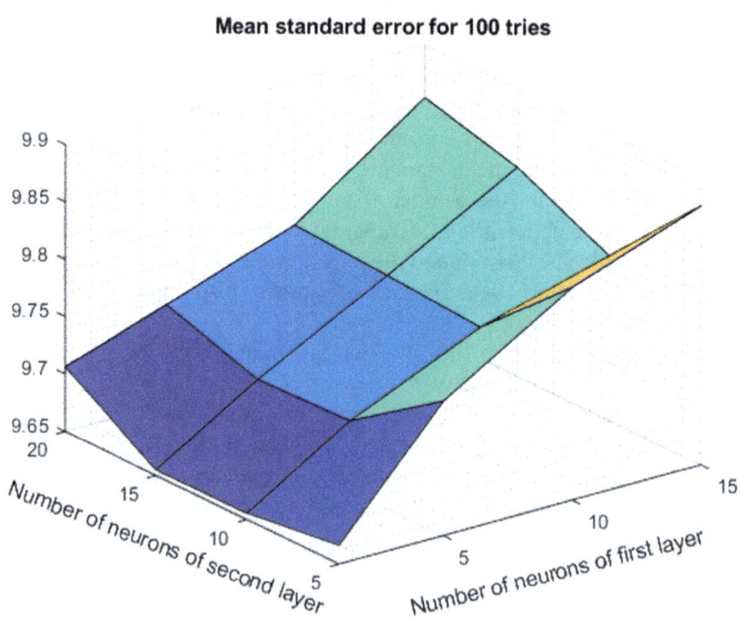

Figure 4: Mean standard error of different configurations of neural networks with data from 2018 for the journey STA1–STA2.

4 RESULT ANALYSIS AND CONCLUSION

If the all results are compared a few conclusions can be reached. First of all, the mean standard error is around a 10–11% which is actually a good precision if all the sources of uncertainty are taken into account. However, when more complex estimators are used (as SVM (Table 5) or neural networks (Table 6)) only a little improvement of around 1–2% is obtained. This means that the estimators are not able to find a correlation between the variations in the total consumption and the input data. This could also mean that the given input data is not valid to estimate the sources of uncertainty (like total mass of the train, behaviour of the train driver…) that have a huge impact in the total consumption

Table 5: SVM estimator performances for the case study.

Estimator	STA 1–STA 2 2016	STA 1–STA 2 2018	STA 2–STA 1 2016	STA 2–STA 1 2018
SVM	9.75% (0.105%)	9.51% (0.132%)	11.43% (0.135%)	10.13% (0.12%)

Table 6: Neural network estimator performances for the case study.

Estimator	STA 1–STA 2 2016	STA 1–STA 2 2018	STA 2–STA 1 2016	STA 2–STA 1 2018
Neural network	9.34% (0.092%)	9.26% (0.098%)	10.71% (0.119%)	9.87% (0.096%)

ACKNOWLEDGEMENTS

We thank Ms. Virginie Dordonnat and Dr. Benjamin Donnot for their help given in the field of AI.

REFERENCES

[1] EU, Single European Railway Directive, 2012. https://eur-lex.europa.eu/legal-content/EN/TXT/PDF/?uri=CELEX:32012L0034&from=en. Accessed on: 19 Nov. 2019.

[2] Alonso, L.M., Energy-meter data validation in railway via physical models. *VPPC2020*, 2020. (Submitted.)

[3] Andre-Philippe, C., Didier, F., Pascal, M., Younes, G., Gwenael, P. & Patrick, H., Energy efficiency for rolling stock, from measuring to savings. *2018 IEEE International Conference on Electrical Systems for Aircraft, Railway, Ship Propulsion and Road Vehicles & International Transportation Electrification Conference (ESARS-ITEC)*, 2019.

[4] Ahmad, N., Specific data model of smart fuel consumption cost estimator. *2009 International Conference on Computer Technology and Development (ICCTD 2009)*, **1**, pp. 429–432, 2009.

[5] Zhang, N. & Liu, K., The predition of SOC based on multiple dimensioned support vector machine. *MACE 2011: Second International Conference on Mechanic Automation and Control Engineering*, **1**, pp. 1786–1788, 2011.

[6] Chen, Y., Long, B. & Lei, X., The battery state of charge estimation based weighted least squares support vector machine. *Asia-Pacific Power & Energy Engineering Conference (APPEEC)*, **1**(2), pp. 1–4, 2011.

[7] Rad, M.K., Torabizadeh, M. & Noshadi, A., Artificial neural network-based fault diagnostics of an electric motor using vibration monitoring. *Proceedings 2011 International Conference on Transportation, Mechanical, and Electrical Engineering, TMEE 2011*, pp. 1512–1516, 2011.
[8] Hannan, M.A., Lipu, M.S.H., Hussain, A., Saad, M.H. & Ayob, A., Neural network approach for estimating state of charge of lithium-ion battery using backtracking search algorithm, *IEEE Access*, **6**, pp. 10069–10079, 2018.
[9] Khan, G.M., Khattak, A.R., Zafari, F. & Mahmud, S.A., Electrical load forecasting using fast learning recurrent neural networks. *Proceedings of International Joint Conference on Neural Networks. International Joint Conference on Neural Networks*, pp. 1–6, 2013.

PASSENGER BEHAVIOUR SIMULATION INSIDE TRAIN CAR COMBINED WITH TRAIN TRAFFIC SIMULATION

TAKETOSHI KUNIMATSU[1], HIRONORI MORIYA[2], HIROAKI KUSAKABE[2] & RYOHEI YAMADA[2]
[1]Railway Technical Research Institute, Japan
[2]JR Souken Information Systems Co., Ltd., Japan

ABSTRACT

The aim of this research is to improve functions of train traffic simulation to evaluate train timetables. We want to improve reproducibility of train delay caused by passengers' boarding and alighting. We also want to estimate boarding passengers' discomforts caused by congestion or collision with other passengers inside the train car. In railway commuter lines in a big city like Tokyo, train delay due to congestion is one of the biggest problems. When the railway company makes a timetable, although it is important to set enough dwell time at stations to prevent train delay, there is no way to confirm the appropriateness of timetable alternatives in advance. In our previous researches, we have developed a train operation and passenger flow simulator, which can reproduce train congestion, train delay, and their interactions, like the "snowball effect". However, the reproducibility of train delay in the simulator is not enough, because dwell time necessary for boarding and alighting is different, along with position of each passenger inside train car. In this research, we proposed a simulation method of passengers' movements to estimate the time necessary for boarding and alighting. The method can consider each passenger's shortest walking path towards the train door and their target position inside the train car, as well as passengers who temporarily alight the train, in which they disturb other alighting passengers. The method also realizes short calculation time enough for evaluating a whole timetable composed of a combination of many trains, stations, and doors. We conducted test calculations for the different movements of passengers. We compared the estimated time for boarding and alighting by the simulator, with the actual dwell time and estimated time by existing method by using a regression formula. As a result, we conclude that the proposed method can estimate each passenger's behaviour in detail without decreasing reproducibility of train delay.

Keywords: train traffic simulation, passenger's behaviour, delay, dwell time, shortest path search.

1 INTRODUCTION

In commuter lines in Tokyo, trains are often delayed due to extension of dwell time caused by congestion [1]. One of the countermeasures for this problem is appropriate timetabling reflecting passenger flow. That is, setting enough dwell time supplements at stations considering number of passengers alighting and boarding, and setting enough buffer time not to propagate train delay from a certain train to the succeeding train.

The authors have developed "Train Operation and Passenger Flow Simulator", which validates a train timetable before it goes into operation [2]. The simulator estimates train congestion and delay by using passenger data collected by automatic ticket gates at stations. Estimation of train delay is conducted by using the number of passengers alighting and boarding at each station, on each train, and from each door. Then the necessary time for alighting and boarding is predicted. However, the necessary time for alighting and boarding is determined not only by the number of passengers, but also the detailed position of passengers, layout inside train car, or the way to line up for the train on a platform. Actually, railway companies take efforts for decreasing the necessary time, and train delay. Moreover, if it is very congested in train car, some passengers near the door may temporarily get out the train to make way for alighting passengers. By using the conventional simulator, we cannot consider such situations or behaviour. So, we cannot estimate effects of reduction of train delay by taking such countermeasures as smooth boarding and alighting.

WIT Transactions on The Built Environment, Vol 199, © 2020 WIT Press
www.witpress.com, ISSN 1743-3509 (on-line)
doi:10.2495/CR200221

In this research, we developed simulation function for passengers' boarding and alighting at stations. It estimates necessary time for boarding and alighting in great detail by simulating each passenger's behaviour at every second. It can also combine with train operation and passenger flow simulator, which enable us to predict effects for reducing train delay by taking a countermeasure for smooth boarding and alighting. At first, we conducted questioner survey about passengers' preference of position inside train car. Then, we developed simulation function for boarding and alighting. Then, we combined the simulation function and passenger's choice model of vehicle into the simulator.

To validate these methods, we compared the estimated dwell time by the developed simulator with the actual ones. We confirmed that accuracy of the devised method is not worse than conventional method. Moreover, we evaluated and compared two plans for smooth boarding and alighting by test calculations for train delay and passengers' comfort. Finally, we concluded that the proposed method is useful enough to evaluate such plans quantitatively.

2 AIM AND APPROACH OF THE RESEARCH

2.1 Problems of rail transportation during peak hours in commuter lines in Japan

In the morning peak hours, as many passengers commute, trains and stations become very congested. Trains tend to be delayed, because the time necessary for boarding and alighting become longer than the planned dwell time at a station. In addition, once a train becomes delayed, train interval between the preceding train and that train become larger, and this in turn increases the number of passengers waiting for that train at the next station. So, the extent of delay for that train at the next station tends to increase as a consequence of longer necessary time for boarding and alighting. This phenomenon is called "snowball effect" of train delay and train congestion. As snowball effect decreases passengers' comfort and punctuality of train operation, railway companies take efforts for avoiding that, by setting as many trains as possible to decrease congestion, or making a robust timetable by including much running time / dwell time supplements and buffer time between trains.

However, if train timetable is changed, passengers take different train paths towards their destination. This in turn affects train congestion of trains and stations, and the necessary time for boarding and alighting. So, it is difficult for us to predict such changes in advance. Thus, there is a problem for railway companies that they have no way for validating robustness of a new timetable, or effectiveness of some counterplans for congestion and delay.

2.2 Related work

There are many related works concerning train delay and congestion. As a previous work for estimating train delay and congestion by simulation is "Train Operation and Passenger Flow Simulator", developed by the authors [2]. This simulator uses passenger data collected by automatic ticket gates at stations. The simulator estimates each passenger's train paths towards his/her destination along with a train timetable (Fig. 1). It then estimates train congestion by aggregating results of all passengers' estimated train paths. It also predicts train delay caused by congestion. That is, it calculates necessary dwell time at a station by using estimated number of passengers boarding and alighting.

When it calculates necessary dwell time, the regression formula introduced by another previous work about passenger flow is used [3]. However, this method has a drawback. That is, although the time for boarding and alighting is different along with the condition inside train car, like position of alighting passengers, it is not considered. In fact, some passengers may temporarily get out of the train, because they disturb other alighting passengers.

On the other hand, there is another work for simulating passengers' behaviour of boarding and alighting [4]. This work treats a passenger as an agent. It estimates necessary time for boarding and alighting by agent-based simulation technique. However, this work does not consider existence of other passengers on each passenger's walking path. That is, if there are some passengers on the shortest walking path of an alighting passenger towards the station platform, he/she will detour to another walking path which few passengers are on.

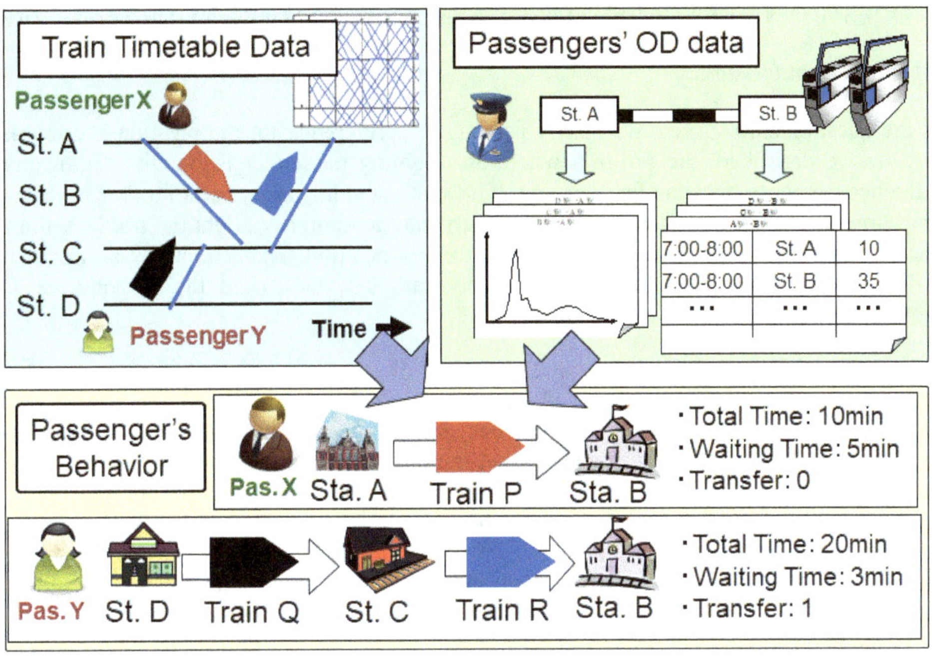

Figure 1: Structure of conventional train operation and passenger flow simulator.

2.3 Purpose of the research

We set the goal of the research as developing a new simulation system for train delay and passengers' comfort, satisfying the following requirements.

- Incorporating detailed passenger behaviour simulation about the situation of boarding and alighting, considering passengers' temporal alighting, and detour walking path for avoiding congested area.
- Estimating each train's delay and each passenger's total comfort under a timetable to evaluate the daily train operation plan within the practical calculation time

We focus on estimating train delay and passengers' comfort under a daily train traffic condition, without any disturbance caused by any external factors like rolling stock failure, humans/animals/vehicles presence on the line, or bad weather. They are rare cases in commuter lines in Japan. In addition, we are going to apply the new simulator for designing and scheduling a better train timetable in the planning stage.

2.4 Procedure of the research

The procedure of the research is as follows. First, we conduct questionaries' survey for passengers' preference of position inside train car. Then, we develop detailed boarding and alighting simulation of passengers for a certain door of a train at a certain station. Finally, we combine the boarding and alighting simulation, passenger's choice behaviour for vehicles, and existing train operation and passenger flow simulator, to realize prediction for train delay and passengers' comfort under a certain timetable on a certain rail line.

3 QUESTIONNAIRE SURVEY FOR PASSENGERS' POSITION INSIDE TRAIN CAR

3.1 Purpose of the survey

In the questionnaire survey, we clarify passengers' preference for the position inside train car. As we described, the position where an alighting passenger is standing/sitting may influence the necessary time for alighting. If he/she is standing inner part of train car, it takes long time to get off, and some other passengers may temporarily alight the train car not to disturb him/her. As passengers' preference for the position inside train car seems to be different according to the congestion inside train car, we asked the respondents for preferences under several conditions.

3.2 Procedure and questions of the survey

The survey was conducted on the websites during 10 days in January 2018. We first conducted preliminary survey to select target respondents who almost always use commuter lines in Tokyo. On the main survey, 1,047 respondents sent the answers for our questions.

In the preliminary survey, we asked his/her time period, rail line, origin and destination stations, and existence of accompany persons in his/her daily use of railway to go to work from home. We picked up 1,748 respondents who use the commuter lines of inbound direction in Tokyo during 7 to 10 AM, and there are no accompany persons.

In the main survey, we first asked his/her standing/sitting position inside train car in the ordinal commuting condition. We also asked his/her riding time, congestion of train cars when boarding or alighting, his/her choice of vehicles, and the reason for the answers. Then, we asked his/her standing position under the imaginary conditions of congestion.

3.3 Outline of the results

In this paper, we describe about ordinally standing position inside train car. In the main survey, we set the alternatives of standing position inside train car (Fig. 2). Each respondent answered the alternative which he/she often stand. The results were shown in Fig. 3. The position most passengers chose was 1. Following this, 3, 2, and 8 tended to be chosen by respondents. The result implies that, many passengers want to stand near the door to get off as fast as possible. It also suggests that the position close to the seats is preferable than that between the seats.

In addition, although this is not described in Fig. 3, when passengers' riding time become longer, position 2 become preferable than position 3. Also, when train congestion become greater, position 4 is preferable than position 7 or 8. So, it can be said that passengers seem to decide their standing position considering not only easiness for get out, but also comfort and duration of standing in the congested train car.

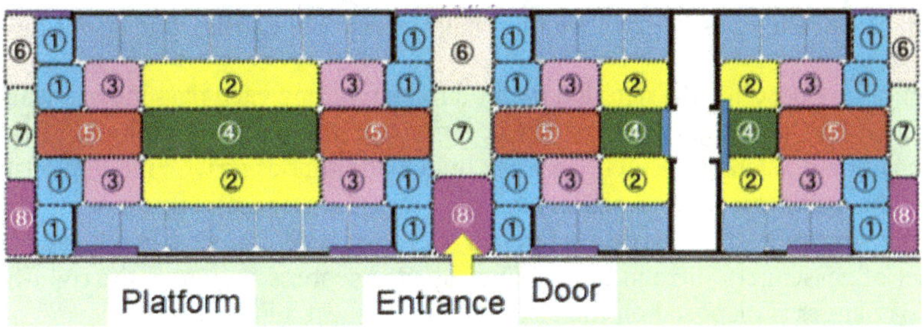

Figure 2: Alternatives for standing position inside train car.

Figure 3: Rate of respondents about ordinally standing position inside train car.

4 PASSENGER SIMULATION IN CASE OF BOARDING AND ALIGHTING

4.1 Outline of the simulation

We developed detailed simulation of boarding and alighting for a certain door in a train at a station. The outline of the simulation is shown in Fig. 4. Passengers' movements for boarding and alighting are represented by transitions between the grids which are set by separating area both inside and outside of train car. Two or more passengers may be assigned on the same grid at the same time.

First, we set the number of passengers boarding, alighting and staying in train car for a certain door on the target situation. We can choose either setting the number of passengers directly, or using the number of passengers estimated by train operation and passenger flow simulator.

Next, initial position of each passenger in train car is decided one by one by using the adjusted score of passengers' preferences about standing/seating condition. The basic score of passengers' preferences is set reflecting the results of questionnaires described in section 3. The score is adjusted reflecting the number of other passengers on the surrounding grids.

For the third procedure, simulation for alighting is conducted until all alighting passengers get off from the train door. Passengers who temporarily get off are also decided in this step.

After that, passengers remaining in train car are relocated before boarding passengers come in. That is, after all alighting passengers get off, if there is more attractive grid for a passenger, he/she moves to the place before the boarding passengers come in.

Then, simulation for boarding is conducted by every second. Each boarding passenger decides his/her destination position by using adjusted score of preferences. When all passengers pass the train door, the boarding simulation ends.

The simulator outputs result of simulation, not only alighting and boarding time, but also all passengers' logs about grid transitions. By using passengers' logs, we can also estimate each passenger's comfort during boarding and alighting.

4.2 Estimation of each passenger's walking path in case of boarding and alighting

In this section, we describe the detailed method for estimating each passenger's walking path in case of boarding and alighting. First, we describe the way for setting initial position in Fig. 5. Initial position of each passenger is set one by one in order of arrival. On the left-hand side of Fig. 5, the basic preference score and position of other passengers who are already disposed are illustrated. The basic preference score for a certain grid is adjusted reflecting the number of passengers surrounding it. The adjusted score is shown on the right-hand side. The next passenger is disposed on the grid with the highest adjusted score.

Then, we describe the method for searching a walking path of alighting passenger towards the door. We conduct shortest path search. The structure of grids is converted into in-directional graph network as described in Fig. 6. The costs of edges are set reflecting the number of other passengers on the neighbouring grids. By estimating the shortest path, passenger's detour behaviour not to pass congested area is reproduced. In case if there is a grid on which more than X shortest paths of alighting passengers pass, the staying passengers on that grid become temporal alighting, because they disturb alighting passengers. In addition, as the shortest paths are changed along with changes of passengers' positions at a certain time, they are recalculated at every simulation clock, by using fast estimation algorithm for the shortest paths [5].

In the simulation for alighting, each passenger proceeds basically a grid per second on his/her shortest path. In case when two or more passengers are located on the same grid at the same time, they have "one-time rest", and do not proceed on the grid in the next time step. In this way, we can represent the situation that it takes time for passengers to pass congested area.

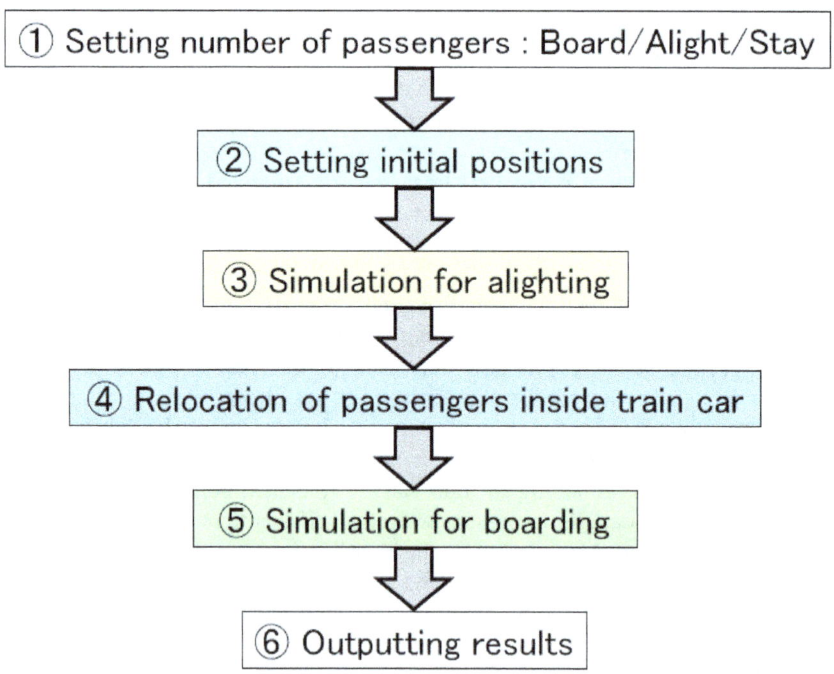

Figure 4: Outline of passenger simulation in case of boarding and alighting.

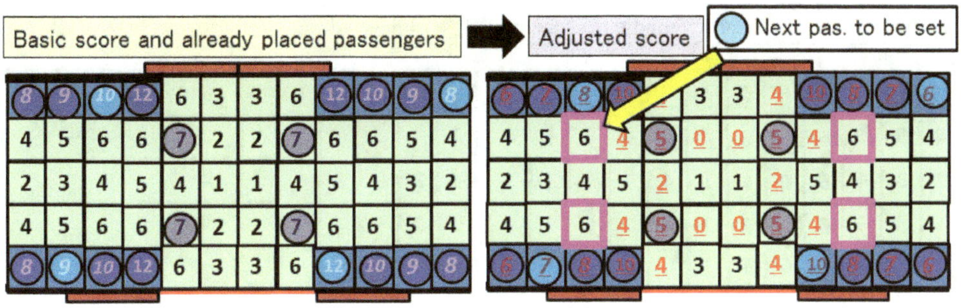

Figure 5: How to set initial position of passengers (Step 2).

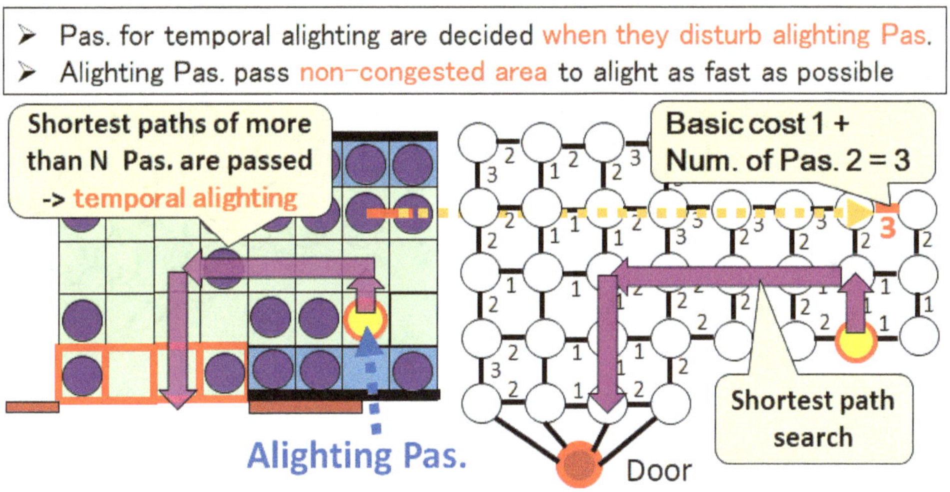

Figure 6: Simulation for alighting (Step 3).

In step 4, passengers remaining in train car are relocated. This is done based on the adjusted preference score for each passenger. Fig. 7 shows an example of adjustment. After all alighting passengers get out, adjusted preference score for each passenger is calculated. It reflects not only basic score and the number of passengers on neighbouring grids, but also the distance from the grid which he/she is now on. It represents the situation that, although a staying passenger in train car want to move more preferable position, as he/she do not want to walk long distance, he/she change his/her position only if more preferable position is occurred near his/her current position.

Fig. 8 illustrates estimation of walking paths for the boarding passengers. The target position inside train car for each boarding passenger is decided along with the adjusted preference score. Then, the shortest path search is conducted by the same way as that for the alighting passengers.

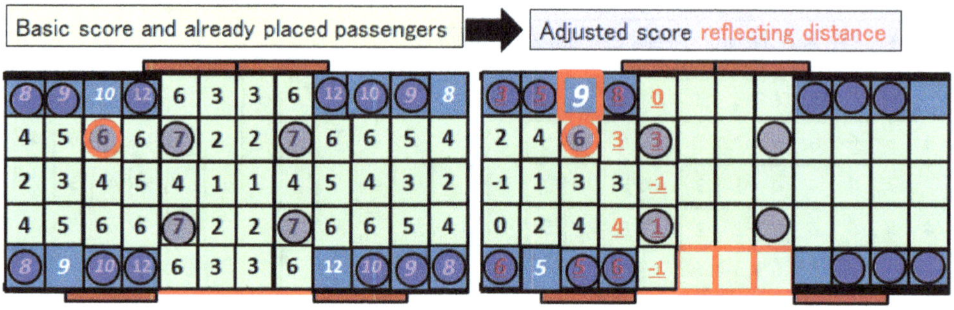

Figure 7: Adjusting preferences of position according to the number of passengers (Step 4).

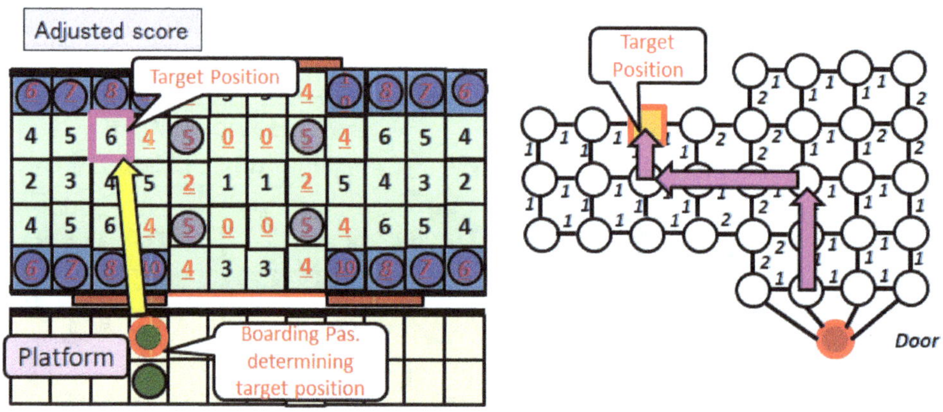

Figure 8: Simulation for boarding (Step 6).

We can consider the influence of some passengers standing near the doors. When we decide a walking path of an alighting or boarding passenger, the shortest path makes a detour for them. In case when there are no detour way, he/she has to walk near them, but when his/her and they are on the same grid, he/she takes "one-time rest", which results additional time for alighting or boarding. As a consequence, the necessary time for boarding and alighting become longer, compared with the situation without such standing passengers near the door.

On the other hand, there are other internal factors affecting boarding and alighting time. For example, the number and size of passenger luggage and their allocation inside the train car may influence on the dwell time, in case of high-speed or inter-city trains. Also, the presence of passengers waiting for the next train on the same platform may also influence on the dwell time. In this research, our target situation is peak hours in commuter rail lines, like metro inside the city. There are no such inter-city trains. Only few passengers have a big luggage. The number of passengers waiting for the next train on the same platform is also limited. So, in this research, we do not consider such influences.

4.3 Case study of boarding and alighting simulation

We will show some an example of boarding and alighting simulation. Fig. 9 illustrates the case that the number of alighting passengers is 30, that of boarding 30, that of staying 30. It firstly determines temporal alighting passengers by using the results of shortest path searches for the alighting passengers. In this case, we suppose that if more than three alighting passengers pass a certain grid, staying passengers on that grid become temporal alighting. For the result of this case, 4 passengers become temporal alighting.

The left-hand side of Fig. 9 illustrate the situation that all alighting passengers start to move for alighting just after the door open. On the right-hand side of Fig. 9 illustrate the situation that all alighting passengers start to move 10 seconds before the door open. Passengers' distributions when the door open are illustrated at the top of figure, and those 18 seconds after the door open are at the bottom. On the left-hand side, the alighting passengers are distributed when the door open. It affects necessary time for alighting, and there are still alighting passengers in case of 18 seconds after the door open. But, on the right-hand side, alighting passengers are concentrated when the door open. This contributes to shorten the necessary time for alighting. In case of 18 seconds after, all alighting passengers have already got off, and the boarding passengers have already started to get on. The total alighting and boarding time were 31 seconds, which is 6 seconds shorter than that in the left-hand side. So, we can conclude that it is effective for shorten dwell time to encourage passengers to start to move several seconds before the door open. In this way, we can quantitatively evaluate effects of countermeasures for decreasing train delay.

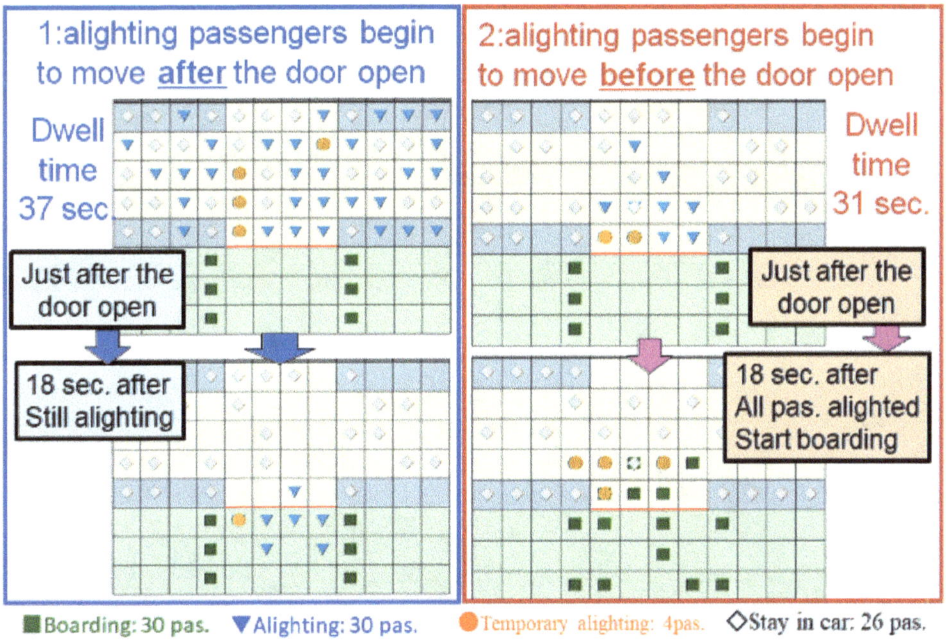

Figure 9: Comparison of boarding and alighting simulations.

4.4 Comparison of dwell time for boarding and alighting

To validate the simulation from the viewpoints of necessary time for boarding and alighting, we compared estimated necessary time with actual data. There were data of necessary time for boarding and alighting in 35 experimental cases. These experiments were conducted in a mock-up train and platform. When we conducted the simulation, we set the same number of boarding, alighting and staying passengers as those in actual data. Because we did not know detailed condition in train car in the experiments, the simulation was conducted tens of times for a set of the number of passengers, varying initial position of alighting passengers probabilistically. We also changed the time when alighting passengers start to move to get off. For a comparison, we conducted regression analysis for actual 35 data, and prepare a regression formula. We calculated the dwell time in case of the conventional method by using the regression formula and the number of passengers in the actual data.

Fig. 10 illustrates the results. The horizontal axis indicates sum of the number of passengers getting on and off. The vertical axis indicates boarding and alighting time. The actual data about boarding and alighting time are indicated by "x". The estimated time by conventional and proposed method are indicated as "■" or "●". For the proposed method, boarding and alighting time are illustrated as interval estimation, because we conducted simulation tens of times varying conditions. The trend of estimation for the three cases are almost the same. By taking rate of cases within 10 seconds error, compared with actual ones, the proposed method become 94%. That for the conventional method was 74%. So, we can conclude that the proposed method has at least the same accuracy with the conventional method.

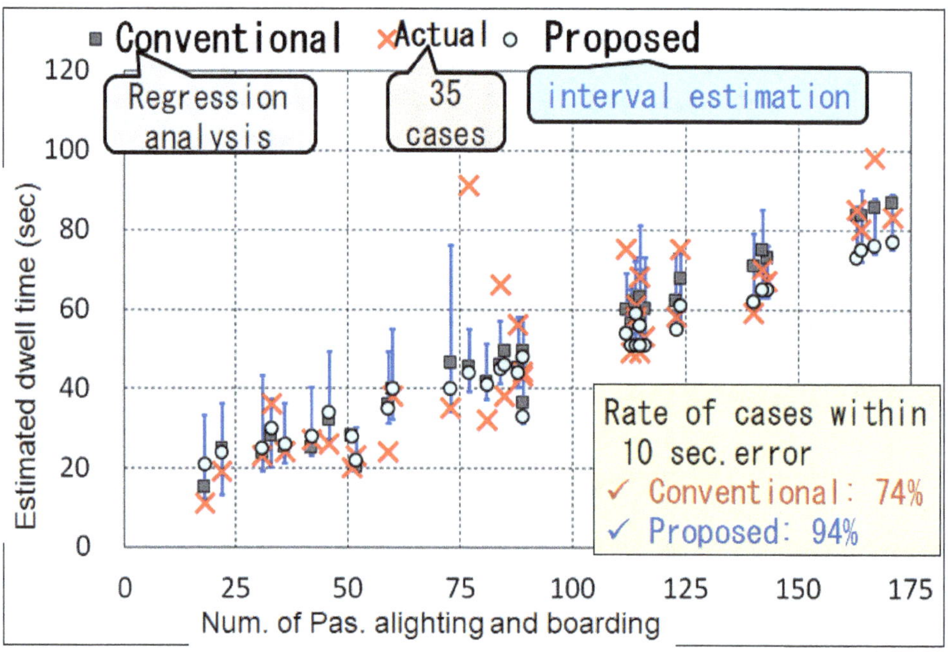

Figure 10: Comparison of estimated dwell time with actual ones.

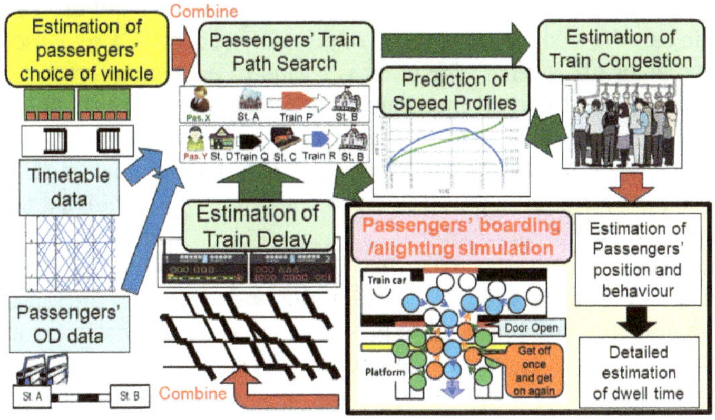

Figure 11: Overall structure of train operation and passenger flow simulator with linkage.

5 LINKAGE WITH TRAIN OPERATION AND PASSENGER FLOW SIMULATOR

To improve accuracy of estimation of train delay in train operation and passenger flow simulator, we combined the boarding and alighting simulation with the simulator. We also combined estimation function for passengers' choice of vehicles. It is decided by not only position of stairs in origin or destination station, but also the number of passengers lining up for the next train [6]. So, our passengers' choice model for vehicles is developed reflecting both structure and congestion of platform in each passenger's origin and destination station.

Fig. 11 shows the new structure with linkage. As train operation and passenger flow simulator can estimate each passenger's train path, reflecting his/her train choice for successive train and left behind, we can estimate dwell time by using more realistic number of boarding and alighting passengers. We can also estimate effects of reduction of train delay directly and quantitatively in a certain rail line by introducing some counter measures for smooth boarding and alighting. Especially in Japan, widening train doors, changing layout inside train car, or changing the way of lining up at station platform are often discussed and enforced to decrease train delay. By using the new simulator, railway companies can know how much delay can be decreased by a certain countermeasure in advance.

In addition, we can also evaluate the effects from the viewpoints of passengers. We can know how many times and how long a certain passenger stands close to other passengers. By estimating them, we can evaluate passengers' comfort quantitatively.

On the other hand, as we use passengers' data collected by automatic ticket gates at stations, some passengers may stop their itinerary between the ticket gates (e.g. for a coffee). In this research, as our target situation is peak hours in commuter rail lines, and few passengers will stop their itinerary during their commuting, we do not consider such break of itinerary.

6 APPLICATION FOR EXISTING RAIL LINE

6.1 Outline of the rail line and preconditions

We applied the updated train operation and passenger flow simulator for an actual commuter line in Japan. The rail line has 19 stations. The number of trains in the whole day is about 1,000. In the peak hours, trains are set about every 4 minutes for one direction.

We simulated and compared two countermeasures for train operation. One is widening doors of the train car. We assumed that width of all doors become 1.8 m, which enables four passengers to get on or off at once. The other is a kind of passengers' manner for boarding and alighting. We assumed that all passengers become polite, and behave cooperative for other passengers' boarding and alighting. That is, when passengers get on a train, they willingly go to inner area of train car, and when they disturb other alighting passengers, they willingly alight temporarily. In addition to these conditions, we simulated train traffic and passenger flow under the current situation.

We calculated two evaluation indexes. One is maximum value of train delay. This is not for a certain train or a certain station, but for all trains and stations in the rail line. The other index is average neighbouring time with other passengers. This is based on the situation in which more than two passengers are on the same grid in the boarding and alighting simulation.

The time during which a certain passenger is involved in such situation is calculated and summed up. Then, we take an average for all passengers in the rail line. The latter index indicates each passenger's comfort during boarding and alighting. By summing up the time for all his/her journey, we can estimate his/her comfort during the trip. We can evaluate the total level of comfort for a rail line by taking average for all passengers.

6.2 Results of test calculations

The results of test calculations are summarized in Table 1. The simulation was conducted from the earliest train in the morning to noon. The maximum values of train delay were calculated by taking all trains running before 10 AM. The average neighbouring time with other passengers are calculated for about 280,000 passengers using the line before 10 AM.

From the viewpoint of punctuality, maximum train delay was estimated to drop by 25 seconds with widening train doors. The effect was estimated to be 24 seconds in case of passengers' cooperative behaviour. But, in the latter case, the condition that all passengers take cooperative behaviour is supposed. So, we have to understand the results that the effects are at most 24 seconds. If we want to realize further decrease of train delay, we have to discuss other countermeasures like widening doors.

On the other hand, from the viewpoints of passengers' comfort, average neighbouring time will drop significantly when widening train doors. When passengers take cooperative behaviour, average neighbouring time is slightly dropped. So, it can be said that, by taking cooperative behaviour, passengers can enjoy better transportation services from the viewpoints of not only punctuality, but also comfort. These results will encourage passengers to take cooperative behaviour during boarding and alighting.

Table 1: Comparison of the simulated results.

	Maximum train delay (seconds)	Average neighbouring time with other passengers (seconds)
1. Current condition	102	25.9
2. Widen train doors	77	21.7
3. Passengers' cooperative behaviour	78	24.3

7 CONCLUSION AND FUTURE WORKS

In this research, to estimate effects of reduction of train delay by taking countermeasures for smooth boarding and alighting, we have developed simulation for boarding and alighting. It can consider detailed situation of boarding and alighting, like temporal alighting not to disturb other alighting passengers. It also reflects passengers' preferences of standing position based on the results of questionaries' survey. Then, it combines with existing train operation and passenger flow simulator, which enable us to estimate changes of train delay or passenger comfort quantitatively. By the test calculation, the effects are estimated in an actual commuter rail line.

For the future works, we will apply the simulator for many other rail lines, and validate the accuracy or usefulness of that. Especially, in rail lines including long-distance transportation like high-speed or inter-city, we have to extend the simulation model to incorporate passenger luggage, and existence of passengers waiting for the successive train on the same platform. They are not so difficult to incorporate such internal factors, because existence of luggage or waiting passengers can easily be added on the boarding and alighting simulation. We also consider passenger's temporal stop of his/her journey. It will become a considerable work, because we first have to clarify passengers' trend for stopping their itinerary. In addition, we are going to model other layouts in train car, like cross seats, and compare the results along with the difference of layouts. By incorporating these further internal factors affecting dwell time, the simulation can become realistic, and be applied for more complicated situation. Moreover, from the viewpoints of further external factors, we will extend the simulation for disrupted condition, like rolling stock failure or bad weather, to evaluate train rescheduling.

REFERENCES

[1] Tomii, N., *Techniques To Make Train Diagrams*, Ohmsha Publishing, 2012. (In Japanese).

[2] Kunimatsu, T., Hirai, C. & Tomii, N., Evaluation of timetables by estimating passengers' personal disutility using micro-simulation. *Proceedings of the 3rd International Seminar on Railway Operations Modelling and Analysis: RailZurich2009*, 2009.

[3] Oto et al., A study on passenger flow in a railway station (in Japanese). *Summaries of technical papers of Annual Meeting Architectural Institute of Japan*, **E-1**, pp. 845–846, 1999.

[4] Iwakura, S. et al., A multi agent simulation model for estimating knock-on train delays under high-frequency urban rail operation. *Transport Policy Studies' Review*, **15**(4), 2013. (In Japanese).

[5] Tatsui, D., Kunimatsu, T. & Sakaguchi, T., Interactive timetabling system with the function of estimating train congestion on timetable planned. *Proceedings of the 14th International conference on Railway Engineering Design and Optimization (COMPRAIL2016)*, 2016.

[6] Fang, J. et al., A model of passenger distribution on metro platforms based on passengers' boarding strategies. *Proceedings of 8th International Conference on Railway Operation, Modelling and Analysis: RailNorrköping2019*, 2019.

LONG TRAIN DYNAMIC SIMULATION BY MEANS OF A NEW IN-HOUSE CODE

NICOLA BOSSO, MATTEO MAGELLI & NICOLÒ ZAMPIERI
Department of Mechanical and Aerospace Engineering, Politecnico di Torino, Italy

ABSTRACT

The need for faster heavy-haul trains with higher axle-loads is responsible for the rise of large in-train forces on wagon connection systems. These forces have a significant impact on longitudinal train dynamics (LTD) from the point of view of performances, running stability and safety during braking and traction operations. Therefore, LTD simulations represent an essential tool to predict in-train forces and to design coupling and braking systems. Long trains are typically modelled as a system of several point masses linked to each other by means of non-linear elements, summarizing the characteristics of the coupling systems. In fact, wagon connection systems present a non-linear force-deflection characteristic, with a hysteresis loop due to the different behaviour in loading and unloading states. The calculation of in train forces is nowadays performed following three main strategies, namely look-up table approaches, mathematical equations based on experimental results and "white-box" models related to physical properties of the connection systems. An International benchmark of LTD simulators was recently proposed to assess the output results obtained by different simulators, following several modelling strategies, in the simulation of the same scenarios. The research group from Politecnico di Torino joined the competition using the multibody commercial software Simpack (ver.9.4). However, inefficiencies appeared because multibody software are usually intended for the simulation of a few vehicles with many degrees of freedom (d.o.fs), rather than many wagons with few d.o.fs each. Therefore, the team developed a new LTD simulator (LTDPoliTO) in MATLAB, totally based on a vector-arithmetic logic. The new simulator can perform both dynamic investigations in time domain and modal analysis in frequency domain. The validation of LTDPoliTO was carried out by performing the four benchmark simulations and a good agreement with the other participants was obtained, in terms of both numerical results and computational speeds.
Keywords: long train simulation, LTD benchmark, draft gear, dynamics modeling, train dynamics.

1 INTRODUCTION

The performances and the safety of railway systems are strongly affected by longitudinal train dynamics, which refer to both the motion of the whole train as well as to the relative motions between adjacent vehicles ("slack actions"). The latter are allowed by clearances in coupling systems [1], usually comprising a coupler and a damping element. The typical solution in Europe is the buffer-hook system, while coupler and draft gears are used in other countries, such as the USA and Australia [2].

Modern freight trains are facing a strong demand for higher axle-loads and speeds, which can eventually lead to the rise of large in-train forces, thus causing a serious threat to the safety and running stability of freight trains during traction and braking operations, with several possible drawbacks, namely wagon climbing, derailments, structural damages and even breaking of coupling elements [3], [4]. Therefore, studying the LTD of long freight trains is a major concern for the design of coupling and braking systems, but also for the optimization of driving cycles and train composition as well as for the forecast of vehicle position and speed during a specific route [5].

Due to the high computational power offered by modern computers, nowadays LTD investigations are performed by means of numerical simulations [6]. Usually, long trains are modelled as a system of several lumped point masses, with a single degree of freedom (d.o.f) in the running direction, linked to each other through nonlinear elements, accounting for the

WIT Transactions on The Built Environment, Vol 199, © 2020 WIT Press
www.witpress.com, ISSN 1743-3509 (on-line)
doi:10.2495/CR200231

mechanical impedance of the coupler and draft gears. The calculation of in-train forces is often the bottleneck of these numerical models, since coupling systems are strongly nonlinear components, with a hysteretic behaviour, as shown in experimental drop-hammer tests [7]. Wheel-rail adhesion phenomena [8] as well as traction/braking control systems [9] are typically not considered in LTD models since even efficient contact codes [10], [11] usually require a high computational effort.

As surveyed by Wu et al. [12], three main strategies are used to model the nonlinear impedance of railway coupling systems, namely (a) look-up table (LUT) approaches, (b) mathematical equations fitting experimental data and (c) white-box models. The LUT approach is based on fixed tables storing the experimental data registered in drop-hammer tests, and it usually requires a logic switch criterion to smooth the transition between loading and unloading conditions. On the other hand, the second strategy relies on analytical equations fitted on experimental data. Finally, white-box models depend on the physical properties of coupling elements.

The in-train forces calculated by numerical models can vary significantly according to the specific strategy used to model the coupling systems. Therefore, an international benchmark ("the benchmark" in the rest of the paper) was established in recent years to compare the results from different LTD simulators [13], [14]. The railway research group from Politecnico di Torino developed an LTD model using the commercial multibody software Simpack (ver. 9.4) [15], [16] and joined the competition, but struggling occurred in the simulation of a system comprising many bodies, each one with few d.o.fs. Therefore, the research group developed a new in-house LTD code in MATLAB (LTDPoliTO) to obtain higher computational efficiencies and a better agreement with the other participants in the output results.

The paper concerns the mathematical description of the new model, the implementation in MATLAB environment and the validation of the proposed in-house tool performed on the same case study as suggested in the benchmark. The novel aspect of the proposed code is that it is totally developed in MATLAB, making the most of built-in high-level functions and vectorization strategies for large array management. The greatest point of merit of this new code is the capability of producing stable output results with high computational speeds.

2 DESCRIPTION OF THE MODEL

The mathematical definition of the new code followed the suggestions from the international benchmark. Each vehicle only has a single d.o.f in the running direction and the track is modelled as straight, while curves and slopes are accounted for by introducing extra retarding forces.

Fig. 1 shows all forces acting on the kth vehicle of the train configuration. Please note that the numbering rule in this section as well as in the code, starts from the tail-end wagon in ascending order. The forces on the vehicle include the inertial term (the product of the total mass M_k times the acceleration $\ddot{x}_k$), the forces due to front ($F_{c,k}$) and rear ($F_{c,k-1}$) coupling elements, the retarding forces ($F_{R,k}$) and the forces due to traction or dynamic braking efforts ($F_{T/DB,k}$).

The equilibrium of forces in the longitudinal direction x for the kth vehicle is expressed by eqn (1):

$$F_{T/DB,k} + F_{c,k} - F_{R,k} - F_{c,k-1} = M_k \ddot{x}_k. \tag{1}$$

Of course, for the first vehicle of the composition, i.e., the tail-end wagon, the force on the back coupler is zero. Similarly, the force on the front coupler of the last vehicle, i.e., the leading loco, is equal to zero, too.

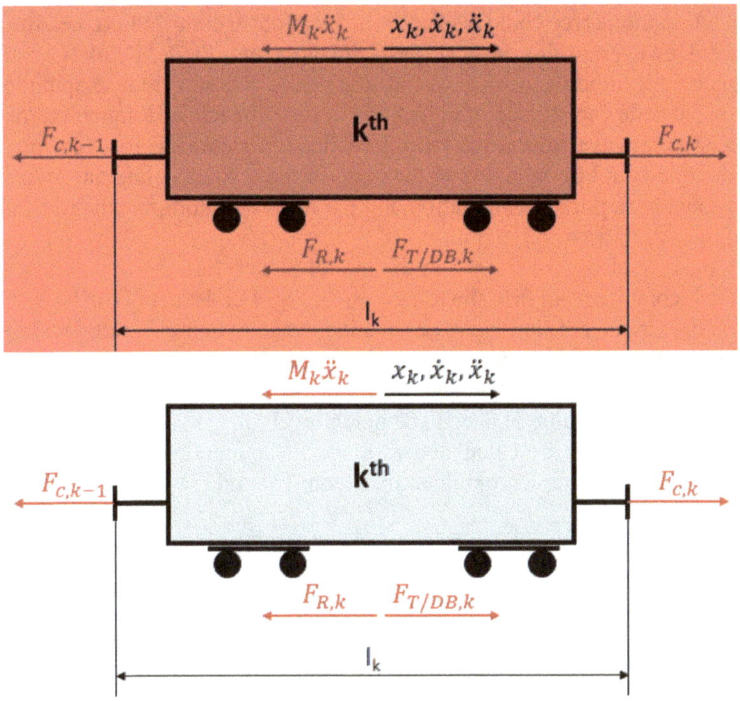

Figure 1: Forces acting on the kth vehicle.

In a vehicle with N_{AX} wheelsets, the total mass of the vehicle (M_k) is calculated as the sum of all the axle-loads (m_{ax}), expressed in kg, according to eqn (2)

$$M_k = m_{ax}N_{AX}. \tag{2}$$

The total retardation force $F_{R,k}$ is the sum of the resistant forces due to rolling resistances $F_{RR,k}$ and the forces due to track curvature $F_{RC,k}$ and grade $F_{RG,k}$. The forces due to curvature and rolling resistances always have positive sign, i.e., they always have an opposite direction with respect to the direction of the vehicle speed $\dot{x}$. On the other hand, forces due to track grade can be in the same direction as the vehicle speed, e.g., in downhill track sections. The resistant forces are modelled using the expressions suggested in the benchmark and shown in eqns (3)–(5), in which the subscript k is omitted for the sake of clarity. Please note that in eqns (3)–(5), the unit for mass is tonne, while the unit for speed is km/h. In eqn (3), Q is an aerodynamic frontal resistant factor, which is equal to 3.2 for the leading loco and equal to 1 for all other vehicles. In eqn (4), R_C is the curve radius expressed in meters and finally in eqn (5) i_G is the track grade in %. Other expressions are witnessed in the literature and are surveyed in a work by Spiryagin et al. [17]

$$F_{RR} = QM\left(2.943 + \frac{89.2}{m_{ax}} + 0.0306\dot{x} + \frac{0.122\dot{x}^2}{M}\right), \tag{3}$$

$$F_{RC} = M\frac{6116}{R_C}, \tag{4}$$

$$F_{RG} = Mgi_G. \tag{5}$$

The total retardation force could also account for a fourth contribution, i.e., the pneumatic braking force. However, at this early stage of development, the LTDPoliTO code does not include a pneumatic module, which would drastically increase the computational effort. Nevertheless, the code can still be validated with a set of benchmarking output results, since the air brake system was neglected in the international benchmark, too. In future work, the pneumatic module could be included in the code in order to calculate air brake forces, but this would probably require co-simulation and parallel computing techniques to reduce the wall time of the simulations.

The only braking effort considered in the model is therefore the dynamic braking force exerted by the locomotives. In fact, the term $F_{T/DB}$ in Fig. 1 and eqn (1) can be either positive, when traction operations are simulated, or negative, when dynamic braking manoeuvres take place. Obviously, this term is set to zero for all non-powered vehicles. On powered vehicles, the tractive/braking effort is calculated from the locomotive mechanical characteristic, depending on the current value of speed and notch level.

Finally, coupling forces are calculated as a nonlinear function of the deflection δx and relative speed δv of each coupler, obtained from eqns (6) and (7)

$$\delta x_k = x_{k+1} - x_k - \frac{l_{k+1} - l_k}{2}, \tag{6}$$

$$\delta v_k = \dot{x}_{k+1} - \dot{x}_k. \tag{7}$$

In eqns (6) and (7), l_k is the total length of the kth vehicle, which includes the car body and both semi-lengths of front and rear coupling systems, see Fig. 1, while $\dot{x}_k$ refers to the speed of the kth vehicle. The transition between loading and unloading conditions is smoothed according to a logic criterion depending on the current value of the relative speed, as suggested by Zhang et al. [18]. The force on each coupler is calculated using eqns (8)–(10), which ensure that a smoothing transition occurs when the absolute value of the relative speed is below a threshold value v_ε, with a coupling force lying between the loading and unloading curves. On the other hand, if the absolute value of the relative speed is above v_ε, the coupling force is equal to either the loading (F_{load}) or the unloading (F_{unload}) force.

$$F_{c,k}(\delta x_k, \delta v_k) = \begin{cases} F_{cM,k}(\delta x_k) + F_{cA,k}(\delta x_k)\,\mathrm{sign}(\delta x_k \cdot \delta v_k), & |\delta v_k| \geq v_\varepsilon, \\ F_{cM,k}(\delta x_k) + \frac{|F_{cA,k}(\delta x_k)|}{v_\varepsilon}\delta v_k, & |\delta v_k| < v_\varepsilon, \end{cases} \tag{8}$$

$$F_{cM,k}(\delta x_k) = \frac{F_{load,k}(\delta x_k) + F_{unload,k}(\delta x_k)}{2}, \tag{9}$$

$$F_{cA,k}(\delta x_k) = \frac{F_{load,k}(\delta x_k) - F_{unload,k}(\delta x_k)}{2}. \tag{10}$$

The next section briefly shows the implementation of the model in MATLAB and shortly describes the LTD scenarios proposed in the benchmark, which were simulated with the new code in order to validate and to assess the output results obtained with LTDPoliTO.

3 IMPLEMENTATION OF THE MODEL IN MATLAB

Eqns (1)–(10) must be written for each of the N_v vehicles in the train composition, thus obtaining a system on nonlinear ordinary differential equations (o.d.e), which needs a numerical integration scheme to be solved. It is not difficult to demonstrate that at each time-step, the LTD o.d.e system can be expressed in the form of eqn (11), where t is time, $\underline{\mathbf{y}}$ is the array storing the dependent variables of the problem, $\underline{\mathbf{y}}'$ is the first derivative of $\underline{\mathbf{y}}$, $\underline{\mathbf{f}}$ is a function of t and $\underline{\mathbf{y}}$ for the calculation of the derivative $\underline{\mathbf{y}}'$ and finally $\underline{\underline{\mathbf{M}}}$ is the mass matrix of the system

$$\underline{\underline{M}}\underline{y}' = \underline{f}(t, \underline{y}). \tag{11}$$

The column vector $\underline{\mathbf{y}}$ stores $2N_v$ elements, corresponding to the positions and speeds of the N_v vehicles. Therefore, $\underline{\mathbf{y}}$' collects the N_v speeds and N_v accelerations of the vehicles. The mass matrix is a constant (i.e., not time dependent) sparse diagonal square matrix $2N_v x 2N_v$ in size, with the first N_v elements on the diagonal all equal to 1, and the following N_v elements equal to the mass of each vehicle. Finally, $\underline{\mathbf{f}}$ is the same size as $\underline{\mathbf{y}}$ and $\underline{\mathbf{y}}$', and its elements 1-N_v correspond to the speed of each vehicle, while each of the elements N_v+1-2N_v is equal to the left-hand side of eqn (1), written for the corresponding vehicle.

Written in the form of eqn (11), the LTD o.d.e system can be easily implemented and solved in MATLAB, using a numerical integration method, once the model inputs are provided. LTDPoliTO inputs include the definition of the train arrangement, in terms of the number and types of vehicle, as well as the following input characteristics:

- Track curvature and elevation along the running direction.
- Driving cycle, i.e., the notch level during the train route, as a function of either time or position along the running direction.
- Mechanical characteristic of locomotives, i.e., the typical curve relating traction/dynamic braking force to speed and notch level.
- Definition of the mechanical impedance of coupling elements.

The train configuration is represented in the code using an array with N_v integer, each one corresponding to a specific type of vehicle. Instead, the input characteristics are imported into LTDPoliTO by means of several text files, which describe the main features of the input data. Forces on coupling elements are calculated using a LUT approach, with two fixed tables storing the experimental force-deflection characteristic for both loading and unloading conditions. Before the simulation starts, the user must define the following 4 re-discretization steps in order to re-interpolate the input characteristics with a fixed step:

1. A step for the discretization of the coupling system deflections, in order to re-interpolate the force-deflection LUT.
2. A re-discretization step corresponding to the shortest track section considered, in order to re-interpolate the track curvature and elevation characteristics.
3. A fixed time-step to discretize the driving cycle input text file, which currently defines the notch level as a function of time.
4. A step for the re-discretization of the vehicle speed, which is needed to re-interpolate the locomotive mechanical curves.

The re-discretization of the input files with fixed steps is needed to allow easy indexing operations throughout the simulation. All indexing and array management operations are handled using MATLAB vectorization strategy and optimized built-in functions, thus increasing the total computational efficiency. At each time-step, the code can perform the calculation of all in-train, retardation and tractive/braking forces with no need for loop flows or Boolean statements, by simply exploiting MATLAB vector logic.

At the end of the simulation, the MATLAB solver outputs the speed and position of each vehicle throughout the simulation. Therefore, deflections, relative speeds and in-train forces on all coupling systems are calculated with a specific post-processing routine. Finally, the whole output dataset can be saved in binary files for further investigations or simply to back up the simulation results.

4 CASE STUDY

The validation of the new code was performed on the four simulation scenarios proposed in the international benchmark questions paper [13], which correspond to four different train compositions, built using 2 different types of locomotives and 2 different types of wagons and schematically illustrated in Fig. 2. As shown in Fig. 2, the number of d.o.fs increases from train 1 to train 3, while train 4 has the same number of d.o.fs as train 3 but it features a different arrangement. Table 1 briefly summarizes the main parameters of each vehicle type.

Two types of wagon connection system are considered in the benchmark, i.e., couplers, with a total 10 mm slack, and drawbars, with no slack. Starting from the front wagon, in all four train configurations, each pair of wagons is connected by means of a drawbar, and wagon pairs are linked to each other with couplers. All locomotives are connected to the other vehicles using couplers.

Among the benchmarking input was a draft gear force-deflection cross plot, representing an experimental drop-hammer test. Fig. 3 shows the loading and unloading characteristics for the series of two draft gears and a coupler/drawbar obtained from the benchmark input data and used in the simulations performed during the numerical activity described in the paper. According to the benchmark rule, tensile deflections and forces have a negative sign, while compression states have positive values. The detailed view in Fig. 3 highlights the total 10 mm coupler slack, with a 2 mm slack in compression and 8 mm slack in tensile conditions.

The benchmarking input also included the mechanical characteristics of two different types of locomotive. Both locomotives have 8 notch levels for traction and dynamic braking, but locomotive #2 guarantees higher power efforts.

Track curvature and elevations as well as driving cycles are once again extracted from the benchmark input dataset. Track features, shown in Fig. 4, are fixed in all four simulations, while the driving cycle is different for each train, and can be easily determined from the benchmarking question paper.

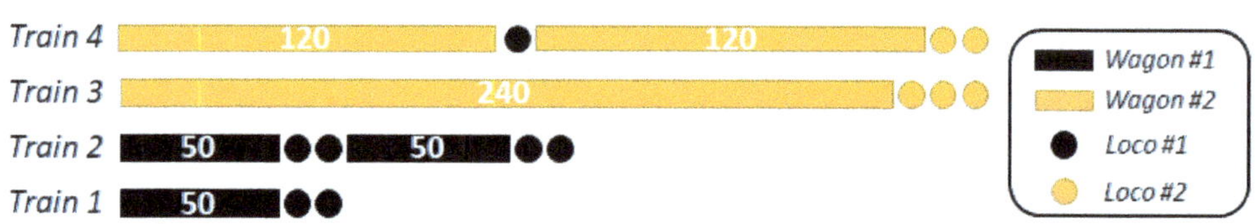

Figure 2: Schematic representation of the four benchmark train configurations. The wagon block rectangles are proportional in length to the number of wagons, which is written in the middle.

Table 1: Main parameters of the four vehicle types presented in the benchmark.

Vehicle type	Axle-load (tonne)	Axle numbers	Total mass (tonne)	Vehicle length (m)
Loco #1	22.33	6	134	22.95
Loco #2	32.50	6	195	24.35
Wagon #1	32.00	4	128	15.00
Wagon #2	40.00	4	160	11.00

As already mentioned in the previous section, all the benchmark input characteristics were summarized into text files and then re-discretized during the simulations presented in the paper using the following values. The coupler deflection was discretized with a step equal to 1e-5 m, the shortest track section was considered equal to 1 m, the fixed time-step for the notch level characteristic was set to 0.001 s and finally the locomotive speed discretization step was assumed equal to 0.001 m/s. The choice of the re-discretization steps must guarantee a good reproduction of the original data without exceeding the computer RAM memory limits.

A deep investigation was carried out to select the proper numerical solver for the case study shown in the previous lines. The best compromise in terms of accuracy of output results and computational efficiency was offered by the MATLAB variable step-size predictor-corrector stiff solver ode15s.

The constant sparsity pattern for the Jacobian matrix was passed to the solver as an input option, in order to reduce the total number of elements to be computed from the ode15s internal routine in failed time-steps.

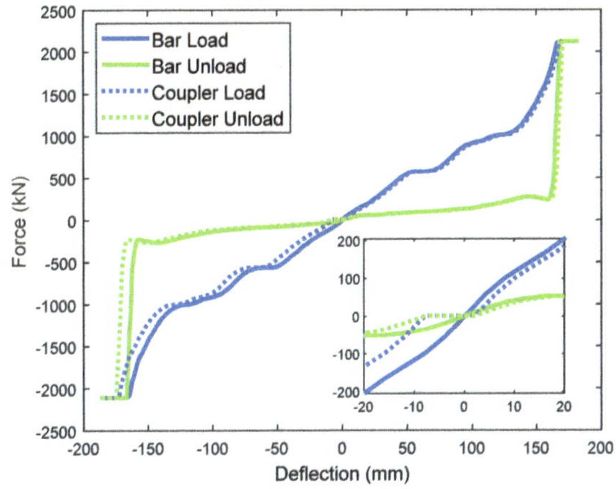

Figure 3: Force-deflection characteristics of drawbars and couplers.

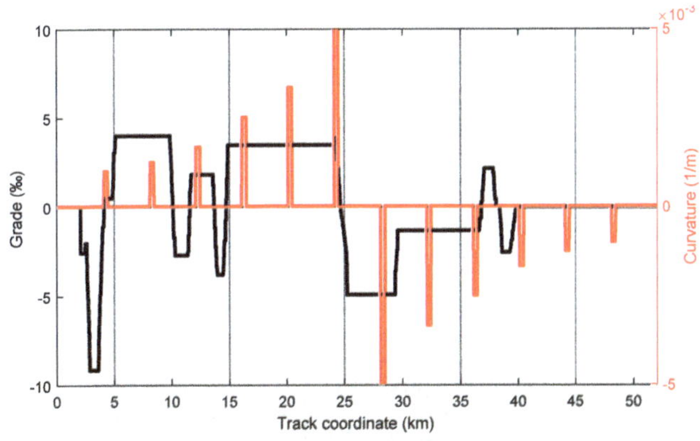

Figure 4: Track curvature and grade.

Finally, a fine tuning was also performed on the relative speed threshold value v_ε, see eqns (8)–(10). In fact, low values of this parameter proved to significantly slow down the computation, while higher values can considerably change the coupling system mechanical impedance. A good compromise was observed for a value $v_\varepsilon = 0.001$ m/s in all four configurations.

5 NUMERICAL RESULTS

This section deals with the validation and assessment of the proposed LTDPoliTO code. The validation of the output results was performed considering the assessment quantities described in the benchmark results paper [14] to compare the results produced by the various simulators, namely:

- The maximum and mean speed value among all vehicles and time steps.
- The largest in-train force and the corresponding position calculated as the maximum value among all time steps and all coupling systems, for both tensile and compressive states. Coupling system positions are numbered in this section in ascending order starting from the one on the rear of the leading loco, in accordance with the benchmark results paper.
- The mean value of the maximum force registered throughout the simulation in each coupling position, for both compressive and tensile states.
- The maximum deflection on a selected coupler position for both tensile and compressive states.

The outputs obtained in all four configurations showed a good agreement with the results from the other benchmark participants, as it can be inferred from the comparison of Table 2 with the analogous table from the international benchmark results paper.

Fig. 5 presents a direct comparison between the new LTDPoliTO code and the previous Simpack model in terms of the forces and deflections calculated on the selected coupling system for train 4, i.e., the one in position 146. As shown in Fig. 5(a), numerical divergences occurring in the previous Simpack model are not present in the new code, which produces numerically stable results. The cross-deflection characteristic in Fig. 5(a) shows the nonlinear transition path predicted by eqns (8)–(10). Discrepancies between LTDPoliTO and Simpack model outputs can be related to slight differences in the re-discretization of the loading and unloading curves as well as to the absence of numerical divergences in the new code. Similar observations are also true for the other simulation scenarios.

Table 2: LTDPoliTO results for the four benchmark train configurations.

	Speed (km/h)		Largest in-train force (kN) @ coupler pos.		Mean in-train force (kN)		Max. deflection (mm) @ selected coupler pos.		
Train	Max	Mean	Tens.	Comp.	Tens	Comp	Pos.	Tens.	Comp.
1	87.30	65.97	-565@2	338@2	-336	195	10	-64.56	45.07
2	89.40	64.55	-769@54	407@54	-444	263	61	-89.22	48.05
3	81.83	44.79	-1636@3	1008@144	-1166	620	147	-169.56	129.00
4	81.80	44.99	-1309@123	867@122	-932	501	146	-161.71	96.35

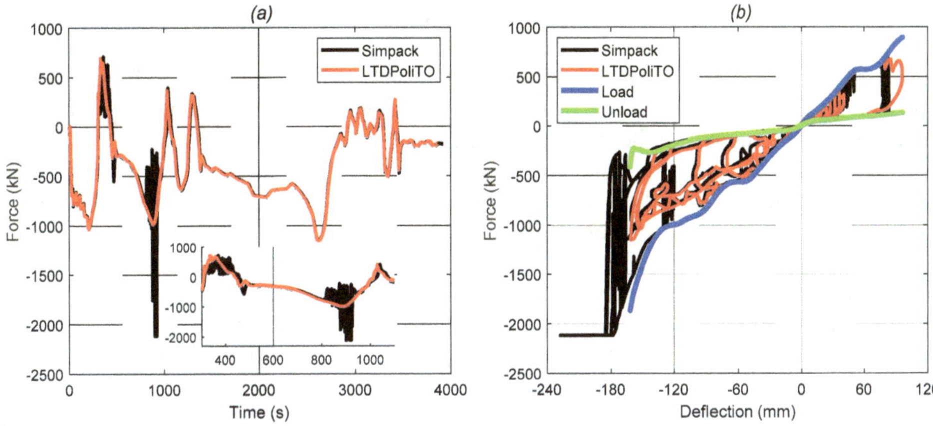

Figure 5: Comparison of output results from LTDPoliTO and previous Simpack model for the selected coupler position on train 4. (a) The in-train force time history; and (b) The force-deflection characteristic.

Shifting focus to the assessment of the computational efficiency of the new code, Table 3 presents the main quantities introduced in the paper for all four simulations, i.e., the wall time T_w, the train operation time T_{op} and their ratio Speed 1. The wall time is the time needed for the code to solve the LTD o.d.e system while the train operation time is the total simulated train route time. Low values of Speed 1 indicate that the time required for the numerical calculation is far lower than the simulated train operation time, thus highlighting a good computational efficiency. The simulations for the code validation were all performed on the same notebook with the following main features: Windows 10 Pro OS, Intel i7 CPU and 16GB RAM memory. Comparing the computational efficiencies of the simulators joining the benchmark is not an easy task since the participants did not run their models on the same computer. In fact, hardware and OS characteristics can drastically modify the computational speeds. However, the juxtaposition of Table 3 with the analogous from the benchmark results paper shows that LTDPoliTO is the fastest code in the four simulation scenarios, apart from TABLDSS. Therefore, LTDPoliTO can surely be considered as a solid and efficient tool for LTD simulations.

Table 3: LTDPoliTO computational times for the four benchmark simulation scenarios.

Train configuration	Train operation time (s)	Wall time (s)	Speed 1 (ms/s)
Train 1	2684	35	13
Train 2	2711	88	33
Train 3	3951	269	68
Train 4	3863	284	74
Average value	-	-	47

6 CONCLUSIONS

The paper deals with the implementation of a new LTD simulator, developed by the railway research group from Politecnico di Torino, in MATLAB, with great focus to its validation performed using the input dataset proposed in the international benchmarking of LTD simulators. The main goal of the work described in the paper was to develop a reliable tool for LTD simulations, able to produce numerically stable outputs with high computing performances, due to the inefficiencies of a previous model realized by the research group using the commercial multibody package Simpack.

The new code is developed in MATLAB and takes advantage of built-in high-level functions and vectorization strategies to handle large array indexing and mathematical operations. This approach guarantees extremely high computing performances and represents the main novel aspect of the proposed code. The simulations performed on the four benchmark scenarios led to a good agreement between the LTDPoliTO outputs and the results from the other simulators joining the benchmark. Moreover, the wall times obtained in the four scenarios make the new code as the second fastest simulator among the benchmark participants, since only TABLDSS was faster in the four configurations. Clearly, a fair comparison of the simulator performances would need all participants to run their codes on the same computer. Anyway, the initial goal of the activity described in the paper was achieved with success.

A lot of work is still needed to further increase LTDPoliTO computational efficiency as well as to widen the application extent of the code. In fact, the proposed simulator currently does not consider the retarding forces due to the air brake system. Therefore, a pneumatic module should be included for the calculation of air brake forces.

Moreover, at the current state of work, no explicit multi-thread computing technique is used, apart from MATLAB internal optimization processes. Future developments could investigate the possibility of parallelizing the code. Parallel computing and co-simulation strategies could be essential to maintain good computational efficiencies especially in case a pneumatic module is included in the code.

REFERENCES

[1] Colin, C., Longitudinal train dynamics. *Handbook of Railway Vehicle Dynamics*, Taylor & Francis: Boca Raton, 2006.

[2] Massa, A., Stronati, L., Aboubakr, A.K., Shabana, A.A. & Bosso, N., Numerical study of the noninertial systems: Application to train coupler systems. *Nonlinear Dynamics*, **68**(1), pp. 215–233, 2012. https://doi.org/10.1007/s11071-011-0220-2.

[3] Wei, L., Zeng, J. & Wang, Q., Investigation of in-train stability and safety assessment for railway vehicles during braking. *Journal of Mechanical Science and Technology*, **30**(4), pp. 1507–1525, 2016. https://doi.org/10.1007/s12206-016-0304-5.

[4] Zhu, T., Yang, B., Yang, C., Xiao, S., Yang, G. & Yang, B., The mechanism for the coupler and draft gear and its influence on safety during a train collision. *Vehicle System Dynamics*, **56**(9), pp. 1375–1393, 2018. https://doi.org/10.1080/00423114.2017.1413198.

[5] Wu, Q., Spiryagin, M. & Cole, C., Longitudinal train dynamics: an overview. *Vehicle System Dynamics*, **54**(12), pp. 1688–1714, 2016. https://doi.org/10.1080/00423114.2016.1228988.

[6] Cole, C., Spiryagin, M., Wu, Q. & Sun, Y.Q., Modelling, simulation and applications of longitudinal train dynamics. *Vehicle System Dynamics*, **55**(10), pp. 1498–1571, 2017. https://doi.org/10.1080/00423114.2017.1330484.

[7] Cole, C., Improvements to wagon connection modelling for longitudinal train simulation. *Conference on Railway Engineering Proceedings: Engineering Innovation For a Competitive Edge*, 1998.
[8] Bosso, N., Gugliotta, A. & Zampieri, N., Study of adhesion and evaluation of the friction forces using a scaled roller-rig C3. *5th World Tribology Congress, WTC 2013*, **3**, pp. 2640–2643, 2014.
[9] Bosso, N. & Zampieri, N., Real-time implementation of a traction control algorithm on a scaled roller rig. *Vehicle System Dynamics*, **51**(4), pp. 517–541, 2013. https://doi.org/10.1080/00423114.2012.750001.
[10] Bosso, N., Gugliotta, A. & Zampieri, N., RTCONTACT: An efficient wheel-rail contact algorithm for real-time dynamic simulations. *2012 Joint Rail Conference*, pp. 195–204, 2012. https://doi.org/10.1115/jrc2012-74044.
[11] Bosso, N. & Zampieri, N., A novel analytical method to calculate wheel-rail tangential forces and validation on a scaled roller-rig. *Advances in Tribology*, 2018. https://doi.org/10.1155/2018/7298236.
[12] Wu, Q., Cole, C., Luo, S. & Spiryagin, M., A review of dynamics modelling of friction draft gear. *Vehicle System Dynamics*, **52**(6), pp. 733–758, 2014. https://doi.org/10.1080/00423114.2014.894199.
[13] Spiryagin, M., Wu, Q. & Cole, C., International benchmarking of longitudinal train dynamics simulators: benchmarking questions. *Vehicle System Dynamics*, **55**(4), pp. 450–463, 2017. https://doi.org/10.1080/00423114.2016.1270457.
[14] Wu, Q. et al., International benchmarking of longitudinal train dynamics simulators: Results. *Vehicle System Dynamics*, **56**(3), pp. 343–365, 2018. https://doi.org/10.1080/00423114.2017.1377840.
[15] Bosso, N. & Zampieri, N., Long train simulation using a multibody code. *Vehicle System Dynamics*, **55**(4), pp. 552–570, 2017. https://doi.org/10.1080/00423114.2016.1267373.
[16] Bosso, N., Gugliotta, A. & Zampieri, N., A mixed numerical approach to evaluate the dynamic behavior of long trains. *Procedia Structural Integrity*, **12**, pp. 330–343, 2018. https://doi.org/10.1016/j.prostr.2018.11.083.
[17] Spiryagin, M., Cole, C., Quan Sun, Y., McClanachan, M., Spiryagin, V. & McSweeney, T., Longitudinal train dynamics. *Design and Simulation of Rail Vehicles*, CRC Press: Boca Raton, 2014.
[18] Zhang, Z., Li, G., Chu, G., Zu, H. & Kennedy, D., Compressed stability analysis of the coupler and buffer system of heavy-haul locomotives. *Vehicle System Dynamics*, **53**(6), pp. 833–855, 2015. https://doi.org/10.1080/00423114.2015.1023318.

SPEED TRAJECTORY OPTIMIZATION FOR A HIGH-SPEED TRAIN

ZHUANG XIAO, ZINING ZHAO, QINGYUAN WANG, PENGFEI SUN & XIAOYUN FENG
School of Electrical Engineering, Southwest Jiaotong University, People's Republic of China

ABSTRACT

The high-speed railway system has been a significant energy consumer due to the rapid increase of operation mileage and speed. Eco-driving is a promising method to reduce energy consumption, which aims at calculating speed profiles with minimization electric energy consumption under temporal and spatial operation constraints. Cruise control is regarded as the most energy-efficient control strategy when the cost function is the mechanical energy on wheels or the efficiency of traction drive system is simplified as a constant value. However, the efficiency of the traction drive system is varies among different traction force and speed pairs. This paper presents an eco-driving cycle computation problem for high-speed trains considering nonlinear loss of traction drive system. On basis of this model, the "traction-and-coasting" driving strategy is proposed to replace the cruise mode, and corresponding bi-level optimization algorithm is designed. The energy saving performance is investigated by comparing optimized results with the solution based on cruise control. Numerical simulation results illustrate that the traction energy consumption can be reduced by the proposed strategy, and the operation efficiency of the traction drive system can be improved. The economy significance of this method is illustrated and analyzed. For application, the proposed method can be used for a driver assistant system (DAS) or an automatic train operation (ATO) system, which can significantly improve the energy economy of the high-speed railway.

Keywords: high-speed railway, eco-driving, energy calculation, driver advisory system.

1 INTRODUCTION

The energy consumption of high-speed trains (HSTs) is increasing rapidly, which has become a significant consumer in transportation sector [1]. Most HSTs have adaptive cruise control system, which can help drivers to maintain a constant speed on the track without steep hills. Although the cruise mode can reduce the energy consumption caused by the basic resistance [2], the load of the traction transmission system (TTS) is low in most cases, which might result in low efficiency of the TTS. On basis of that, the electrical energy from the traction power system is not the smallest [3]. To reduce energy consumption of the HSTs from the catenary and improve efficiency of the TTS, a new optimization algorithm to generate energy-efficient strategies should be carried out.

Various studies about calculating energy saving speed profiles have already been applied to railway vehicles [4], [5]. These methods can be categorized into two groups:

1. Different numerical search algorithms based on discrete state space are widely used to solve the energy saving optimization problem directly.
2. The construction of speed profile is achieved by the linkage of the optimal modes that are derived from the maximum principle.

For the first kind of method, an nonlinear programming model is established according to the operation constraints such as speed limitations, trip time and traction forces, and then different algorithms are applied to obtain the eco-driving trajectory: brute force search [6], Dynamic Programming [7], Genetic Algorithms [8] and pseudo-spectral method [9]. With the increase of operation speed and distance between two stations, the search space would increase exponentially, and the computation burden would increase significantly. In addition,

WIT Transactions on The Built Environment, Vol 199, © 2020 WIT Press
www.witpress.com, ISSN 1743-3509 (on-line)
doi:10.2495/CR200241

the HSTs are driven by drivers manually. The frequent changing of operation modes would be infeasible to guide drivers. The second kind of method can avoid this problem due to specific operation modes. The summary of the switching principle among different optimal modes is presented in [10], [11]. An effective dual-loop algorithm is applied to locate the switching position and the sequence of different optimal modes [12].

The above studies aim at obtaining minimum mechanical energy on wheels, and the nonlinear power loss of the traction transmission system (TTS) is ignored [13]. The low load of the TTS would result in the low energy efficiency of the TTS when the train is in cruise mode. The maximum traction-coasting pair to replace cruise mode can improve the system efficiency in some situations.

This paper presents a bi-level optimization method that decreases energy consumption and improves energy efficiency. Specifically, the speed profile with the minimum mechanical energy is calculated first based on the Pontryagin Maximum Principle (PMP), and then traction-coasting pairs are used to replace cruise mode when the energy saving can be achieved. In addition, the quantitative analysis about the factors on energy efficiency of the traction-coasting pairs is discussed.

This paper is organized as follows. The system model and problem statement are described in Section 2. Section 3 illustrates the optimization method. The quantitative analysis of the energy efficiency of the traction-coasting pair is presented in Section 4. Case studies are carried out in Section 5.

2 SYSTEM MODEL AND PROBLEM STATEMENT

The acceleration/deceleration of the HST is controlled by the mechanical power on wheels. The electric energy is provided from the catenary. The energy conversion between electric and mechanical energy is achieved by the TTS, and the efficiency of the TTS changes with traction force and velocity pairs. The system model is the foundation to describe the train movement.

2.1 System model

The movement of the HST is driven by the interaction of traction or braking force, aerodynamic drag resistance $R(v)$ and the additional resistance $Mg\sin(\alpha)$ caused by the gradients of the track. The train dynamic model is presented as eqns (1) and (2)

$$\frac{\mathrm{d}v}{\mathrm{d}s}=\frac{F_t\text{-}F_b\text{-}R(v)\text{-}Mg\sin(\alpha)}{Mv}, \tag{1}$$

$$\frac{\mathrm{d}t}{\mathrm{d}s}=\frac{1}{v}, \tag{2}$$

- F_t is the traction force.
- F_b is the braking force.
- v is the train speed.
- s is the train position.
- t is the train operation time.
- M is the train mass.
- α is the angle of the gradient.

The calculation of $R(v)$ is given as eqn (3), which is based on field tests.

$$R(v) = M(a + bv + cv^2), \tag{3}$$

- a, b and c are constant values.

The traction power of the HST on traction motors P_w is calculated as

$$P_w = F_t \cdot v. \tag{4}$$

An accurate power consumption model should be introduced in this paper. The power of the rectifier P_t can be calculated as

$$P_t = \begin{cases} P_w + P_l & (P_w + P_l > 0 \ and \ P_w \neq 0) \\ 0 & P_w = 0 \end{cases}, \tag{5}$$

- P_l is the power loss of the TTS.

The power losses of the TTS for different speed and traction force pairs are shown in Fig. 1.

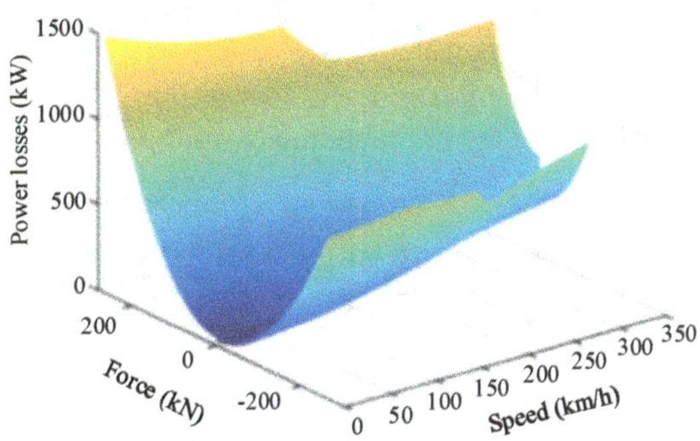

Figure 1: Power loss for the TTS.

2.2 Optimization problem

To minimize energy consumption, the cost function is written as

$$J = \int_{s_0}^{s_f} \frac{P_t}{v} ds, \tag{6}$$

- s_0 *and* s_f are the start and terminal position of the planned journey.

The operation constraints of the HST can be summarized as two groups: equality constraints and inequality constraints.

1. Equality constraints: the punctuality of railway is important, which can be regarded as

$$t(s_f) - t(s_0) = T, \tag{7}$$

$$v(s_0) = 0 \ and \ v(s_f) = 0, \tag{8}$$

- T is the planned trip time.

2. Inequality constraints: the control forces are bounded by the output characteristics of the TTS, expressed as

$$0 \leq F_{\mathrm{t}} \leq F_{\mathrm{t,max}}(v), \tag{9}$$

$$0 \leq F_{\mathrm{b}} \leq F_{\mathrm{b,max}}(v). \tag{10}$$

In addition, the train speed cannot exceed the speed limitations of the track

$$v \leq v_{\mathrm{lim}}(s), \tag{11}$$

- $F_{t,max}(v)$ and $F_{b,max}(v)$ are the maximum traction force and braking force, respectively.
- $v_{\mathrm{lim}}(s)$ is the speed limitation of the track.

The optimization problem for minimizing the electrical energy consumption of the HST is a typical nonlinear programming problem, and the optimization method is illustrated in next section.

3 OPTIMIZATION METHOD

3.1 Structure of optimization method

For the optimization model with nonlinear power loss of the TTS, the optimized operation strategies can be found by dividing the problem into two subproblems, shown in Fig. 2.

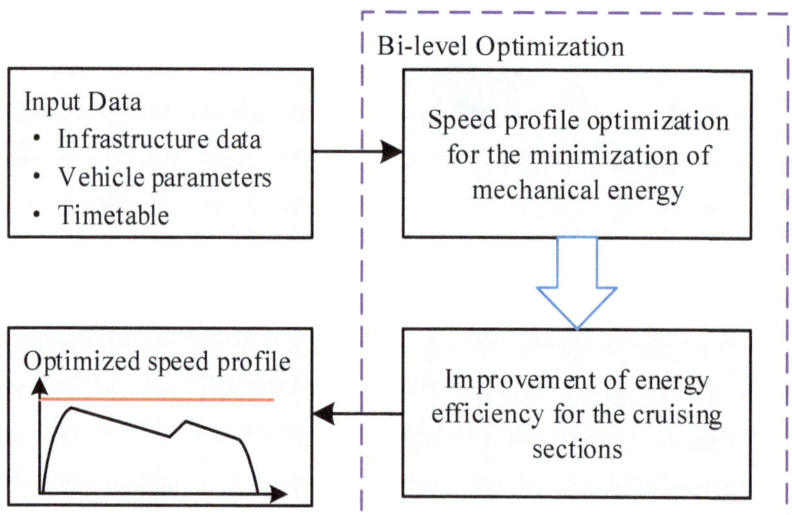

Figure 2: Structure of the bi-level optimization method.

For the optimization of mechanical energy, the planned timetable (i.e., the start and stop stations, the planned operation time), infrastructure data (i.e., the track gradients, curves and speed limits) and vehicle parameters (i.e., traction/braking characteristics, nonlinear power loss model and running resistance coefficient) are sent to the optimization subsystem. The PMP method is used to generate the optimal speed profile with minimal energy consumption on wheels.

For the improvement of energy efficiency on the cruising sections, the cost function is the electric energy considering the nonlinear power loss of the TTS. In the cruising sections of the speed profile with minimization mechanical energy, the energy efficiency of the TTS might be low. The traction and coasting (TnC) strategy is used to replace the cruising sections under the same operation time and distance constraints, which aims to improve the energy efficiency and reduce energy consumption.

3.2 Speed profile optimization with the minimization of mechanical energy

To minimize the mechanical energy, the cost function is written as

$$J' = \int_{s_0}^{s_f} \frac{P_w}{v} ds. \tag{12}$$

The optimal modes with minimum mechanical energy can be derived from the Maximum Principle [2], and the Hamiltonian function is defined as

$$H = -\frac{P_w}{v} + \frac{\lambda_1}{v} + \lambda_2 \cdot \frac{F_t\text{-}F_b\text{-}R(v)\text{-}Mg\sin(\alpha)}{Mv}. \tag{13}$$

The two adjoint variables λ_1 and λ_2 are determined by

$$\frac{d\lambda_1}{ds} = -\frac{\partial H}{\partial v}, \tag{14}$$

$$\frac{d\lambda_2}{ds} = -\frac{\partial H}{\partial t} = 0. \tag{15}$$

To maximum the Hamiltonian function, the four optimal modes are:

- Full Traction ($F_t = F_{t,max}(v)$), which is limited by the maximum traction force.
- Cruising at constant speed
- Coasting
- Full Braking ($F_b = F_{b,max}(v)$), which is limited by the maximum deceleration.

The speed profile can be constructed based on these modes. The typical sequence is shown in Fig. 3. The key technologies are the calculation methods to determine the cruising speed v_c and the end speed v_b of coasting with the constraint of operation time.

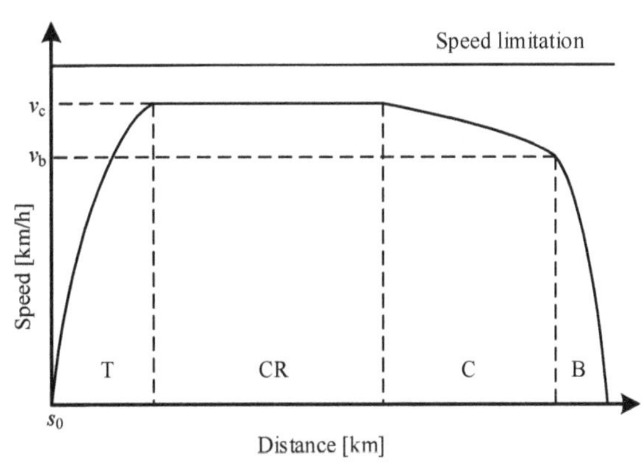

Figure 3: The typical optimal sequence for minimizing mechanical energy.

When the coating phase is on the same slope values, the relationship of the two speed values can be derived based on eqns (1), (13) and (14)

$$v_\mathrm{b} = \frac{\psi(v_\mathrm{c})}{R(v_\mathrm{c})+\frac{\psi(v_\mathrm{c})}{v_\mathrm{c}}-\sin\alpha}, \tag{16}$$

$$\psi(v_\mathrm{c}) = v_\mathrm{c}^{\,2}R'(v_\mathrm{c}). \tag{17}$$

By the iteration of cruising speed v_c, the optimized speed profile can be obtained under the constraint of trip time.

3.3 Improvement of energy efficiency on the cruising sections

The cruising mode is the most energy saving strategy under the distance and time constraints when the cost function is the mechanical energy on wheels [2]. However, the load of the cruising mode is lower than the full traction, and the power loss of the TTS would increase. The TnC strategy is proposed to reduce the power loss on the TTS. The main idea of the TnC is to increase the load of the TTS by the full traction mode and improve the energy efficiency, and the coast mode does not consume energy. Two situations would occur when the TnC strategy is used, shown as Fig. 4(a) and Fig. 4(b).

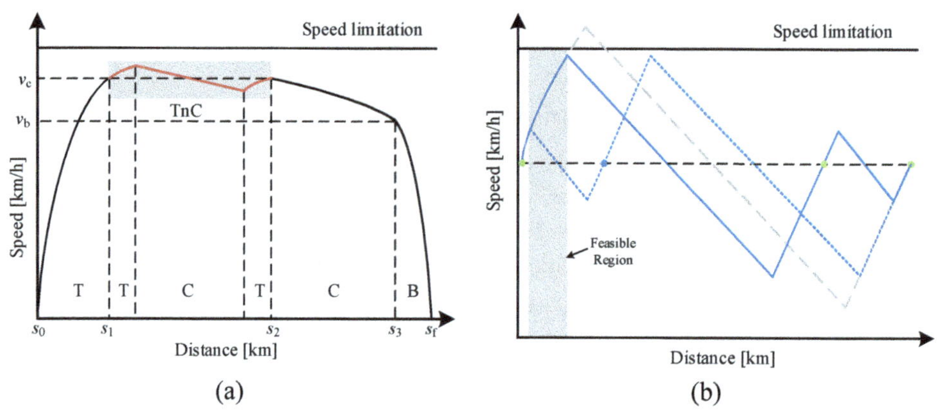

Figure 4: The TnC strategy for minimizing electrical energy.

In Fig. 4(a), the maximum speed of the TnC strategy is under the speed limitations, and one pair of traction and coast mode can replace the cruise mode under same time and distance constraints. In addition, the maximum speed of the TnC strategy might exceed the speed limitations when one pair of traction and coast mode is used, shown as the grey dotted line in Fig. 4(b). In this case, more TnC pairs should be utilized. Nowadays, the HSTs are mainly operated by drivers. To reduce the switching frequency of the operation mode, we should try to reduce the TnC pairs. Fig. 4(b) illustrated the feasible region determined by the two boundary linkage sequences as the green and blue lines. Then the energy-optimal operation strategy can be determined by the brute force search in the feasible region.

To summarize, the proposed method is shown as the following steps:

1. Denote $v_c = \frac{s_f}{T}$, then the value of v_b is determined, and the optimized speed profile with the minimum mechanical energy can be obtained. Comparing the operation time T_c with the given trip time T. If $|T_c - T| < \varepsilon$, turn to step 2. Otherwise, adjust the speed v_c.
2. Find the start and end positions of the cruise mode, one pair of the TnC strategy is used to replace the cruise mode.
3. Compare the maximum speed of the TnC with speed limitations, if the maximum speed is lower than the speed limitation, the optimal speed profile is the strategy with smaller energy consumption, either the cruise strategy or the TnC strategy. if the maximum speed is larger than the speed limitation, turn to step 4.
4. Find the feasible region show in Fig. 4(b) using the multiple TnC pairs. In the feasible region, the energy-optimal strategy can be found by brute force search.

4 QUANTITATIVE ANALYSIS

Although the TnC strategy can reduce the energy loss on the TTS, the loss would increase caused by the fluctuation of speed profile. The quantitative analysis of the energy consumption for the TnC and the cruising strategies should be discussed. The load on the TTS has a great impact on the energy performance of the HST, which determines the energy-efficient operation strategy that is the cruising or the TnC. The constant speed of the cruising mode and the gradients of the track are two parameters to influence the TTS load.

4.1 Train characteristics

The CRH3 high-speed train is selected to evaluate the performance of the TnC and cruising strategies. The maximum traction force and aerodynamic drag resistance are presented in Fig. 5.

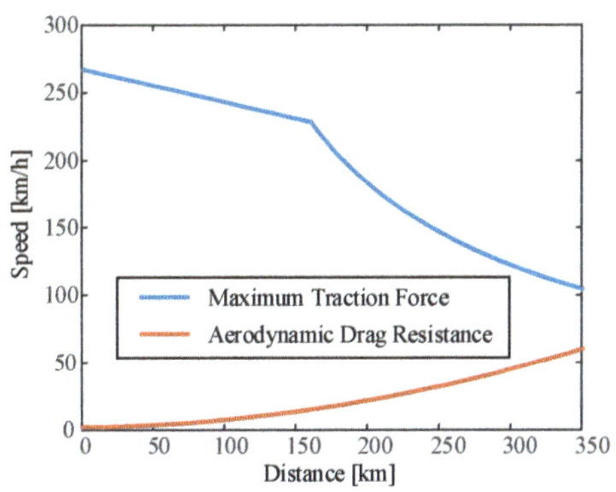

Figure 5: The maximum traction force and aerodynamic drag resistance.

4.2 Analysis for different constant speed values

The aerodynamic drag resistance is proportional to the square of the train speed, and it becomes the main energy consumer at high speed [7]. With the increase of the constant speed, the bigger traction force is required to overcoming the aerodynamic drag resistance, and the energy efficiency of the TTS would increase. The flat track is chosen in this case, and the range of investigated speed is from 150 km/h to 300 km/h.

Fig. 6(a) illustrates the speed profiles of the TnC and the cruising strategies at 150 km/h, 225 km/h and 300 km/h. Moreover, the corresponding energy efficiency profiles are shown in Fig. 6(b). It can be observed that the coasting phase becomes shorter with the increase of constant speed values. The energy efficiency of the TnC strategy changes slightly at different constant speed, but the efficiency of the cruising strategy improves significantly when the constant speed values are increased.

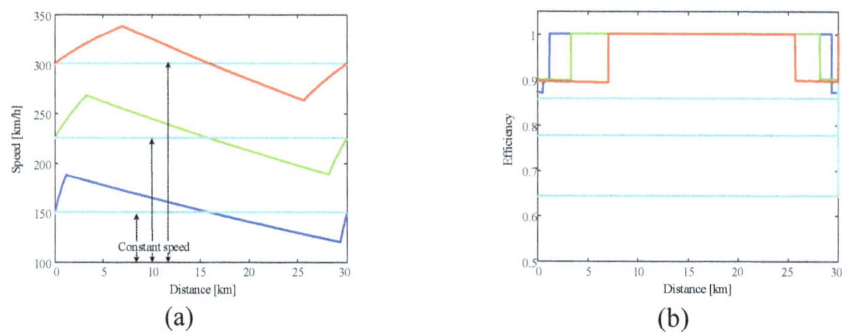

Figure 6: The results for different speed values. (a) Speed profile; and (b) Efficiency profile.

The energy consumption for different constant speed values is shown in Table 1. From the table, we can see that the more energy is needed to drive the HST either of the cruising mode or the TnC strategy when the constant speed value is risen. In addition, the TnC strategy can save energy under different constant speed values, and the energy saving rate is decreased with the increase of the constant speed value.

Table 1: Energy consumption for different constant speed values.

Constant value (km/h)	Cruising mode (kW·h)	The TnC (kW·h)	Energy saving rate (%)
150	175.1973	133.4892	23.81
175	212.2834	169.7276	20.05
200	251.6512	211.0614	16.13
225	291.0044	258.5505	11.15
250	336.5951	311.8014	7.37
275	388.3206	370.269	4.65
300	446.2018	433.8835	2.76

4.3 Analysis for different gradients

The gradient of the track is another important factor that influences the load of the train at cruising mode. The energy values for different constant speed and gradients pairs are presented in Fig. 7. When the cruise mode occurs on the track with slope less than 2‰, the TnC strategy can save energy under different constant speed values, and the energy saving rate decreases with the increase of the TTS load. The energy consumption of the TnC is larger than the cruising mode when the speed is 150 km/h and the gradient is 6‰. This indicates that the fluctuation of speed will increase the energy loss caused by aerodynamic drag resistance, although the power loss on the TTS is low using the TnC.

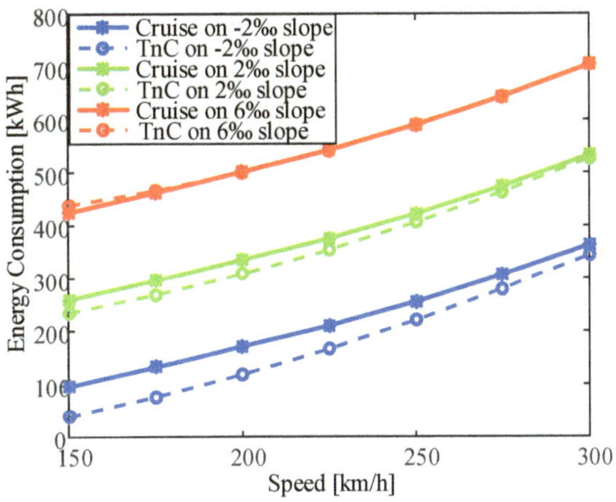

Figure 7: The energy consumption for different constant speed and gradients.

5 CASE STUDY

The distance of the simulation high-speed railway line is 70 km, and the planned trip time is 1200 s. The characteristics of the HST are shown in Fig. 5. Two cases are designed to evaluate the proposed TnC strategy.

- Case 1: the speed limitation is not considered, which means one pair of the TnC can replace the cruising mode.
- Case 2: the speed limitation is considered, and two pairs of the TnC should be adopted.

5.1 Case 1

This case aims to evaluate the effectiveness of the TnC strategy without considering the speed limitations. Firstly, the speed profile with the cruising mode based on the PMP is obtained in Fig. 8(a), and then the TnC strategy is used to replace the cruise mode. The energy efficiency profiles of the two strategies are shown in Fig. 8(b).

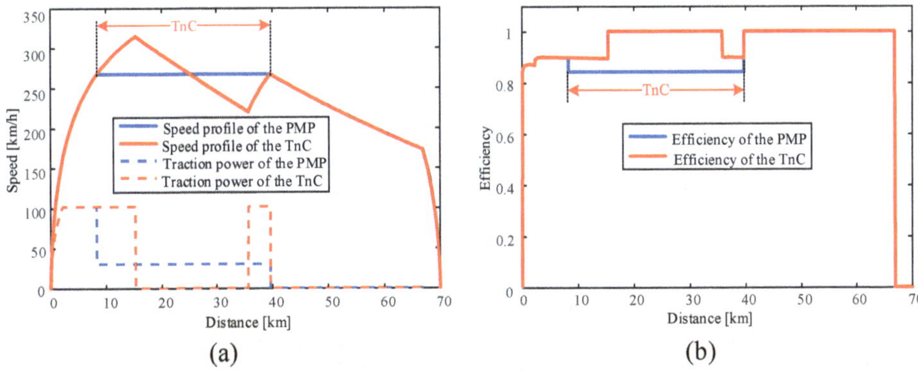

Figure 8: The speed profiles and energy efficiency. (a) Speed profile; and (b) Energy efficiency.

The cruising speed obtained by the PMP is 266.8 km/h, and the energy efficiency of the cruising mode is 0.8425. The maximum speed of the TnC strategy is 314 km/h, and the energy efficiency of the TnC is improved significantly. For the TnC operation section, the electrical energy consumption of the PMP and the TnC are 474.8873 kWh and 462.7282 kWh, respectively. The energy saving rate is 2.56% of the TnC strategy.

Table 2 illustrates the energy loss caused by the basic resistance and the TTS. It shows that the reduced loss of the TTS is larger than the increased energy consumption of the basic resistance, and the energy efficiency can be improved by the TnC strategy when the nonlinear power loss model of the TTS is considered.

Table 2: Energy consumption for different devices.

Operation strategy	Energy loss of resistance (kW·h)	Energy loss of the TTS (kW·h)
The TnC	404.5647	58.1635
The Cruising	395.7458	79.1415

5.2 Case 2

This case aims to find the switching sequence of the multiple TnC pairs when the maximum traction speed of one TnC pair exceeds the speed limitations. The speed profiles of different switching positions are shown in Fig. 9(a). The operation time and distance for theses profiles are the same. Furthermore, the relationship between the energy consumption and the switching position of the first coast mode is shown in Fig. 9(b). It can be observed that the energy consumption decreases with the increase of the position before 3.19 km, and then the energy consumption increases with the position. A minimum energy consumption can be found, and the energy-optimal switching sequence of multiple TnC pairs can be determined.

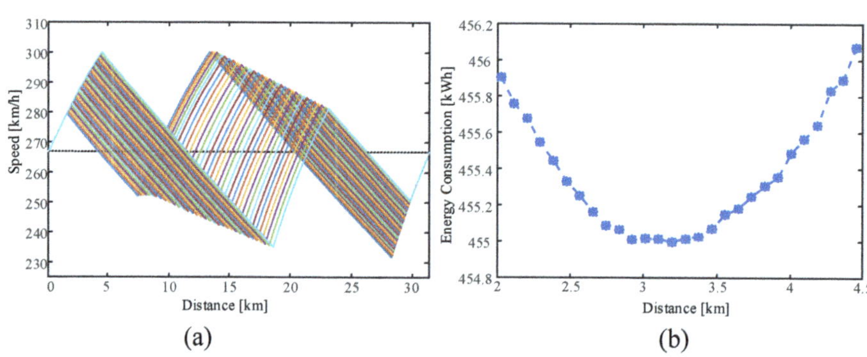

Figure 9: The performance of different TnC pairs switching position. (a) Speed profile; and (b) Energy consumption

In this case, the speed profile based on the PMP is the same as Fig. 8(a). The strategy with two pairs TnC is shown in Fig. 10(a), and the related energy efficiency profiles are shown in Fig. 10(b). The maximum speed of the TnC strategy is 289 km/h, and the energy consumption of the TnC is 455 kWh. The two pairs TnC strategy can reduce electrical energy consumption further compared to the one pair TnC strategy. It is obvious that the two pairs TnC strategy is more complex than the only one pair TnC strategy for drivers.

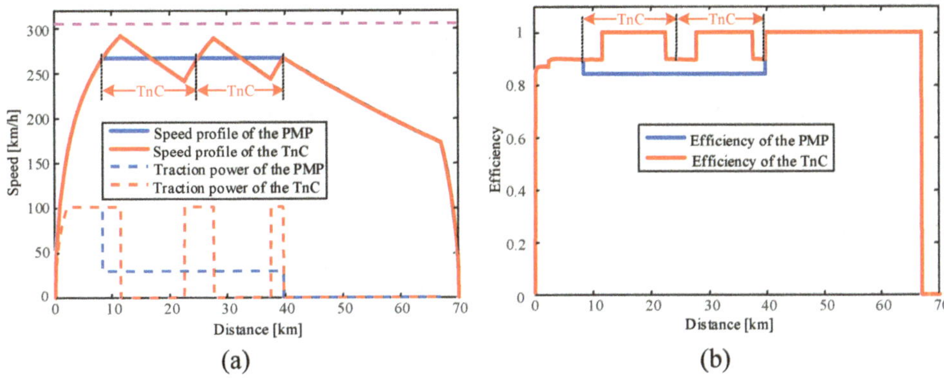

Figure 10: The speed profiles and energy efficiency. (a) Speed profile (b) Energy efficiency.

6 CONCLUSION

This paper presents a bi-level optimization method to minimize the electrical energy consumption with the nonlinear power loss model of the TTS for the HSTs. Firstly, the minimum mechanical energy on wheel is obtained based on the PMP, and the speed profile is constructed with four optimal modes such as full traction, cruising, coasting and full braking. Then, the TnC driving strategy is proposed to improve the energy efficiency of the TTS when the train is in cruising mode with low load. Comparing optimized results with the solution based on cruise control, the energy saving rate can be achieved without and with speed limitations. In addition, the performance of proposed TnC strategy is illustrated and analyzed under different constant speed values and gradients. For application, the proposed method can be used for a driver assistant system (DAS) or an automatic train operation (ATO) system to improve the energy economy of the high-speed railway.

ACKNOWLEDGEMENT

This research was supported by the National Key Research and Development Program of China under the Grant 2017YFB1201302 and Grant 2016YFB1200502.

REFERENCES

[1] Wang, K., Hu, H., Chen, J., Zhu, J., Zhong, X. & He, Z., System-level dynamic energy consumption evaluation for high-speed railway. *IEEE Transactions on Transportation Electrification*, **5**(3), pp. 745–757, 2019.

[2] Howlett, P.G., Milroy, I.P. & Pudney, P.J., Energy-efficient train control. *Control Engineering Practice*, **2**(2), pp. 193–200, 1994.

[3] Zhao, X.-H., Ke, B.-R. & Lian, K.-L., Optimization of train speed curve for energy saving using efficient and accurate electric traction models on the mass rapid transit system. *IEEE Transactions on Transportation Electrification*, **4**(4), pp. 922–935, 2018.

[4] Yang, X., Li, X., Ning, B. & Tang, T., A survey on energy-efficient train operation for urban rail transit. *IEEE Transactions on Intelligent Transportation Systems*, **17**(1), pp. 2–13, 2016.

[5] Scheepmaker, G.M., Goverde, R.M. & Kroon, L.G., Review of energy-efficient train control and timetabling. *European Journal of Operational Research*, **257**(2), pp. 355–376, 2017.

[6] Tian, Z., Zhao, N., Hillmansen, S., Roberts, C., Dowens, T. & Kerr, C., Smartdrive: Traction energy optimization and applications in rail systems. *IEEE Transactions on Intelligent Transportation Systems*, **20**(7), pp. 2764–2773, 2019.

[7] Xiao, Z., Feng, X., Wang, Q. & Sun, P., Eco-driving control for hybrid electric trams on a signalized route. *IET Intelligent Transport Systems*, **14**(1), pp. 36–44, 2020.

[8] Wong, K.K. & Ho, T.K., Dynamic coast control of train movement with genetic algorithm. *International Journal of Systems Science*, **35**(13), pp. 835–846, 2004.

[9] Wang, Y., De Schutter, B., van den Boom, T.J. & Ning, B., Optimal trajectory planning for trains–a pseudospectral method and a mixed integer linear programming approach. *Transportation Research Part C: Emerging Technology*, **29**, pp. 97–114, 2013.

[10] Albrecht, A., Howlett, P., Pudney, P., Vu, X. & Zhou, P., The key principles of optimal train control–part 1: Formulation of the model, strategies of optimal type, evolutionary lines, location of optimal switching points. *Transportation Research Part B: Methodologies*, **94**, pp. 482–508, 2016.

[11] Albrecht, A., Howlett, P., Pudney, P., Vu, X. & Zhou, P., The key principles of optimal train control – Part 2: Existence of an optimal strategy, the local energy minimization principle, uniqueness, computational techniques. *Transportation Research Part B: Methodologies*, **94**, pp. 482–508, 2016.

[12] Liu, R.R. & Golovitcher, I.M., Energy-efficient operation of rail vehicles. *Transportation Research Part A: Policy Practice*, **37**(10), pp. 917–932, 2003.

[13] Frank, P.T. & Meyer, M., An algorithm for the optimal control of the driving of trains. *Proceedings of the 39th IEEE Conference on Decision and Control*, pp. 2123–2127, 2000.

CALCULATION METHOD OF LOAD CONDITIONS FOR PERFORMANCE TEST OF RAIL FASTENINGS OF A FLOATING SLAB TRACK

SHINGO TAMAGAWA
Department of Track Structures and Components, Railway Technical Research Institute, Japan

ABSTRACT
This study aims to establish a calculation method of load conditions for a performance test of rail fastenings of a floating slab track. A floating slab track is a type of vibration-reducing tracks. It is composed of rails, rail fastenings and concrete slabs supported by elastic layers such as coil springs. Among these components, the design of rail fastenings is an important factor in determining the performance of the floating slab track. In the design of rail fastenings, the Japanese design standard for railway structures requires that the performance of rail fastenings should be checked by the static loading test at a laboratory. In the performance test, the test load is applied to the rail in consideration of the load dispersion by the track. In this study, an FEM model of the floating slab track is developed to estimate the dispersed forces. It also shows the calculation method of load conditions for a performance test of rail fastening of the floating slab track. Furthermore, the performance test of the rail fastening is performed.
Keywords: rail fastening, floating slab track, performance test, load condition, FEM.

1 INTRODUCTION

A floating slab track is a type of vibration-reducing tracks and is composed of rail, rail fastenings and concrete slabs supported by elastic layers such as coil springs (Fig. 1). Even among these components, the design of rail fastenings is an important factor in the determining the performance of the floating slab track. In the design of rail fastenings, the Japanese design standard for railway structures (RTRI [2]) requires that the performance of rail fastenings should be checked by the static loading test at a laboratory (hereafter, performance test). Fig. 2 shows examples of the test arrangement for a general rail fastenings. In the test, the test load is applied to the rail in consideration of the load dispersion by the track (Youichi [3]). The dispersed design load, namely vertical rail pressure, lateral rail pressure, and rail tilting moment should be applied to the rail with a single assembled rail fastening. Vertical and lateral rail pressures refer to the dispersed forces that act on a single assembled rail fastening when the wheel load and lateral load act on the track. Rail tilting moment refers to the moment by rail tilting that acts on a single assembled rail fastening and is defined around the central point of the rail bottom surface. These dispersed forces have been estimated using the practical solution based on the torsion theory (Yutaka [4], Takeshi et al. [5]). On the other hand, the author has developed the FEM model of a ballasted track or a general slab track to estimate the dispersed forces mentioned above and has proposed the calculation method of the load conditions for the performance test of rail fastenings based on the FEM results (Shingo et al. [6], Tadashi et al. [7]). However, the conventional FEM model cannot be applied to the floating slab track.

In this study, a FEM model of the floating slab track is developed to estimate the dispersed forces and the calculation method of load conditions for a performance test of rail fastening of the floating slab track is shown. Furthermore, the performance test of the rail fastening is performed.

WIT Transactions on The Built Environment, Vol 199, © 2020 WIT Press
www.witpress.com, ISSN 1743-3509 (on-line)
doi:10.2495/CR200251

Figure 1: Floating slab track [1].

Figure 2: Performance test arrangement for rail fastenings.

2 FEM MODEL OF A FLOATING SLAB TRACK

2.1 FEM model

Fig. 3 shows the FEM model of a floating slab track. The FEM model is composed of solid elements, plate elements and spring elements. The rail is modelled by solid elements and is supported by 64 rail fastenings, which are spaced at 625 mm intervals. Two slabs are modelled by plate elements and are supported by 32 coil springs, which are spaced at 1,250 mm intervals. The rail clips composed of tip springs and lateral springs are modelled by non-linear elastic spring elements. The tie plates are modelled by plate elements. The springs under the rail and under the tie plates are modelled by the non-linear spring elements as well as the rail clips. The following are the constraining conditions of the respective elements.

- All elements: v=0 on the cross section to which loads are applied
- Slabs: $u=\theta_y=0$ on the side opposite to the rail
- Coil springs: $u=v=w=\theta_x=\theta_y=\theta_z=0$ on the side opposite to the slabs.

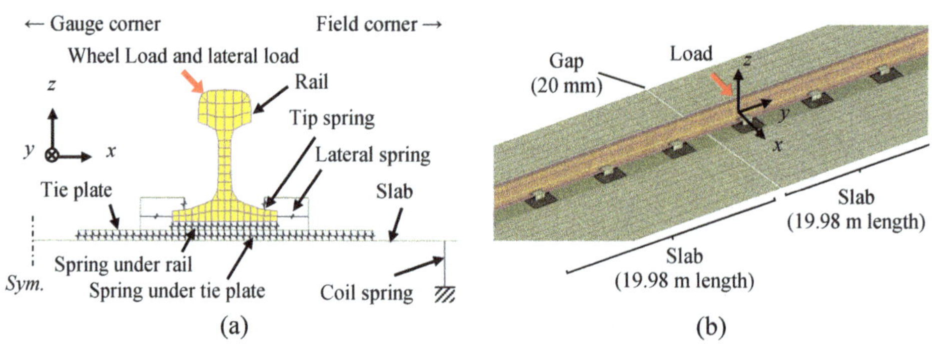

Figure 3: FEM model of a floating slab track. (a) Rail cross section; and (b) Longitudinal direction.

Here, u, v, and w indicates the displacement in x, y, and z directions and θ_x, θ_y, and θ_z indicates the rotation around x, y, and z axes. The loads are applied to the side of rail head assuming the train loads, which are composed of the wheel load and the lateral load.

In this study, FEM analyses executed are non-linear elastic analyses using NASTRAN (ver. 10) software (Siemens [8]). The total values of applied loads are divided into 10 steps. The convergence calculation at each step is performed with the quasi-Newton method.

2.2 Spring coefficients

The spring coefficients were measured prior to the performance test and incorporated into the FEM model. This study focuses on the rail fastening as shown in Fig. 4. The coefficient of the spring under the rail and the tie plate was measured by applying the compression force from 0 kN to 100 kN to the rail as shown in Fig. 5. The coefficient of the tip spring was measured by applying the vertical force to the rail as shown in Fig. 6. The test procedure was done as follows: The initial displacement of rail was set to 0 mm. The rail was pulled up and the rail pad was extracted from under the rail. The force applied to the rail at this time was defined as Pmax. The force was decreased until 0 kN and was increased until Pmax once again. As a result, the relationship between the force and the displacement was measured and was defined as the coefficient of the tip springs. The force at the time of displacement was decreased until 0 mm was defined as the initial clamping force. The coefficient of the lateral springs was measured by applying the lateral force from 0 kN to 60 kN to the rail base as shown in Fig. 7. The coefficient of the coil spring was presented by manufacturer. The relationships between the force and the displacement obtained above were approximated by some lines and the coefficient of each spring was determined as shown in Fig. 8. The spring coefficients of the lateral springs and the tip springs are the values per a clip. The initial clamping force was 4 kN per a clip.

2.3 FEM results

FEM analyses were carried out in two cases (case A and case B). The wheel loads and lateral loads which apply on the FEM model are shown in Table 1. The design load A is applied to the rail from the gauge corner and the design load B is applied from the field corner alternately. These loads are generally used at the sharp curved track. As FEM results, Fig. 9 shows the distribution of rail head displacement, rail base displacement and rail tilting angle.

Figure 4: Rail fastening.

Figure 5: Test for spring under rail and tie plate.

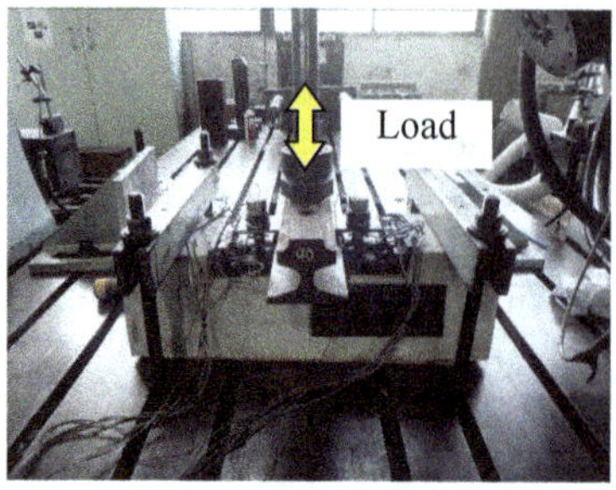

Figure 6: Test for tip spring.

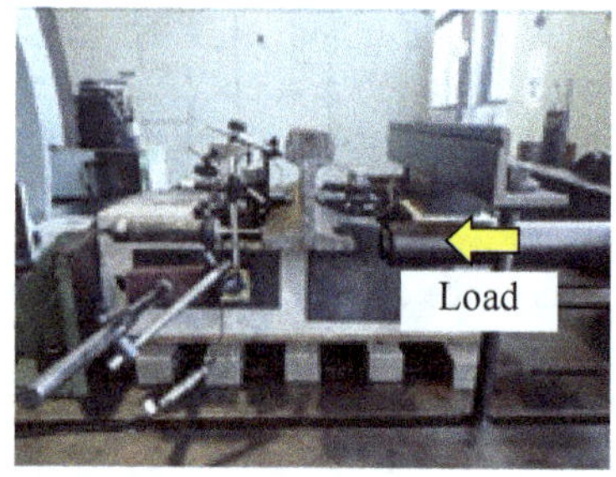

Figure 7: Test for lateral spring.

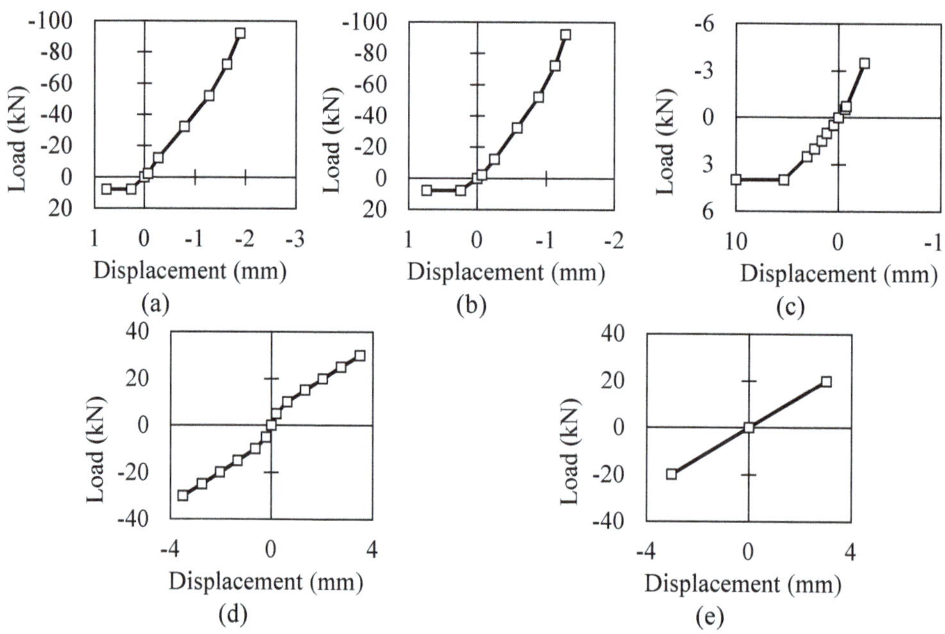

Figure 8: Spring coefficients of each spring. (a) Spring under rail; (b) Spring under tie plate; (c) Tip spring; (d) Lateral spring; and (e) Coil spring.

Table 1: Wheel load and lateral load (Meter-gauged line, R < 600 m).

Load	Case A (Design load A)	Case B (Design load B)
Wheel load P	98kN	86kN
Lateral load Q	60kN	30kN

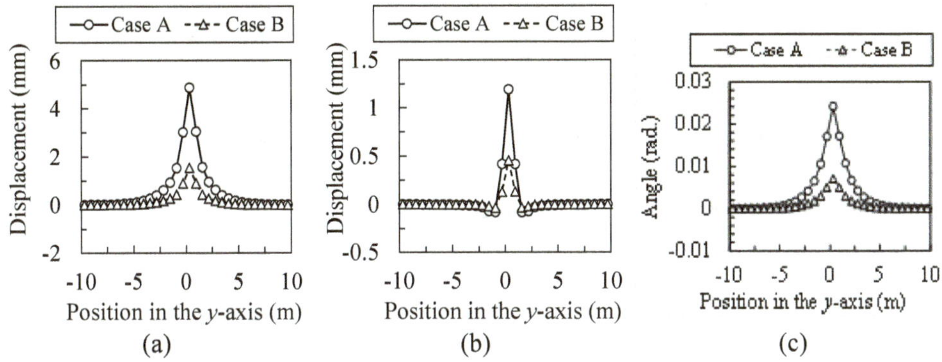

Figure 9: FEM results of rail displacements and rail tilting angle. (a) Rail head displacement; (b) Rail base displacement; and (c) Rail tilting angle.

On the basis on the FEM results, vertical rail pressure W, lateral rail pressure H and rail tilting moment M was calculated by the following equations (wherein each symbol is as described in Fig. 10)

$$W = \sum_{i=-n}^{n} (W_i) - (P_l + P_r), \tag{1}$$

$$H = R_l + R_r, \tag{2}$$

$$M = \sum_{i=1}^{n} (W_i \cdot b_i) - \sum_{i=-n}^{-1} (W_i \cdot b_i) + (P_l \cdot d_l - P_r \cdot d_r) + (R_l + R_r) \cdot c. \tag{3}$$

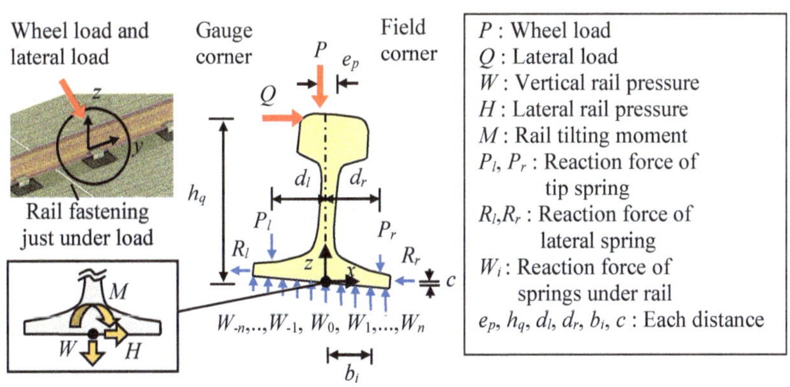

Figure 10: FEM model balanced under wheel load and lateral load.

Fig. 11 shows the distribution of vertical rail pressure, lateral rail pressure and rail tilting moment. Table 2 shows the rail displacements and dispersed load at the point of the load application.

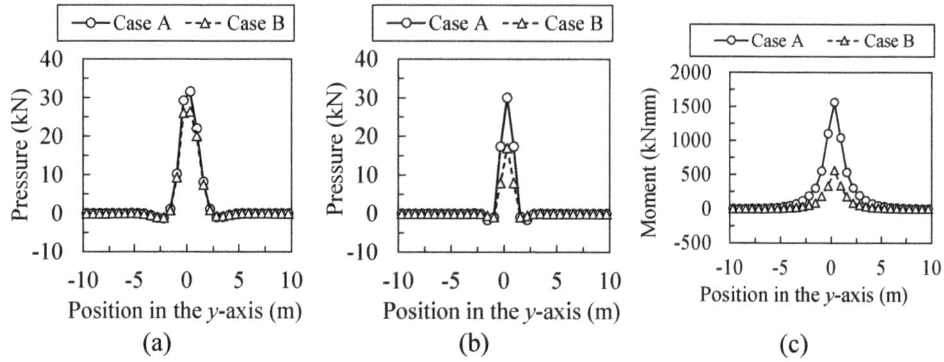

Figure 11: Dispersed design loads estimated from FEM results. (a) Vertical rail pressure; (b) Lateral rail pressure; and (c) Rail tilting moment.

Table 2: Rail displacements and dispersed loads at the point of the load application.

Item		Unit	Case A	Case B
Rail displacements	Rail head displacement	mm	4.88	1.54
	Rail base displacement	mm	1.2	0.45
	Rail tilting angle	rad.	0.0241	0.0071
Dispersed loads	Vertical rail pressure W	kN	31.6	26.3
	Lateral rail pressure H	kN	30.1	17
	Rail tilting moment M	kNmm	1567	564

3 CALCULATION OF LOAD CONDITIONS FOR PERFORMANCE TEST

This section shows the method for determining the loading conditions of the performance test based on the FEM results.

3.1 Calculation method

The performance test is performed under bi-axial loading condition in which the loads are applied alternately from the gauge corner and the field corner of test rail as shown in Fig. 12. Test loads L_A and L_B, the angles of load application θ_A and θ_B, and the height of load application h are expressed by the following equations

$$L_A = \sqrt{(W_A - L_0 \sin\theta_B)^2 + (H_A + L_0 \cos\theta_B)^2}, \tag{4}$$

$$L_B = \sqrt{(W_B - L_0 \sin\theta_A)^2 + (H_B + L_0 \cos\theta_A)^2}, \tag{5}$$

$$\theta_A = \tan^{-1}\left(\frac{W_A - L_0 \sin\theta_B}{H_A + L_0 \cos\theta_B}\right), \tag{6}$$

$$\theta_B = \tan^{-1}\left(\frac{W_B - L_0 \sin\theta_A}{H_B + L_0 \cos\theta_A}\right), \tag{7}$$

$$h = \frac{M_A + e\left(W_A - 2L_0 \sin\theta_B\right)}{H_A}, \tag{8}$$

$$h = \frac{M_B + e\left(W_B - 2L_0 \sin\theta_A\right)}{H_B}. \tag{9}$$

Here, W_A, H_A, and M_A or W_B, H_B, and M_B are vertical rail pressure, lateral rail pressure, and rail tilting moment calculated by eqns (1)–(3) when design load A or B are applied. The subscript A or B means Case A or B. L_0 is the minimum load applied in the opposite direction to L_A or L_B and is set at 5 kN–10 kN in consideration of stability of the test. The detail of the solution of eqns (4)–(9) is described in the paper (Shingo et al. [6]).

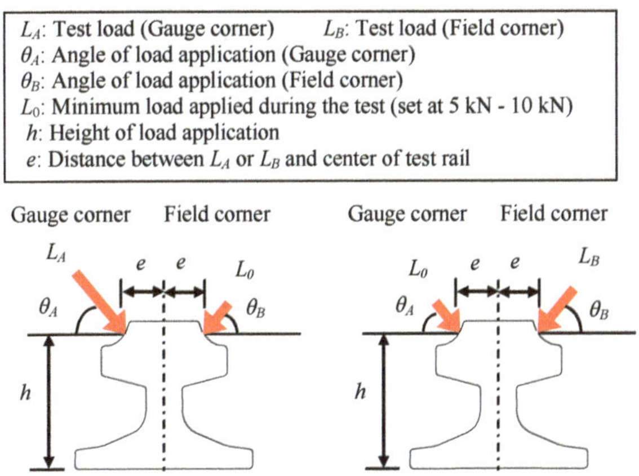

Figure 12: Bi-axial loading condition for performance test. (a) Design load A; and (b) Design load B.

3.2 Calculation results

Loading condition of the performance test was calculated on the basis of the FEM results. Substituting the values of vertical rail pressure, lateral rail pressure and rail tilting moment shown in Table 2 into eqns (4)–(9), we calculate test loads L_A and L_B, the angles of load application θ_A and θ_B, and the height of load application h as shown in Table 3. L_0 and e is set at 5 kN and 30 mm.

Table 3: Load condition of performance test.

Item	L_A	L_B	θ_A	θ_B	h	L_0	e
Unit	kN	kN	Degrees	Degrees	mm	kN	mm
Value	42.0	29.3	41.2	51.6	80	5	30

4 PERFORMANCE TEST OF THE RAIL FASTENING

In order to verify the validity of the calculation of loading condition based on the FEM model, the performance test was carried out. The test result was compared with the FEM results.

4.1 Test procedure

Fig. 13 shows the test specimen. A concrete block cut to the length of 700 mm was set on the testing machine. JIS-50 kgN test rail (height 80 mm) was set on the block and was fastened to the block by the rail fastenings of the floating slab track. Rail clips were fastened by the bolts with 140 Nm, the torque specified in the specifications. The block was fastened to the testing machine firmly without elastic layers. Table 4 shows the process of loading. The load condition refers to Table 3. Fig. 14 shows the measuring points of rail displacement. Rail tilting angle was calculated by dividing the difference between the displacement at measuring point 1 and 2 by the distance between two measuring points (115 mm). Rail base displacement was measured at point 3. We converted rail head displacement into a value of the top of JIS-50 kgN rail by multiplying rail tilting angle by a height of JIS-50 kgN rail (153 mm) and by adding rail base displacement. The direction of lateral displacement from gauge corner to field corner was defined as positive and clockwise rail tilting angle was defined as positive.

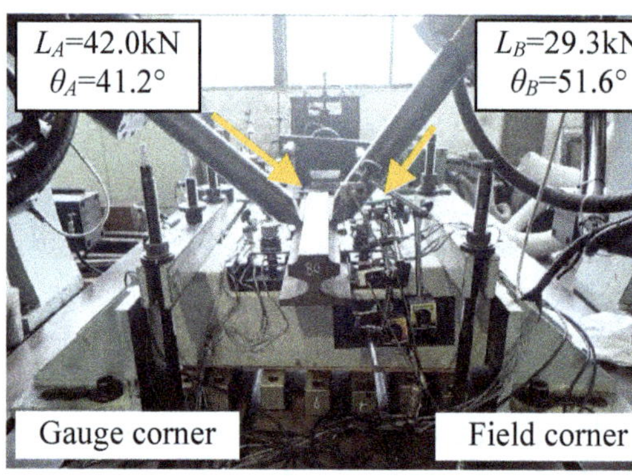

Figure 13: Performance test specimen.

Table 4: Load procedure.

Load procedure	Test load (kN)
Preload	$L_A: 0 \rightarrow 5.0$
	$L_B: 0 \rightarrow 5.0$
L_A	$L_A: 5.0 \rightarrow 42.0 \rightarrow 5.0$
	L_B: 5.0 (fixed)
L_B	L_A: 5.0 (fixed)
	$L_B: 5.0 \rightarrow 29.3 \rightarrow 5.0$

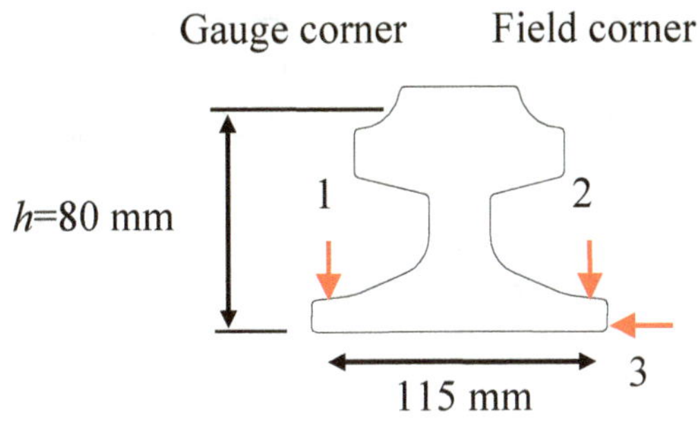

Figure 14: Measuring point of test rail displacement.

4.2 Test results

Fig. 15 shows the relationship between the test load, and the rail tilting angle and rail displacement. In Fig. 16, the test results were compared with the FEM results described in Table 2. Regarding rail head displacement, FEM result was in good agreement with test result when L_A and L_B was applied. As regards rail base displacement, FEM result was 0.6 times of test result when L_A was applied and was 0.4 times of test result when L_B was applied. As for rail tilting angle, FEM result was 1.4 times of test result when L_A was applied and was 1.9 times of test result when L_B was applied.

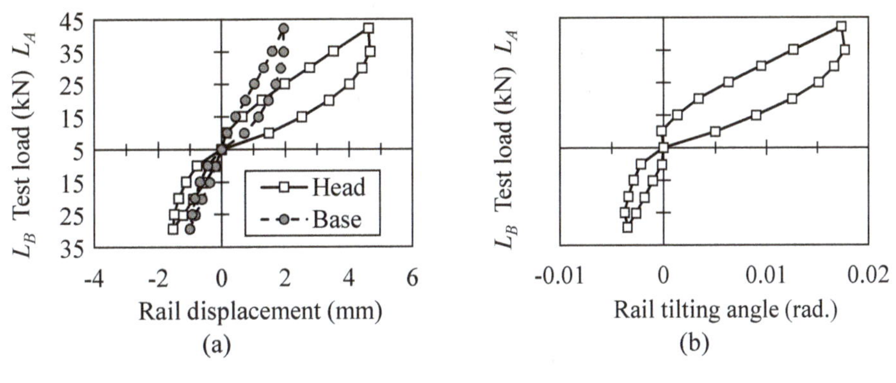

Figure 15: Rail head and base displacement and rail tilting angle obtained by the performance test. (a) Rail head and base displacement; and (b) Rail tilting angle.

5 CONCLUSION

This study examines the calculation method of load conditions for performance test of rail fastenings of a floating slab track. The obtained main results are as follows:

1. A FEM model of a floating slab track is developed to estimate the dispersed forces and the calculation method of load conditions based on the FEM results are proposed.
2. The effectiveness of the calculation method of load conditions was verified by comparing FEM results with test results.

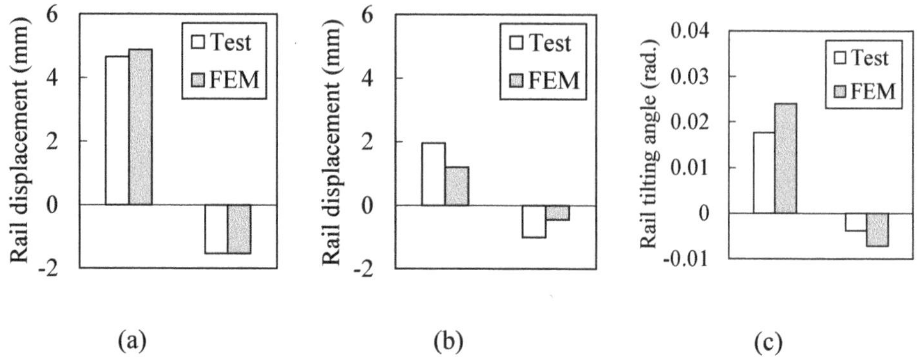

Figure 16: Comparison between test results and FEM results. (a) Rail head displacement; (b) Rail base displacement; and (c) Rail tilting angle.

From the above results, FEM results were substantially coincident with test results. Therefore, it is concluded that the effectiveness of the calculation method of load conditions for performance test based on the developed FEM model is verified.

The calculation method based on the FEM model can evaluate the performance of rail fastenings of the floating slab track with higher accuracy than before. Therefore the rational design of the rail fastenings can be expected and the further adoption of the floating slab track can be expected. In addition, the proposed method may become the Japanese standard of the performance test for the rail fastenings of the slab track. In recent years, in order to respond to the globalization of the railway business, Japan have participated in the international standard development of the railway track (Katsumi [9]). In particular, international standard for the rail fastenings are now discussed at ISO/ TC269/SC1/WG7. The author will continue efforts to reflect the Japanese ideas in the international standards.

ACKNOWLEDGEMENT

The author wishes to thank Kita-Osaka Kyuko Railway Co. Ltd. for providing the specimens and supporting the tests.

REFERENCES

[1] Japan Railway Construction Public Corporation, Construction of Rinkai Line, 2003.

[2] Railway Technical Research Institute, Design Standards for Railway Structures and Commentary (Track Structures), Maruzen, 2012.

[3] Youichi, H., A practical solution for the torsion (tilting) of rail. *Journal of Japan Society of Civil Engineering*, **210**, pp. 33–46, 1973.

[4] Yutaka, S., On the lateral strength of railway track. *Railway Technical Research Report, Railway Technical Research Institute*, **110**, 1960.

[5] Takeshi, Y., Shizuya, U. & Toshiyuki, K., Relationship between spring coefficient of fastening device and rail overturning angle. *Quarterly Reports, Railway Technical Research Institute*, **22**(4), pp. 153–156, 1981.

[6] Shingo, T., Hiroo, K. & Tadashi, D., Practical model for rail tilting and its application to performance test of rail fastening system. *Journal of JSCE*, **7**, pp. 1–14, 2019.

[7] Tadashi, D., Shingo, T., Masato, N. & Hiroo, K., Improving the performance of rail fastening system evaluation. *Quarterly Report, Railway Technical Research Institute*, **59**(3), 2018.

[8] NX Nastran User's Guide, Siemens PLM software, 2014.
[9] Katsumi, M., Development status of International standards on the rail track and the trend of the related research of RTRI. *RTRI Report*, **31**(12), pp. 1–4, 2017.

APPLICATION EXPLORATION OF B METHOD IN THE DEVELOPMENT OF SAFETY-CRITICAL CONTROL SYSTEMS

LIU NING[1], WANG KEMING[2], HOU XILI[3], WANG XIA[2] & CHENG PENG[2]
[1]Graduate School of Tangshan, Southwest Jiaotong University, China
[2]School of Information Science and Technology, Southwest Jiaotong University, China
[3]TongHao GBA (GuangZhou) Smart Control Co., Ltd., China

ABSTRACT
This article presents our experience in developing a tram control system using the B formal method. We find that there are some notable issues when using an abstract machine model to express software systems and in automatic code generation, for which we have summarized the solutions. In this paper, we illustrate how to use the B module to develop complex systems, the rational choice of implication relations of the invariants, as well as the correctness of the variable definition in the model for code generation. The solutions to these issues can help developers with less experience to better use the B method to develop the reliable systems.
Keywords: B method, invariant, modelling, code generation, application exploration.

1 INTRODUCTION

Using formal methods to develop the safety-critical system can greatly reduce the design defects caused by the system requirements specification and the misinterpretation by developers, which consequently improves the safety of the system [1].

The B method has been successfully applied to the industry of transportation [2], and there are some successful cases [3]. Our group used the B method on the development of the tram control system of Guangzhou Huangpu Line 1 in China. To the best of our knowledge, this is the first time for the B method to be applied to develop the tram control system in China.

It remains a major challenge for both researchers and engineers to develop a complex industrial system with formal methods. Scientific theoretical knowledge and engineering empirical knowledge are equally important for applying formal methods. The former helps developers apply formal methods to develop software systems, and the latter helps developers better apply formal methods, but few articles share the experience.

This article focuses on the experience of using B method modelling and code generation. Due to their lack of background knowledge and engineering experiences of formal methods, some developers encounter many notable problems with their applying B method of engineering practice. A period of time for the past we spent on developing the Guangzhou tram project and encountered many difficulties in the using of the B method. We also met some typical issues such as developing complex software systems with the B module, the choice of implication relations of the invariants as well as the definition of variables in the model for code generation, etc. We believe that how to deal with these issues mentioned above is critical to the successful application of the B method on engineering projects.

The remainder of this paper is organized as follows: The B method is briefly introduced in Section 2. Section 3 demonstrates how to use the B module to develop complex systems and the rational choice of implication relations of the invariants by cases. Section 4 discusses two inappropriate forms of modelling. In Sections 5 and 6, we discuss some related work and give the conclusion of this paper.

WIT Transactions on The Built Environment, Vol 199, © 2020 WIT Press
www.witpress.com, ISSN 1743-3509 (on-line)
doi:10.2495/CR200261

2 B METHOD

As a software development method, the basic idea of the B method is to provide a representation method that uses the Abstract Machine Notation (AMN) and the Generalized Substitution Language (GSL) to describe software. The software described by the B method can eventually be implemented in imperative programming languages, or even in assembly language [4]. The method provides a framework in which specifications can be refined through to implementations, which can be translated into a programming language. The B method is the first single formal method that covers all stages of the software development process from specification to design to implementation [5].

The basic unit for describing software with the B method is an abstract machine, which includes: data description (constants, variables); operation description (a set of operations on data); invariants (a set of relationships that the data must satisfy) [6], these abstract machines describe the most basic requirements. Developers understand software systems in the manner of an abstract machine model, that is, a state and various operations that can modify this state. This analysis method constitutes a study of their static and dynamic. The static corresponds to the definition of states, while dynamic correspond to various operations [4].

Besides, the B method supports refinement, that is, the process of modifying from an abstract specification to a concrete implementation through some data or operation substitution. Implementation is a special case of refinement and can be done at any stage in refinement, but may be done only once. The implementation cannot have any private state, and have to implement these operations using the specified operations of other machines imported into the implementation [5]. When using the executable code of the B method to implement the system, the system must be refined into implementation [7].

The supporting platform tool for the B method is Atelier B [8]. After the developer builds and saves the B model, the model must pass the type-checker before it can generate proof obligations and prove. The ProB [9] also provides an interface for Atelier B to link.

3 PROPERTIES VERIFICATION BASED ON INVARIANTS

B method states that after describing the operation, developers need to prove the invariant to ensure that this operation has not broken the consistency of the abstract machine. This requires developers have to understand how to use the abstract machine model to express software systems, as well as be able to choose the implication relations of the invariants correctly. In the B method, we use the B module to describe a subsystem. The combination of some B modules constitutes a complex software system. The following describes how to develop the complex software systems with the B module, and the rational choice of implication relations of the invariants.

3.1 Develop complex software systems with the B module

Developers use the B method to develop the complex software systems, that is, the developer builds software systems according to the concept of the B method, and ensures the safety of the developed software systems [6]. Developers can use the B modules to describe a subsystem of the software, therefore, the combination of several B modules constitutes a complete system. Each module has its specifications and implementation, their development is independent. The B module must have an abstract machine, which is used to describe the specifications of the system. The content of the abstract machine is an objective fact and describes the system "what to do". The B module have an implementation, which is used to describe the concrete implementation of the corresponding specification. In other words, the content of the implementation is the details and describes the system "how to do". There may be several transition machines between the system specification and the implementation, which are refinement machines.

Sometimes the specifications that need to be described are complex and need to be decomposed. The abstract machine can use INCLUDES, USES, and SEES to construct more complex abstract specifications, like *M* INCLUDES *N*, *M* regards *N* as a part of itself. SEES can be used to refer to information in another abstract machine. The implementation can use IMPORTS and SEES to construct more complex software systems. For example, an implementation needs to import other abstract machines as the basis. It should be noted here that IMPORTS can only be used to import abstract machine specifications, but its connection relationship can be reflected to the implementation [6].

Developers must describe the static laws of the system, that is, invariants. The invariant describes the properties that must be maintained between variables, constants, etc., the properties that must be not violated at the run-time of the operations. When the value of variables is updated, the invariant should be preserved [4]. For some developers, they may think that the relevant properties of the system need to be verified in the implementation because what is described in the implementation is the final implementation, although they know that the specification has described in the abstract machine. This is a misunderstanding.

In fact, after describing an operation in the abstract machine, developers need to ensure that this operation preserves the invariant of the abstract machine. The invariant here can not only describe the value range and mutual relationship of variables but also can be used to describe certain safety properties that must be satisfied at the run-time. So how to ensure that the concrete operations of the implementation also preserves the invariant? The developer only needs to ensure that the operations in the implementation satisfy the specifications of the abstract machine, then the concrete operations in the implementation preserve the invariant of the abstract machine at the run-time, thus ensuring the safety of the system, as shown in Fig. 1.

Without understanding how to develop complex software systems with the B module, developers may not be able to verify safety-critical properties, or even determine where to verify safety-critical properties, abstract machine or implementation?

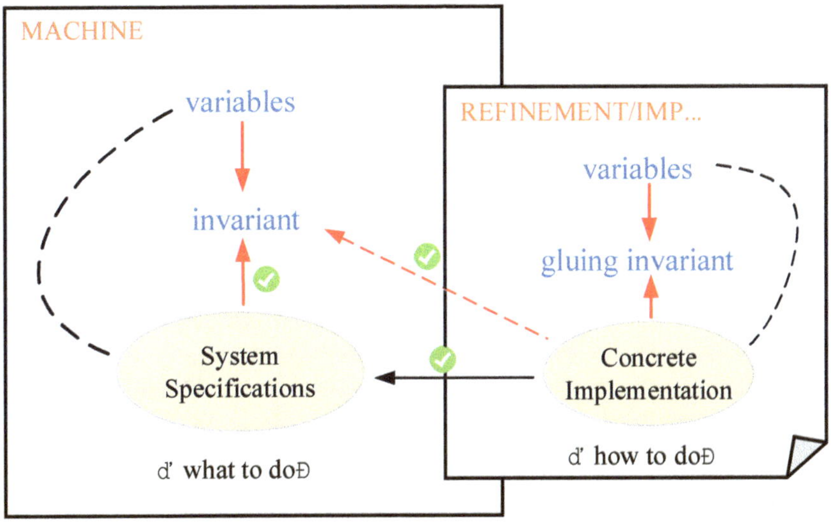

Figure 1: The B module development framework.

3.2 Implication in invariants

There are two forms of implication in invariants: $THEN \rightarrow IF$, it is consequent driven implication in this invariant; $IF \rightarrow THEN$, it is antecedent driven implication in this invariant. When the developer verifies the system properties, the appropriate form of invariant should be chosen to ensure the accuracy of implication of invariants for the verification of the safety properties.

For the rational choice of implication relations of the invariants, we conclude three following scenarios:

1. When there is only one unique transition between state A and state B, the invariants in the form of $Antecedent \rightarrow Consequent$ and $Consequent \rightarrow Antecedent$ can be both used to verify the properties of the system, as shown in the following Fig. 2(a).
2. When temporal property is involved in the *IF* statement, even if there is one unique transition between A and B, $p \rightarrow q$ may not hold, but at this time $q \rightarrow p$ holds.
3. If there are two or more transition paths between state A and state B, when $p' \rightarrow q'$ holds, $q' \rightarrow p'$ may not hold, in the case shown in Fig. 2(b).

When the form of invariant derived from the safety properties is $THEN \rightarrow IF$, the invariant may not be able to verify this safety property of the system. A case is shown in Fig. 2(c), Whether the value of the *checkPoints_pointInPosition* is TRUE or FALSE, both of these contradictory invariants are true at the same time. Why do the two invariants hold at the same time? As illustrated by Fig. 2(b), the *checkPoints_pointInPosition* is TRUE in state A', FALSE in state A", and in state B the value of *PCS_routeSectionCmd_lock* is equal to 15. But when *checkPoints_pointInPosition* is TRUE, it does not represent the state A', nor does its FALSE value represent A". Therefore, the invariant in Fig. 2(c) cannot be used to verify the properties of state transitions between A'-B, A"-B. If the description of the system states in the invariant is incomplete, verification process will produce unexpected errors because this type of invariant has no constraint on the state of the system.

When writing the invariants for the verification of safety properties, the developer should distinguish the implication of different forms invariant, then choose the appropriate and correct form of invariant for safety-critical properties verification, especially the case as shown in Fig. 2(b).

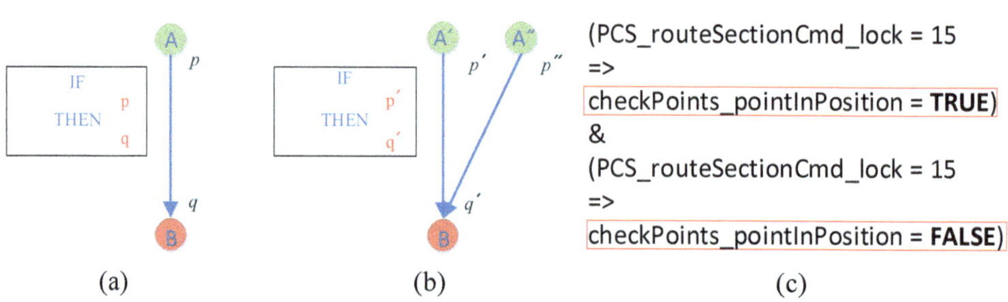

Figure 2: The implication in invariants.

4 DEFINITION OF THE VARIABLE FOR CODE GENERATION

Atelier B supports the transformation of the B models into high-level programming languages, like C codes. In this section, we focus on two forms of modelling that may be inappropriate. The C codes generated by this modelling form will cause unnecessary troublesome when used, then show how to avoid them.

4.1 The influence of the output parameters of the B model on the generated C code

Considering the following example, First, we define *tt* as a variable, and the value of operation *plusStart*(*aa*,*bb*) is assigned to *tt* , as shown in Fig. 3(a). Next, instead of defining a variable, we only define *nn* as an output parameter, as shown in Fig. 3(b). The function of the operation is the same.

We found that if the operation has the output parameters, the generated C codes are difficult to understand and need to be modified (except adding the *main* function) because the output parameter *nn* of the C codes does not get the value, the developer needs to modify the *Function* types in C codes, thus destroying the integrity of the generated code. If the operation has no output parameters but with a variable which gets the output value of the operation, the generated C codes are more standardized and easier to use, the variable *Observer_tt* can get the value, this issue is shown in Fig. 4.

Developers should draw the lesson from this above case when defining the operation in the B model if they would like to get the better C codes, they may use the way shown in Fig. 3(a) to build the B model.

```
IMPLEMENTATION Observer_i
REFINES Observer
IMPORTS
  Plus
CONCRETE_VARIABLES
  tt
INITIALISATION
  tt := 0
OPERATIONS
  ObserverStart(aa,bb)=
    tt <-- plusStart(aa,bb)
END
```
(a)

```
IMPLEMENTATION Observer_i
REFINES Observer
IMPORTS
  Plus
OPERATIONS
  nn <-- ObserverStart(aa,bb)=
    nn <-- plusStart(aa,bb)
END
```
(b)

Figure 3: Two different forms of operation definition in B model.

```
/* Clause CONCRETE_VARIABLES */
static int32_t Observer__tt;
void Observer__ObserverStart(int32_t aa, int32_t bb)
{
  Plus__plusStart(aa, bb, &Observer__tt);
}
```
(a)

```
void Observer__ObserverStart(int32_t aa, int32_t bb, int32_t *nn)
{
  Plus__plusStart(aa, bb, nn);
}
```
(b)

Figure 4: C code generated from the models in Fig. 3.

4.2 Difference between VARIABLES and CONCRETE_VARIABLES in generating code

The variables used in the B method can be divided into abstract variables and concrete variables. The abstract variables are defined in the abstract machine, and the concrete variables are usually defined in the refinement and the implementation. For example, if an abstract variable defined in the abstract machine needs to be implemented, it has to be changed to a concrete variable in the refinement or the implementation. Here we only discuss the case where the variable name does not need to be changed when the variable implements.

Some developers might directly define the variable as a concrete variable in the abstract machine. When they compile the C codes generated by the B model, an error will occur. The error message is as follows: "[Error] static declaration of 'xxx' follows non-static declaration".

Finally, conclusions are summarized as follows: It causes errors in compiling C codes due to the inconsistency of the variable type in the ".h" file and the ".c" file which generated by the B model, if the variable is defined as a concrete variable directly in the abstract machine. Specifically, this variable is the external storage type in the ".h" file but the static storage type in the ".c" file. Remarkably, to make the C codes generated by the model more standardized and easier to understand, the variable should not be directly defined as a concrete variable in the abstract machine. The variable should be defined as an abstract variable firstly, then changed to as a concrete variable in the refinement or the implementation, although it is feasible to directly define as concrete variables in the abstract machine.

5 RELATED WORK

Formal methods have taken more attention in the industry in recent years, EN-50128 suggests the use of formal methods to develop systems with high safety integrity levels. However, the formal method faces various obstacles in the application process. For example, formal methods are based on mathematical logic, the verification process requires strong logical reasoning knowledge, reasoning process are complicated, reasoning methods and strategies are flexible. These reasons lead to higher learning cost of formal methods and application difficult. Moreover, the analysis of the final verification results relies on relevant experience and there lacks a formal verification platform for specific industry applications [10]. Abrial [11] mentioned that the B method is essentially used in train systems. Although this modelling method can also be used in energy, automotive and other industries, the engineer needs to change the long-established engineering method, which is almost impossible. Also, some developers believe that the use of formal methods in industrial software development means that extra effort is invested in formal modelling, and there are potential safety risks from the process of the specification to implementation, these problems can be solved by automatically generating codes from the formal model [12].

At present, the application of formal methods in the field of rail transportation is mainly concentrated in railway systems, such as train control systems and interlocking systems. Formal methods are rarely applied to the safety research of tram control system. With the development of trams, the safety of tram control systems in particular needs to be guaranteed. We use the B method to develop the tram control system to improve the quality of its software system and improve its safety. Moreover, the C codes are generated from the formal requirement model to improve the automation ability.

We have summarized the content that is likely to errors in the formal modelling process, which can help developers better use the B method correctly. Once the model does not satisfy the expectation, it is meaningless to perform formal verification on the model. Besides, there are some interesting points when the B model generates the C codes, such as the influence of output parameters on the generated code. We consider that developers need to notice when generating C codes, although the B method does not specify which practice is wrong.

6 CONCLUSION

Based on our experience in developing tram control systems with the B method, we have summarized some of the issues and solutions that are likely to be encountered when developing the system using the B method, which should be useful to developers with less experience in this field.

In this paper, we illustrated (1) Using the B module to develop complex systems. It is different from the general method of expressing software systems, developers should understand the software system according to the abstract machine model;(2) Rationally choose the implication relation of the invariant. We summed three cases, developers can choose the appropriate invariant form by comparison; (3) The influence of output parameters on the generated code. We generated code for the B model with and without output parameters and found that the code generated by the model without output parameters is easier to understand and use; (4) The influence of directly defining concrete variables in the abstract machine on code generation. Although the B method allows concrete variables to be defined in the abstract machine, the error related this concrete variable will appear when the generated C code is compiled, and the codes must be modified before they can be used. These points are critical for developers to develop systems with the B method. These can help developers avoid unnecessary work and enhance their confidence while reducing learning costs.

ACKNOWLEDGEMENTS

This work was supported by the National Natural Science Foundation of China (No. 71502146, 61673320). We would like to thank Dr. Colin Snook for his discussion in this work.

REFERENCES

[1] Pressman, R.S., *Software Engineering: A Practitioner's Approach*, McGraw-Hill: New York, 2005.
[2] Lecomte, T., Deharbe, D., Prun, E. & Mottin, E., Applying a formal method in industry: A 25-year trajectory. *Brazilian Symposium on Formal Methods*, pp. 70–87, 2017.
[3] Bicarregui, J.C. et al., Formal methods into practice: Case studies in the application of the B method. *IEEE Proceedings Software Engineering*, **144**(2), pp. 119–133, 1997.
[4] Abrial, J.R., *The B-Book: Assigning Programs to Meanings*, Cambridge University Press: Cambridge, 1996.
[5] Robinson, K., The B Method and the B Toolkit. *International Conference on Algebraic Methodology and Software Technology*, pp. 576–580, 1997.
[6] School of Mathematical Sciences, Peking University. www.math.pku.edu.cn/teachers/qiuzy/fm_B/slides. Accessed on: 13 Mar. 2020.
[7] Lano, K., *The B Language and Method: A Guide to Practical Formal Development*, Springer-Verlag: New York, 1996.
[8] Atelier B, www.atelierb.eu/en/atelier-b-tools/. Accessed on: 11 Feb. 2020.
[9] ProB, www3.hhu.de/stups/prob/. Accessed on: 15 Feb. 2020.

[10] Bjørner, D. & Havelund, K., 40 Years of formal methods. *International Symposium on Formal Methods*, pp. 42–61, 2014.
[11] Abrial, J.R., On B and Event-B: Principles, success and challenges. *International Conference on Abstract State Machines, Alloy, B, TLA, VDM, and Z*, pp. 31–35, 2018.
[12] Mashkoor, A., Kossak, F., Biró, M. & Egyed, A., Model-driven re-engineering of a pressure sensing system: An experience report. *European Conference on Modelling Foundations and Applications*, pp. 264–278, 2018.

FORMAL SOFTWARE INSPECTIONS: AN INDUSTRIAL APPLICATION OF FUNCTION TABLES AND EVENT-B TO SOFTWARE OF A WAYSIDE TRAIN MONITORING SYSTEM

ROBERT ESCHBACH
ITK Engineering GmbH, Germamy

ABSTRACT

The experience gained in the industrial application of software inspections using Function Tables and Event-B to a subsystem of a Wayside Train Monitoring system (WTMS) is presented in this paper. The WTMS Configuration Management System (CMS) supports the creation and management of configuration data for the WTMS. The correct and reliable implementation of the required system functions, especially those dealing with data handling and data management, is of particular importance for the overall quality of the system since faults in these functions may lead to critical failures and malfunctioning. Therefore, the development of the data handling part of a CMS requires the use of high integrity methods like systematic software inspections in order to ensure the highest quality. Function Tables have been successfully applied for the inspection of safety-critical software. In our industrial project, a special variant of Function Tables was defined that can be easily mapped to formal Event-B specifications. Event-B with its set-theoretic basis for modeling, its concept of refinement and the use of formal proof to ensure correctness of refinement steps, is used to formally analyze the derived Function Tables. The systematic derivation of Function Tables is done by a verification-based inspection using reading technique "stepwise abstraction".

Keywords: software inspection, Function Table, Event-B, stepwise abstraction.

1 INTRODUCTION

The purpose of a Wayside Train Monitoring System (WTMS) is to detect early threats that may lead to hazards and damages by monitoring trains and environmental conditions [1]. Typical examples are the detection of hot box, brake-locking or load displacements. A WTMS is a highly distributed system that, from a system data perspective, gathers, analyzes and exchanges different kinds of measurement and status data. A central part of a WTMS is the Configuration Management System (CMS). It is responsible for the correct data exchange between connected devices and control and management service points. The goal of our project was the development of a server-based subsystem of the CMS. In Eschbach [2] the formal conceptual data analysis of CMS management data using Event-B has been presented. In this paper the experiences gained in the systematics inspection of software using Function Tables and Event-B will be described.

In Basili and Selby [3] the authors compared three software quality assurance techniques: (a) code reading by stepwise abstraction, (b) functional testing using equivalence partitioning and boundary value analysis, and (c) structural testing using 100% statement coverage criteria. These techniques were compared w.r.t. three aspects: fault detection effectiveness, fault detection cost, and classes of faults detected. One major outcome of this study is summarized by the authors as:

> With the professional programmers, code reading detected more software faults and had a higher fault detection rate than did functional or structural testing, while functional testing detected more faults than did structural testing, but functional and structural testing were not different in fault detection rate.

WIT Transactions on The Built Environment, Vol 199, © 2020 WIT Press
www.witpress.com, ISSN 1743-3509 (on-line)
doi:10.2495/CR200271

In Linger et al. [4] this reading technique is characterized as follows:

> The process of reading a poorly documented program bottom up is called *stepwise abstraction*. Stepwise abstraction may be required in either the verification of correctness or the determination of program function.

Based on these results we chosen the reading technique stepwise abstraction for software inspection in our project. The applied inspection technique was strongly influenced by the cleanroom inspection methods (see Mills et al. [5] and Mills [6]). The cleanroom inspection methods are verification-based methods with the goal of finding functional defects in software through mathematical verification building upon the box-structured method (see Mills [7]). In our project we have used Function Tables in order to abstract from concrete clear-box software structures to more abstract state box and black box structures.

The results of Elberzhager et al. [8] were used to integrate the software inspection with our testing processes. For this purpose, project-specific assumptions have been identified and contextfactors have been derived in order to integrate our inspection processes based on stepwise abstraction with our testing processes. Furthermore, so called Goal Indicator Trees (GITs), as defined in Kloos et al. [9], which support inspectors in performing an inspection, thereby assuring the quality of work products and improving the quality of the overall system, were used to identify the relevant quality objectives and refine them into appropriate quality indicators. One outcome of this analysis was to focus formally on software functions that may cause database inconsistencies.

In Powell [10] the authors present a verification-based software inspection technique using tool support. The proposed tool extracts and generates proof obligations directly from software and supports and guides an inspector by his semi-formal verification. The proposed approach does not lead to a formal model of software functions as intended in our project. How an inspection process can be improved by focusing and tailoring the inspection on experiences and expertise of the reviewers is described in Parnas and Weiss [11]. The proposed approach does not directly address formal specification and verification but is of great practical value. In Singh et al. [12] the authors present a refinement strategy to generate formal Event-B models from tabular expressions. The refinement strategy, especially the generation of models is based on a correct-by-construction approach. The proposed approach will be illustrated by an insulin infusion pump case study. The approach addresses formal specification and verification but is not used for formal software inspection, which requires an appropriate abstraction strategy. Module interface specification enriched with assertions and semi-formal proofs can be of great advantage, when inspecting code w.r.t. contracts given by pre- and postconditions. These results are described in Jackson and Hoffman [13]. The paper covers some of the underlying ideas of our verification-based software inspections technique, but does not address suitable abstraction strategies as well as the formal specification and verification aspects. The verification-based inspection technique presented in this paper is based on a systematic abstraction strategy that allows for stepwise abstraction and the creation of a formal Event-B model. Furthermore, the presented inspection technique is scalable w.r.t. the degree of formality, i.e. depending on the concrete context it is possible to work with an informal model specified with natural language statements as well as with a complete formal Event-B model. Techniques like the GITs, as defined in Kloos et al. [9], can be used to derive criteria, that supports the decision which degree of formality is appropriate in a concrete context.

Event-B [14], [15] with its set-theoretic basis for modeling, its concept of refinement and the use of formal proofs ensuring correctness of refinement steps has been used for the formal code inspection based on Function Tables and reading technique stepwise abstraction.

Formal methods, especially Event-B have been applied successfully in many safety-critical railway systems [16]–[18]. The success of Event-B and the B method has even an influence on the definition of CENELEC standard EN50128 [19]. A state-of-the-art of methods and tools for the verification of interlocking systems can be found in Haxthausen and Peleska [20].

In the following it will be described how the application of Function Tables and Event-B together with the reading technique stepwise abstraction can be successfully used to systematically inspect and analyze critical software.

2 WAYSIDE TRAIN MONITORING SYSTEM (WTMS)

2.1 Architecture

The CMS can be used to create configuration data for the WTMS. On an abstract level the CMS consists of data and operations related to this data (see Fig. 1). CMS data can be further divided into data for the ticket system and technical configuration data related to the WTMS. The CDM and FCDM has been created for the ticket system.

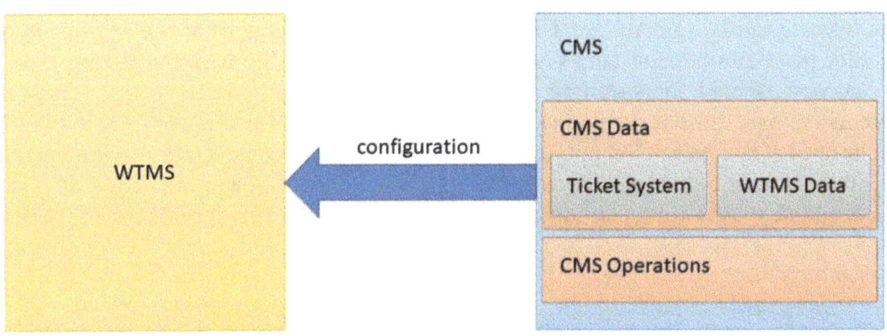

Figure 1: Conceptual view of WTMS and CMS.

2.2 Challenges and solutions

Since a CMS is responsible for correct data exchange, data integrity is a very important quality attribute. Furthermore, a systematic handling of faults is very important. From an engineering point of view the main focus was on fault prevention rather then fault identification and removal. Another important concept is fault tolerance. Strategies for fault tolerance help to avoid failures during system execution by means of error detection and recovery mechanisms. Using database constraints and triggers based on formal models as well as SQL statements involved in transactions with roll-back mechanisms, a specific kind of fault tolerance has been achieved.

Data integrity and especially data consistency of system data is one of the most important quality properties of a CMS since inconsistency may lead to critical malfunctioning. Therefore, the development of a CMS requires the use of high integrity methods in order to ensure the highest quality.

During system development, it is important to establish techniques for *fault prevention.* A fault may cause errors during system execution and may lead to critical situations. Different techniques for fault prevention exist. For example, coding guidelines or systematic code inspections may help software developers to prevent faults. The overall goal should be to prevent almost all faults in order to deliver (almost) zero-fault software.

Another important quality attribute required for the CMS is *fault tolerance*. Since each non-trivial system may have faults, it is very important to deal with faults in a systematic way. Fault tolerance covers different techniques for error detection as well as for system recovery. A special way to recover a consistent system state in a CMS is to remove inconsistent data and roll back to an earlier consistent and saved data restoration point. In our project, a sequence of database statements was embedded into a transaction with associated rollback mechanisms. Whenever an exception occurs (for example, by violating a unique or foreign key constraint), the database will roll back to the last saved restoration point. The function *ewt* will check whether the execution of the transaction was successful or not and will trigger a warning in the latter case.

3 EVENT-B

Event-B is a formal method that can be used to model and analyze state transition systems (see Abrial [14]). Event-B is part of the B-Method (see Abrial [15]). Event-B supports refinement-based development of transition systems. It is based on of first-order logic and typed set theory. Tool support for Event-B is given by the Rodin Platform (http://www.event-b.org). An Event-B specification consists of a static and a dynamic part. The static part describes constants (context), the dynamic part behavior (machine).

A context is used for specifying sets, constants, axioms and theorems. Axioms and theorems are used to state properties of sets and constants. In this paper execution modes were specified as constants among others (see Fig. 2). The context $ctx - ewt$ depicted in Fig. 2 defines a set *ExecMode*, four constants *start*, *execute*, *wait* and *exit*. The axiom states that $ExecMode = \{start, execute, wait, exit\}$ and that set *ExecMode* has exactly four elements.

```
context
   ctx−ewt
sets
   ExecMode
   ...
constants
   start
   execute
   wait
   exit
   ...
axioms
   partition (ExecMode , { start } , { execute } , { wait } , { exit })
   ...
end
```

Figure 2: Context ewt.

A machine describes transitions (events) and may use the information of a connected context. A machine specifies variables, invariants and events. An event describes a transition and consists of a possibly empty set of event parameters, guards, that describes conditions under which an event is enabled and actions, that describe how parameters will change, if the event occurs (see Fig. 3).

```
machine
   ewt
sees
   ctx−ewt
variables
   phase
   current_db_state
   consistent_db_state
   transaction
   . . .
invariants
   phase ∈ {ewt , ef } → { start , execute , exit , wait }
   phase ( ewt ) = e x i t ⇒
      consistent_db_state ( current_db_state) = TRUE
   . . .
end
```

Figure 3: Machine ewt.

The machine *ewt* depicted in 3 can use context $ctx - ewt$ (sees clause) and defines the variables *phase*, *current_db_state*, *consistent_db_state* and *transaction*. The invariant $phase \in \{ewt, ef\} \rightarrow \{start, execute, exit, wait\}$ states that *phase* is a total function with domain $\{ewt, ef\}$ and range $\{start, execute, exit, wait\}$. The second invariant states, that when $phase(ewt) = exit$ holds, the current db state *current_db_state* is consistent.

4 SOFTWARE INSPECTIONS

Since data consistency is one of the most important quality attributes of the CMS, the decision was made to systematically inspect almost all software providing functionalities with database access. The reading technique stepwise abstraction was chosen based on the results given in Basili and Selby [3].

In our project, the execution of a sequence of database statements are always embedded into a SQL transaction with rollback mechanism. A simplified skeleton of the function that embeds the execution of multiple SQL-instructions into a transaction is shown in Fig. 4.

Function *ewt* has function pointer *ef* as argument. In the first step, relevant program variables of *ewt* will be identified. In this example these are the following variables:

1. *rid* (return value of function *ewt*),
2. *rc* (assigned to the return value of function *ef*).

We use so-called meta-variables for expressing abstract properties. The status of a transaction will be abstractly represented by meta variable *transaction*

$$transaction \in \{tr\_begin, tr\_commit, tr\_rollback\}. \tag{1}$$

If function *ewt* starts a transaction, we simply write

$$transaction := tr\_begin. \tag{2}$$

```
int ewt ( const std : : function <bool ()> &ef ) {
    try {
        tr_begin ( db ) ;
        bool rc = ef ( ) ;
        if ( ! rc ) { throw Exc ( " ef␣. . . " ) ; }
        tr_commit ( db ) ;
        rid = . . . ;
        return rid ; }
    catch ( const C_Exc &exc ) {
        tr_rollback ( db ) ;
        throw ; } . . . }
```

Figure 4: C++-Function ewt.

In the same way, a transaction commit or rollback wil be treated. Since function *ef* may throw an exception, we introduce a meta variable *exception*, which is either *TRUE* or *FALSE*. The following condition describes, that *ef* has thrown an exception:

$$exception = TRUE. \tag{3}$$

Furthermore, we define a meta variable for the current database state *current_db_state* and the current restoration point *restoration_point*. Since we just want to distinguish database states, we model both meta variables as natural numbers, i.e.

$$current\_db\_state, restoration\_point \in \mathbb{N}. \tag{4}$$

The consistency of db states is represented by meta variable *consistent*, which is either *TRUE* or *FALSE*. The execution phases of functions *ewt* and *ef* will be represented by variable

$$phase : \{ewt, ef\} \rightarrow \{start, execute, wait, exit\}. \tag{5}$$

In phase *start* variables and meta variables will be initialized. Phase *execute* covers the execution and phase *wait* is used to model the exection of other Function Tables. The execution ends in phase *exit*.

During the stepwise construction of Function Table FT-EWT (see eqn (5)), the explicit scheduling of function *ewt* will be replaced by an implicit scheduling based on conditions. The table is created bottom up using the reading technique stepwise abstraction as defined in Linger et al. [4]. The Function Tables presented in this paper are a variant of those proposed by Dave Parnas (see [21], for example).

Each column of the derived Function Table, as shown in Fig. 5, describes a specific condition and associated actions, i.e. a parallel assignment of values to variables. In our example, a condition is related to:

- the current execution phase $phase \in \{start, execute, wait, exit\}$,
- the existence of an exception $exception \in \{TRUE, FALSE\}$ and
- the return code $rc \in \{undef, ok, nok\}$.

Initially, the execution starts with *phase*(*ewt*) = *start* and *phase*(*ef*) = *wait*. The conditions are organized hierarchically. The Function Table can be read column by column starting with the column condition, then going to the values that will be assigned to the specified variables

Event Name	ewt-start	ewt-execute-1	ewt-execute-2	ewt-execute-3	ewt-execute-4	ewt-exit
Conditions	phase =					
	start	execute				exit
		exception =				
		FALSE			TRUE	
		rc =				
Variables		undef	ok	nok		
rid	-1/ok	nc	last_insert_rowid/ok	nc	nc	nc
rc	undef	ef()	nc	nc	nc	nc
transaction	undef	tr_begin	tr_commit	nc	tr_rollback	nc
exception	FALSE	nc	undef	TRUE	nc	nc
restoration_point	undef	current_db_state	undef	nc	undef	nc
current_db_state	nc	nc	new_db_state	new_db_state	restoration_point	nc
consistent	nc	nc	TRUE	FALSE	nc	nc
phase	execute	nc	exit	nc	exit	nc

Values

Figure 5: Function Table FT-EWT.

in parallel. A value cell marked with *nc* (no change) does not cause a change of the corresponding variable.

Each column of this table is associated with an event in Event-B with guards related to the column conditions and actions defined as parallel assignment of the column values to variables. For example, event *ewt-execute-1*, as shown in Fig. 6, has been derived from column with *Event Name = ewt-execute-1*. The column condition is

$$phase(ewt) = execute \land exception = FALSE \land rc = undef \land ef \in \{ok, nok\}. \quad (6)$$

The parallel assignment can be specified as

$$rc := ef, transaction := tr\_begin, restoration\_point := current\_db\_state. \quad (7)$$

The corresponding event defined in Event-B can be seen in Fig. 6.

Ewt_execute_1
any
  ef
where
  phase (ewt) = execute
  exception = FALSE
  rc = undef
  ef ∈ {ok , nok}
then
  rc := ef
  transaction := tr_begin
  restoration_point := current_db_state
end

Figure 6: Event ewt_execute_1.

The main objective of the inspection was to verify that in phase *exit* the current data base state is consistent. This can be formalized as following invariant:

$$phase(ewt) = exit \Rightarrow consistent(current\_db\_state) = TRUE. \quad (8)$$

The formal analysis was performed with the eclipse-based tool Rodin. The verification of all invariants, especially invariant 8, was performed with theorem provers, SMT solvers available in Rodin (using appropriate plugins). Furthermore, the model checker ProB was also used to prove the deadlock freeness of the model. The state space has been explored by stepwise execution of the model. The inspection of *ewt* und four instances of *ef* revealed several serious incompleteness faults.

5 CONCLUSION

The knowledge gained in the application of systematic software inspections based on Function Tables and Event-B to the configuration management system of a wayside train monitoring system has been presented in this paper. The configuration management system supports the creation and management of configuration data for the wayside train monitoring system. The correctness and reliability of all software functions dealing with data handling and data management is of utmost importance. Systematic software inspections of these functions have been successfully used for fault detection. The reading technique stepwise abstraction was used to create Function Tables that were later on mapped to Event-B specifications for critical functions. In this way a formal analysis based on several formal verification techniques like theorem proving, SMT solving, model checking and constraint solving was possible. Furthermore animation of the functional behavior was used for validation purposes. Several critical incompleteness faults could be identified and corrected. The Function Tables, especially informal tables, were of great practical value since they were used not only for fault detection but also for creating common understanding within the development team. The reading technique stepwise abstractions helps to get the analyzed functionalities under intellectual control. Event-B as underlying formal method has greatly improved the analysis of critical functions. Especially the integration of formal specifications and formal verification, as well as the refinement concept, makes Event-B to a practical and powerful formal technique useful in industrial projects.

REFERENCES

[1] Bracciali, A., Wayside train monitoring systems: A state-of-the-art and running safety implications. *International Journal of Railway Technology*, **1**(1), pp. 231–247, 2012.

[2] Eschbach, R., Industrial application of Event-B to a wayside train monitoring system: Formal conceptual data analysis. *Formal Methods – The Next 30 Years*, eds M.H. ter Beek, A. McIver & J.N. Oliveira, Springer International Publishing, pp. 738–745, 2019.

[3] Basili, V. & Selby, R., Comparing the effectiveness of software testing strategies. *IEEE Transactions on Software Engineering*, **SE-13**(12), pp. 1278–1296, 1987.

[4] Linger, R.C., Mills, H.D. & Witt, B.I., *Structured Programming, Theory and Practice*, The Systems Programming Series, Addison-Wesley, 1979.

[5] Mills, H., Dyer, M. & Linger, R., Cleanroom software engineering. *IEEE Software*, **4**(5), pp. 19–25, 1987.

[6] Mills, H.D., Zero defect software: Cleanroom engineering. *Advances in Computing*, **36**, pp. 1–41, 1993.

[7] Mills, H., Stepwise refinement and verification in box-structured systems. *Computer*, **21**(6), pp. 23–36, 1988.

[8] Elberzhager, F., Eschbach, R. & Munch, J., The relevance of assumptions and context factors for the integration of inspections and testing. *2011 37th EUROMICRO Conference on Software Engineering and Advanced Applications*, IEEE, pp. 388–391, 2011.

[9] Kloos, J., Elberzhager, F. & Eschbach, R., Systematic construction of goal indicator trees for indicator-based dependability inspections. *2010 36th EUROMICRO Conference on Software Engineering and Advanced Applications*, IEEE, pp. 279–282, 2010.

[10] Powell, D., Tool support for verification-based software inspection. *2004 Australian Software Engineering Conference Proceedings*, IEEE, pp. 232–240, 2004.

[11] Parnas, D. & Weiss, D., Active design reviews: Principles and practices. *Journal of Systems and Software*, **7**(4), pp. 259–265, 1987.

[12] Singh, N.K., Lawford, M., Maibaum, T.S.E. & Wassyng, A., Use of tabular expressions for refinement automation. *Model and Data Engineering*, eds. Y. Ouhammou, M. Ivanovic, A. Abello & L. Bellatreche, Springer International Publishing, Vol. 10563 of *Lecture Notes in Computer Science*, pp. 167–182, 2017.

[13] Jackson, A. & Hoffman, D., Inspecting module interface specifications. *Software Testing, Verification and Reliability*, **4**(2), pp. 101–117, 1994.

[14] Abrial, J.R., *Modeling in Event-B: System and Software Engineering*, Cambridge University Press: Cambridge, 2010.

[15] Abrial, J.R., *The B-Book: Assigning Programs to Meanings*, Cambridge University Press: Cambridge, 1996.

[16] Lecomte, T., Servat, T. & Pouzancre, G., Formal methods in safety-critical railway systems. *10th Brazilian Symposium on Formal Methods*, 2007.

[17] ter Beek, M.H., Fantechi, A., Ferrari, A., Gnesi, S. & Scopigno, R., Formal methods for the railway sector. *ERCIM News*, **112**, 2018.

[18] Ferrari, A. et al., Survey on formal methods and tools in railways: The astrail approach. *Reliability, Safety, and Security of Railway Systems. Modelling, Analysis, Verification, and Certification*, eds S. Collart-Dutilleul, T. Lecomte & A. Romanovsky, Springer International Publishing: Lille, pp. 226–241, 2019.

[19] Fantechi, A., Fokkink, W. & Morzenti, A., *Some Trends in Formal Methods Applications to Railway Signaling*, John Wiley & Sons: Hoboken, pp. 61–84, 2012.

[20] Haxthausen, A.E. & Peleska, J., *Model Checking and Model-Based Testing in the Railway Domain*, Springer Fachmedien Wiesbaden, pp. 82–121, 2015.

[21] Parnas, D.L., Inspection of safety-critical software using program-function tables. *Proceedings of the IFIP 13th World Computer Congress, Hamburg, Germany,* pp. 270–277, 1994.

TRAFFIC FLOW ANALYSIS CONSIDERING THE EFFECTS OF FEATURES OF LEVEL CROSSING AND ROAD PASSERS

SHINGO NAKAGAWA
Signalling and Transport Information Technology Division, Railway Technical Research Institute, Japan

ABSTRACT

At level crossings in urban area where trains and cars pass frequently, traffic jams are sometimes social issues. One way to improve this situation is to control and shorten the warning time of a level crossing by controlling train running. So far, several methods based on this idea have been proposed, however most of them aimed at shortening the active warning time, and they do not fully take into account the effect on the road traffic flow. On the basis of this situation, a traffic flow simulator has been developed which can consider the behavior of road passers including cars, bicycles and pedestrians within a level crossing area. The simulator can especially consider interaction among the road passers at a crowded level crossing, such as mutual avoidance behaviors. With this simulator, we conduct experimental calculations to evaluate how the traffic flow varies with factors, such as the size of the level crossing, active time of the warning bell and numbers and speed of road passers. This paper analyzes the results of the calculations and quantifies how the active warning time and the number of bicycles and pedestrians affect the car traffic capacity.
Keywords: level crossing, timetable, traffic capacity, avoidance behavior, experimental calculation.

1 INTRODUCTION

In Japan, level crossing accidents have been decreasing in recent years, but traffic jams at level crossings, as shown in Fig. 1, are still social issues, especially in urban area where trains and cars pass frequently. This problem reduces convenience for road passers and can lead to accidents. To control and shorten the warning time of a level crossing by controlling train running can be a solution to improve this situation [1]. Under such problem awareness, we have developed a traffic flow simulator which can consider the behavior of road passers within a level crossing area. This paper d escribes the experimental calculations which we have conducted with this simulator. By analysing the results, we quantify how the active warning time and the number of bicycles and pedestrians affect the car traffic capacity.

1.1 Abbreviations and terminologies

In this article, we use some abbreviations and terminologies described as below:

- LC(s): level crossing(s).
- A LC: active when the level crossing's bells are ringing. It means that the active level crossing warns the road passers not to cross.
- TTLC: timetable of level crossing. It includes when the level crossing's bell starts to ring and when the gate arms start to descend, finish descending and start to ascend.
- TCC (of a level crossing): traffic capacity of cars. The number of cars which can pass a level crossing during certain time.
- Inside/outside of a LC: inside of a LC is the area surrounded by gate arms of the LC (regardless of state of the gate arms.) Other area is outside of the LC. In other words, all road passers are prohibited to be inside of a LC during the LC closes.

WIT Transactions on The Built Environment, Vol 199, © 2020 WIT Press
www.witpress.com, ISSN 1743-3509 (on-line)
doi:10.2495/CR200281

Figure 1: A crowded level crossing in Japan.

2 TRAFFIC FLOW SIMULATOR AT A LEVEL CROSSING

The screenshot of the traffic flow simulator we have developed is as shown in Fig. 2. It can simulate the behavior of road passers including cars, bicycles and pedestrians within a LC area, based on structure of the LC, the number of road passers and distribution of their speed and TTLC. The simulator calculates all road passers' position every 0.5 second as one step; it repeatedly simulates their behavior based on the previous step every 0.5 second.

All road passers basically go straight and cross the LC at their original constant speed; they also follow the rules described below.

- If they go straight and collide or get very close with other one, they take avoidance behavior; turning left or right, going diagonally or slowing down. If they cannot take any avoidance behavior, they stop and stay there.
- If pedestrians and bicycles are forced out to the car lane, they try to return sidewalk, if possible.
- When the LC activates, pedestrians and bicycles that are within the LC will take action either hurrying up or returning back to the outside of the LC. The rate of such road passers and the acceleration rate can be set by users.
- Some pedestrians and bicycles go into the LC, even if the LC is active. The rate of such road passers can be set by users.
- No road passers go into the completely closed LC.

The simulator calculates where each road passer is under these rules and output the results as CSV files.

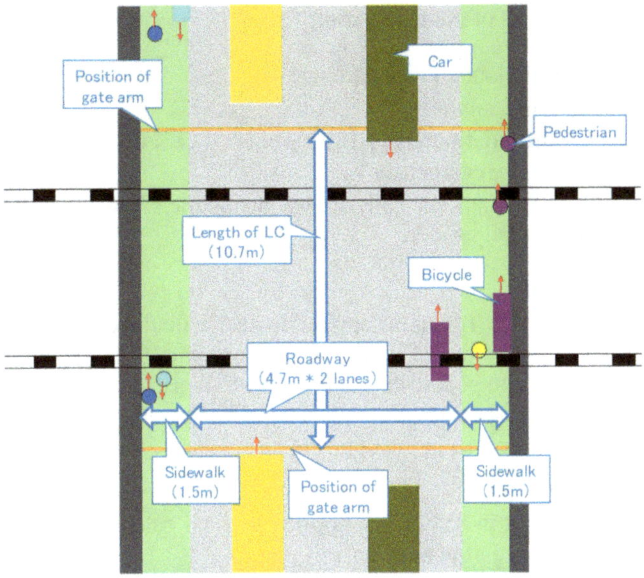

Figure 2: Example of a screenshot of the simulator. (Balloons, arrows and railways are added to the actual screenshot; Lengths in the figure are those of LC-A described later in Table 1.)

Table 1: Target LCs.

	Car Tracks	Sidewalks	Length
LC-A	4.7 m width, 2 lanes	1.5 m width, 2 lanes	10.7m
LC-B	4.0 m width, 2 lanes	1.4 m width, 2 lanes	16.2m
LC-C	3.9 m width, 2 lanes	1.7 m width, 2 lanes	10.9m
LC-D	2.8 m width, 1 lane (One-way traffic)	0.9 m width, 1 lane	11.8m

3 OUTLINE OF EXPERIMENTAL CALCULATION

3.1 Target level crossings

We choose four LCs as the targets of experiment, as described in Table 1. All of them are based on but not exactly same as the real LCs in Tokyo metropolitan area, because a railway line and a road cross diagonally at these LCs. Our simulator cannot deal with such LCs so that on our experiment we regard they cross vertically at the target LCs.

At LC-A, B and C, there are two car lanes (single lane for traffic in both directions) and two sidewalks, as shown in Fig. 2. LC-D is one-way traffic for cars so that there are one car track and one sidewalk, which are narrower than those of LC-A, B and C.

LC-B is expected to be more crowded than LC-A, because LC-B is narrower and the length is longer than LC-A. With respect to LC-C, car tracks are narrower and sidewalks are wider than LC-A and B, therefore it is difficult to predict the grade of congestion compared to LC-A. LC-D has only one car lane and is narrower than other LCs, therefore TCC at LC-D is estimated to be less than half of those at other LCs.

3.2 Premises and parameters of our calculation

1. Common settings throughout all calculation
   - Simulation time is 5 minutes.
   - Cars arrive the LC continuously. The number of cars which can cross over the LC in simulation time is regarded as TCC.
   - Bicycles and pedestrians basically follow inside the sidewalk but may walk or run inside the car tracks if needed.

2. Number of pedestrians
   One of five cases shown in Table 2 is applied to each calculation. They are total number of pedestrians who come to the LC in the simulation time (five minutes) per a sidewalk. For example, on case 1, 40 pedestrians appear at LC-A, B and C and 20 at LC-D. All of them appear randomly. Note that on case 5 at LC-D, there are 30 pedestrians in the same direction as that of cars and 10 in the opposite direction.

3. Number of bicycles
   One of two cases shown in Table 3 is applied to each calculation. Meanings of these numbers are same as those of pedestrian.

4. Original speed of road passer
   We assume that there are 10 types of pedestrians whose speed are different. On our calculation, one of four cases shown in Table 4, which are different in terms of the component ratio of these 10 types, is applied. On the basis of a multiple regression analysis described in Suzuki et al. [2] and considering that LC-B is longer than other LCs, we determined median speed of pedestrian 140 cm/s at LC-B and 135 cm/s at the rest LCs.

   Regarding bicycles, 10 speeds are set in the same way as pedestrian and one of four cases is applied to each calculation.

   Case 1 is uniform distribution. Case 2/3 is the case whose average speed is fast/slow. Note that the original speed of cars is uniformly 530 cm/s.

Table 2: The number of pedestrians per one sidewalk.

	Number of pedestrians per one sidewalk
Case 1	10 in each direction
Case 2	20 in each direction
Case 3	30 in each direction
Case 4	40 in each direction
Case 5	30 in one direction, 10 in the opposit direction

Table 3: The number of bicycles per one sidewalk.

	Number of bicycles per one sidewalk
Case 1	10 in each direction
Case 2	20 in each direction

5. Behavior selection of the road passer when LC is activated while the passers are crossing
When a LC starts warning, some road passers can be within the LC. Then they will change their behavior. On our calculation, three choices of their behavior are assumed: hurrying up, returning back and waiting at outside of the gate arm and keeping their previous behavior unchanged. Regarding the selection rate, one of two cases shown in Tables 5 and 6 is applied to each calculation.
It is natural that these rates vary depending on the positions of the road passers in the inside of the LC. For example, when they are almost going through the LC, they will not return back because passing through the LC is obviously safer for them than returning. For this reason, applied selection rates change depending on the positions of the passers.

Note that cars choose to always keep going, because to accelerate or to return is illegal or dangerous for them.

Table 4: The composition ratio of original speed of pedestrian and bicycle.

Pedestrian (LC-A, C and D) (cm/s)	Pedestrian (LC-B) (cm/s)	Bicycle (cm/s)	Case 1	Case 2	Case 3	Case 4
180	185	410	10%	20%	-	20%
170	175	390	10%	10%	10%	-
160	165	370	10%	20%	-	20%
150	155	350	10%	20%	-	-
140	145	330	10%	10%	10%	10%
130	135	310	10%	10%	10%	10%
120	125	290	10%	-	20%	-
110	115	270	10%	-	20%	20%
100	105	250	10%	10%	10%	-
90	95	230	10%	-	20%	20%

Table 5: Behavior selection of the road passer when LC start to warn. (When the road passer has moved 20% or less of the LC's length.)

	Pedestrian	Bicycle
Case 1	Hurry: 50% Return: 20% Keep going: 30%	Hurry: 10% Return: 20% Keep going: 70%
Case 2	Hurry: 30% Return: 20% Keep going: 50%	Return: 20% Keep going: 80%

Table 6: Behavior selection of the road passer when LC start to warn. (When the road passer has moved more than 20% of the LC's length.)

	Pedestrian	Bicycle
Case 1	Hurry: 50% Keep going: 50%	Hurry: 10% Keep going: 90%
Case 2	Hurry: 30% Keep going: 70%	Keep going: 100%

Table 7: The rate of pedestrians and bicycles who go into the LC after its bell starts ringing.

	While the bell are ringing (before the gate arms start to descend)	While the gate arms are descending
Case 1	30%	10%
Case 2	For pedestrians and bicycles whose original speed are: - fast: 70% - medium: 50% - slow: 30%	10%

6. Penetration inside active LC
   While a LC is warning, road passers should not start to cross; they have to stop at in front of the gate arm. However, some of them who are in hurry go into the LC which has not been completely shut by the gate. Regarding the rate of such pedestrians and bicycles, one of two cases shown in Table 7 is applied to each calculation. In case 2, the rate for faster pedestrians is set to be higher, because they tend to think they can pass through the LC before the gates shut.
   Note that cars do not go into active LC. In addition, after the gates are completely shut, no road passers including pedestrians and bicycles go into the LC.
7. TTLC (timetable of level crossing)
   We define 10 TTLCs shown in Fig. 3. One of them is applied to each calculation. In all TTLCs, the LC closes once or twice for five minutes and each closure time is 40 or 70 seconds. Therefore total closure time is 40 seconds at the shortest and 140 seconds at the longest (see Table 8). The bell of the LC starts to ring four seconds before when the gate arms start to descend taking 12 seconds; in other words, the LC activates 16 seconds before its closure. After the closure, the gate arms ascend in three seconds.

As described in (6), pedestrians and bicycles go into the inside of the LC unless it is completely shut. So the time they cannot go into the LC is 40 or 70 seconds per closure. For cars, the time is 56 or 86 seconds, because they cannot go into the LC after it activates.

We conducted our experiments with 1600 settings, which are the combinations of the cases described in (2)–(7). We conducted five calculations under each of 1600 settings for each of four LCs listed in 3.1; in other words, we conducted totally 32000 calculations.

3.3 Propriety of the simulations

Rules of behavior of road passers described in Section 2 and parameters described in 3.2 are defined to depict natural behavior inside the LC. To check the propriety of them, we compared the results of our calculation to behavior within real LC.

We have recorded the traffic at LC-B. From this video, each pedestrian's walking distance in 0.5 second is measured. We also obtain such walking distance from our calculation regarding LC-B. Note that the target of comparison is right after the LC opens, when the congestion occurs.

The results are shown in Table 9. Differences of average for both directions are only about 1 cm, which do not matter. On the other hand, standard deviations of results of our simulation are larger than those of real LC. On our simulation, 8% of the obtained data do not go forward due to the congestion; they walk aside or stop. To revise the rule to make them walk forward or diagonal may reduce the standard deviation.

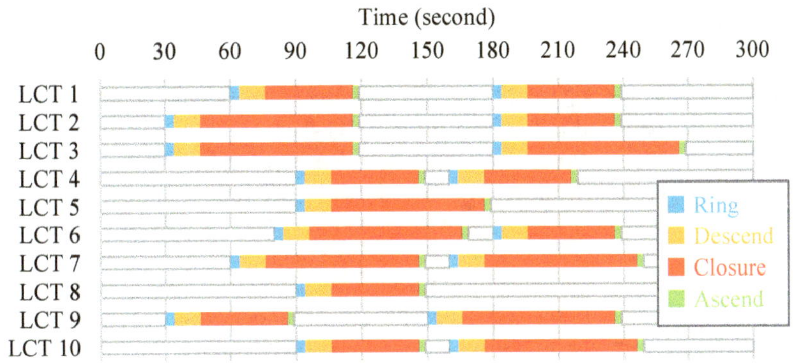

Figure 3: Timetables of level crossing.

Table 8: Closure time of each TTLC.

	Closure time (second)			Interval (second)
	Total	1st closure	2nd closure	
TTLC 3	140	70	70	61
TTLC 7	140	70	70	11
TTLC 2	110	70	40	61
TTLC 6	110	70	40	11
TTLC 9	110	40	70	61
TTLC 10	110	40	70	11
TTLC 1	80	40	40	61
TTLC 4	80	40	40	11
TTLC 5	70	70	-	-
TTLC 8	40	40	-	-

Interval is the duration from the gate arms finish to ascend to the bell starts to ring again, which is painted white between two closures in Fig. 3.

Table 9: Walking distance of pedestrian in 0.5 second.

	Number of data	Direction of travel (cm)		Lateral direction (cm)	
		Average	Standard deviation	Average	Standard deviation
Recorded video	13848	51.9	16.6	8.3	7.7
Simulation	34523	53.2	23.9	9.2	17.6

4 ANALYSIS ON TRAFFIC CAPACITY OF CARS

4.1 Analysis for LC-A

As described in Section 3.2, 8000 calculations have been conducted for LC-A. The distribution of TCC obtained in these calculations is shown in Fig. 4. They ranged from 18 to 140 and their average is 90.2. The distribution has five peaks and does not resemble to normal distribution, about which we will consider latter section.

Note that in some calculations a very small number of TCC is obtained. It is because heavy congestion has occurred just after the LC has opened and some road passers obstruct other people each other. Such passers block the car lanes. In a real situation, under such congestion, each road passers settle the matter by mutual concessions, such as walking back to let other people go. These results show that our simulator cannot perform such cooperation, which is our future issue.

Table 10 shows the summarization of obtained TCC classified by each parameter. The difference of TCC due to TTLC is larger than other factors. Five peaks observed in Fig. 4 are almost same as average TCC under each TTLC. Among the TTLCs whose total closure time are same, average TCC under them are almost the same. We can say that the TTLC affects TCC mostly and the number of pedestrians and bicycles also makes some difference to TCC. Other parameters have only little effect.

We conducted significance tests to obtained TCC; checking whether or not the obtained TCC have significantly different among each case/TTLC. The results are shown in Table 11. We conducted the tests for pairs of case/TTLC; for example, "case 1&2" compares the TCC obtained under case 1 and under case 2.

Significant differences due to the number of pedestrians, the number of bicycles and the original speed are observed. Furthermore, regarding TTLC, the p-value is less than 0.01 for 41 out of 45 pairs. The remaining four are the pairs of TTLC whose total closure time are same, but some of the pairs which have such relation are included in the 41 pairs mentioned above. We can also say that for all the pairs of TTLC whose total closure time are same, more TCC has obtained under the TTLC whose interval of two closures is shorter. The reason for these different is not clear, but it is presumed that to make the closure time closer and open the LC continuously may lead the traffic flow smoothly and effective to increase the TCC (note that to shorten the closure time is obviously more effective).

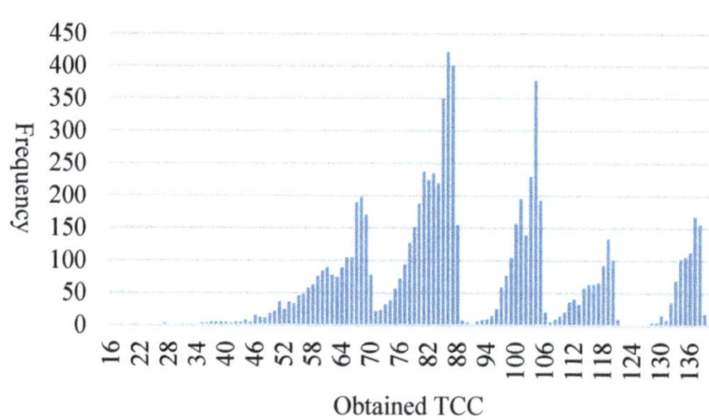

Figure 4: Distribution of obtained TCC at LC-A.

Table 10: The obtained TCC at LC-A classified by case/TTLC.

Parameters	Case/TTLC	TCC		
		Average	Max.	Min.
Number of pedestrians	Case 1	92.5	139	27
	Case 2	91.0	139	35
	Case 3	89.3	139	24
	Case 4	87.2	140	18
	Case 5	91.0	139	32
Number of bicycles	Case 1	93.9	140	42
	Case 2	86.4	137	18
Original speed	Case 1	90.2	139	18
	Case 2	91.3	140	25
	Case 3	89.3	139	18
	Case 4	89.9	139	21
Behavior selection when LC activates	Case 1	90.2	140	21
	Case 2	90.1	139	18
Penetration into active LC	Case 1	90.1	139	18
	Case 2	90.2	140	18
TTLC	TTLC 1	101.4	107	66
	TTLC 2	80.6	88	38
	TTLC 3	61.2	71	18
	TTLC 4	101.8	106	62
	TTLC 5	113.4	121	50
	TTLC 6	83.2	89	50
	TTLC 7	62.7	71	36
	TTLC 8	135.3	140	51
	TTLC 9	80.8	88	18
	TTLC 10	81.5	89	55

Table 11: The results of significance test among each case/TTLC.

Parameters	Significance	Pair of cases/TTLCs
Number of pedestrians	$p \leq 0.01$**	Case 1&3, 1&4, 2&4, 4&5
	$p \leq 0.05$*	Case 2&3, 3&4, 3&5
	$p > 0.05$	Case 1&2, 1&5, 2&5
Number of bicycles	$p \leq 0.01$**	Case 1&2
Original speed	$p \leq 0.01$**	Case 2&3
	$p \leq 0.05$*	Case 2&4
	$p > 0.05$	Case 1&2, 1&3, 1&4, 3&4
Behavior selection when LC activates	$p > 0.05$	Case 1&2
Penetration into active LC	$p > 0.05$	Case 1&2
TTLC	$p \leq 0.01$**	41 pairs (all pairs except the following)
	$p \leq 0.05$*	TTLC 2&10, 9&10
	$p > 0.05$	TTLC 1&4, 2&9

Table 12: Average of obtained TCC at each LC.

Parameter	Case/TTLC	LC-A	LC-B	LC-C	LC-D
Number of pedestrians	Case 1	92.5	87.1	86.2	37.0
	Case 2	91.0	85.6	84.1	34.0
	Case 3	89.3	83.3	81.8	31.0
	Case 4	87.2	80.6	79.0	28.6
	Case 5	91.0	85.3	84.2	35.6
Number of bicycles	Case 1	93.9	88.4	87.2	39.1
	Case 2	86.4	80.4	78.9	27.3
Original speed	Case 1	90.2	84.7	83.2	33.3
	Case 2	91.3	85.3	84.2	35.5
	Case 3	89.3	83.3	81.8	31.4
	Case 4	89.9	84.3	83.1	32.7
Behavior selection when LC activates	Case 1	90.2	84.4	83.0	33.4
	Case 2	90.1	84.4	83.1	33.1
Penetration into active LC	Case 1	90.1	84.4	82.9	33.3
	Case 2	90.2	84.4	83.1	33.2
TTLC	TTLC 1	101.4	96.1	94.5	39.4
	TTLC 2	80.6	75.0	74.1	25.3
	TTLC 3	61.2	56.2	55.5	18.0
	TTLC 4	101.8	95.9	94.8	41.4
	TTLC 5	113.4	104.7	102.6	40.0
	TTLC 6	83.2	78.5	77.0	32.6
	TTLC 7	62.7	59.0	57.2	22.2
	TTLC 8	135.3	126.4	125.3	54.6
	TTLC 9	80.8	75.6	74.5	27.0
	TTLC 10	81.5	76.4	75.0	31.7

4.2 Comparison of four LCs

The average TCC at four LCs are shown in Table 12. The obtained TCC are in the order of LC-A>B>C>D for all of 25 parameters. Obtained TCC on LC-D are less than half of those on other LCs, as we have estimated in 3.1. The trends of difference between the obtained TCC under each parameter are almost same on all LCs.

4.3 Regression analysis on TCC

We examined multiple regression analysis to obtain regression formula which explains TCC by using parameters described above as descriptive variables. We set 22 dummy variables which are correspond to each parameter as candidate of descriptive variables and apply forward-backward stepwise selection method for variable selection. Note that each of "number of bicycles," "behavior selection when LC activates" and "penetration into active LC" can be expressed by one dummy variable, because there are only two selections.

The results are shown in Table 13. Selected variables and their significances are almost the same through all LCs.

Table 13: Obtained regression formula which explains TCC.

Parameter	Case/TTLC	LC- A		LC- B		LC- C		LC- D	
(Constant term)		81.97	**	76.51	**	75.17	**	32.42	**
Number of pedestrians	Case 1	5.29	**	6.55	**	7.20	**	8.39	**
	Case 2	3.87	**	5.00	**	5.10	**	5.40	**
	Case 3	2.12	**	2.77	**	2.81	**	2.46	**
	Case 4	-		-		-		-	
	Case 5	3.82	**	4.73	**	5.22	**	6.99	**
Number of bicycles	Case 2	-7.51	**	-8.01	**	-8.33	**	-11.83	**
Original speed	Case 1	0.30		0.41	*	-		0.67	*
	Case 2	1.42	**	0.99	**	1.07	**	2.86	**
	Case 3	-0.63	**	-0.98	**	-1.41	**	-1.21	**
	Case 4	-		-		-		-	
Behavior selection when LC activates	Case 2	-		-		-		-	
Penetration into active LC	Case 2	-		-		-		-	
TTLC	TTLC 1	19.93	**	19.65	**	19.50	**	7.66	**
	TTLC 2	-0.91	**	-1.38	**	-0.92	**	-6.40	**
	TTLC 3	-20.36	**	-20.21	**	-19.46	**	-13.71	**
	TTLC 4	20.26	**	19.44	**	19.79	**	9.70	**
	TTLC 5	31.92	**	28.25	**	27.59	**	8.24	**
	TTLC 6	1.69	**	2.06	**	2.04	**	0.90	*
	TTLC 7	-18.81	**	-17.43	**	-17.82	**	-9.58	**
	TTLC 8	53.76	**	50.03	**	50.33	**	22.90	**
	TTLC 9	-0.76	**	-0.79	*	-0.52		-4.76	**
	TTLC 10	-		-		-		-	
Adjusted R^2		0.94		0.91		0.90		0.65	

The mark ** and * mean the significance of selected variables, whose meanings are same as those of Table 11. The mark - means the variable is not selected.

As for LC-A, B and C (i.e. LCs with two car lanes,) coefficients for each variable are almost the same. It means that differences of constant terms correspond to those of obtained TCC. Coefficients of determination (adjusted R^2) are over 0.90 for these LCs, which show the obtained regression formulas are appropriate.

Regarding LC-D, constant term is less than those of other LCs because it has only one car lane. Magnitude relationship of coefficients is slightly different from that of other LCs, while their signs are same as those of other LCs. But the adjusted R^2 is 0.65, which means less appropriate than others.

Next, to estimate the effect of shortening the active time of the LC, we tried to obtain regression formula which explains TCC by using the time car can go through the LC (i.e. 300 seconds – active time of the LC) as descriptive variables. The formula is as eqn (1) below, where y is obtained TCC, x is the time and α, β are coefficients.

$$y = \alpha x + \beta. \tag{1}$$

The results are shown in Table 14. As for LC-A, B and C, adjusted R^2 is sufficiently large and all α, β are sufficiently significant, which mean obtained formulas are appropriate. Note that all α are about 0.6. It means that if the time the car can go through the LC extends one second, the TCC will increase about 0.6. This result is well match with the previous paper by

Iwasaki et al. [3], which estimates the traffic capacity of cars at the LC with straight road as about 1000 per an hour and one car lane.

Regarding LC-D, α is about half of α for other LCs and α, β are sufficiently significant, both of which are reasonable result. But considering the adjusted R^2, this formula has weak propriety.

Table 14: The estimated relationship between the TCC and the time car can go through the LC.

	LC-A	LC-B	LC-C	LC-D
α (coefficient for the time)	0.624	0.582	0.580	0.284
β (constant term)	-17.26	-15.90	-16.84	-15.64
Adjusted R^2	0.90	0.86	0.84	0.41

All α, β are sufficiently significant ($p \leq 0.01$**).

5 CONCLUSION

We have described about our traffic flow simulator and the experimental calculations to evaluate how the traffic flow varies with factors, such as the size of the level crossing, active time of the warning bell and numbers and speed of road passers. By analysing the results, we can show that the active warning time of level crossing affects the traffic capacity of cars, whose effect is about 0.6 car/second at level crossing with two car lanes. We also show that the number of pedestrians and bicycles and pedestrians affect traffic capacity of cars significantly. Moreover, it is suggested that making the closure time of level crossing closer and opening the level crossing continuously may lead the traffic flow smoothly and effective to increase the traffic capacity of cars. Note that this may force road passers to wait longer and reduce convenience and safety.

From our calculation, we also analysed the number of pedestrians who are left inside the closed level crossing. But we could not obtain significant regression formula. The main reason is considered the obstruction described in 4.1. We aim to depict the road passers' behavior more precise and improve the validity of evaluation to solve these problems including we show in this paper.

REFERENCES

[1] Fukuda, M., Study of the control method for Level Crossing with CBTC System in case of train delay and congested traffic. *IEE Japan*, pp. 1–6, 2014, (in Japanese).

[2] Suzuki, D., Endoh, H., Saito, A., Akiu, N. & Mizukami, N., Effects on walking velocity by length, warning and congestion of railroad crossing, *Proceedings of the 59th Conference of Japan Ergonomics Society*, 2018, (in Japanese).

[3] Iwasaki, M., Watanabe, T. and Miyazawa, T., Traffic flow characteristics and delay at urban railroad crossings. *Proceedings of Japan Society of Civil Engineers*, **401**, pp. 61–67, 1989, (in Japanese).

DEVELOPING THE TRAM CONTROL SYSTEM BASED ON SIMULINK/STATEFLOW AND B METHOD

CHENG PENG[1,2], WANG KEMING[1,2], HOU XILI[3], LIU NING[2,4] & WANG ZHENG[1,2]
[1]School of Information Science and Technology, Southwest Jiaotong University, China
[2]National-Local Joint Engineering Laboratory of System Credibility Automatic Verification, Southwest Jiaotong University, China
[3]TongHao GBA (Guangzhou) Smart Control Co., Ltd., China
[4]Graduate School of Tangshan, Southwest Jiaotong University, China

ABSTRACT

The huge gap between the requirements and the system model is an obstacle to the application of formal methods in industry. To reduce this gap, as well as to enhance the consistency and completeness before implementation, we proposed an approach integrating Simulink/Stateflow and Atelier B for developing the tram control system. The Simulink/Stateflow was used to quickly model the requirements for the system. Moreover, we analysed and debugged the logic of the system with the fast-iterative simulation of Simulink/Stateflow. Afterwards, we outlined the analysis in which the B architecture is chosen and manually built the B model according the process flows of the Stateflow model, to further explore through proof of safety invariants. Finally, we introduced the approach by developing the point controlling module of our projects. In this paper, following the approach we presented, not only can the consistency between the requirements and formal specification be improved, but the safety of system model is strengthened.

Keywords: Simulink/Stateflow, B method, formal verification, simulation, safety-critical, tram control system.

1 INTRODUCTION

In recent years, the formal method, a technology with rigorous mathematics foundation, has been successfully applied to the development of the railway safety-critical system. For example, ClearSy [1], a French Company, developed the driverless system of Paris metro line 14 using Atelier B [2], a B method-based IDE tool [3].

However, there are remaining challenges on the application of the B method. The challenges consist of correctness of specifications, correctness of implementation and correctness of proofs, etc. [4]. To revise the incorrect specifications, which lead to the inconsistency between the informal requirement specifications and formal modelling, the UML [5] is used to explore the potential designs before modelling. It can help the coder to understand the design of the system, but is unable to validate the correctness of the functional flow analysis of the requirements. The Matlab Simulink/Stateflow [6] provides the graphical diagrams, which not only can quickly model the behaviors of the control system, but also enable the developer to analyze and debug the logic flow of the model. Nevertheless, the correctness of the model cannot be guaranteed by simulation testing alone. Although the Simulink Design Verifier [7] can use Bounded Model Checking [8] algorithm to generate test cases to improve test coverage, it lacks the sufficient support for formal verification (e.g. The state space explosion of the model). Therefore, we propose an approach which combines the advantages of both Simulink/Stateflow and B method.

We apply the integrated approach to the developing of the tram control system (TCS) of Guangzhou Huangpu Line 1 in China. As far as we know, this is the first application of B method on the tram control system in China. The approach boosts our confidence in design consistency and completeness before implementation.

WIT Transactions on The Built Environment, Vol 199, © 2020 WIT Press
www.witpress.com, ISSN 1743-3509 (on-line)
doi:10.2495/CR200291

The rest of this paper is organized as follows. The developing map of our project presented in this paper follows the discussion of Section 1. By introducing a case, point controlling module (PCM) of TCS, the processes of Stateflow modelling and translating that are described in Section 1 are developed into a concrete activity in Section 2. In Section 3, the simulation steps are listed to explain how to do iterative simulating; the example of formal verification is given to illustrate the benefits between simulation testing and formal technologies. Finally, we discuss the related work and conclude our work.

2 THE PROCESSES OF OUR APPROACH

Developing the safety-critical systems in the railway domain usually follows the V model. Here we define an enhanced V model in our approach, as depicted in Fig. 1.

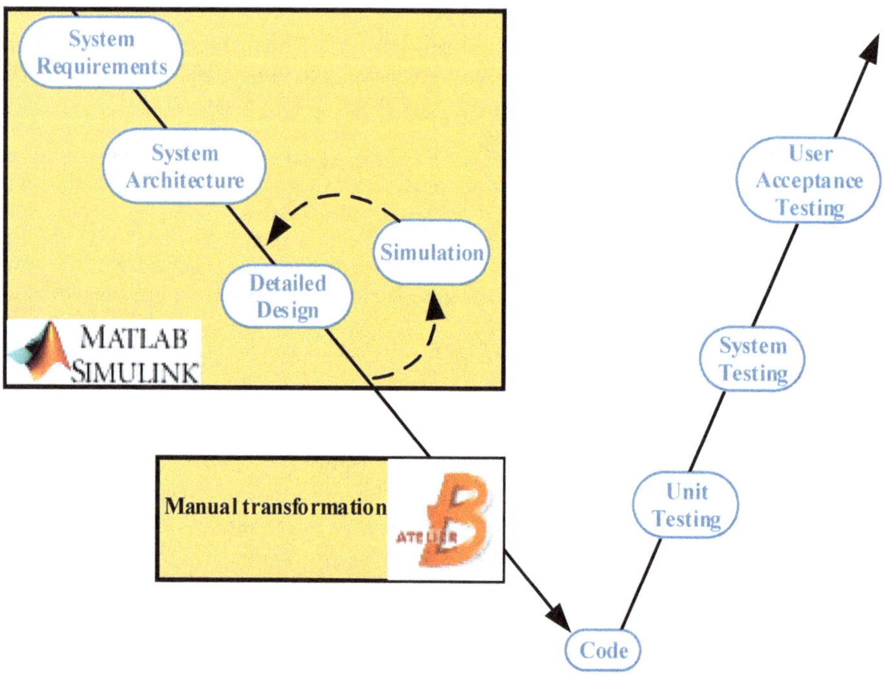

Figure 1: The developing map.

In Fig. 1, the approach consists of two processes, which are described below:

1. Modelling and validation based on Simulink/Stateflow: the flow charts and state machines of Stateflow are used to model the system functions, which are derived from the requirements. After the Stateflow model was established, the correctness of its function flow is validated through fast iterative simulation, which is a cyclic control represented by the dashed line in Fig. 1.

2. Modelling and verification based on Atelier B: we manually build a B model following the architecture and logic flows of Stateflow model, which will be proved in the proof environments of Atelier B.

3 MODELLING AND TRANSLATING THE PCM

In this section, along with the steps of this integration approach, the PCM and its development will be illustrated.

3.1 The brief introduction of PCM and its simulink/stateflow

The TCS includes many modules such as *PCM*, *Route*, *Manual*, etc., which are required to achieve safety integrity level 4 (SIL4) and comply with the European standard for railway software, CENELEC EN50128 [9]. Fig. 2 shows the connections between PCM and other modules, which cooperate with the data-flows to send and receive the commands.

Simulink contains a set of blocks, sub-systems, and wires. Stateflow can be encapsulated as blocks or sub-systems, where Stateflow can be modelled by a hierarchical structure. Fig. 3 is presented to offer a real hierarchy snippet of our Stateflow model and it is also an extension of the POL diagram. The wires connect among the different blocks or sub-systems to cooperate with the data-flows. To provide the actual connections, Fig. 4 draws a Simulink project on the basis of the Fig. 2.

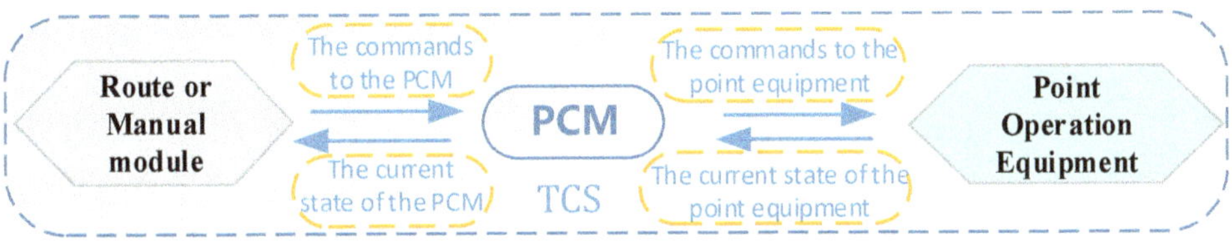

Figure 2: The framework of the PCM.

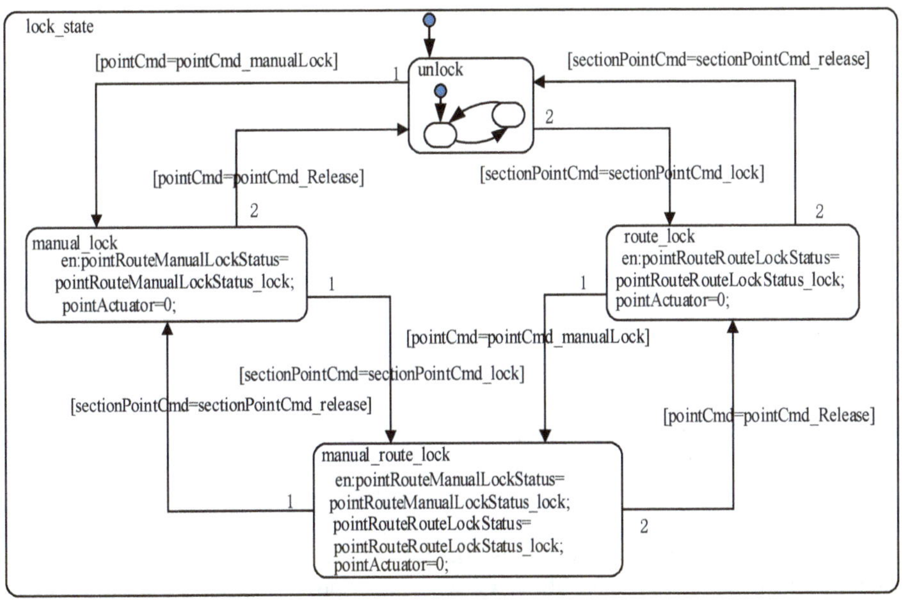

Figure 3: The picture depicts the *unlock* state, it has a sub-diagram as well as the extension of POL.

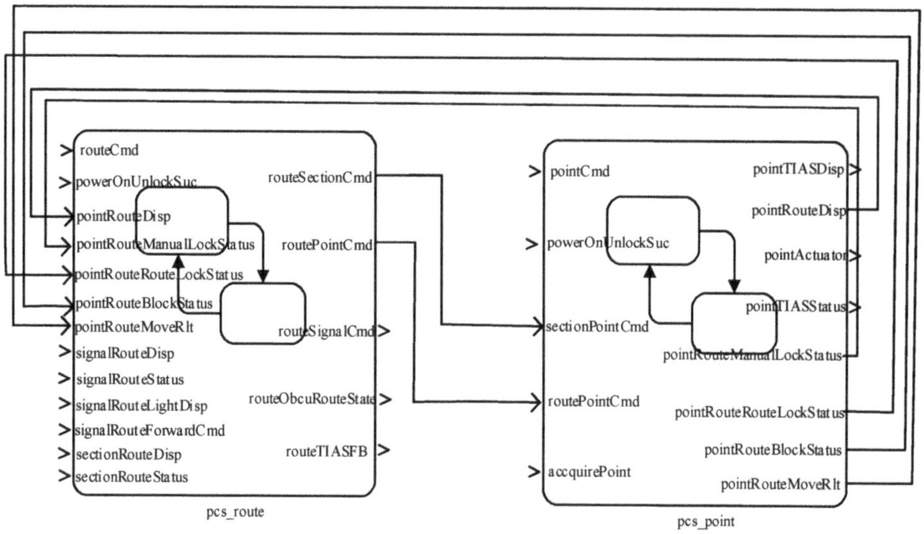

Figure 4: The connections between Route and PCM.

Fig. 2 ignores the command *PowerOnUnlockSuc*, which is the input of *pcs_point* of Fig. 5. In this paper, *pcs_point* is approximately equivalent to PCM.

3.2 Modelling Stateflow of PCM

In Section 1, the first step of our approach is modelling and validation based on Simulink/Stateflow. As a correspondence to the first step, modelling PCM is dedicated in this part and the validation of PCM is discussed in Section 3.

Stateflow acts as a bridge between the requirements and system design. We use Stateflow to create a multiple-level hierarchical model to represent the system functions that is decomposed by the domain experts. The different levels indicate the different development stages of the model, which are portrayed as a development views, from abstract to concrete.

The Stateflow diagram of each level can be an AND diagram, for which states are distributed in parallel and all of them are actually executed in sequential order according to their priority once AND is triggered; or an OR diagram, where only one of the states is active when OR is activated. Fig. 5 gives the Stateflow diagram of the *pcs_point*, containing the *PowerOnlock* module and PCM module. The Stateflow of PCM is an AND diagram, which is composed of three parallel states: point operating and locking (POL), point blocking (PB), and the display of the train integrated automated system (TIAS), respectively. Because of hierarchy, POL is modelled as a sub-diagram and it is also an OR diagram, comprising four states, *unlock*, *manual_lock*, *route_lock*, and *manual_route_lock* respectively.

After the Stateflow model was established, the correctness of its functional flow was validated in the simulation environments. In Fig. 4, the data-flows that are connected between the PCM and Route constitutes closed loop control. The loop can be iteratively simulated to debug and analyze the correctness of function of the model, which are discussed in Section 3.

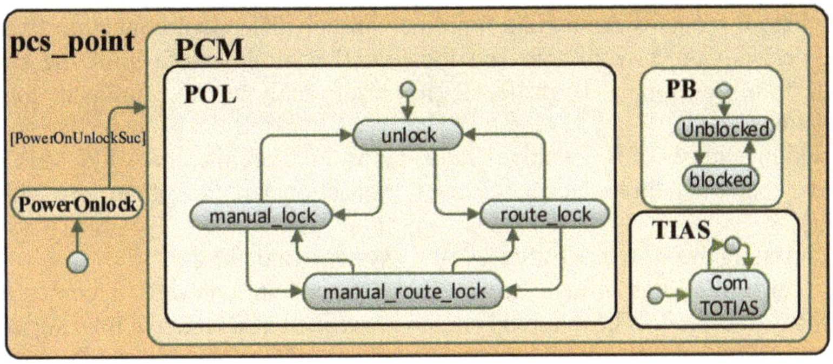

Figure 5: The overview of Stateflow of pcs_point.

3.3 Translating Stateflow to B

Owing to the correctness of the Stateflow model cannot be guaranteed by simulation alone, formal technologies are introduced into the development process to deal with this issue. In Section 1, the formal activity is the second step of our approach, which is divided into the modelling and verification based on Atelier B. Modelling PCM is also the translating of PCM. As a correspondence to the second step, translating is analyzed in this part and the verification is discussed in Section 3.

In terms of translating of PCM, two aspects should be inherited by B modules, which are the structure of the hierarchical diagram of Stateflow model and its detailed design respectively.

3.3.1 Translating the structure of Stateflow to B module

The first aspect concerns the inheritance of the architecture of Stateflow. Fig. 6 provides the mathematical relation which represents a mapping, from Stateflow diagram to B modules. The mapping is defined as a total surjection:

$$Transition \in \{D1, D2, ..., Dn\} -->> \{B1, B2, ..., Bn\}.$$

In Fig. 6, the set $\{D1,\ D2,\ ...,\ Dn\}$ denotes all diagrams of this module and the set $\{B1,\ B2,\ ...,\ Bn\}$ denotes the all B modules, where a B module can be made up of three components – abstract machine specification, refinement, and implementation [1].

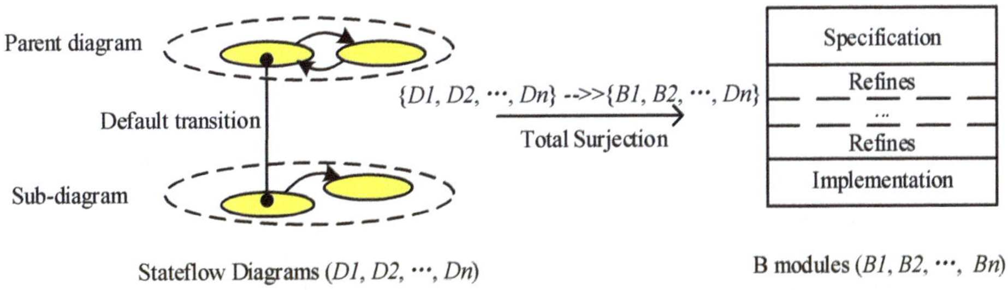

Figure 6: The mathematical relation between Stateflow and B.

Since these B modules can build a B project through different architectures, hand in hand with the decision on how to translate the Stateflow architecture goes the choice of B architecture to develop a B project. The brief analyses are outlined for choosing the translations.

Translation analysis of B development: as far as the B architecture of PCM is concerned, we believe there are three mechanisms of translation (In fact, there are only two for this paper).

(1) Translating the structure of Stateflow to system modelling option.

System modelling has always been a development option when it comes to using B method for system level. The mapping should be defined as a concrete total surjection:

$$\mathrm{Re}\,\mathit{finement\_structure} \in \{D1, D2, ..., Dn\} -->> \{B1\}.$$

On the basis of the definition, all diagrams of PCM are coupled into a B module. It actually adopts a coupling translation paradigm. To construct final software systems, one issue should be taken which is that the specification of each B module needs comprise the inputs and outputs of all diagrams of a Stateflow module. However, this mechanism can demonstrate system but does not generate codes for system. To this end, B method provides software development option for code generation. Software development mainly adopts INCLUDES and IMPORTS mechanisms to develop system [1].

As a preliminary statement below, the mapping of (2) and (3) both are defined as a bijection in this module:

$$\mathit{INCLUDES\_IMPORTS} \in \{D1, D2, ..., Dn\} >->> \{B1, B2, ..., Bn\}.$$

(2) Translating the structure of Stateflow to INCLUDES mechanism (software develops). INCLUDES mechanism is used to bring together the components, where the information of the *included* component can be obtained by the *inclusion* component. It is much fitter for developing specification text. According to the bijection, each diagram of PCM corresponds to a B module that includes a specification. INCLUDES mechanism is first used to link among the specifications, then they are implemented separately.

(3) Translating the structure of Stateflow to IMPORTS mechanism (software develops).

IMPORTS mechanism links between an implementation and a specification. The implementation can use specification's data to implement its own data and operation. It is more suitable for constructing software systems. Based on the bijection, these specifications are first independently developed into the implementations, where each implementation uses IMPORTS mechanism to link corresponding specification.

Result for developing PCM: In terms of Stateflow of PCM, the specifications have been divided into different hierarchical diagrams. Intuitively, such a Stateflow specification offers exceptional advantages for software development. From our opinion, for choosing (2) or (3), the main criteria are how the relations between different hierarchical diagram in PCM are linked. By the composition of the specifications, INCLUDES mechanism can build a refinement structure of PCM. For instance, if POL *includes pcs_point*, the guard conditions of each transition of POL should *include* the condition *PowerOnUnlockSuc* in *pcs_point*. IMPORTS mechanism divide the guard conditions of the refinement structure into different implementations through decomposing an implementation into a number of specifications. For example, the *PowerOnUnlockSuc* is not *included* POL in case of *pcs_point imports* POL, but the *PowerOnUnlockSuc* is still the guard condition of the transitions of POL as the substitution of the implementation of *pcs_point* is decomposed into the operation of the specification of POL. Consequently, both (2) and (3) can be used to develop PCM.

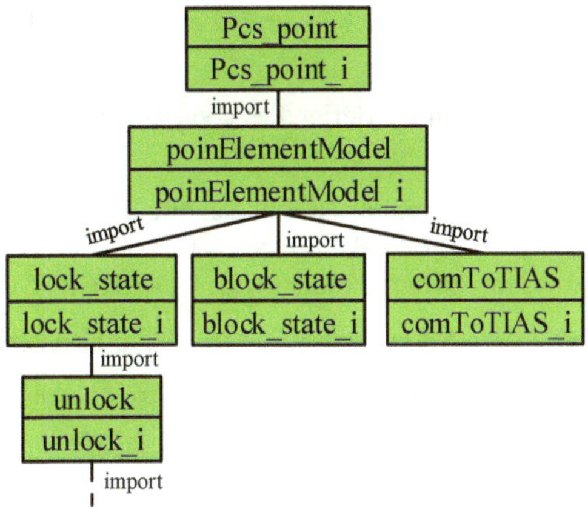

Figure 7: Snippets of the B architecture of PCM.

Fig. 7 shows that (3) is used to develop the architecture of PCM in this paper.

In Fig. 7, the four abstract machines – *pointElementModel*, *lock_state*, *block_state*, and *comToTIAS* respectively denote PCM, POL, PB, and TIAS. The *unlock* abstract machine denotes the state *unlock* of POL, which has a sub-diagram.

3.3.2 Translating the contents of Stateflow to B module

Once the architecture is determined, another aspect will concern how the contents of each specification and its refinement and implementation (*B1*, *B2*, …) should be developed. The main elements of a Stateflow diagram include a finite set of states, transitions, and variables.

States: except for having the hierarchy like *unlock* state in POL, a state may be involved with the three optional types of actions: An *entry action* executed when the state is activated; A *during action* executed when no transition is satisfiable; An *exit action* executed immediately before the source state is marked inactive.

Transitions: to select an active state of the OR diagram on condition that the OR diagram was activated, default transitions with no source states or source junctions are allowed. For instance, the *unlock* state becomes active as the POL diagram was triggered. A transition might consist of several transitions joined by junctions such as in TIAS diagram. A transition label is a line with an arrowed, from source state to target state; It can contain an *event*, a *condition*, a *condition action*, a *transition action*, which are formed as follows:

$$transition\_label = event[condition]\{condition\_action\} / transition\_action \ [17].$$

The valid transition label can be an *event* only; or a *condition* only; or an *event* and a *condition*; or NOT specified; etc. The *condition_action* will be executed when the *event* is triggered and the *condition* becomes true. The *transition_action* is executed after the relevant transitions were determined to be valid. For instance, the *transition_action* is triggered after the *exit action* of source state was exectuted.

Variables: the variables may be a set of input data-flows, output data-flows, or local, and them exist in the *condition*, *action*, etc.

We translate the main elements of a Stateflow diagram into the OPERATIONS CLAUSE and the INITIALISATION CLAUSE of an implementation of a B module. The relation between them and ELSIF definition are depicted in Fig. 8. Apart from the ELSIF structure, other substitution structures are allowed as long as the substitution can be implemented.

Following the previous definition that each diagram corresponds a B module, the translation items are presented as shown below:

1. The States in each diagram are defined as a variable to denote the states such as the variable *pointLockstate* in Fig. 9;
2. The Variables of each diagram are defined as the input and output parameters (e.g. the *pointCmd* in Fig. 9) of OPERATIONS. The parameters will be replaced by the temporary variables to verify the model with INVARIANT CLAUSE;
3. INITIALISATION is coded by translating the default transitions of Stateflow.
4. The predicates of OPERATIONS may be expressed by the source state itself (e.g. the variable *pointLockState* in Fig. 10) and *event* or *condition* in the transition label;
5. The substitution of OPERATION can be *condition_action*; or *transition_action*; or relevant *entry action*, *during action*, and *exit action*; or a target state itself. Besides the substitutions above, the OPERATIONS of the specifications to be *imported* by the implementation are also substituted.

Based on the translation, the snippets of code of the PCM are given to explain the above discussion. The snippets of code of *lock_state* as shown in Fig. 9, where the sets like *pointCmdSets* are defined in the Context file, in which the data-flows such as *pointCmd* are derived.

The snippets of code of *lock_state_i* as shown in Fig. 10, where the transitions and states in POL diagram are inherited. In Fig. 10, the operation *unlockStart* comes from the unlock machine. There is no refinement machine as an interim from *lock_state* to *lock_state_i*.

INITIALISATION : BEGIN **S** END;
OPERATION: IF **P** THEN **S** ELEIF **Q** THEN **T** ELSE **U** END;
Where P and Q are predicate and S, T , and U are substitution

Figure 8: The relation and ESLIF definition in the implementation of B module.

```
INITIALISATION
    pointLockState := pointLockState_unlock
OPERATIONS
    pointRouteManualLockStatus_out , pointRouteRouteLockStatus_out ,
    pointActuator_out, pointRouteMoveRlt_out <-- pointLockStateStart ( pointCmd , sectionPointCmd ,
        getRoutePointCmd_cmd , accquirePoint ) =
    PRE
        pointLockState : pointLockStateSets &
        pointCmd : pointCmdSets &
        sectionPointCmd : sectionPointCmdSets &
        getRoutePointCmd_cmd : routePointCmdSets &
        accquirePoint : accquirePointSets
    THEN
        pointLockState :: pointLockStateSets ||
        pointRouteManualLockStatus_out :: pointRouteManualLockStatusSets ||
        pointRouteRouteLockStatus_out :: pointRouteRouteLockStatusSets ||
        pointActuator_out :: pointActuatorSets ||
        pointRouteMoveRlt_out :: pointRouteMoveRltSets
    END
```

Figure 9: Snippets of the B code of *lock_state*.

```
OPERATIONS
    pointRouteManuallockStatus_out , pointRouteRouteLockStatus_out ,
    pointActuator_out, pointRouteMoveRlt_out <-- pointLockStateStart ( pointCmd , sectionPointCmd ,
        getRoutePointCmd_cmd , accquirePoint ) =
    IF
        pointLockState = pointLockState_unlock & pointCmd /= pointCmd_manualLock & sectionPointCmd /= sectionPointCmd_lock
    THEN
        pointRouteMoveRlt_out,pointActuator_out <-- unlockStart ( pointCmd , getRoutePointCmd_cmd , accquirePoint );
        pointRouteRouteLockStatus_out:=pointRouteRouteLockStatus_release;
        pointRouteManuallockStatus_out := pointRouteManualLockStatus_release
    ELSIF
        pointLockState = pointLockState_unlock & pointCmd = pointCmd_manualLock
    THEN
        pointLockState := pointLockState_manual_lock ;
        pointRouteManuallockStatus_out := pointRouteManualLockStatus_lock ;
        pointRouteRouteLockStatus_out := pointRouteRouteLockStatus_lock;
        pointRouteMoveRlt_out := pointRouteMoveRlt_null;
        pointActuator_out:=pointActuator_null
```

Figure 10: Snippets of the B code of *lock_state_i.*

Section summaries: Stateflow can quickly generate the logic model of the specification. Moreover, a well-defined architecture of the software system and detailed design are established by Stateflow, which helps to better understand the behaviors of the system during the formal modeling activity.

4 SIMULATION AND FORMAL VERIFICATION

4.1 Simulation

In this part, by integrating PCM and the Test Sequence within the Test Harness, the test-specific simulation environments are built as shown in Fig. 11.

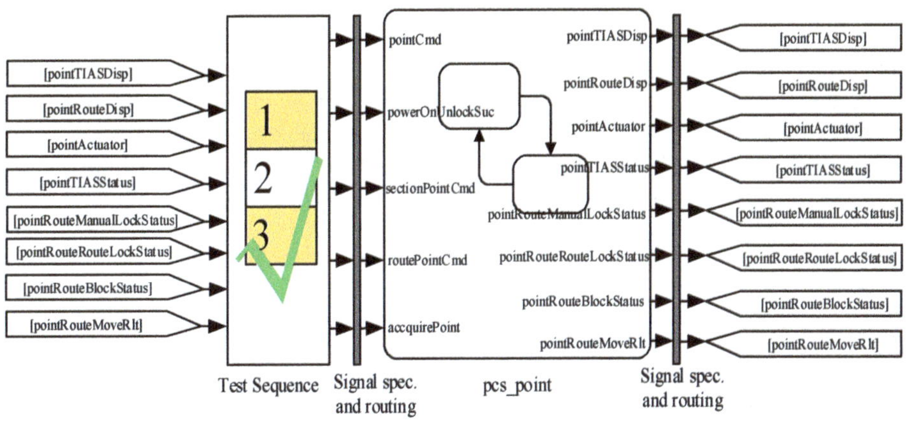

Figure 11: The simulating environment of PCM.

In Fig. 11, the outputs of PCM is linked with the inputs of the Test Sequence. A Test Sequence contains test steps, where a test step consists of actions and transitions. Action executes at the starting of the step. Transition defines when the step stops executing. The entire simulation progresses observe the logical transformations within PCM by periodically performing the steps defined in the Test Sequence. A step comprises the following three processes:

1. The first is to assign values to the inputs of PCM through executing the outputs of Test Sequence; The outputs are the actions of a step, which can be edited in the Test Sequence Editor.
2. After the outputs of PCM was returned to the Test Sequence, it can be assessed whether to comply with the expectation;
3. The current transition condition is then validated. If it is satisfiable, it jumps to the next step to execute. Otherwise, it stays at the current step.

When the simulation model is running, the results are sent to the Test Sequence. Based on the results, the model is analyzed to validate whether it meets the requirements.

One concern for our work is how does the result of simulation conducive to assist formal proving. Indeed, the simulation information only are used to correct the Stateflow model while not work on proving. However, based on the checking of the simulation results, the accuracy of the formal modelling can be improved.

4.2 Formal verification

Verification with invariants: after the B model was built, the hidden safe defects can be further explored by the obligation proving of the safety invariants.

Note that there is no gluing invariant in *lock_state_i* since the variables *imported* from the other abstract machines are not refined relation for the variables in *lock_state*. In addition, the variable name is consistent from abstract machine to its implementation.

Example: we assume that the safety requirement is whether the reachability of the state *manual_route_lock* in Fig. 3 can satisfy the requirement, then the invariant is:

$$\begin{aligned} & pointLockState = pointLockState\_manual\_route\_lock => \\ & verify\_section = sectionPointCmd\_lock \wedge \\ & verify\_manual = pointCmd\_manualLock \end{aligned}$$

Notice also that the two temporary variables *verify_section* and *verify_manual* separately represent the two input commands *sectionPointCmd* and *pointCmd*, as the two input commands are parameter that INVARIANT CLAUSE is unable to identify. In Fig. 3, the three variables *pointRouteManualLockStatus*, *pointRouteRouteLockStatus*, and *pointActuator* respectively become equal to *pointRouteManualLockStatus_lock*, *pointRouteRouteLockStatus_lock*, and 0 when the entry actions of the state *manual_route_lock* are executed. In the invariant, the value of the variable *pointLockState* is to represent the values of two variables *pointRouteManualLockStatus_lock* and *pointRouteRouteLockStatus_lock*.

In the proof window, the deduction result of the invariant does not involve violations which indicates that the value of the variable *pointLockState* becomes equal to the state *pointLockState_manual_route_lock* when the *sectionPointCmd* and *pointCmd* hold true.

Section analyses – Benefits analysis between simulation and formal verification: model-based simulation technologies can validate the functional flow correctness of the model from requirements by editing test cases that are executed logically. Thus, we conclude that there is an improvement for the correctness of requirement specifications under the above simulation testing. Nevertheless, the effectiveness of them cannot satisfy the criterions of high-test coverage and completeness specifications in safety-critical applications. For this, formal technologies are utilized to fill the gap and complement verification mechanism.

5 RELATED WORK

The huge gap between the informal requirements and formal specifications has been an issue of interest in the safety-critical domains over past decades. It promotes quite a few works to reduce this gap. For example, two approaches are presented to integrate the structured analysis and VDM in Fraser et al. [8], and Object Constraint Language specifications from UML cases are used as a guidance for customer-side to develop formal specifications in Giese and Heldal [9], and so on. For B method, using UML has been a de-facto standard for modelling software systems in informal stage. Many researches have been done on the transition from the UML to B language such as Laleau and Mammer [10], [14]. In Fotso et al. [12], SysML/KAOS, a requirement engineering method of extending SysML UML profile, is used to derive B system specifications; Being apart from extracting the elements of the model, it translates the ontologies, refinement structure etc., of the model into B system specifications to build more links between models.

Besides the concern of bridging the gap, two issues have been widely discussed at the intersection of the software development community and formal method. The first one is related to the benefits of software ecosystems. Historically, software engineering has been also seeking a high level of accuracy at earlier stages of development. According to the statistics [13], about one-third of the errors in the product are buried in the requirement stages, and the earlier they are corrected, the lower the cost. In our work, by modelling the requirements with Stateflow and using the simulation in Simulink environments to correct the errors of the informal model, not only is the gap filled but also the accuracy of formal modelling activities has an improvement.

Another issue attaches the industry development challenge of formal methods, and different formal technologies have their own challenges. In this paper, we focused on the challenge of introducing the B method into the industry. In Lecomte [3], although the B method are successfully used in the railway domain, the main challenge is poor spreading in the safety-critical industrial community; One main reason is that it is too difficult to modify the developing approaches which has been established over many years in the aeronautics, energy, etc. [14]. As a possible future direction, the approach we presented would be interesting to extend formal methods to more safety-critical domains by combining the advantages of Simulink project and B method.

6 CONCLUSION

Industrial developing including formal method exist a huge gap between requirements and formal specifications. In this paper, aimed at this gap, we proposed an approach integrating the advantages of Simulink/Stateflow and B method on the developing of the safety-critical system. The process of this approach was introduced in an improved V model, with its function being illustrated by developing PCM of TCS. After the functions of PCM from requirements were modelled by Stateflow, two aspects are concerned towards the transformation from Stateflow to B: (1) the analyzes are outlined to choose suitable mechanisms for the structure transformation; and (2) the elements of Stateflow diagram are described to explain the corresponding translation relation. Afterwards, orienting the higher industry standard, we clarified the advantages of formal verification over simulation.

While bridging the gap, our approach contributes to increasing the benefits of software development. According to our developing experiences, the approach we proposed not only can check the correctness of the requirements, but also significantly improves the accuracy of the formal modeling processes which consequently enhance safety of the system.

Model-based engineering involving formal method still are not common. As a spread, the integration may enable more industrial applications to apply formal method. The possible future work is to translate Stateflow model automatically into the B model, so as to establish a highly practical approach based formal method for the development of the industrial control systems.

ACKNOWLEDGEMENT

We would like to thank the National Natural Science Foundation of China (No. 71502146, 61673320).

REFERENCES

[1] Clearsy. www.clearsy.com/en. Accessed on: 6 Apr. 2020.

[2] Lecomte, T., Deharbe, D., Prun, E. & Mottin, E., Applying a formal method in industry: a 25-year trajectory. *SBMF 2017, LNCS*, eds. S. Cavalheiro & J. Fiadeiro, Springer: Cham, pp. 70–87, 2017. Clearsy. www.clearsy.com/en. Accessed on: 6 Apr. 2020.

[3] Abrial, J.R., *The Book title: Assigning Programs to Meanings*, Cambridge University Press: Cambridge, 1996.

[4] Batra, M., Formal methods: Benefits, challenges and future direction. *Journal of Global Research in Computer Science*, **4**(5), pp. 21–25, 2013.

[5] Rational Software Corporation: OMG Unified Modeling Language Specification – version 1.4., Sep. 2001.

[6] Mathworks Automotive Advisory Board (MAAB): Control Algorithm Modeling Guidelines Using Matlab, Simulink and Stateflow, Version 2.0, 2007.

[7] The Mathworks, Simulink Design Verifier. www.mathworks.com/products/simulink-design-verifier.html. Assessed on: 6 Apr. 2020.

[8] Clarke, E. et al., Bounded model checking using satisfiability solving. *Formal Methods in System Design*, **19**(1), pp. 7–34, 2001.

[9] CENELEC – EN 50128, Railway applications-communication, signaling and processing systems-software for railway control and protection systems, 2011.

[10] The Mathworks, Help Center, Transitions. www.mathworks.com/help/stateflow/ug/transitions.html. Assessed on: 6 Apr. 2020.

[11] Fraser, M.D., Kumar, K. & Vaishnavi, V.K., Informal and formal requirements specification languages: bridging the gap. *IEEE transactions on Software Engineering*, **17**(5), pp. 454–466, 1991.

[12] Giese, M. & Heldal, R., From informal to formal specifications in UML. *International Conference on the Unified Modeling Language*, pp. 197–211, 2004.

[13] Laleau, R. & Mammar, A., An overview of a method and its support tool for generating B specifications from UML notations, *ICS*, pp. 269–272, 2000.

[14] Snook, C. & Bulter, M., UML-B: formal modeling and design aided by UML. *ACM Transactions on Software Engineering and Methodolgy*, **15**(1), pp. 92–122, 2006.

[15] Fotso, S.J.T. et al., Event-B expression and verification of translation rules between SysML/KAOS domain models and B system specifications. *International Conference on Abstract State Machines, Alloy, B, TLA, VDM, and Z*, Cham, pp. 55–70, 2018.

[16] Leffingwell, D. & Widrig, D., *Managing Software Requirements: A Use Case Approach*, Addison-Wesley: New Jersey, pp. 10–40, 10–40.

[17] Abrial, J.R., On B and event-B: Principles, success and challenges. *International Conference on Abstract State Machines, Alloy, B, TLA, VDM, and Z*, Cham, pp. 31–35, 2018.

SECTION 6
INTELLIGENT RAILWAY SYSTEMS

DEVELOPMENT OF A ROUTE GUIDANCE SUPPORT SYSTEM FOR THE VISUALLY IMPAIRED IN RAILWAY STATIONS

JONG-GYU HWANG, KYEONG-HEE KIM & TAE-KI AHN
Korea Railroad Research Institute (KRRI), South Korea

ABSTRACT

It is difficult for a commuter to find a desired path, which can be very complex in the case of navigating an urban railway station where there are several routes. The visually impaired, especially, have more difficulty finding their desired paths or points within a railway station. They have even more difficulty finding available paths, including stairs, to reach a platform where they want to go than the general public do. Therefore, the satisfaction level on mobility for the handicapped, including the visually impaired, when using railway stations is very low, in comparison with that of the general public. In this paper, a route guidance support system is provided that can guide by smart Braille block-based routes in railway stations is proposed, to enhance the mobility satisfaction level of the visually impaired.

Keywords: mobility, blind, Braille, handicapped, visually impaired, route guidance support system, railway station, smart Braille block.

1 INTRODUCTION

The mobility handicapped form a user group that has temporary or continuous restrictions or inconvenience in movement when using public transportation facilities, due to their behavioural inconvenience. In the case of Korea, the portion of these people has increased remarkably. Thus, inconvenience will also be increased, due to the decline in their cognitive ability to judge, physical conditions, etc., when they make use of the railway stations. Among the mobility handicapped, especially the visually impaired are experiencing inconvenience, because of their blindness or being weak-eyed. Addressing this mobility enhancement for the mobility handicapped, including the visually impaired, the Mobility Enhancement for the Mobility Handicapped Act (abbreviated as MEMH ACT) was enacted in Korea, under the vision of realization of society having traffic welfare become convenient for all [1].

Table 1 shows the result of surveying the utilization satisfaction of the railway station and installation rate prescribed by the MEMH ACT. The satisfaction rate compared to the installation rate is relatively low in all of the railway stations. In particular, it can be seen that the satisfaction rate of railway station utilization for the mobility handicapped is about 10% lower than that of the general public [2]. It can also be seen that the transportation handicapped feel much inconvenience and difficulty in using public transportation facilities such as railway stations, compared with the general public [4]–[6]. As such, the satisfaction rate is lower than the installation rate of the mobility facility in the railway stations. To overcome these problems, this paper proposes a smart Braille block-based route guidance system in railway stations that can enhance the satisfaction of visually impaired people using the railway stations (Fig. 1).

2 OVERVIEW OF THE PROPOSED SYSTEM

The mobility handicapped, according to the MEMH ACT, are defined as people who are experiencing difficulties in moving about while living their lives, such as the visually impaired, deaf, pregnant, elderly, children, infants, etc. Among them, the visually impaired

WIT Transactions on The Built Environment, Vol 199, © 2020 WIT Press
www.witpress.com, ISSN 1743-3509 (on-line)
doi:10.2495/CR200301

Table 1: Installation and satisfaction rate for the railways and other passenger facilities.

Classification		The end of 2017		
		Installation rate	Satisfaction rate	
			Public	Mobility handicapped
Passenger facility	Passenger car terminal	60	64	56
	Urban railway and subway station	89	72	64
	Railway station	81	70	63
	Airport passenger terminal	80	72	64
	Passenger ship terminal	87	66	54
	Bus terminal	53	64	48

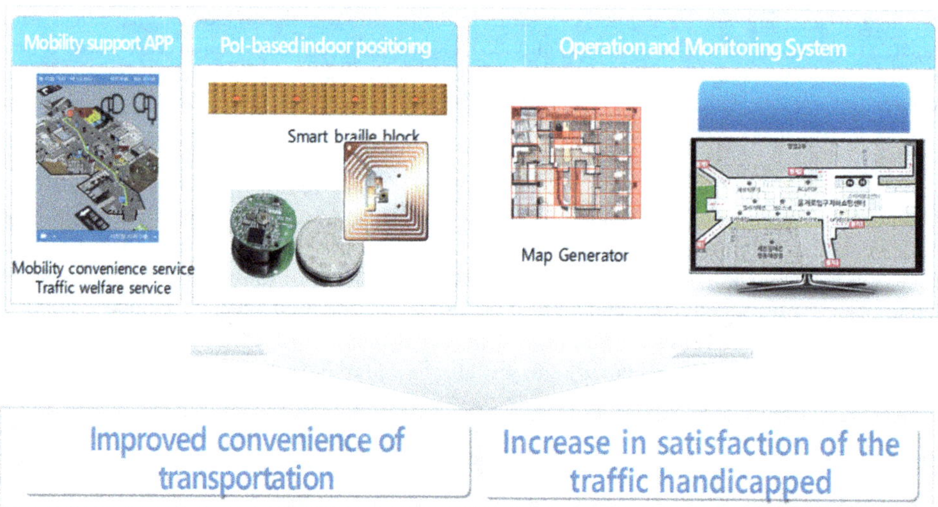

Figure 1: Overview of the proposed mobility-enhancing system.

have many difficulties. Therefore, it is necessary to establish measures to improve the mobility satisfaction of the visually impaired. A Braille block application-based system for supporting the visually impaired is proposed and developed in this paper (Fig. 1) [3].

Average satisfaction of the railway station was 72 points for the general public. On the other hand, the average for mobility handicapped was 63, which means that their dissatisfaction rate is high, and satisfaction rate is low [2]. Traffic disadvantages are being analysed, as they were the most complained-about problem in utilizing the railway stations. In order to improve these rates, an additional installation or improvement is required at station facilities, such as the vertical facilities. These needed changes could be very many and costly, so it is a difficult problem to solve.

The cause providing the most dissatisfaction for the impaired is complaints about facility infrastructure such as stairways, elevators and such, related to the limitations of the informational equipment, when commuters are looking for their destination within railway stations. The methodology which can be used to solve this problem using information and communication technology (ICT) was considered in our research. The Braille blocks have been installed at most station floors. We propose a smart Braille block system, with an Internet of Things (IoT) sensor added to it [3], and propose a route guidance system based on this Braille block that can be used through a mobile phone application (APP). Fig. 1 shows the overall concept of the proposed system, while Fig. 2 shows the architecture of this system. It is necessary to provide these services, such as station map-based convenience route guidance services, and location information on facilities for a mobility-supported application system. Also, it would be better if safety information were provided, but providing that safety information will be the subject of further research.

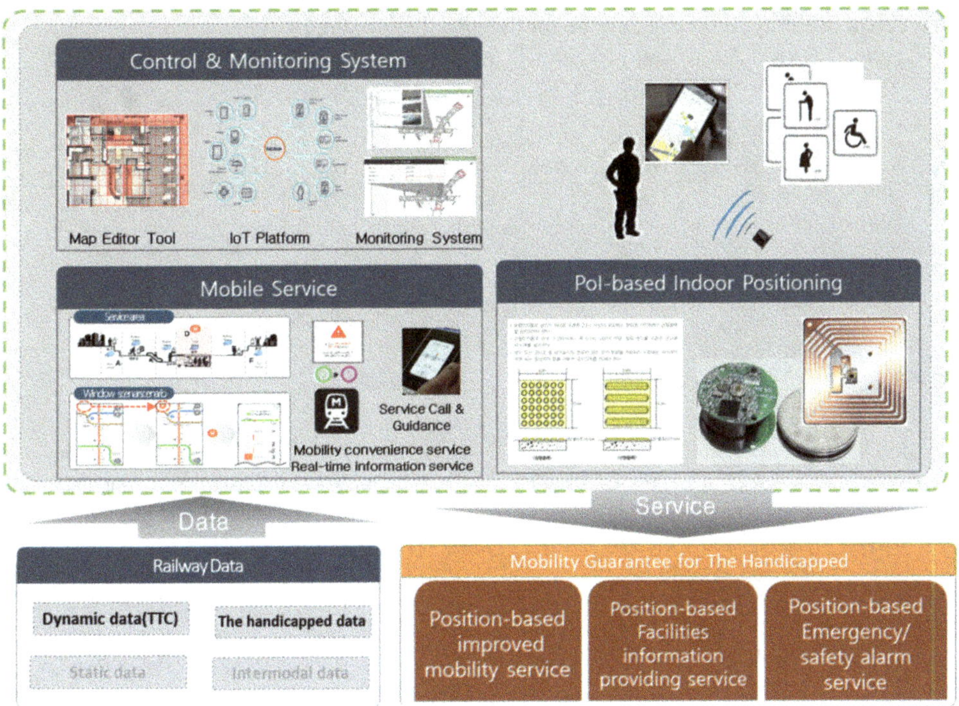

Figure 2: Architecture of the proposed Braille block-based mobile phone application system for guiding the visually impaired through train stations.

3 DEVELOPMENT OF PROPOSED SYSTEM

The visually impaired, when at train stations, need a system that can connect with a server anytime and anywhere, to provide convenient route information and facilities location information. This paper proposes the architecture of the system to support the impaired (Fig. 2). This system consists of a monitoring and control system, smart Braille block-based positioning system, and the mobile service via a mobile APP. The detailed architecture and operation mechanism is represented in Hwang et al. [3]. Fig. 2 shows the route guidance and support system for the visually impaired, a system designed to provide the visually impaired with the convenience of mobility and as a traffic welfare service, based on positioning sensors within the station. To support this service, a monitoring and control system including a station map generation module, by which the control system can create the optimal convenient path for the impaired to follow using a mobile terminal with the service APP.

Although this system has been basically proposed for the visually impaired, this system was configured to have extendibility, so that even general users or the other mobility handicapped can use it, through generation of dynamic and static data. As shown in Fig. 3, our developed service flows for the visually impaired will be provided on the mobile terminal with a service APP. Basically, this mobile APP is used as a terminal for both the service request and receiving path information. It is designed for the visually impaired and basically provides voice; and also, an additionally standardized user interface/user experience (UI/UX) is provided for the weak-eyed. The developed standardized UI/UX will be verified by its convenient usability, by organizations for the visually impaired.

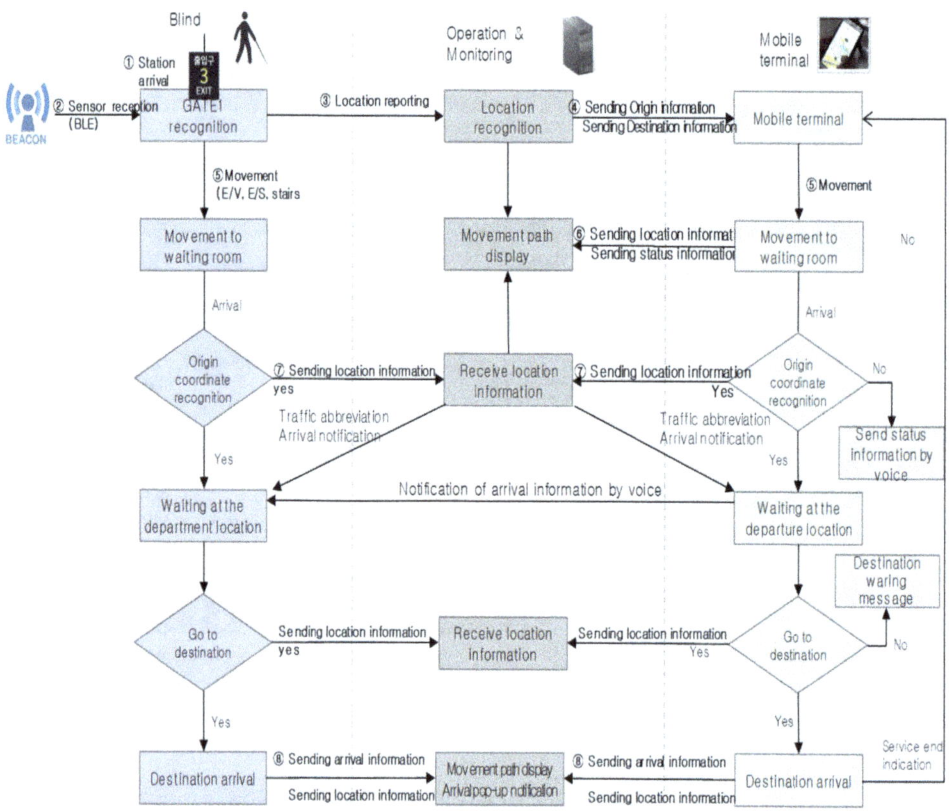

Figure 3: Designed service flow for the visually impaired.

As shown in Fig. 3, the system consists of a mobile APP for a support service, operating and monitoring system, with the visually impaired as a service user. The mobile APP will be requested for services within a station; and then the user will receive feedback on the location and route information, as a result of requests. The operation and monitoring system performs functions such as searching for the optimal travel route for each request and transmitting it to the impaired by the mobile APP. Also, the module functions as the location monitor for service users within the station and in station map generating.

As shown in Fig. 3, the operation and monitoring system receives the service request with information on the type of impaired user, their requested destination, and their current location when the impaired person utilizes the service, when approaching or entering the station. And the user will be provided a customized optimal route, as a response to his/her requests. If the impaired user submits the request inside or near the station, the system receives the request through the mobile terminal of the applicant. If the applicant requests help outside the station, the information on the gate to access that station, with stairs or elevators, is provided to the impaired applicant. And when the impaired arrives at the station, that is detected through the BLE sensor installed at the top of the gate, so that the system automatically recognizes that the user has arrived at the station gate. Those functions can be improved to mitigate the navigation dissatisfaction rate without changing the infrastructure.

Upon recognition of arrival at the station, the applicant's mobile phone will be provided with information such as the best route to navigate the destination, the expected arrival time to the destination, and the predicted remaining time, through the mobile APP. And also, the applicant's position is periodically transmitted to the system, so that applicants are monitored in real time for where they are within the station. If the help of station staff is needed for requested services, then a staff member will receive these, and the applicant receives information on the predicted arrival time of helpful staff. The termination process: the service can be automatically terminated when the visually impaired user arrives at the destination point, and can also be terminated by inputting service termination by their mobile phone.

Fig. 4 shows windows for map generation of the service's target station, registering the detailed information of the specific area of the station to the operation and monitoring system. For this proposed service, a digitalized station map is required. For these, a station map generation module was developed. On the right side of Fig. 4, the UI/UX is designed to monitor environmental information such as the number of mobility handicapped people in the specific area of the station, temperature in that area, IoT sensors' indoor locations, etc.

As mentioned above, there are also the weak-eyed, within the visually impaired users. Therefore, the mobile service application is based on the voice interface, but has even been developed to provide information on a simplified application screen. An addition to the impaired user interfaces, such as bone conduction headsets suitable for the visually impaired, are under review. Fig. 5 shows the concept of the impaired using railway station services: the mobile phone provides information via voice or screen, and it guides the appropriate route in the station along the smart Braille block.

The screen has a standard design, so it also works for the weak-eyed. This designed screen will soon be certified for ease of use by a certification authority, in the near future. An example of a screen designed by this system is shown in Fig. 6. As shown in Fig. 6, there are several PoI (Point of Interest) indication screens, route guide screens, and more that are necessary for route guidance. Designed in each railway station, this designed UI was implemented as an APP. The several screens desired are provided to the user along with a sound for information guidance, during the route guidance by mobile phone.

The system we designed and developed was installed and tested in the Busan city hall metro station. For the test, various visually impaired people, from amblyopia to blind, were included (Fig. 7). Then the survey to determine the user satisfaction rate was conducted. The result: the satisfaction rate was significantly improved, as it was 68.7 points before the application of this system, but after use of this system it improved 6.8%, to 73.4 points.

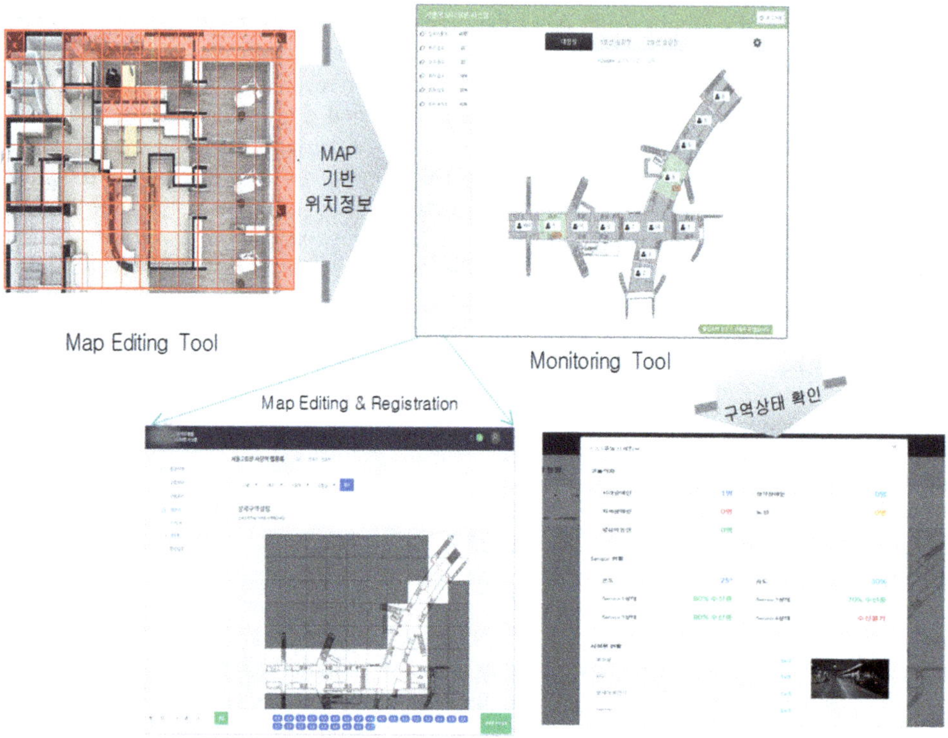

Figure 4: Station map editing tool and identification of other information windows.

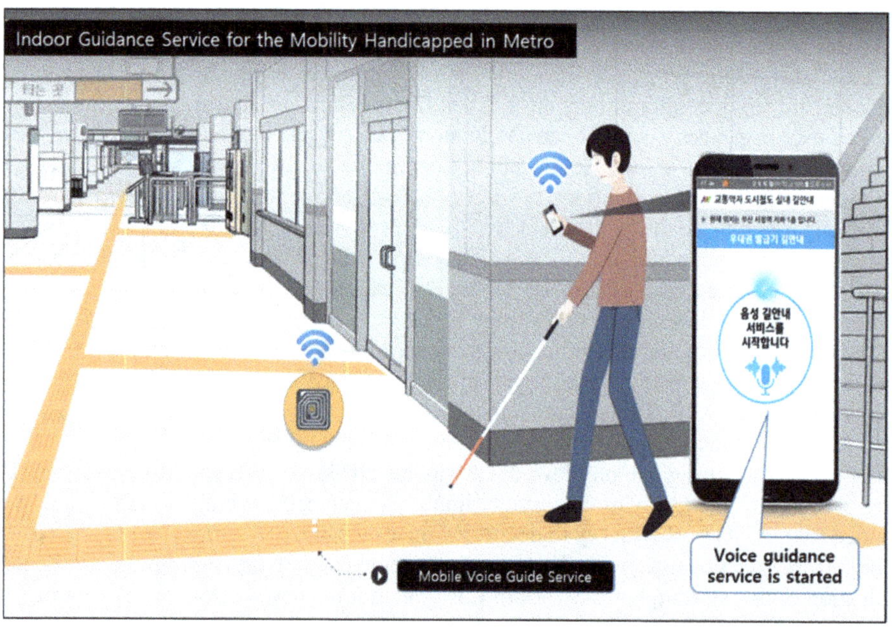

Figure 5: The concept of the impaired using our services.

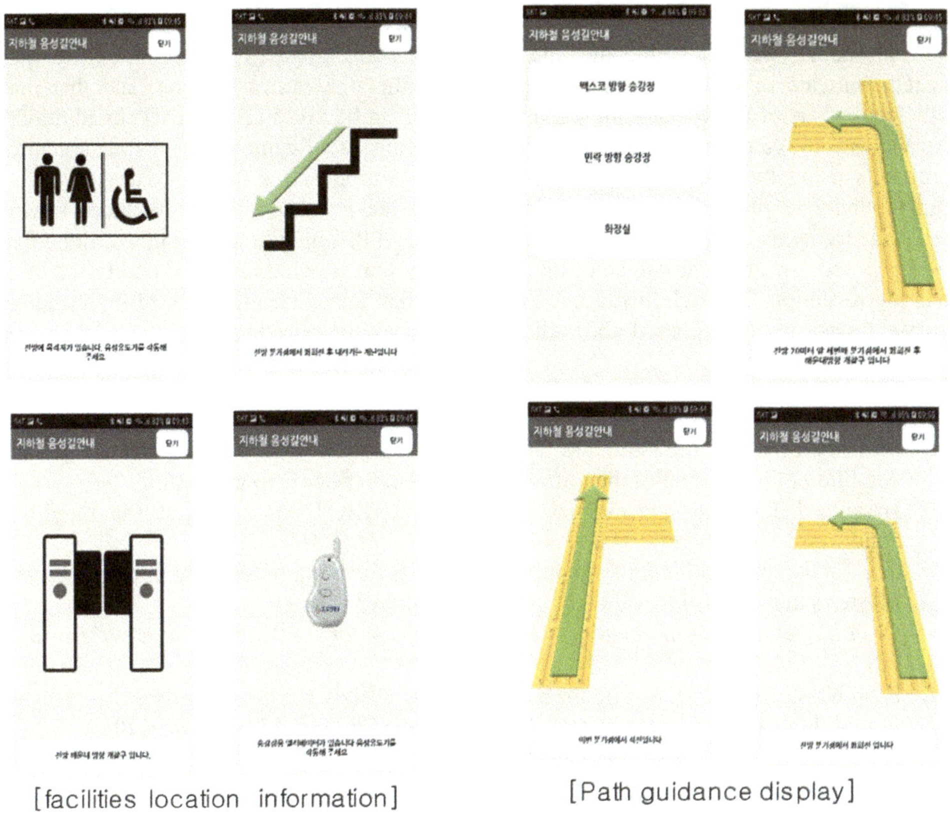

Figure 6: Developed application display for the visually impaired, showing screens for facility location information and the path guidance display.

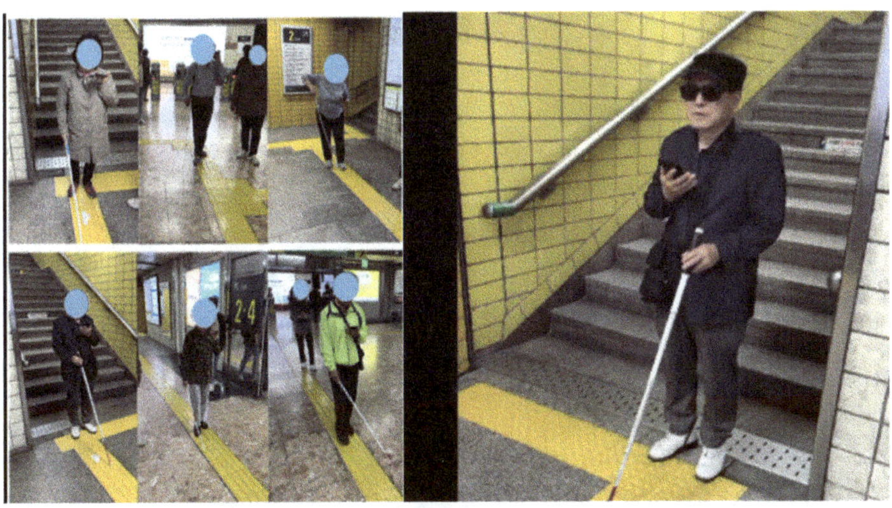

Figure 7: Field testing by the visually impaired.

4 CONCLUSION

For this paper, we first found that the user satisfaction rate was low compared to the installation rate of mobility-supported facilities within the railway station, and that the satisfaction rate of the visually impaired was lower than that of a general user. In addition, we analysed the causes of the low satisfaction rate by the visually impaired: we found out the problems using the railway station. In addition, we proposed a system to improve the satisfaction rate of the visually impaired, through the provision of an enhanced travel convenience by use of a mobile phone with a special APP: a mobile UI design was done for the proposed system. A prototype of the proposed system is being produced and tested in a real metro station. Through this test, we confirmed that the satisfaction rate of the visually impaired was greatly improved, for finding their way within train stations.

REFERENCES

[1] Korean Act, Mobility Enhancement for the Mobility Handicapped Act.

[2] Ministry of Land, Infrastructure and Transport, In year 2017, a study on the actual condition of movement of the traffic handicapped. Service Report, 2018.

[3] Hwang, J.G. et al., Operational scenario of COBOT for escort of the mobility handicapped at railway station. *Proceedings of COMPRAIL*, 2018.

[4] Ge, T., Indoor positioning system based on BLE for blind or visually impaired users. Master's thesis, KTH Royal Institute of Technology, 2015.

[5] Co, E.-Y. et al., *Requirement of Navigation Service for the Mobility Impaired Person*, 2017.

[6] Ryu, K.-S., Lee, S.-H. & Choim, S.-H., The basic study for the mobility enhancement of mobility handicapped in metro. *Proceeding of KSR Spring Conference*, 2017.

EXPERIMENTAL STUDY ON THE VIABILITY OF DECENTRALIZED RAILWAY TRAFFIC MANAGEMENT

QUENTIN PERRACHON[1], RÉMY CHEVRIER[1] & PAOLA PELLEGRINI[2]
[1]SNCF Innovation and Research, France
[2]COSYS-LEOST, University Gustave Eiffel, IFSTTAR, University Lille, France

ABSTRACT

In this paper we propose an experimental study aimed at assessing a new paradigm for railway traffic management. In current practice and academic state of the art, train priority decisions in the case of perturbation are made centrally. In this work, train traffic is modelled as a multi-agent system in which trains are decision makers. We implement a microscopic simulator for assessing the viability of this decentralized traffic management and for comparing it with the centralized one. Our results show that it is indeed possible to let trains decide their own priority and it is possible to achieve results which are comparable to simple centralized approaches in terms of passenger travel time in the mass-transit area of Paris, France.

Keywords: railway dispatching, decentralisation, multi-agent system.

1 INTRODUCTION

Railway traffic management (RTM) follows a centralized approach in which dispatchers regulate trains from an operational control centre. Their role is to manage trains with a global vision of traffic, current and future, by anticipating conflicts. In conflicts, one or more trains have to slow down because the track section they need to traverse is occupied. Dispatchers' decisions lead to the interlock of itineraries for trains to pass.

Thanks to novel technologies in the fields of, for example, communication, computing and artificial intelligence, it is now possible to reconsider the way to manage traffic. In relation with the recent works on autonomous vehicles in road traffic and the SNCF autonomous train project, our idea is to not only distribute RTM but also to decentralize dispatching decisions directly on trains.

This research theme is quite unexplored [1]. Very few papers in the literature focus on non-centralized approaches [2]–[4]. The first paper proposing a decentralized approach [5] describes a model in which the train behaviours are conditioned by the impact of their traveling on other trains. However, many details are missing.

Globally, the few existing approaches are rather focused on the distribution of the operational decision centres, i.e., on the organization of physical decision centres in the network, than on the decentralization of the decision-making. Compared to the distributed approach which keeps local decision centres, the decentralized one is fundamentally different because of the change of paradigm: there is no more "decision centre". Decisions are made by trains themselves, which are thus no more passive executors but become active actors of the system.

In this paper, we present a study on this very prospective research theme. We bring results obtained on simulations inspired from real existing lines in the Paris area: the L and RER A lines.

The rest of the paper is organized as follows. Section 2 details the rationale and the objective of this study. Section 3 describes the characteristics of the railway system which we consider in the analysis, and Section 4 the ones of the multi-agent system we use for representing traffic. Sections 5 and 6 report the simulation set-up and results we achieve. Section 7 concludes the paper.

WIT Transactions on The Built Environment, Vol 199, © 2020 WIT Press
www.witpress.com, ISSN 1743-3509 (on-line)
doi:10.2495/CR200311

2 RATIONALE AND OBJECTIVES

With regards to the evolution of computer science or other fields, there is a progressive evolution from centralized approaches to decentralized ones. This evolution responds to needs such as robustness face to the risks of system failures and resource availability.

Transposed to the railway system, these characteristics incite us to reconsider the way of managing traffic and to introduce changes within it. For that, AI and communication technologies make it possible to change the role of trains from passive, i.e., receiving and executing orders, to active.

As new research theme, a lot of underlying research topics will have to be investigated. Our objective is to make a first step towards the experimental assessment of the fully decentralized RTM concept, in terms of viability and ability to reach results of comparable quality to the centralized ones. As a first step, we focus on conflict resolution at convergences in the network and on the assessment of the impact of communication disruptions.

3 MODELING OF THE RAILWAY SYSTEM

We model train traffic as a multi-agent system. To do so, train-agents need a specific railway environment, that commercial railway simulators do not offer. Then, we develop a simulator for this study. With the representation of track sections and the block signals, it is microscopic and simulates actual train protection systems.

Given that the accuracy of the physical models used to represent train runs is not necessary to prove the feasibility of the decentralized approach, we choose a simplified railway modelling. We use the motion formula with constant acceleration and braking. Hence, we calculate speed profiles of train runs considering distance and speed as functions of time. Tracks are divided into sets of block sections. A block is delimited by two signals and defined by a length and a speed limit. To ease computations and make them faster, parameters such as gradient or curve radius are not taken into account in the modelling.

Blocks are delimited by three-aspect signals which are very common in railways: (i) freeway (green): the train can run at the authorized maximum speed; (ii) closed (red): the train has to stop in front of the signal; and (iii) warning (yellow): the train driver is notified that the next signal is closed (red) and has to follow a speed control curve to be able to stop at the next signal if it is still closed when the train arrives in front of it.

Hence, signalling separates trains and protects specific areas which may see the emergence of conflicts such as switches in junction areas. Indeed, when a train enters a block, the latter is locked and all signals protecting the corresponding track section turn to red.

The example in Fig. 1 shows that train A is late whereas train B is on time, and they arrive at the junction simultaneously. To respect the planned order, train B has to stop before the block including the switch to let train A pass. Hence, also train B will suffer a delay for the rest of its route.

To match the reality of our case study, we model the French KVB device (beacon-based speed control) [6].

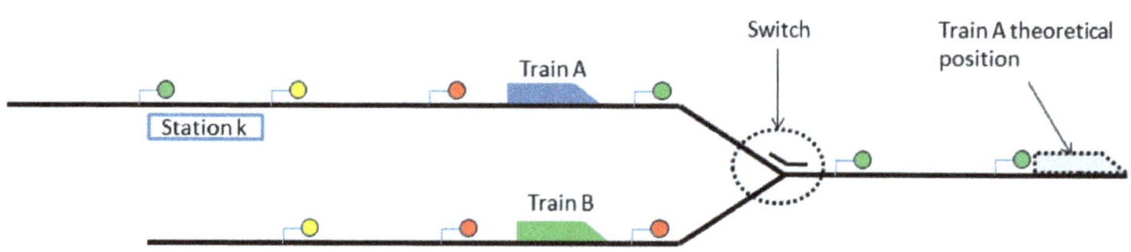

Figure 1: Conflict occurring with the simultaneous arrival of two trains at a junction.

4 MODELING OF TRAIN-AGENTS

In this section, we describe the system we put in place to assess the performance of decentralized RTM. In particular, each train is an agent. Each train is defined by a timetable which gives the departure times from the stations it serves and a utility function is associated to possible delays. The timetable is conflict-free, but conflicts may emerge due to perturbations, especially at junctions as explained in Section 3.

When conflicts emerge, train-agents do not have this global vision of the traffic situation and must coordinate to properly schedule their arrival at the junction. In our modelling, trains know the characteristics of the infrastructure on which they run and the timetable of neighbour trains, and do trajectory projections from their current position.

When a train detects a conflict, it informs the neighbour trains. Then, they decide how to solve the conflict and schedule their passing order by exchanging their utility values regularly. We consider different decision-making strategies as explained in Section 5. Along the exchanges, the trains can possibly update their utility values. The final decision is made at the latest when the train which has to pass first according to the current passing order reaches the junction (Fig. 2).

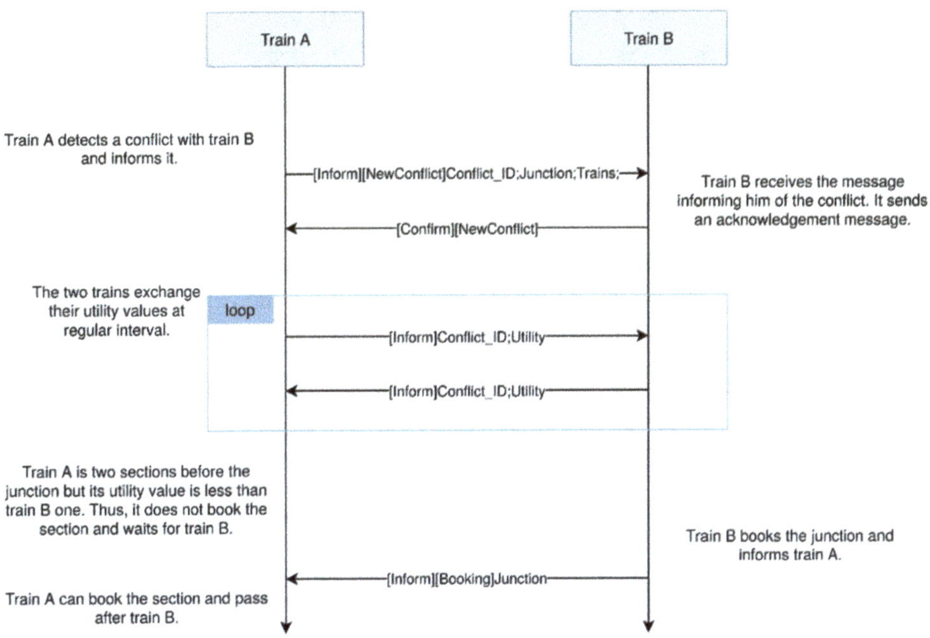

Figure 2: Conflict solution procedure.

5 EXPERIMENTAL SET-UP

In this section, we describe the instances on which we assess the decentralized approach. We also explain the criteria chosen to evaluate and compare the decentralized and the centralized approaches, as well as the decision making strategies considered.

At the end of a simulation run, we assess the traffic in terms of passenger travel time. In particular, we consider the ratio between the actual travel time and the planned one, i.e., the minimum time necessary to go from a station to another in nominal traffic. We consider the total of these times over all passengers.

5.1 Instances

We consider a network inspired from a perimeter operated by both SNCF Transilien and RATP, the two main operators in the Paris area: the western RER A and L lines, as depicted in Fig. 3. The SNCF part of the line has two terminal stations: Cergy and Poissy. Trains run towards Paris through the Nanterre-Préfecture station, from which the line is operated by RATP. Line L also starts from Cergy station and shares its infrastructure with the RER A branch from Cergy to Maisons-Laffitte and with two RER A branches from Maisons-Laffitte to the Nanterre divergence. From this divergence, the L trains run towards the Paris Saint-Lazare station (terminus). The third branch of the RER A line, operated by RATP, departs from Saint-Germain-en-Laye and joins the other branches before the Nanterre-Préfecture station.

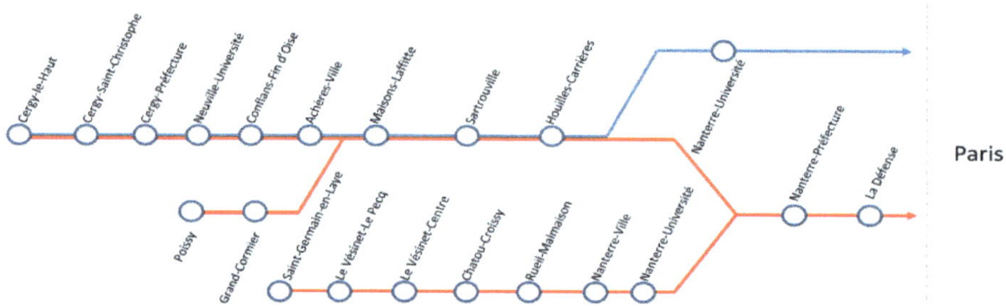

Figure 3: Schematic description of the western RER A (red) and L (blue) lines.

In the morning peak hours, the nominal timetable is cyclic and periodic and plans (per hour): 6 L trains departing from Cergy-Le-Haut towards Nanterre-Université; 6 RER A trains departing from Cergy-Le-Haut; 6 RER A trains departing from Poissy; 6 RER A trains departing from Saint-Germain-en-Laye. One train from each line and branch departs every ten minutes and the trains gather in a common flow between Maisons-Laffitte and Houilles-Carrières stations, after which line L goes to Nanterre-Université and RER A to Nanterre-Préfecture. Before reaching this station, the traffic flow of the three RER A branches merge. It has to be noted that the RER A trains from Cergy are not omnibus and do not stop neither at Maisons-Laffite nor at Houilles-Carrières stations.

All other trains are omnibus. All stops theoretically last 30 seconds, but they often end up to be longer due to large passenger flows. This is a frequent source of perturbation. The infrastructure topology includes double tracks and multiple platforms in some stations, allowing trains to overtake others if necessary. Trains traveling in opposite directions never share tracks, hence we consider only the ones going to Paris.

We consider three 6-hour simulation scenarios. The simulations differ in the passengers' arrival pattern. In the proportional simulation the number of passengers uniformly increases at all stations at peak time. The passenger ratio is constant through lines, and the trains originating at Cergy are more crowded that those coming from Poissy, since the former cross more stations. In the inverse simulation, the number of passengers is inversely proportional to the number of stations traversed: trains from Poissy are more crowded than those from Cergy. Finally, in the average simulation the number of passengers from both origins is in average the same. In these simulations, we are particularly interested in observing the interaction between trains originating in Cergy and Poissy: the junction between the two branches has a strong impact on the rest of the line, due in particular to the different services performed. Passengers arrive at stations in a continuous flow and try alight in the first train which stops, provided that capacity is available. The destination station of each passenger is fixed in coherence with the observed demand on the lines.

5.2 Traffic management strategies

To define the train-agents' decision making, first we compare the system performance when traffic is managed according to three simple strategies: on-time, according to which on-time trains have priority over delayed ones; delay, according to which, on the contrary, delayed trains have priority; passengers, according to which train priorities are proportional to the number of on-board passengers when the conflict is detected. Indeed, being the strategy decentralized, only the trains directly involved in a conflict are considered.

Then, to make a step forward toward the definition of a finer strategy, we consider the implication of the choices after the conflict. To do so, each train makes a projection of the delay it will generate for all trains in its neighbourhood if it passes first, taking into account all trains with which it has a conflict. This delay is typically named secondary delay in the railway community. Since it is the train perspective, the projection can be local and no computational issues emerge. Considering this generated delay, we introduce two more strategies: secondary, according to which the train generating the smallest secondary delay has priority; weighted secondary, according to which the train which generate the smallest secondary delay weighted by the number of passengers on each train has priority.

For comparison, we consider two centralized traffic management strategies: *fcfs* (first-come-first-served), in which the fist train claiming the use of the junction passes first, and plan, in which the train priority follows the original timetable disregarding delays. The latter strategy is the one currently implemented on this line. Finally, we take into account the centralized strategy: we use the same criteria as the best decentralized strategy, but the decision is made centrally. The difference between decentralized and centralized strategies is the larger scope of the latter, in which we do not consider only the immediate neighbourhood of each train in the utility assessment of a decision.

As an example of the difference between the centralized and decentralized strategy, we consider the following situation in the network shown in Fig. 3. One train departing from Cergy arrives at the junction together with one departing from Poissy. Another train departing from Cergy follows closely. In a decentralized traffic management strategy, the first two trains will detect their conflict and they will assess the impact of the two possible entry orders in the common track. The order with the lower impact will be privileged. If the train from Cergy passes second, it may possibly penalize the following train, too, but this is not considered given the local nature of the decision. In a centralized strategy, instead, all trains are considered and a globally good solution is sought. Indeed, this comes at the cost of a higher computational complexity of the centralised case.

6 SIMULATION RESULTS

In this section we report the results of the simulations performed. We start by comparing different decentralized traffic management strategies. Then we select the best performing one and we compare it with centralized ones.

6.1 Decentralised strategies

Fig. 4 reports the comparison of the *proportional*, *inverse* and *average* simulations described in Section 5.2 for each simulation scenario depicted in Section 5.1. Specifically, we analyse the evolution over time of the ratio between the actual total travel time and the planned one. As it is evident, the *passengers* strategy appears effective in the first two simulations. This was indeed expectable, since a train delay impacts the total travel time proportionally to its number of passengers. Instead, if trains transport equivalent numbers of passengers (*average* simulation), this strategy becomes less effective. For what concerns the two other strategies, *on-time* and *delay*, their relative performance changes in the different simulations. This is due, again, to the different numbers of passenger in the trains originating from Cergy and Poissy and to the characteristics of the infrastructure. Indeed, as it can be seen in Fig. 3, the line branch from Cergy is longer than the one from Poissy. Trains coming from the former face more perturbations and cross more stations than the others. Hence, they get to the junction with higher delay and are penalized. When the passenger flow is higher from Cergy in the *proportional* simulation, the *delay* strategy prioritizing trains with higher delay implies that these most crowded trains are favoured. Thus, the total travel time is lower than in the case in which the *on-time* strategy is applied and the trains from Poissy have priority. The opposite holds in the *inverse* simulation, as expected.

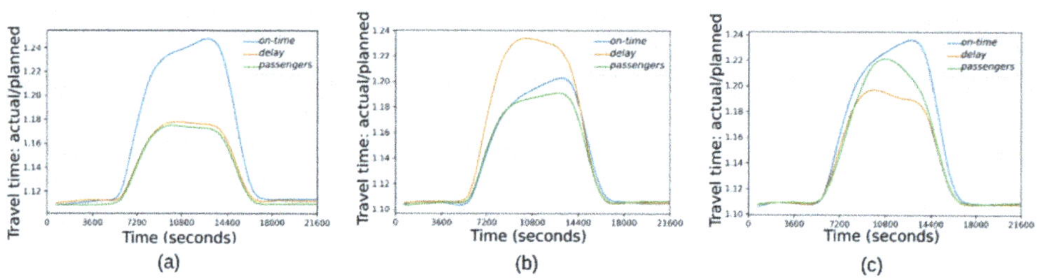

Figure 4: Assessment of traffic management strategies in the *proportional* (a); *inverse* (b); and *average* (c) simulations: *on-time*, *delay* and *passengers* strategies.

This analysis shows that these three simple strategies are too naive: if the number of passengers does not make obvious the appropriate priority to affect to trains, these three strategies are not capable of properly capturing the system performance associated to traffic management choices.

Hence, we compare the performance of the two finer *secondary* and *weighted secondary* strategies with the *passengers* one, which had the best results in the first round of experiments, on the three simulation scenarios. The results are shown in Fig. 5.

As it could be expected, the *weighted secondary* strategy appears the best performing in almost every case. Only in the *average* simulation, the *secondary* strategy performs slightly better at peak time. The fact that *weighted secondary* does not always minimize the actual total travel time depends on the lack of knowledge on the passengers' destination when the secondary delay projection is made. Indeed, if there is at least one stop between the junction where the conflict emerges and the station where the delay is suffered, some passengers may get on and off the train and this will slightly change the impact of the traffic management choice with respect to the expectation. Indeed, it may be possible to use information on

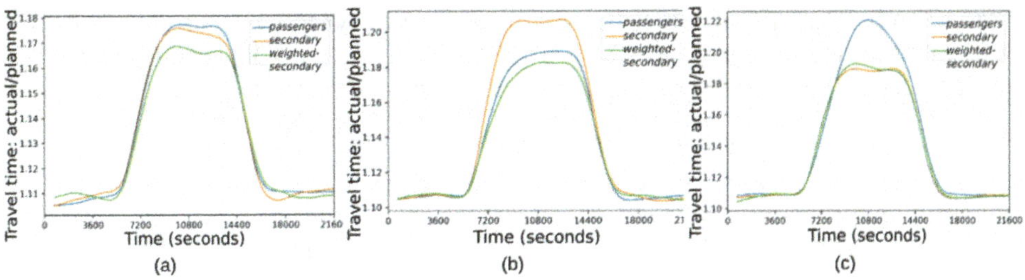

Figure 5: Assessment of traffic management strategies in the *proportional* (a); *inverse* (b); and *average* (c) simulations: *passengers*, *secondary* and *weighted secondary* strategies.

passengers' trips if available, for example in case the train tickets include destination data. However, this piece of information is not often available, as in our case study. Another option may consist in the inclusion of an analysis of the infrastructure and of the services planned when a train needs to assess the secondary delay it would generate by passing first. Based on this analysis, an educated guess on the number of passengers suffering this secondary delay may replace the straight computation of the *weighted secondary* strategy. Nonetheless, this would complicate quite significantly the computation and we consider the performance of the proposed strategy good enough not to justify the additional complication.

In the following, we will consider the *weighted secondary* strategy and we will simply name it *decentralized.*

6.2 Centralized vs decentralized approach

In this section, we report the experimental analysis which we obtain by comparing the *decentralized* strategy with two centralized ones mentioned in Section 5.2. We consider the *proportional* simulation presented in Section 5.1. Fig. 6(a) reports these results. They show that the *decentralized* strategy achieves the best performance in this comparison. The *plan* one is quite competitive with it, but especially at peak time the difference is quite remarkable. Indeed, the *fcfs* strategy is definitely outperformed.

In Fig. 6(b), we show the results of a further comparison: we add to the *decentralized* and the *plan* strategy the results achieved with the *centralized* one.

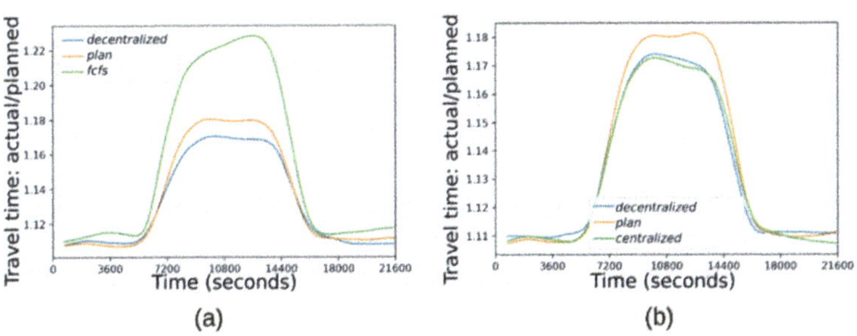

Figure 6: Assessment of centralized and decentralized traffic management strategies. (a) *Decentralized*, *fcfs* and *plan* in the *proportional* simulation; and (b) *Decentralized*, *plan* and *centralized* in the *average* simulation.

As it can be seen, the difference between these two strategies is very small. In our eyes, these results show that managing traffic in a decentralized way is a viable option, which does not significantly worsen the performance from the system perspective.

7 CONCLUSIONS AND PERSPECTIVES

In this paper, we proposed an experimental study aiming to be a first step towards the assessment of the feasibility of a new paradigm for railway traffic management. In this new paradigm, decisions are made in a decentralized framework: traffic is modeled as a multi-agent system. We implemented a simulator to perform this study.

We tested different decentralized strategies for traffic management, and we compared them with centralized ones. The simulation results show that traffic is indeed manageable in a decentralized way, without any remarkable loss of service quality. We measured service quality as the ratio between the total passengers' travel time computed after simulation and the theoretically possible one.

In future research, we will study more articulated decentralized strategies, focusing on state of the art approaches for decision making in multi-agent systems. These strategies will allow assessing the performance of the new paradigm on more complex networks with mixed traffic and different traffic characteristics. On these networks, we will compare the decentralized system with a centralized one in which traffic is managed in an optimized fashion, using well performing algorithms existing in the literature. Moreover, we will assess the impact of communication failures on the performance of the system, to quantify the difference in the robustness of centralized and decentralized RTM.

ACKNOWLEDGEMENT

The work of Paola Pellegrini was partially funded by the CPER ELSAT2020 project co-financed by the European Union with the European Regional Development Fund, the French state and the Hauts-de-France Region Council.

REFERENCES

[1] Marcelli, E. & Pellegrini, P., Literature review toward decentralized railway traffic management. *IEEE Intelligent Transportation Systems Magazine*, 2020.

[2] Haxthausen, A.E. & Peleska, J., Formal development and verification of a distributed railway control system. *IEEE Transactions on Software Engineering*, **8**, pp. 687–701, 2000.

[3] Parkes, D.C. & Ungar, L.H., An auction-based method for decentralised train scheduling. *Proceedings of the 5th International Conference on Autonomous Agents*, pp. 43–50, 2001.

[4] Vernazza, G. & Zunino, R., A distributed intelligence methodology for railway traffic control. *IEEE Transactions on Vehicular Technology*, **39**(3), pp. 263–270, 1990.

[5] Cui, Y., Martin, U. & Liang, J., Decentralized, autonomous train dispatching using swarm intelligence in railway operations and control. *7th International Conference on Railway Operations Modelling and Analysis, RailLille2017*, pp. 521–540, 2017.

[6] Chevrier, R. & Ramond, F., Improving the mass-transit flow by train-holding in station with a simple and efficient heuristic. *International Journal of Transport Development and Integration*, **2**(4), pp. 330–341, 2018.

PRINCIPLES OF THE NEWDIMENSIONS SOFTWARE CREATION FOR A CONTROL CENTRE OF THE FUTURE: CLOUD COMPUTING AND SOFTWARE ARCHITECTURE

RÚBEN ARAÚJO[1], JOAQUIM NUNES[1], AFONSO FERNANDES[2] & ROLANDO MARTINS[2]
[1]EFACEC, Portugal
[2]INESC TEC – Instituto de Engenharia de Sistemas e Computadores, Tecnologia e Ciência, Portugal

ABSTRACT
The ongoing technological evolution implies a continuous effort to update OCC (Operation Control Centre) solutions to match market needs and directives. Given that, EFACEC, a long-time player on the transportation business domain, is actively seeking novel solutions to correspond to customer requirements. From this context NewDimensions was born, a project that aims to make use of the current scientific state-of-the-art technologies and architectures for a new generation of company's software solutions for the railway market, through seamless integration of cloud computing, machine learning and cybersecurity. A particular product (INOSS PIS Rail) is becoming a pilot project for the ongoing modernization that is taking place within EFACEC's transportation solutions. A modular microservice oriented approach was chosen to drive the development onwards. Moving from legacy architecture to microservices allows us to build more reliable products, that are easier to scale up and down while having simpler deployment processes. In this paper we reveal the first steps taken to achieve a modern "OCC-as-a-service" solution. It is analysed the impact of the new architectures and technologies implemented, why they were chosen and their practical application. We also address the principles followed during the progressive modernization process. Finally, we present the next steps to continue this modernization and the direction that shall be taken.
Keywords: OCC, cloud computing, microservices, machine-learning, cybersecurity, NewDimensions.

1 INTRODUCTION

Being involved in the transportation business for a long time, EFACEC (with its Transportation Business Unit) is facing challenges coming from its long longevity and legacy systems. With the continuous technological evolution, EFACEC's OCC (Operation Control Centre) solutions are being restructured to respond adequately to customers' needs. These include fast release of new features and use of emerging technologies associated to the cloud and artificial intelligence, without neglecting cybersecurity threats.

From this context NewDimensions emerges. This project aims to take advantage of current scientific state-of-the-art technologies and software architectures for the railway market. Cloud computing, new software architectures, artificial intelligence and cybersecurity are part of the range of subjects in question.

With this modernization, EFACEC intends to present the market with a new and modern generation of OCC solutions and keep up with the latest trends on the business field.

In this paper we intend to reveal the steps that EFACEC is taking in the context of this modernization; explaining the decisions taken and what they aim to achieve after implementation. We also look to the results of the implementation that we achieved at the time of writing the paper.

WIT Transactions on The Built Environment, Vol 199, © 2020 WIT Press
www.witpress.com, ISSN 1743-3509 (on-line)
doi:10.2495/CR200321

2 METHODOLOGY

2.1 Software architecture

The modernization process carried out by the NewDimensions project relies primarily on the implementation of a new software architecture, in this case a microservice architecture. Microservices is a software architecture approach that encourages for many granulated services, each of them responsible for a single business purpose. Each microservice is a self-contained, independently deployable part of a big application that communicate with other elements, usually via well-defined REST APIs. REST, or Representational State Transfer, is a standard for communication between systems via web [1].

Microservices are often recognized as an architecture for building scalable applications running in the cloud. Additionally, they also guarantee high maintainability due to smaller code bases and solid component separation. These characteristics make microservices an interesting option for software modernization [2]. Additionally, the team autonomy promoted by microservices is likely to reduce coordination effort and improve team productivity.

For business applications, which are typically in service for many years, maintainability is also of major importance. Therefore, microservices promise an improvement over conventional monolithic applications. The major reason for modernizing the application architecture is to bring more efficiency to the system and allow the deployment of new features with a proper time-to-market response.

The goals of modernizing the software architecture are:

- To establish a well-defined and independent interface based on the bordered contexts of the underlying platform. The independent interfaces will give the developers the capability of evolving the product without affecting the existing system.
- An incremental migration of programming languages, from C++ to Java.
- An incremental migration from TAO to REST (The ACE ORB (TAO) is a C++ implementation of CORBA).

Microservices is a suitable architecture for achieving these goals due to their organization around business capabilities. The ability to evolve easily, strong component separation, and focus on cross-platform interaction is of extreme importance. Microservices architectures enable a gradual modernization of the legacy applications.

2.2 Modernization process of the software architecture

Modernization of legacy systems creates many challenges due to their size and complexity. Size and complexity issues frequently dictate that these systems are incrementally modernized, and new modules could be deployed before the modernization process is finished. This implies that legacy components work side by side with modernized components in the same system.

The idea of a progressive modernization involves a modernization piece by piece. Adding new modules and modernizing the old ones one by one [3], until the system becomes completely renovated. Progressive modernization of a legacy system is illustrated in Fig. 1.

Since modernized modules and new components are being deployed in a gradual process, it is required to merge components from the legacy system along with the modernized ones. This is necessary to preserve the existing functionality during the development phase. Interface adapters and other wrapping methods may be necessary to provide a communication mechanism between the legacy system and the modernized one. This interaction is shown in Fig. 2.

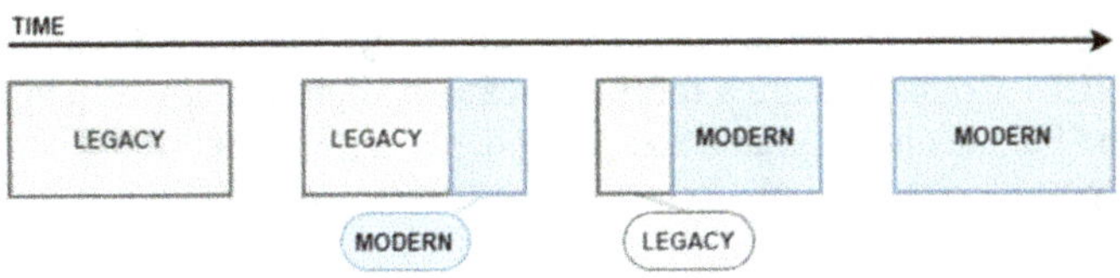

Figure 1: Progressive modernization over time.

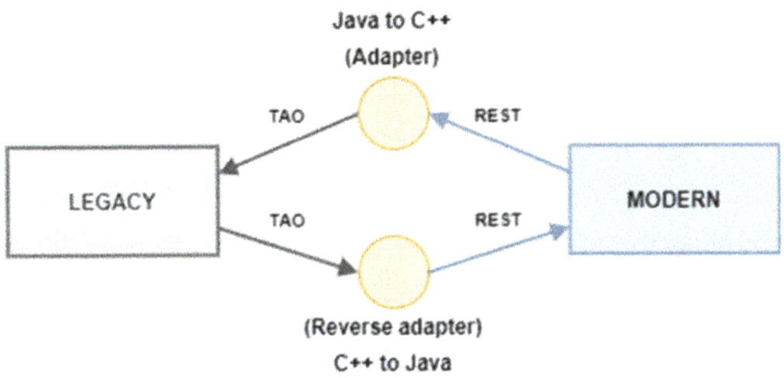

Figure 2: Interoperation of modernized and legacy systems.

A progressive modernization process always aims to keep the system fully operational. This minimizes the amount of rework and technical risk during the upgrade of existing modules or addition of new ones.

2.3 Cloud computing

Today, cloud computing appears as one of the most used and matured technological environment for businesses, hosting and provisioning software systems from all around the world. A continuously growing set of cloud-based solutions in the cloud stack is available to application owners and developers. There are quite a lot of solutions to build applications and exploit the advanced features of this paradigm for elasticity, high availability and performance [4]. These solutions provide many advantages for new applications, but they also create constraints to the modernization and migration of legacy applications. EFACEC's legacy applications, as software not developed with the Cloud in consideration and implemented based on traditional architectural concepts, cannot be as easily scaled and do not share resources beyond infrastructure (e.g. database, memory). A semi-monolithic architecture design style may be deprecated or cannot simply work with the concept of "as a service". Another reason is the fact that being implemented on-premises increases the maintenance costs of the system.

Despite of the reasons for migrating to the cloud, most part of the applications could not profit from the cloud environment, if their main intent is simply to dump the existing architecture to a virtual machine and call it a cloud application. Application scalability would not be achieved without a scalable architecture. Cloud-native architectures like microservices are the ones that have these characteristics, therefore our choices as presented in Section 2.1.

At a high level, cloud native applications should be containerized, segmented into microservices and designed to be dynamically deployed. Consequently, it will allow them to be efficiently managed by orchestration systems like Kubernetes. Kubernetes is a container orchestrator, designed by Google and maintained by the Cloud Native Computing Foundation (CNCF). Containers are a form of packaging applications with all their required dependencies and libraries in a portable and easily deployable format [5]. When implemented, these packages provide a stable, platform independent and predictable environment for the containerized application.

The steps to approach a cloud native/semi-native system are:

- Start containerizing the various components into portable and discretely deployable pieces in the new modules.
- Decomposing the existing business functions of the application into microservices and containerize them.
- Deploy the containers to the cloud, using a container orchestration tool such as Kubernetes.

3 IMPLEMENTATION

Although the NewDimensions project aims to address numerous technological areas, such as: cloud computing, software architectures, artificial intelligence and cybersecurity; this paper only addresses two of these points: cloud computing and software architectures, with its applicability in the development of the INOSS PIS Rail product.

INOSS PIS Rail (Integrated Network Operations Support System, Public Information System for Rail) is a pilot for the ongoing modernization that is taking place within NewDimensions project. Very briefly, there is a whole process of importing timetables and managing trains' circulation information, together with automatic, semi-automatic and manual modes of operation that needs to be improved. The PIS Rail service, as a new module, is the responsible for implementing the associated business logic.

Due to the high degree of complexity of the legacy system, and based on the principle of progressive modernization, the new modules were developed as individual structures that communicate with the existing system by well-defined interfaces. Therefore, following a modernization process like the one presented in the Section 2.2.

As can be seen in Fig. 3, between the new module and the legacy system two interfaces are placed. These interfaces make a bridge between the modules, adapting the languages and communication protocols, used by the legacy system, with the new module. This architecture allows the communication between modules with different languages and communication protocols.

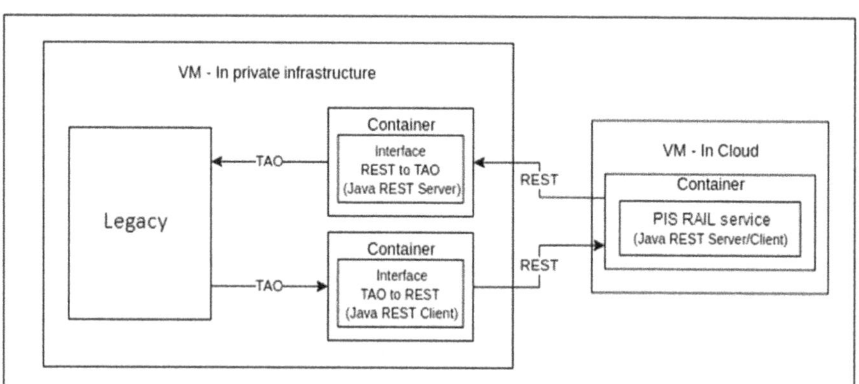

Figure 3: System architecture.

The new modules are packed in containers and, although initially hosted locally, in a final stage they can be hosted in the cloud, as can be seen in Fig. 3. The use of containers allows a better resource efficiency as compared to virtual machines, and an increased flexibility as an application management framework. Containerisation improves development and deployment aspects, such as testing and monitoring of container-based applications. The use of containers also gives us the capability to adjust cluster sizes for the deployment, have an easier maintenance and quicker deployments.

4 DISCUSSION

The implementation described in Section 3, allows the evolution of the legacy system without compromising anything already implemented. Using interfaces like the ones described, we can gradually replace old modules with new ones or keep using them as always.

The adoption of Java as programming language follows the actual development trends. Using microservices, teams can also work on the existing system without having to dive into legacy code. This also facilitates the integration of new developers into the software development cycle.

The product INOSS PIS, and in the scope of the project NewDimensions, is the target of a modernization process focusing on the protocols and technologies used in communication between modules. REST web services are the direct answer to an easy, clean and simpler maintenance communication between the system modules. The use of HTTP (Hypertext Transfer Protocol), in conjunction with REST, transforms the modules into easy pluggable independent services that can be invoked and may evolve easily. Additionally, by using this kind of communication protocols, it enables future integration with external services using APIs (Application Programming Interfaces).

Cloud-based architectures have taken an increasingly important role in building elastic systems with effective fault tolerance. Control and supervision systems, such as those intended to be modernized within NewDimensions project, need guarantees in its execution performance 24/7.

Cloud-based technologies do, in fact, offer quite a lot of benefits. We are able to take advantage of some of them from the moment the new modules are deployed in the cloud. Among several advantages, the cloud allows us to have an easier and more automatic scaling process and turning infrastructure management effortless to the new modules. Since it does not rely on hardware on-premises, or installed software in company servers, the cloud is essentially unlimited in its storage potential, turning the lack of storage space a problem of the past.

Applications have mainly evolved from huge monolithic structures, with fully coupled functions that only work on large physical machines with large computational resources, to fully running decoupled microservices, in a minimal container with high horizontal growth capacity. Therefore, as we can see in Fig. 3, every module or communication interface implemented is packed into an individual container each.

Containerisation granted a lightweight virtualisation through the creation of containers, as application packages that need less machine resources and time of deployment. They additionally address a more interoperable application packaging needed for portable software applications in the cloud [6]. Containerised modules give us the capability to develop, test and deploy applications to a large number of servers and to interconnect those containers.

These were the first steps to a new generation of EFACEC Transportation products. Products that will make use of cutting-edge software technology applied to the transportation business area. Taking this into account, a lot more is yet to come from the NewDimensions project. This evolution will continue, not only in the scope of the INOSS PIS Rail product development but also in other projects. The next steps include implementing automation features, using artificial intelligence, and improvements that need to be addressed related to cybersecurity.

5 CONCLUSION

A modernization process such as the one explained in this the paper, must be a well-defined one. Such process is of extremely importance for the impact of changes on the entire system. We have shown that with a progressive modernization, besides the positive aspects, frequently one may face some drawbacks. As we have argued, technological incompatibilities may occur. In order to overcome the problem, we concluded that system architecture design techniques are the answer.

Our research also concludes that shifting to the cloud should be the next step to large business applications like EFACEC's operational control centre solutions. As a result, as we explained, a cloud-ready and good system's design is the key. The research also indicates that a system implemented using microservices and using containers can take more advantage of the cloud infrastructure.

Having this architecture implemented, we can build around the existing system without changing the legacy modules. This architecture allows us to make new and old modules communicate as if they were developed using the same technologies. Also, we can modernize the entire system, step by step, replacing old modules by new ones.

Future research on artificial intelligence and cybersecurity might shine the light on how to increase the capabilities and quality of EFACEC's OCC solutions. Therefore, those are the steps already underway in the NewDimensions project.

ACKNOWLEDGEMENTS

The work presented is under the scope of the project NewDimensions, being developed by EFACEC in cooperation with INESC TEC (Institute for Systems and Computer Engineering, Technology and Science) for state-of-the-art studies.

NewDimensions is supported by the operational programme COMPETE 2020, financed by PORTUGAL 2020, a partnership agreement between the Portuguese government and the European Commission.

REFERENCES

[1] Massé, M.H., *REST API Design Rulebook*, 2011.
[2] Newman, S., *Building Microservices: Designing Fine-Grained Systems*, 25 Dec. 2014.
[3] Advances in service-oriented and cloud computing, Workshops of *ESOCC 2015*, Taormina, Italy, 15–17 Sep. 2015.
[4] Seacord, R.C., Comella-Dorda, S. Lewis, G., Place, P. & Plakosh, D., *Legacy System Modernization Strategies*, Jul. 2001.
[5] Jetha, H. & Tagliaferri, L., *Running Cloud Native Applications on Digital Ocean Kubernetes*, White Paper, Jul. 2018.
[6] Pahl, C., Brogi, A., Soldani, J. & Jamshidi, P., *Cloud Container Technologies: A State-of-the-Art Review*, May 2017.

SECTION 7
ENVIRONMENTAL IMPACT AND CO_2 REDUCTION

FEASIBILITY VALIDATION OF AN ECO-SUSTAINABLE PHOTOVOLTAIC/ENERGY STORAGE SYSTEM INTEGRATED AC TWO-PHASE TRACTION POWER SUPPLY SYSTEM OF ELECTRIFIED RAILWAY

WENLI DENG[1], CHAOHUA DAI[1], HANBO ZHANG[2] & MI LI[1]
[1]School of Electrical Engineering, Southwest Jiaotong University, China
[2]Southwest Municipal Engineering Design and Research Institute, China

ABSTRACT

With the rapid development of high-speed and heavy-haul railways around the world, how to realize energy saving and emission reduction has been a huge challenge to promote the green development of the railway system by fully utilizing existing resources (such as idle-land resources and solar energy along the railway). This paper proposes an eco-sustainable photovoltaic/energy storage system (PV/ESS) integrated into AC two-phase traction power supply system (TPSS) of electrified railway. Based on the proposed topology, a hierarchical optimization control method is also presented. Concretely, in the system layer, the energy management strategy considering triple random fluctuations (i.e., the real-time power of PV system and two-phase independent traction loads) is designed to prevent PV power flow reverse transmission and improve the utilization rate of PV energy. In the converter layer, besides the conventional maximum power point tracking (MPPT) control of boost converter, a dual-loop state-decoupling control strategy of PV inverter is adopted to ensure the effective power exchange between PV side and traction side. At the same time, the internal coordination control strategy of bidirectional DC/DC converter is detailed elaborately to enhance the long-term operational ability of ESS. Thereafter, to assess the utilization rate of PV energy and carbon emission reduction effect of TPSS under the long-time scale, the mathematical models of PV power estimation and carbon emission reduction are established, respectively. Finally, based on the typical working conditions and the measured loads/external meteorological data from a traction substation in China railway, the technical feasibility of the proposed system and hierarchical optimization control is validated.

Keywords: energy storage system, feasibility verification, photovoltaic power generation, traction power supply system, hierarchical optimization control.

1 INTRODUCTION

In recent years, in order to realize the energy saving and emission reduction in the field of electrified railway, more and more scholars and related departments are beginning to explore the interconnection of new energy network and transportation network [1], [2]. In especial, there are abundant solar energy resources available for exploitation along the railway [3]. However, traction power supply system (TPSS) of electrified railway is a special load of power system with randomness and volatility, especially for AC two-phase TPSS with asymmetric power supply structure, which will greatly increase the difficulty of photovoltaic (PV) access [4]. In fact, in China, there is no precedent for new energy to be connected to TPSS of more than 100,000 km of electrified railway lines.

On the whole, the existing extensive researches mainly focus on the non-traction field (e.g., communication systems, locomotive inspection sites, stations along the railway, etc.) [5]–[9], whereas the traction field with higher energy consumption is still in the stage of theoretical exploration and lack of systematic and in-depth research. Concretely, for AC co-phase TPSS, there are already some PV engineering applications on German railways thanks to the convenience of railway-owned independent power grid [10]. However, for AC two-phase TPSS, due to the particularity of power supply structure, it is difficult for conventional

WIT Transactions on The Built Environment, Vol 199, © 2020 WIT Press
www.witpress.com, ISSN 1743-3509 (on-line)
doi:10.2495/CR200331

PV system to connect directly to TPSS with a transformation from three-phase (on utility grid side) and single-phase (on a power supply arm) independent operation. Additionally, most of the existing literatures only roughly evaluate the economics of new energy access [11], the technical framework of multi-energy complementary structures [12], [13], etc. The detailed analysis of connection topology, energy management and operational control of each subsystem, consumption of PV energy under long-time scale, etc., has not been carried out.

To address these issues, an eco-sustainable PV/energy storage system (ESS) is proposed to adapt the particularity of AC two-phase TPSSs in this article. In terms of control of PV/ESS, a hierarchical control architecture constituted by the system layer and the converter layer is presented. To prevent PV power flow reverse transmission and improve the utilization rate of PV energy, the comprehensive energy management strategy is elaborated in the system layer. To ensure stable and rapid dynamic performance, a dual-loop control strategy of VSC based on state decoupling, and the internal coordination control strategy of bidirectional DC/DC converter is detailed illustrated. In addition, the mathematical analysis models are established to evaluate the effect of PV energy consumption and carbon emission reduction. Finally, the technical feasibility of the proposed system and control strategy is validated, which is expected to provide some technical references for the future demonstration projects or commercial applications. Such as, the construction of intelligent traction substation, rail transit energy internet, etc.

2 TOPOPOGY OF PV/ESS INTEGRATED AC TWO-PHASE TPSS

Fig. 1 illustrates the topology of PV/ESS integrated AC TPSS. 110 kV utility grid is stepped down two-phase 27.5 kV voltage sources for α and β phase feeders by V/v traction transformer. PV/ESS is mainly composed of PV power generation unit and energy storage unit. In PV power generation unit, PV array is connected with DC/DC converter to achieve the function of boost and maximum power point tracking (MPPT), and then makes use of three-phase voltage source converter (VSC) and LC filter to implement the effective transmission of PV energy. In energy storage unit, the battery is connected onto DC-link capacitance via the bidirectional DC/DC converter. Thereafter, PV power generation unit is respectively connected with α and β phase feeders by Dyn11 step-down transformer to establish a sinusoidal, symmetrical, and stable three-phase grid-connected voltage for VSC. That is, the proposed system can effectively cooperate with original TPSS supplying power for locomotives together.

In Fig. 1, $\boldsymbol{U}_\alpha$ and $\boldsymbol{U}_\beta$ are the voltages of α and β phase power supply arms, respectively. $\boldsymbol{I}_A$, $\boldsymbol{I}_B$, and $\boldsymbol{I}_C$ are the three-phase primary currents of V/v transformer, respectively. $\boldsymbol{I}_{\alpha l}$ and $\boldsymbol{I}_{\beta l}$ are the equivalent currents of α and β phase traction loads, respectively. $\boldsymbol{I}_\alpha$ and $\boldsymbol{I}_\beta$ are the equivalent currents of α and β phase traction feeders, respectively. $\boldsymbol{I}_{\alpha c}$ and $\boldsymbol{I}_{\beta c}$ are the equivalent currents provided by PV system to α and β phase power supply arms, respectively. $\boldsymbol{I}_a$, $\boldsymbol{I}_b$, and $\boldsymbol{I}_c$ are the three-phase primary currents of Dyn11 step-up transformer, respectively. U_{ESS} and I_{ESS} are the output voltage and current of ESS, respectively. Additionally, for the sake of illustration, define the direction of currents and voltages depicted in Fig. 1 as positive direction.

3 HIERARCHICAL OPTIMIZATION CONTROL METHOD OF PV/ESS

3.1 Comprehensive energy management strategy in the system layer

In view of the fact that there is no relevant policy in railway systems of most countries to support the reverse transmission of PV power to unity grid through traction substations, some

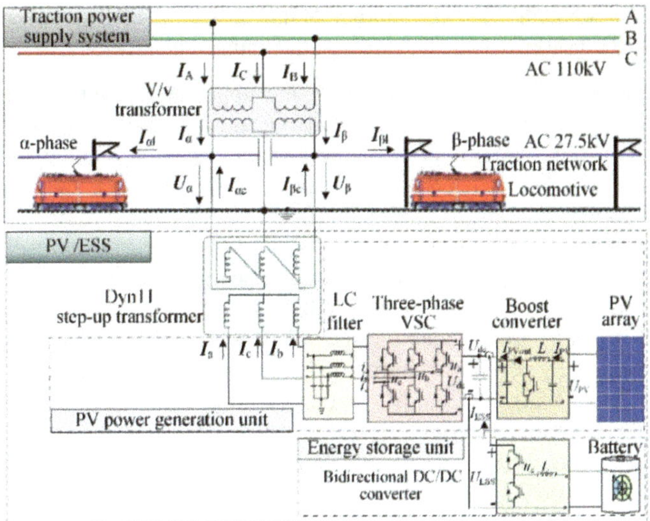

Figure 1: The topology of PV/ESS integrated AC two-phase TPSS.

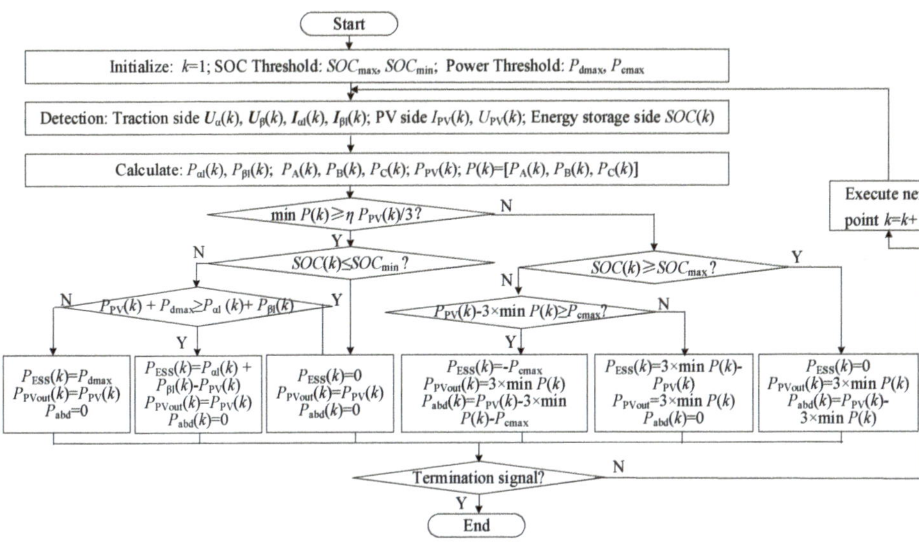

Figure 2: The flowchart of comprehensive energy management strategy of PV/ESS.

treatment measures should be adopted in application. Routinely, adopting the power-limiting control strategy can prevent the reverse transmission of PV power, however, the utilization rate of PV energy will be sacrificed. Therefore, cooperating with some ESSs to improve the utilization rate of PV energy by reasonable charging and discharging behaviour is an optionally and effectively solution.

The flowchart of comprehensive energy management strategy described in Fig. 2 shows the allocation rules of reference power instructions of PV/ESS, which incorporates the real-time output power of ESS $P_{ESS}(k)$, external output power of PV power generation unit $P_{PVout}(k)$, unused PV power $P_{abd}(k)$. Concretely, the main steps include:

1. Initialize variables: Iterative coefficient k; the upper and lower limit value of SOC SOC_{min}, SOC_{max}; the maximum values of charge and discharge power P_{cmax}, P_{dmax}.
2. Detection: Real-time voltages of two power supply arms on traction side $\boldsymbol{U}_{\alpha}(k)$, $\boldsymbol{U}_{\beta}(k)$ and currents $\boldsymbol{I}_{\alpha l}(k)$, $\boldsymbol{I}_{\beta l}(k)$; meteorological data on PV side $S_{ref}(k)$, $T_{ref}(k)$, $FF(k)$; state of charge on ESS side $SOC(k)$.
3. Calculate: Active power of loads on traction side $P_{\alpha l}(k)$, $P_{\beta l}(k)$; three-phase power on system side $P_A(k)$, $P_B(k)$, $P_C(k)$; real-time output power on PV side $P_{PV}(k)$.
4. Judge whether $P(k) \geq \eta \cdot P_{PV}(k)/3$ is satisfied: If yes, the traction loads can completely consume PV power, further judge the state of ESS, proceed to step 5. If no, PV power cannot be completely consumed, further judge the state of ESS, proceed to step 6.
5. Judge whether $SOC(k) \leq SOC_{min}$ is satisfied: If yes, ESS enters into standby mode $P_{ESS}(k)=0$; the external output power of PV system $P_{PVout}(k)=P_{PV}(k)$; unused PV power $P_{abd}(k)=0$. If no, ESS enters into discharge state, further judge whether $P_{PV}(k)+P_{dmax} \geq P_{\alpha}(k)+P_{\beta}(k)$ is satisfied: if yes, let ESS discharge at $P_{ESS}(k)=P_{\alpha}(k)+P_{\beta}(k)-P_{PV}(k)$, external output power of PV system $P_{PVout}(k)=P_{PV}(k)$, unused PV power $P_{abd}(k)=0$; if no, let ESS discharge at $P_{ESS}(k)=P_{dmax}$, $P_{PVout}(k)=P_{PV}(k)$, $P_{abd}(k)=0$.
6. Judge whether $SOC(k) \geq SOC_{max}$ is satisfied: If yes, ESS enters into standby mode $P_{ESS}(k)=0$; the external output power of PV system $P_{PVout}(k)=3\times\min P(k)$; unused PV power $P_{abd}(k)=P_{PV}(k)-3\times\min P(k)$. If no, ESS enters into charge state, further judge whether $P_{PV}(k)-3\times\min P(k) \geq P_{cmax}$ is satisfied: if yes, let ESS charge at $P_{ESS}(k)=P_{cmax}$, external output power of PV system $P_{PVout}(k)=3\times\min P(k)$, unused PV power $P_{abd}(k)=P_{PV}(k)-3\times\min P(k)-P_{cmax}$; if no, let ESS charge at $P_{ESS}(k)=3\times\min P(k)-P_{cmax}$, $P_{PVout}(k)=3\times\min P(k)$, $P_{abd}(k)=0$.
7. Judge whether the end condition is satisfied: If yes, end the run. Otherwise, execute next point $k=k+1$, jump to step 2.

3.2 Modeling and optimization control in the converter layer

Since DC/DC converter of PV power generation unit achieving the functions of boost and MPPT is similar to the control strategy of conventional micro-grid PV power generation system. The disturbance observation method with more engineering applications and mature technology is adopted in this paper. Consequently, this section mainly focuses on the analysis of modeling and optimization controls of VSC on PV side and bidirectional DC/DC converter on ESS side.

3.2.1 Modeling and optimization control of VSC on PV side

As the core component, VSC guarantees the safe and stable operation of PV power generation system. The topology of the adopted three-phase VSC and its control unit is shown in Fig. 3.

The mathematical model of three-phase VSC in dq coordinate system can be described as

$$\begin{bmatrix} U_d \\ U_q \end{bmatrix} = L\frac{d}{dt}\begin{bmatrix} I_d \\ I_q \end{bmatrix} + R\begin{bmatrix} I_d \\ I_q \end{bmatrix} + \omega L\begin{bmatrix} -I_q \\ I_d \end{bmatrix} + \begin{bmatrix} e_d \\ e_q \end{bmatrix}, \tag{1}$$

where U_d, U_q, I_d, I_q are the dq components of VSC output voltage and current, respectively; L is the filter inductor; R is the line equivalent resistance; e_d, e_q are the dq components of grid voltage, respectively.

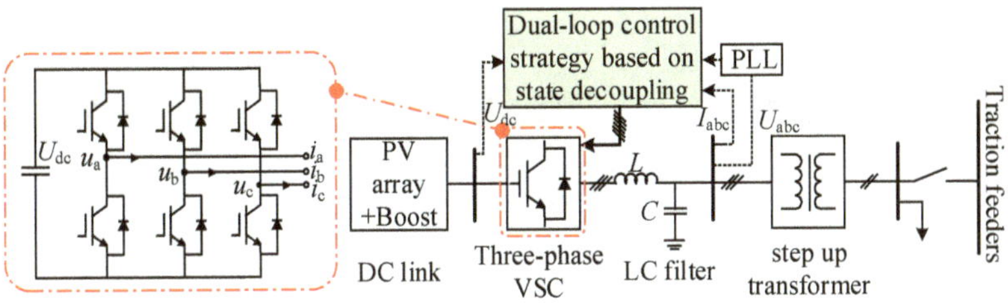

Figure 3: The topology of three-phase VSC and its control unit.

The active power P and reactive power Q injected into traction side through VSC can be calculated as

$$\begin{cases} P = (3EI_d)/2, \\ Q = (3EI_q)/2, \end{cases} \tag{2}$$

where E is the amplitude of grid voltage synthesis vector.

Based on this, the decoupling control of input active power and reactive power can be realized through independent control of I_d and I_q in the dq synchronous rotating coordinate system. Hence, a dual-loop control strategy based on state decoupling shown in Fig. 4 is adopted to ensure the effective power exchange and establish a stable DC-link voltage, which mainly includes status detection, voltage outer loop, current inner loop, PWM generator, etc.

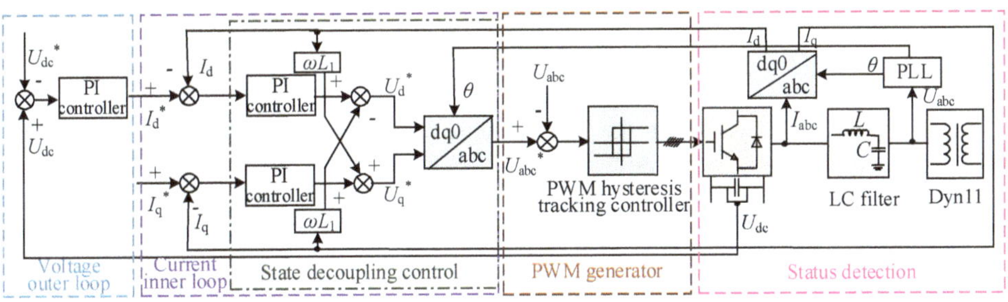

Figure 4: The principle block diagram of dual-loop state-decoupling control.

3.2.2 Modeling and optimization control of bidirectional DC/DC converter on ESS side

As depicted in Fig. 5(a), the bidirectional DC/DC converter is comprised of two IGBTs S_1 and S_2, two diodes D_1 and D_2, a filtering inductor L, which mainly operates in buck and boost mode. In this way, the battery can assist PV power generation unit to adjust power balance. In detail, the battery absorbs energy from DC bus by triggering S_1 in buck mode, and releases energy to DC bus to power two-phase locomotives by triggering S_2 in boost mode.

Since three-phase VSC introduced in Section 3.2.1 is mainly responsible for maintaining the stability of DC-link voltage, ESS controller adopts the control strategy of power outer-loop and current inner-loop presented in Fig. 5(b), which mainly incorporates power allocation unit and charge–discharge control unit of battery. The threshold values in power allocation unit are obtained as follows

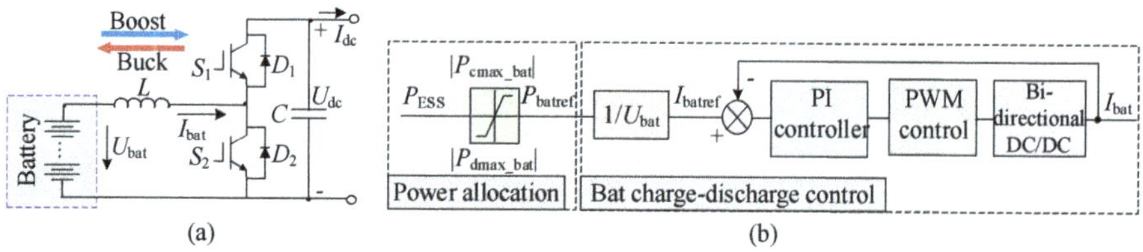

Figure 5: Principle and control block diagram of bidirectional DC/DC converter on ESS side. (a) The principle block diagram of bidirectional DC/DC converter; and (b) The principle block diagram of internal coordination control strategy.

$$\begin{cases} P_{\text{cmax\_bat}}(t) = R_{\text{c\_bat}} Cap_{\text{bat}}, & \text{if } P_{\text{ESS}}(t) \geq 0, \\ P_{\text{dmax\_bat}}(t) = R_{\text{d\_bat}} Cap_{\text{bat}}, & \text{if } P_{\text{ESS}}(t) < 0, \end{cases} \tag{3}$$

where $R_{\text{c\_bat}}$ and $R_{\text{d\_bat}}$ represent the maximum charging and discharge rate of battery, respectively; Cap_{bat} is the rated capacity of battery.

Considering the efficiency loss of energy storage medium and converter, the charging and discharging efficiency of ESS should be considered. Set η_c and η_d represent the charging and discharging efficiency of ESS, respectively. Then, the SOC change caused by charging and discharging process can be described as

$$SOC(t+1)=\begin{cases} SOC(t) - \dfrac{P_{\text{ESS}}(t)\cdot \Delta t}{Cap_{\text{ESS}}\cdot \eta_d}, & \text{if } P_{\text{ESS}}(t) \geq 0, \\ SOC(t) - \dfrac{P_{\text{ESS}}(t)\cdot \Delta t \cdot \eta_c}{Cap_{\text{ESS}}}, & \text{if } P_{\text{ESS}}(t) < 0, \end{cases} \tag{4}$$

where $SOC(t)$ and $SOC(t+1)$ are the SOC at time t and time t+1, respectively; Δt is the sampling interval.

3.3 Estimation models of PV power and carbon emission reduction under the long-time scale

3.3.1 Estimation model of PV power

In order to carry out large-scale and long-term case study, it is necessary to establish suitable estimation model of PV power. Obviously, the output power of PV system is affected by the surface temperature and irradiation of PV modules, while the daily irradiation, temperature, wind speed and other meteorological data are all changing in real-time. It is assumed that PV system operates normally within the estimated time period and the performance of each control unit is well. Thus, the dynamic output power of PV system can be estimated as

$$P_{\text{PV}} = \eta \cdot P_{\text{STC}} \cdot \frac{S_{\text{ref}}[1+k(T_{\text{STC}} - T_{\text{ref}})]}{S_{\text{STC}}}, \tag{5}$$

where P_{PV} is the real-time output power of PV power generation system; P_{STC} is the maximum output power under standard test conditions, that is, installed PV capacity; η is the system efficiency; k is the maximum power temperature coefficient of PV modules under standard test conditions; S_{ref} is the surface irradiance of PV module; T_{ref} is the surface temperature of PV module.

In view of the actual situation, wind speed will indirectly affect the surface temperature of PV modules. To make the model more accurate, T_{ref} can be calculated as

$$T_{\text{ref}} = T_{\text{air}} + S_{\text{ref}} \cdot e^{-3.473-0.0594FF}, \tag{6}$$

where FF is the wind speed near PV system; T_{air} is the air temperature.

In practical application, traction loads of electrified railway will operate periodically on a daily basis according to the traffic operation chart, meanwhile PV output will be affected by annual meteorological conditions. In order to evaluate the consumption capacity of TPSS to PV system, it is necessary to combine the spatial and temporal distribution characteristics of traction loads and solar resources to investigate the overall consumption situation on an annual basis. Define R_{abd} to evaluate the consumption capacity of traction loads to PV/ESS system, that is, the smaller R_{abd}, the stronger the consumption capacity, as follows

$$R_{\text{abd}} = \frac{E_{\text{abd}}}{E_{\text{PV}}} \times 100\% = \frac{\int_0^T P_{\text{abd}}(t)dt}{\int_0^T P_{\text{PV}}(t)dt} \times 100\%, \tag{7}$$

where E_{abd} is the unused PV energy within the time period T; E_{PV} is the total PV energy within the time period T.

3.3.2 Estimation model of carbon emission reduction throughout the life cycle

In order to evaluate the effect of carbon emission reduction of the proposed system throughout its life cycle, the power consumption E_u of PV power supply chain, which covers the processes of production, transportation and waste treatment of PV system, can be calculated as

$$E_u = N\sum_{1}^{n}(DE_n / c + B_n)f_n, \tag{8}$$

where N is the total installed capacity of PV system; DE_n is the direct carbon emission per unit product; c is the total emission of PV system; B_n is the amount of electricity consumed by the production unit product; f_n is the influence factor. To simplify the calculation, according to literature [14], set E_u= 3786.364N kWh.

On the other hand, with the increase of operation time, PV modules will emerge the phenomenon of power attenuation, which includes initial photoinduced attenuation, material aging attenuation and power attenuation caused by external environment or destructive factors. In this paper, the power attenuation rate of PV modules is assumed to be 3% in the first year and 0.7% per year thereafter. Assume that the operating life of PV system is 20 years. The total power E_g produced by PV system throughout its life cycle is evaluated as

$$E_g = \sum_{1}^{20} m_i E_{\text{PV}}, \tag{9}$$

where m_i is the power attenuation rate of PV modules in i year; E_{PV} is the total annual PV energy under ideal conditions.

Hence, the estimation model of carbon emission reduction throughout the life cycle is obtained as

$$C_{\text{total}} = \left(E_g - E_u\right)C, \tag{10}$$

where C_{total} is the total carbon emission reduction; C is large power generation group unit power supply carbon dioxide emissions, where $C = 550$ g CO_2/kWh.

For E_u, literature [14] already assessed that the total power consumption of 1 kW PV system equalized 3,786.364 kWh. Based on this, this paper takes $E_u = 3786.364 \cdot P_{STC}$.

4 SIMULATION VERIFICATION AND CASE STUDY

4.1 Simulation verification

In order to verify the technical feasibility of the proposed system and hierarchical optimization control method, the simulation model of the system presented in Fig. 1 are constructed, where traction loads adopts power following control strategy, and PV installed capacity is set to 1 MWp. Table 1 lists the main simulation parameters based on a system under planning and construction in China. In detail, the parameters of TPSS comes from an actual TPSS already operated, PV modules adopts the nominal data of Suntech STP270S-24_Vb, the parameters like PV installed capacity, Dyn11, battery and other modules, mainly take from system planning data and part of them provided by equipment manufacturers. Considering that PV installed capacity is small relative to traction loads, a set of small active power is respectively given to α and β phase traction loads to simulate different scenarios, as

Table 1: Main simulation parameters of the system.

Item	Parameter	Value
Three-phase power system	Rated voltage level	110 kV
	Short circuit capacity	750 MVA
V/v traction transformer	Nominal capacity	20 MVA
	Ratio of transformation	110/27.5
Dyn11 step-up transformer	Nominal capacity	1 MVA
	Ratio of transformation	27.5/0.5
PV power unit	Installed capacity	1 MWp
	PV module rated power	(STC) 270 Wp
	PV module short circuit current	(STC) 8.2 A
	PV module open circuit voltage	(STC) 44.5 V
	PV maximum power point current	(STC) 7.7 A
	PV module maximum power point voltage	(STC) 35 V
Three-phase VSC	Nominal capacity	1 MVA
	DC side reference voltage	1000 V
	DC-link capacitance	20 mF
LC filter	Filter inductance	0.5 mH
	Filter capacitor	120 pF
Battery	Rated voltage	750 V
	Nominal capacity	420 Ah
	Initial state of charge	50%
	Charge and discharge efficiency	90%
	Charge and discharge ratio	1 C
Bidirectional DC/DC converter	Inductance	1 mH
	Capacitor	100 pF

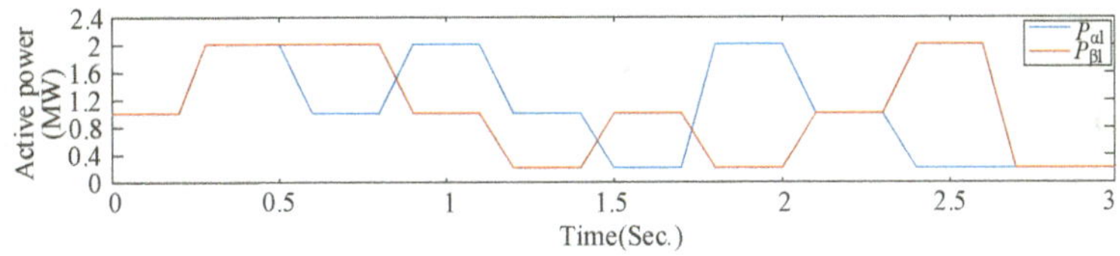

Figure 6: The given working conditions of α and β phase traction loads.

shown in Fig. 6. The external working conditions of temperature and irradiance of PV system are set as standard test conditions, i.e., S_{STC} = 1,000 W/m^2, T_{STC} = 25°C.

Fig. 7(a) compares and analyzes the changes of three-phase active power P_A, P_B, P_C on system side before and after PV/ESS access. The results show that P_A, P_B and P_C all present a certain degree of decline after PV/ESS access, and the degree is roughly the same. Additionally, the phenomenon of reverse transmission of active power in A, B and C phase is not appeared. Combined with Fig. 7(b), It is easy to know that when two arms traction loads are small and ESS has reached the upper limit of maximum charging power, the excess PV power needs to undergo necessary power abandoning treatment (as displayed in the curve of P_{abd}) to ensure that PV power flow does not reversely transmit to 110 kV unity grid. From the dynamic distribution curve of P_{ESS} in Fig. 7(b), ESS can be conducted reasonable charging and discharging behavior based on the energy management strategy mentioned in Section 3.1. That is, the above simulation results verify the effectiveness of the proposed system and hierarchical optimization control strategy.

4.2 Case study

In order to evaluate the consumption capacity of TPSS to PV/ESS system, the typical daily measured load data of a traction substation in China railway is selected as shown in Fig. 8(a)

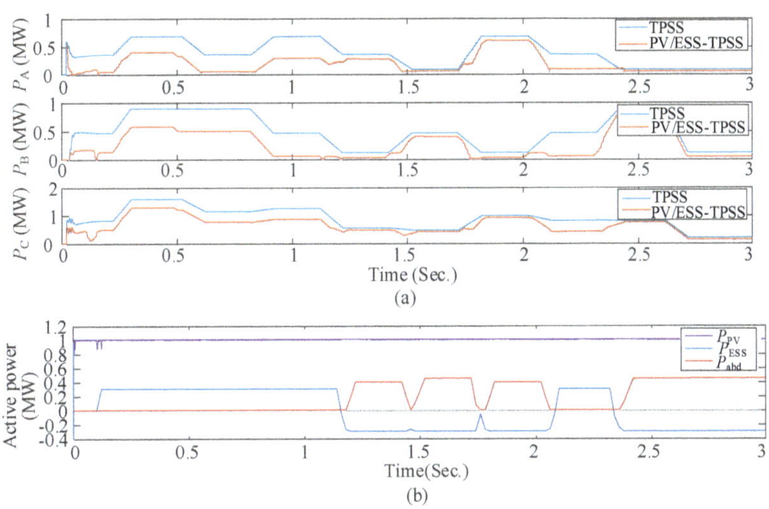

Figure 7: The results of simulation verification. (a) Three-phase active power on system side before and after PV/ESS access; and (b) Active power distribution among different subsystems.

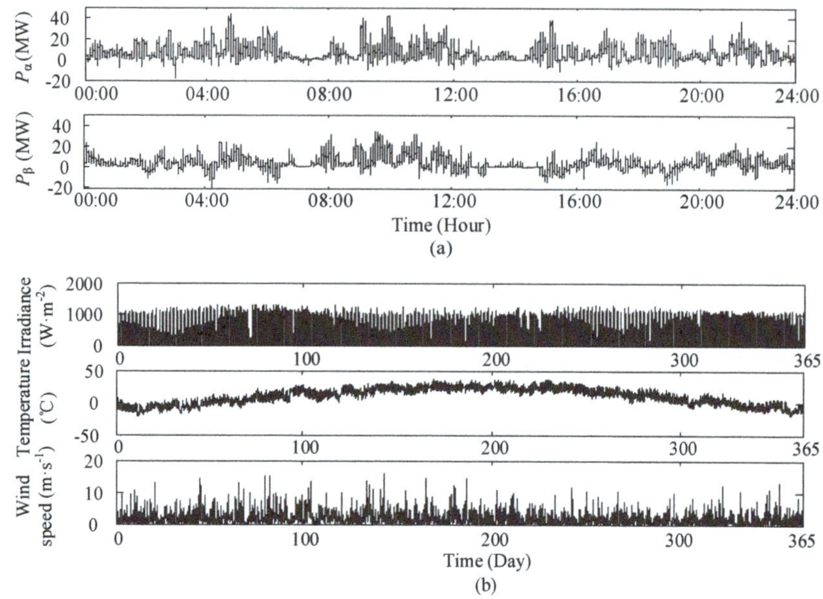

Figure 8: The measured data of case study. (a) Typical daily measured load curve of a traction substation (α and β phase); and (b) The main annual meteorological data in this area.

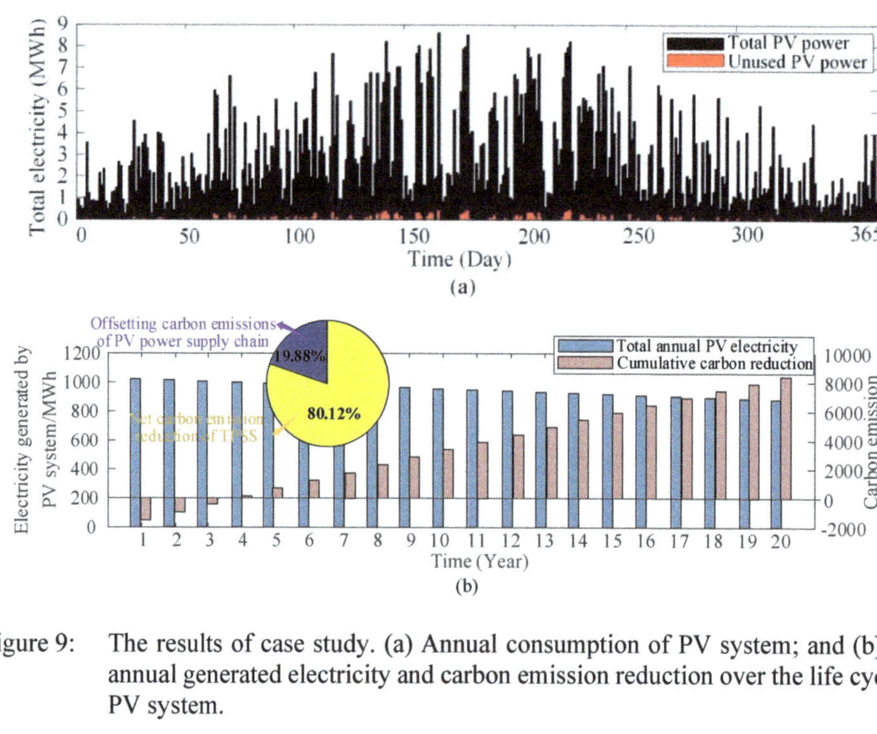

Figure 9: The results of case study. (a) Annual consumption of PV system; and (b) The annual generated electricity and carbon emission reduction over the life cycle of PV system.

on traction side (sampling interval: 3 s/time), and the annual measured meteorological data in this area is also selected as shown in Fig. 8(b) on PV side (Sampling interval: 1min/ time). By means of the hierarchical optimization control method and the estimation model given in Section 3, the model of case study under long-time scale is established based on Matlab. The results of case study are illustrated in Fig. 9.

Fig. 9(a) shows the distribution of annual consumption of PV system, where statistical data indicate that the annual electricity generated by PV system is 1054.2 MWh, the annual consumption electricity of traction loads is 1,034.642 MWh, unused electricity is 19.558 MWh, and the unused rate R_{abd} is 1.86%. On the whole, the consumption situation of the proposed system is easily acceptable.

Fig. 9(b) shows the annual generated electricity and carbon emission reduction over the life cycle of PV system, where the life of PV system is set to 20 years. The results clearly presented that annual electricity generated by PV system is declining year by year, concretely, it is 1,054.2 MWh in the first year and 882.4 MWh in the 20th year. Additionally, statistical data also indicate that the total carbon emissions of 1 MW PV power supply chain is 2,082.5 kg CO_2, which includes the processes of production, transportation and waste treatment of 1MW PV system. Based on the distribution of cumulative carbon emission reduction of TPSS, the proposed system can offset the carbon emission of PV power supply chain in the 4th year, that is, the green and low-carbon effect truly starts to carry out. After 20 years, the cumulative carbon emission reduction of TPSS reaches 8,394.5 kg CO_2. In detail, the carbon emission reduction of PV system in its life cycle, where 19.88% is used to offset the carbon emission of PV power supply chain, and 80.12% truly plays an environmental protection role.

5 CONCLUSIONS

In this paper, an eco-sustainable PV/ESS integrated AC two-phase TPSS of electrified railway is proposed, which can be used to promote the interconnection of new energy network and transportation network, and is conducive to forming a novel green development mode of TPSS. On this basis, a hierarchical optimization control method composed of the system layer and the converter layer is also presented. The system layer determines the energy transfer relationship between different subsystems and sends power instructions to corresponding controllers in different subsystems, which can be used to prevent PV power flow reverse transmission and improve the utilization rate of PV energy. In the converter layer, a dual-loop control strategy of PV inverter based on state decoupling, and the internal coordination control strategy of bidirectional DC/DC converter is designed in detail, respectively. In addition, the mathematical models of PV power estimation and carbon emission are established to assess the effect of new energy utilization and carbon emission reduction. Finally, both dynamic simulations and case study validate the technical feasibility of the proposed system and hierarchical optimization control method.

ACKNOWLEDGEMENTS

The authors would like to thank the reviewers for their helpful suggestions. This work was supported in part by the National Key Research and Development Program of China (2017YFB1201003), and Sichuan Science and Technology Program (20YYJC4428), China.

REFERENCES

[1] Khayyam, S., Ponci, F., Goikoetxea, J., Recagno, V., Bagliano, V. & Monti, A., Railway energy management system: Centralized-decentralized automation architecture. *IEEE Transactions on Smart Grid*, 7(2), pp. 1164–1175, 2016.

[2] Rifkin, J., *The Third Industrial Revolution: How Lateral Power is Transforming Energy, the Economy, and the World*, Palgrave Macmillan: New York, 2011.

[3] Institute of Energy Internet Innovation, Tsinghua University, 2035 report on photovoltaic development for all. www.sohu.com/a/221649580_100049846. Accessed on: 1 Feb. 2020. (In Chinese.)

[4] Li, Q.Z., Xie, S.F., Zhang, L. & Zhang, L.Y., *Electrified Railway Power Supply System and its Power Quality Control Technology*, China Electric Power Press: Beijing, 2015.

[5] Jaffery, S.H.I. et al., The potential of solar powered transportation and the case for solar powered railway in Pakistan. *Renewable and Sustainable Energy Reviews*, **39**, pp. 270–276, 2014.

[6] Morita, Y. et al., Analysis of local smoothing effect on the PV on Tokyo station. *International Conference on Renewable Energy Research and Applications, IEEE*, 2012.

[7] Hayashiya, H. et al., Potentials, peculiarities and prospects of solar power generation on the railway premises. *International Conference on Renewable Energy Research and Applications, IEEE*, 2012.

[8] Miyagawa, T. et al., Cooperative control of reactive power of distributed PV systems to suppress voltage of distribution line along railroad track. *European Conference on Power Electronics and Applications, IEEE*, 2015.

[9] Alam, K.S. et al., Modeling and computation of a solar-piezoelectric hybrid power plant for railway stations. *International Conference on Informatics, Electronic and Vision. IEEE*, pp. 155–159, 2012.

[10] Méndez-Hernández, Y. et al., Design aspects for high voltage MW PV systems for railway power supply. *Proceedings of the 29th European Photovoltaic Solar Energy Conference and Exhibition*, 2014.

[11] Wu, M.L., Gao, Y., Deng, W.L., Chen, H.B., Dai, C.H. & Chen, W.R., Study of PV generation application on AC powered traction system. *Proceedings of 2017 Chinese Automation Congress, IEEE*, pp. 1641–1646, 2017.

[12] Dai, C.H., Deng, W.L. & Wu, M.L., A multi-energy complementary grid connection system for electrified railway and its control method. Chinese invention patent no. CN107749637A, 2017.

[13] Aguado, J.A., Racero, A.J.S. & de la Torre, S., Optimal operation of electric railways with renewable energy and electric storage systems. *IEEE Transactions on Smart Grid*, **9**(2), pp. 993–1001, 2018.

[14] Guo, X.P., Lin, K., Huang, H. & Li, Y., Carbon footprint of the photovoltaic power supply chain in China. *Journal of Cleaner Production*, pp. 626–633, 2019.

IMPACT OF DRIVING STYLE ON THE EXHAUST EMISSION OF A DIESEL MULTIPLE UNIT

MACIEJ ANDRZEJEWSKI[1], PAWEŁ FUĆ[2], DAWID GALLAS[1], ANDRZEJ ZIÓŁKOWSKI[2] & PAWEŁ DASZKIEWICZ[1]
[1]Łukasiewicz Research Network IPS TABOR, Poland
[2]Poznan University of Technology, Poland

ABSTRACT

This paper considers aspects regarding exhaust emissions from vehicles, the reduction of which is one of the main trends of current worldwide development efforts. These efforts resulted in numerous legal tools being created to tackle the problem of toxic gas emissions and global warming. These tools include emission limits and type approval procedures. Such norms for rail vehicles are simplistic and fail to provide a reliable method of assessing the real exhaust emission from rail vehicles operating on the tracks. Hence, PEMS (portable emission measurement system) type apparatus was used to perform measurements of exhaust emissions of a diesel multiple unit designed for short–mid range passenger transport in real operating conditions. The vehicle tested belonged to a widely and commonly used group of passenger vehicles, and the testing methods were similar to current RDE (real driving emissions) test requirements for road vehicles. The selected vehicle was powered by an internal combustion engine in which ecological and economic properties depend on the operating ranges and parameters used. These types of train lines are characterized by frequent stops at stations, which confirms the need to investigate this subject. The vehicle exhaust emissions were measured for different driving styles, regular driving style typical for the driver and a drive more closely in line with the concepts of eco-driving, with reduced maximum acceleration values. Since most previous studies on the eco-driving principles used road vehicles for testing, this resulted in a gap in knowledge in this field regarding rail vehicles. The aim of the paper is to assess the impact of the choice of driving style on the exhaust emission values of a selected passenger rail vehicle.
Keywords: RDE, eco-driving, railbus, exhaust emissions.

1 INTRODUCTION

The transport sector is responsible for 24% of all CO_2 emissions produced by burning fossil fuels according to IEA (International Energy Agency) [1]. This equates to approximately 8 Gt of CO_2, of which 60% is produced by passenger transport and 40% by freight transport. This poses a significant problem to tackle in lieu of reducing global CO_2 emissions and preventing global warming effects. This article focuses on the environmental impact of passenger rail transport, as in addition to their CO_2 contribution combustion engines also produce a significant amount of other toxic compounds. While freight transport is often used over longer distances and between areas focused on industry or commerce, passenger transport is more often nestled in the most densely populated areas. As a result the exhaust emissions released by passenger transport is one of the primary concerns for the health and wellbeing of the human population [2] as well as having significant impact on the long-term air pollution levels in inhabited areas [3]. Transport related air pollution has long been well known to have a strong impact on human health [4]. A significant aspect of this impact relates to the emission of particulates, mainly measured as PM_{10} and $PM_{2.5}$, which is known to be carcinogenic as well as cause lung damage and upper respiratory infections [5]. As a result of its impact, PM emissions from vehicles have been tested and researched intensively in the last few decades, and other sources can provide fairly comprehensible data with regards to PM emissions from rail transport [6]. The environmental and health impact of rail vehicles can be reduced through multiple means, such as design modifications, engine computer

WIT Transactions on The Built Environment, Vol 199, © 2020 WIT Press
www.witpress.com, ISSN 1743-3509 (on-line)
doi:10.2495/CR200341

modeling [7], combustion process shaping [8], and fuel chemical engineering [9]. This article investigates the effect that can be achieved by using more effective driving styles, known as eco-driving, when performing the vehicle regular operations. Thus it discusses RDE style exhaust emission tests of a passenger rail vehicle, specifically a diesel multiple unit, measuring the emission values of CO_2, CO, HC, and NO_x. A comparison of two test drives performed in a normal driving style and eco-driving was made, to assess the potential impact of what can be considered a simple non-invasive solution to exhaust emissions reduction. The popularity of adopting eco-driving principles in vehicles with combustion engines stems from its known effectiveness in road vehicles [10]. Even though rail vehicles typically contribute much less to the overall emissions compared to rail transport, much of their exhaust emission is still released in urban agglomerations, magnifying its impact on human health. Prokolej estimates the average CO_2 emissions for the passenger rail vehicles to be about 60 g/pkm while for road vehicles this value is over 140 g/pkm [11]. The investigation of its effects in the rail sector, as well as potential applications has been lagging behind, however. This article aims to provide an impact assessment of eco-driving on exhaust emissions measured in real driving conditions.

2 TEST VEHICLE AND METHOD

The tested vehicle belonged to the NRMM (non-road mobile machinery) group – the SA108 rail bus (Fig. 1). It is powered by two MAN D2866 LUH21 diesel engines with a power output of 257 kW (Fig. 2). Other major technical and operational parameters of the SA108 rail bus engine were provided in Table 1.

Figure 1: Rail bus with mounted measuring apparatus for testing in RDE conditions.

Figure 2: Source of propulsion of the test vehicle – the MAN D2866 LUH21 engine.

Table 1: Selected technical and operational parameters of the MAN D2866 LUH21 engine.

Engine design	In-line, 6 cylinders horizontal configuration
Ignition type	Compression
Maximum power output	257 kW (350 KM) at 2,000 rpm
Maximum torque	1500 N·m in the range 1,000–1,500 rpm
Stroke volume	11.96 dm^3
Compression ratio	17:1
Piston diameter	128 mm
Piston stroke	155 mm
Average piston velocity	10.85 m/s
Exhaust emission norm	Euro II

The assessment of pollutant exhaust emissions of the SA108 rail bus engines was carried out in the conditions of its real operation during the inspections carried out on the 36-km section of the railway line from Krzyż Wielkopolski to Trzcianka (Figs 3 and 4, points A and B). This route was chosen as it was almost completely straight and located on flat terrain. For the drive one way the route was covered while maintaining the typical way of controlling rail bus engines, while the drive on the return was performed while taking into account the principles of economic and ecological driving (so-called eco-driving). It mainly boiled down to the use of less "aggressive" engine loadings and frequent coasting. Fig. 5 shows an example of the vehicle speed profile recorded during the tests (standard travel). The test drive max speed was 100 km/h, while the mean travel speed was about 62 km/h.

The tests were performed with the use of mobile analyzers from the PEMS (portable emission measurement system) group to determine the composition of engine exhaust gases, including components of the Semtech-DS device and their software (exhaust flow meter, Figs 6 and 7).

Figure 3: View from the driver's cab during tests in real driving conditions.

Figure 4: The test route between Krzyż Wielkopolski–Trzcianka.

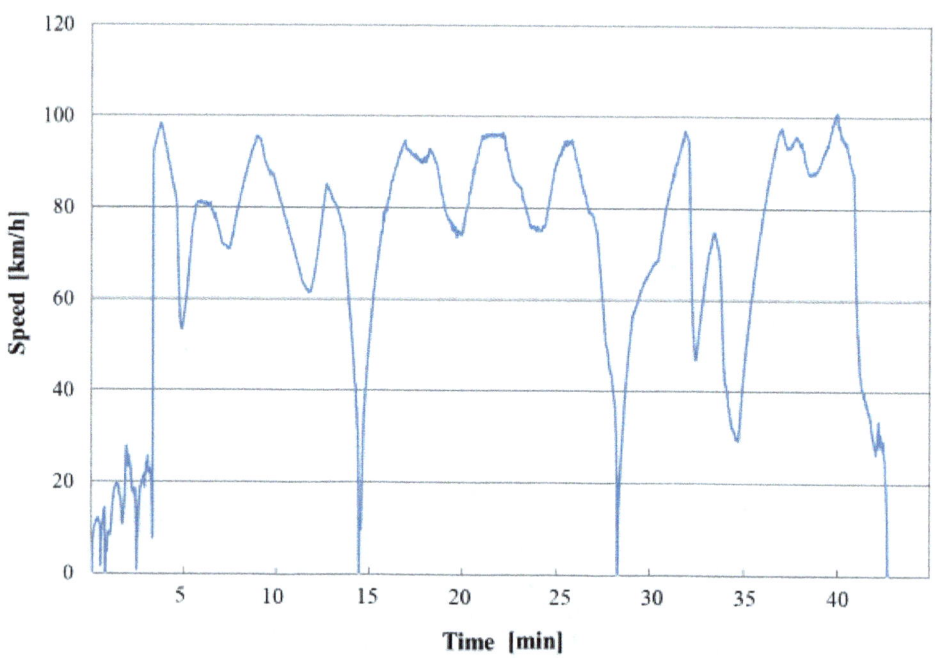

Figure 5: The speed profile of the rail bus SA108 determined for a standard drive.

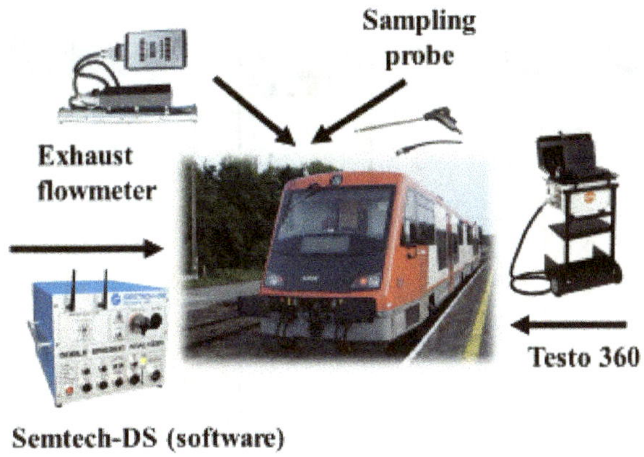

Figure 6: Measuring apparatus mounted on the test vehicle.

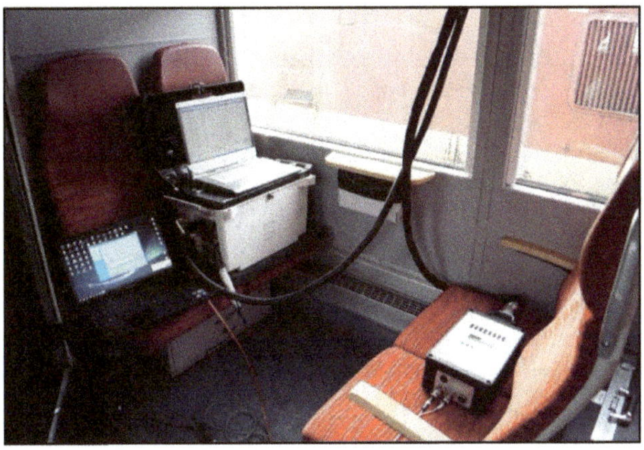

Figure 7: The measuring apparatus placed inside the test vehicle.

3 RESULTS AND ANALYSIS

In order to carry out the analyses that are the subject of this study, measurements of exhaust gas flow from the engines of the tested rail bus were performed beforehand (Fig. 8). Due to the fact that the first part of the test route consisted mainly of manoeuvres in the vicinity of the railway station in Krzyż Wielkopolski, the nature of the exhaust gas flow is significantly different (much lower flow rate) from the other recorded data. This also had an impact on the exhaust gas temperature (Fig. 9).

A significant drop in exhaust gas temperature can be observed when using eco-driving as opposed to the standard drive. Lower temperature in the combustion engine reduces the formation of nitrogen oxides species, which are among the most toxic exhaust components. When using mobile apparatus for exhaust emissions testing, the concentration of individual gaseous compounds in exhaust gas can be determined. The measurement results were used to calculate the emission intensity of these compounds (Figs 10–13). In all cases, higher emission levels were observed for the standard drive. This is primarily due to the much more dynamic and liberal application of engine loads.

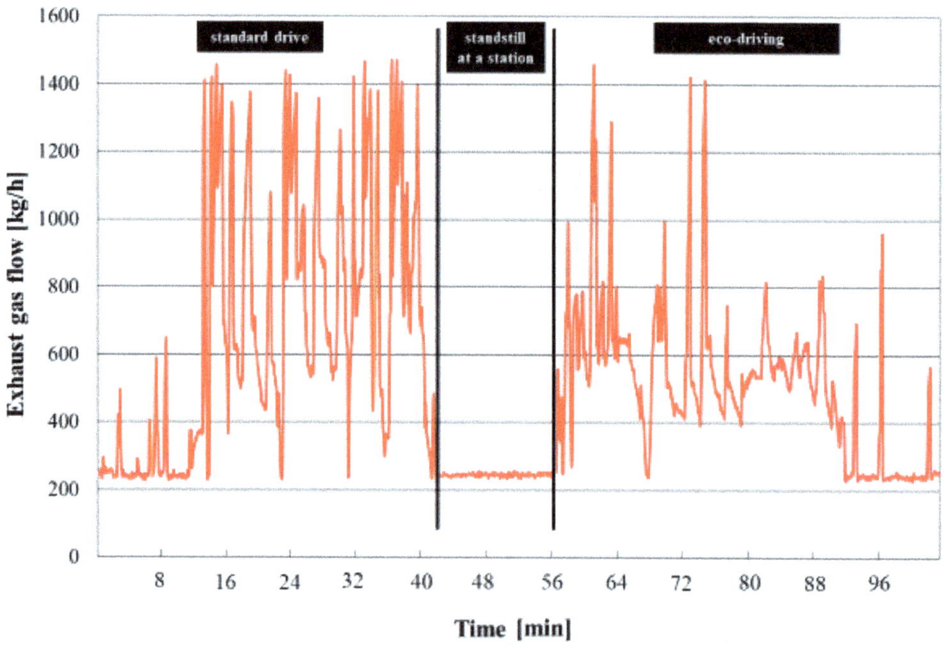

Figure 8: Exhaust mass flow rate recorded during testing – one vehicle engine.

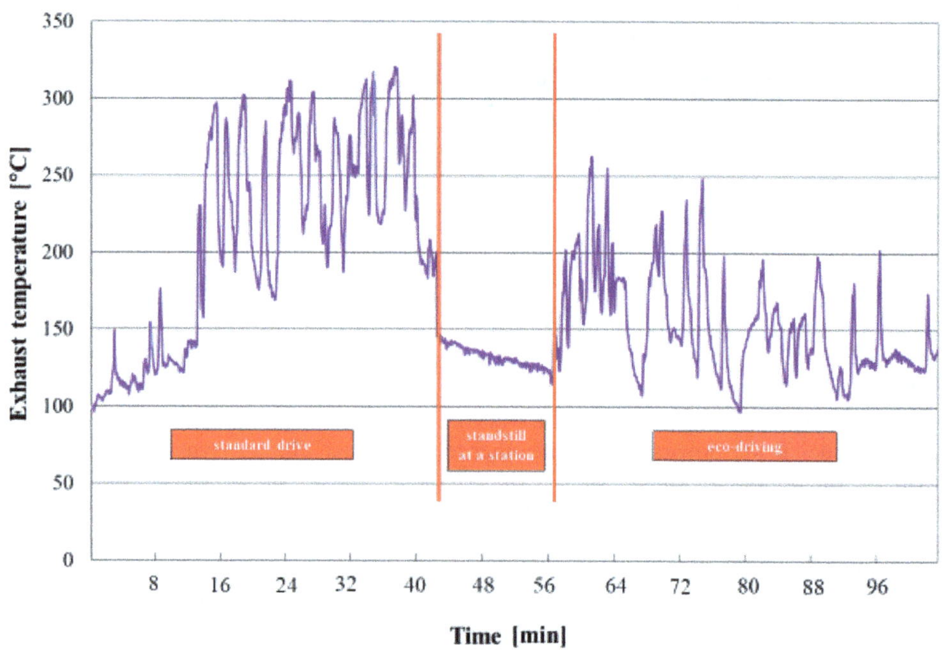

Figure 9: Exhaust gas temperature recorded during tests – one vehicle engine.

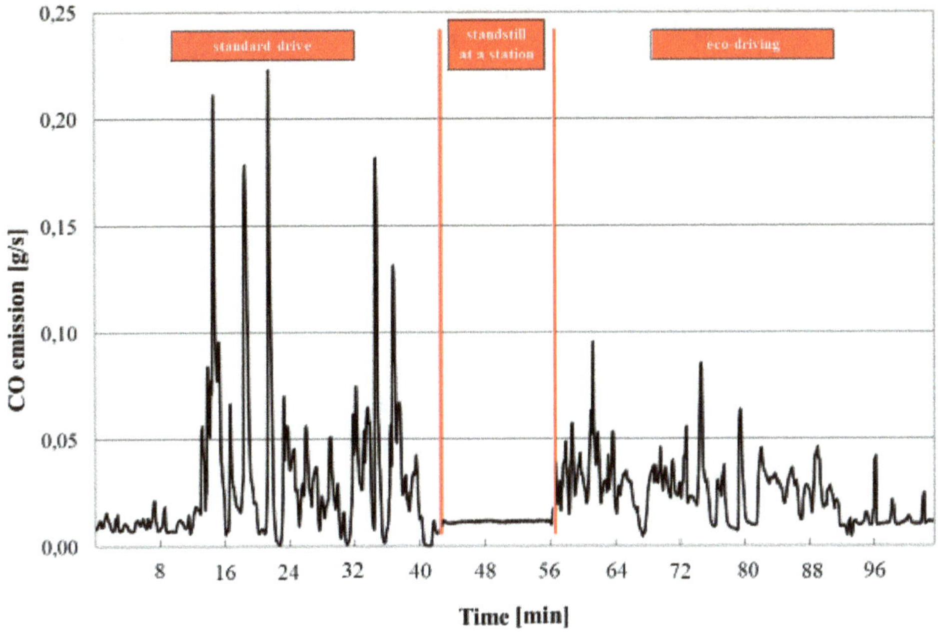

Figure 10: Carbon monoxide exhaust emission rate from the SA108 rail bus exhaust system – one vehicle engine.

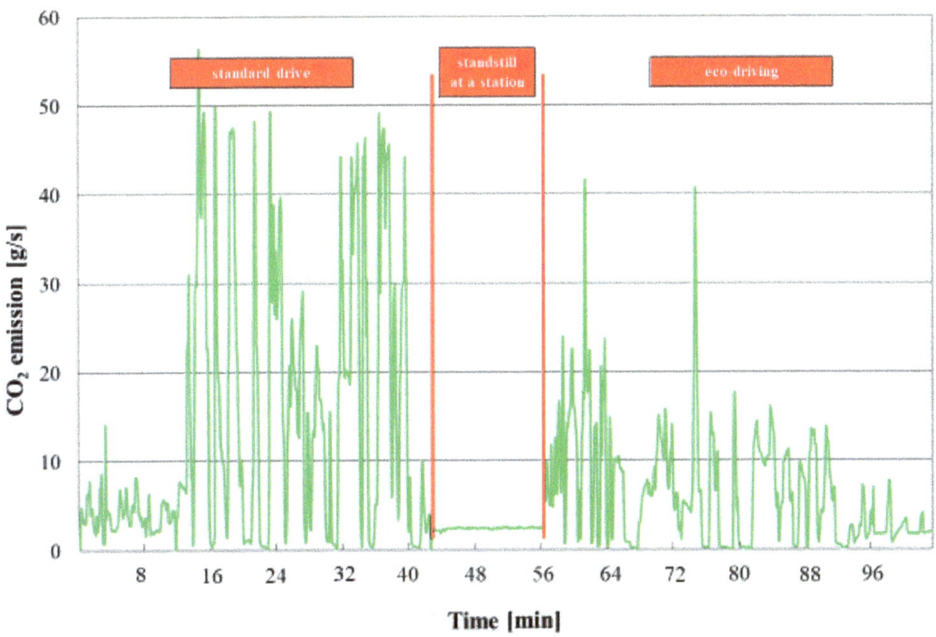

Figure 11: Carbon dioxide exhaust emission rate from the SA108 rail bus exhaust system – one vehicle engine.

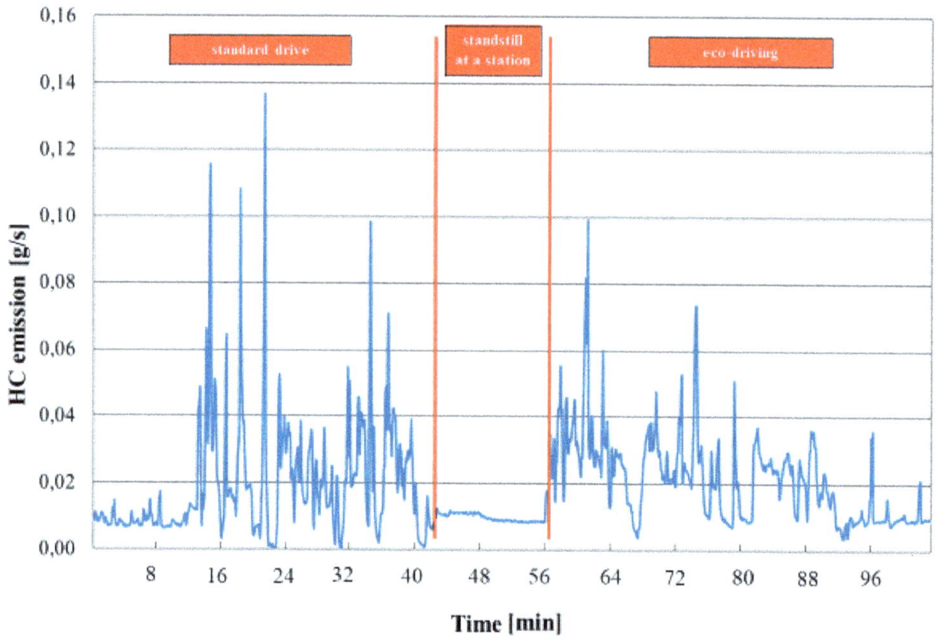

Figure 12: Hydrocarbons exhaust emission rate from the SA108 rail bus exhaust system – one vehicle engine.

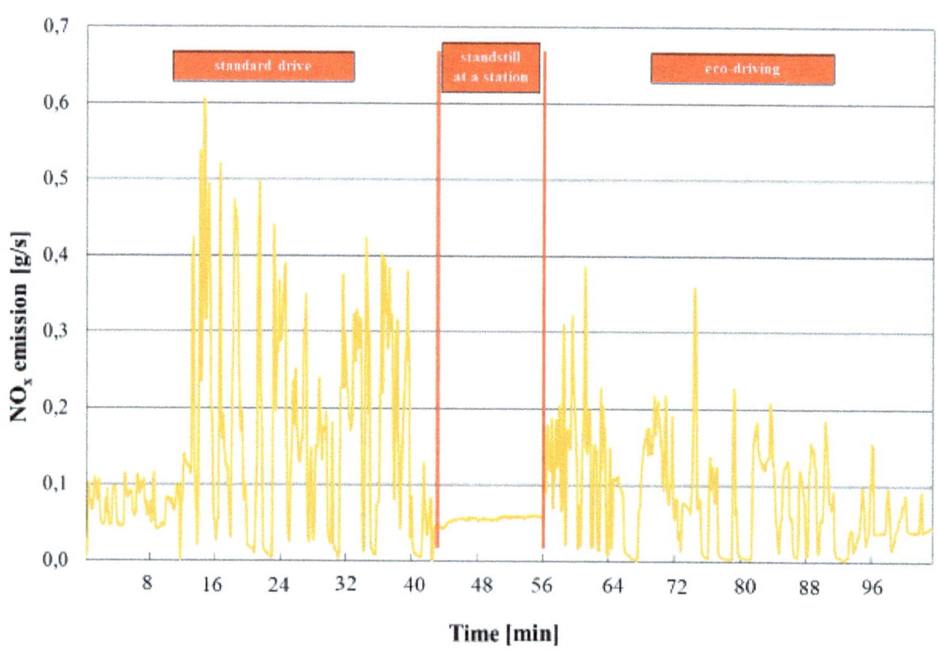

Figure 13: Nitrogen oxides exhaust emission rate from the SA108 rail bus exhaust system – one vehicle engine.

A noticeable reduction in the exhaust emissions of each of the measured compounds (CO, CO_2, HC, NO_x) has been observed. This indicates that the reduction of acceleration and deceleration values in the eco-driving test portion had a positive impact on the exhaust emissions of the vehicle. The observed changes in exhaust emissions are in line with what is often found for tests on road vehicles when applying the eco-driving style. Through the performed analyses, the exhaust emission of each harmful exhaust component for each entire drive was then calculated (Fig. 14). As in the previous case of exhaust emission intensity, a significant diversification of values was observed (Fig. 15).

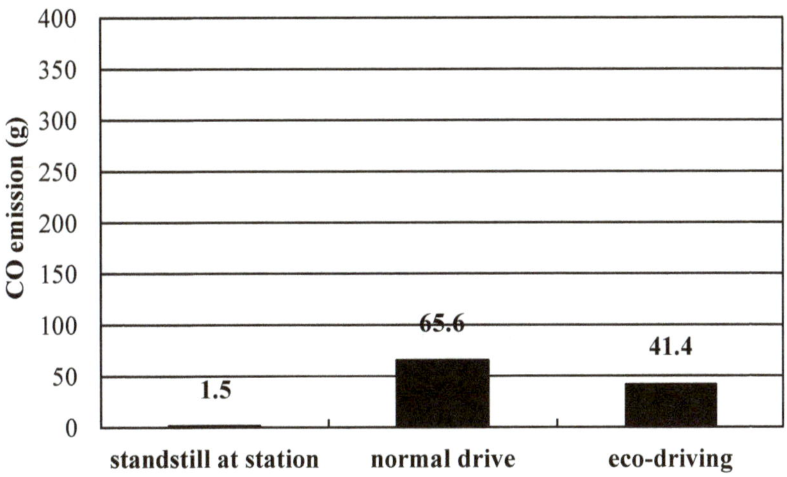

(a)

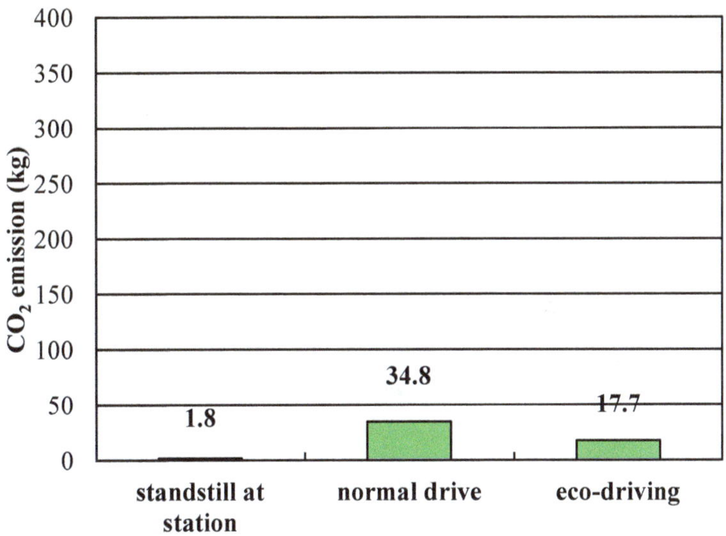

(b)

Figure 14: Exhaust emission value at standstill and in the two individual drives (one vehicle engine) for: (a) carbon monoxide; (b) carbon dioxide; (c) hydrocarbons; and (d) nitrogen oxides.

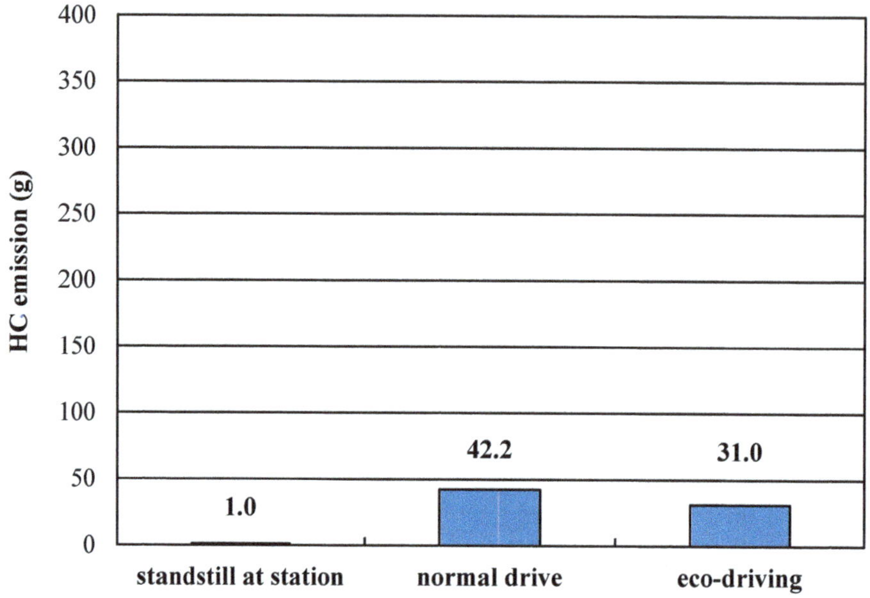

(c)

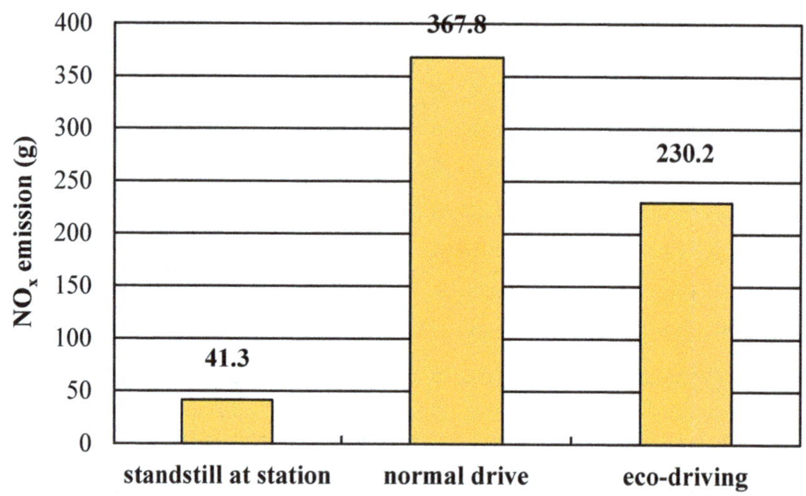

(d)

Figure 14: Continued.

Using eco-driving had the largest impact (49% reduction) on exhaust emissions of CO_2, which could be explained by the fact that this driving style is mostly used to maximize fuel efficiency, which is proportional to CO_2 emitted. The least affected was the emission of HC, with a relative change of less than 27%. This may be due to the fact that with lower

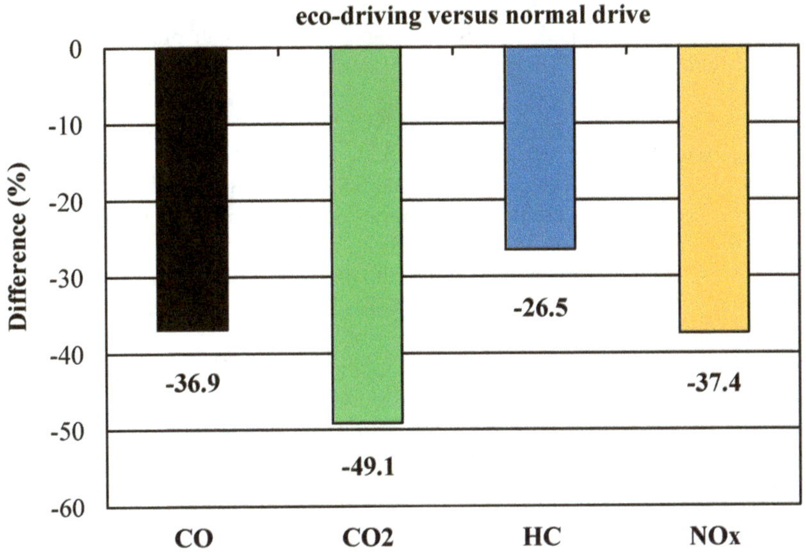

Figure 15: The relative change in exhaust emissions of individual exhaust components when using the eco-driving principles.

acceleration and deceleration values, the engine operates at a lower average rpm, and lower loads over increased periods of time. This results in lower temperature in the combustion engine which in turn reduces the rate of oxidation of hydrocarbons.

4 CONCLUSIONS

The performed exhaust emissions tests and analysis for the standard drive and eco-driving indicate a significant change in exhaust emission values of each of the measured exhaust components. The most significant impact was observed in the case of CO_2 emissions, which is directly correlated with fuel consumption. This should be expected, as eco-driving principles have been developed primarily to help improve the fuel efficiency of vehicles with combustion engines. The overall change reached up to about 50% of CO_2 emission reduction for the second drive. In addition the ecological impact of eco-driving for the rail bus was as follows:

- 37% exhaust emission reduction of carbon monoxide;
- 26% exhaust emission reduction of hydrocarbons;
- 37% exhaust emission reduction of nitrogen oxides.

This result indicates an average exhaust emission reduction of toxic exhaust components by about a third. This result will vary depending on the vehicle type, engine type and age, and the type of work performed (route, stops, slope, speeds etc.). Despite these variables, the obtained results indicate an opportunity to significantly reduce exhaust emissions from rail vehicles through the use of driver training, instead of invasive repairs and modifications, such as retrofitting or engine unit replacement.

ACKNOWLEDGEMENT

The authors would like to thank POLREGIO sp. z o.o. Poznan branch for their help in conducting this research.

REFERENCES

[1] International Energy Agency (IEA), Tracking transport report.

[2] Chester, M.V., Horvath, A. & Madanat, S., Comparison of life-cycle energy and emissions footprints of passenger transportation in metropolitan regions. *Atmospheric Environment*, **44**(8), pp. 1071–1079, 2010.

[3] Monzon, A. & Guerrero, M.J., Valuation of social and health effects of transport-related air pollution in Madrid (Spain). *Science of The Total Environment*, **334–335**, pp. 427–434, 2004.

[4] Krzyzanowski, M., Kuna-Dibbert, B. & Schneider, J., *Health Effects of Transport-Related Air Pollution*, WHO, 2005.

[5] Shaughnessy, W.J., Venigalla, M.M. & Trump, D., Health effects of ambient levels of respirable particulate matter (PM) on healthy, young-adult population. *Atmospheric Environment*, **123**, pp. 102–111, 2015.

[6] Abbasi, S., Jansson, A., Sellgren, U. & Olofsson, U., Particle emissions from rail traffic: A literature review. *Critical Reviews in Environmental Science and Technology*, **43**(23), pp. 2511–2544, 2013.

[7] Wickman, D.D., Senecal, P.K. & Reitz, R.D., Diesel engine combustion chamber geometry optimization using genetic algorithms and multi-dimensional spray and combustion modeling. *SAE Transactions*, **110**(3), pp. 487–507, 2001.

[8] Pielecha, I., Borowski, P., Czajka, J., Wisłocki, K. & Kaźmierowski, J., Combustion process shaping by use of different strategies of multiple fuel injection in a CI model engine. *Journal of Thermal Analysis and Calorimetry*, **119**, pp. 695–703, 2015.

[9] Ingham, M. & Warden, R., Cost-effectiveness of diesel fuel modifications for particulate control. SAE Technical Paper 870556, 1987. https://doi.org/10.4271/870556.

[10] Barkenbus, J.N., Eco-driving: An overlooked climate change initiative. *Energy Policy*, **38**(2), pp. 762–769, 2010.

[11] Prokolej Foundation, www.prokolej.org/pl/infografiki. Accessed on: 9 Jun. 2020.

SECTION 8
RAILWAY LOGISTICS

DATA-DRIVEN JOINT OPTIMIZATION OF PRICING AND SEAT ALLOCATION IN TRAINS

NAOTO KONNO[1] & ARVIND U. RAGHUNATHAN[2]
[1]Mitsubishi Electric Corp, Japan
[2]Mitsubishi Electric Research Laboratories Inc., USA

ABSTRACT

Revenue management (RM), a management science method, employs demand predictions to maximize revenue. Since its introduction in the airline industry in the 1980s, it is now widely used in various industries such as hotel, retail, and railway, among others. Compared to the airline industry, penetration has been slow in the railway industry. In recent years, RM has seen increased adoption in the railway industry. Existing approaches to RM in trains have mostly considered the determination of price and capacity separately. In this paper, we consider the problem of maximizing revenue by optimizing over price and capacity simultaneously. We make three important contributions to this problem. For the first time, we consider the problem of train sizing, i.e. number of coaches in each fare class. Second, instead of logit-based models estimated from price-demand data, we consider data-driven models that alleviate the problem of modelling errors. Finally, we formulate the optimization as an integer linear program as opposed to the nonlinear formulations that result from logit-based models. Thus, we present data-driven revenue maximization to jointly determine pricing, seat allocation, train sizing for a single fare class over multiple legs and train lines. Our simulation results targeting Nozomi on the Tokaido-Sanyo Shinkansen in Japan show that it is possible to formulate a railway operation plan to maximize the operation profit by applying our proposed method. In our numerical experiments, the proposed approach can identify solutions that increase profit per day by 50%, with a computation time of less than 1 second.

Keywords: data-driven, optimization, pricing, capacity allocation, train sizing, revenue management, railways, integer linear program.

1 INTRODUCTION

Revenue management (RM) (or yield management) is a fundamental problem in service operations such as airlines, bus companies, hotel industry, manufacturing companies, and railways among others. RM studies the problem of allocating a firm's resources at different rates to maximize revenues. Seat inventory control, a key concept in RM, is concerned with the problem of determining whether to accept a request at a certain fare under the condition that its value is lost after the sales season. The fares vary based on the class (first class or economy class) and the date of purchase. For example, the fare is more expensive when purchased closer to the date of travel.

RM has been widely employed in the context of airline operations. Seat inventory control has been investigated in Glover et al. [1], Belobaba [2], Davis [3], Weatherford and Bodily [4], McGill and van Ryzin [5] and Talluri and van Ryzin [6]. In the airlines industry, a leg refers to a single take-off and landing. Hence, seat booking for passengers with one or more stops from origin to destination it is necessary to consider the seat allocation by considering the airline's entire network operations. In the context of airlines, the single leg and multi-leg inventory control have been considered.

Train line refers to a train running between terminal stations that stops at multiple intervening stations. A leg (or a segment) in rail operations refers to journey between successive stations at which the train makes a stop. Train lines are inherently multi-leg operations. Hence, the seat allocation for all possible origin–destinations (stations on the train line) has to be considered simultaneously. The application of RM in freight railways can be

www.witpress.com, ISSN 1743-3509 (on-line)
doi:10.2495/CR200351

traced to the work of Strasser [7]. To the best of our knowledge, Ciancimino et al. [8] was the first work on RM for passenger railways. Ciancimino et al. [8] considered the problem of determining seat allocations on a single fare class, multiple legs of the train line to maximize the revenue. A deterministic linear programming formulation was proposed for the case of deterministic demand and a nonlinear programming formulation was proposed to handle demand uncertainty. The authors ignored the integrality of the seat allocation decisions. You [9] formulated the seat allocation among two fare-classes, multiple legs on a train line as a mixed integer nonlinear program and proposed a heuristic solution approach. Armstrong and Meissner [10] present an overview of RM approaches and models in the context of railways. An extension of the two-fare class model of You [9] to the multi-fare class was presented in [10]. Hetrakul and Cirillo [11] developed a nonlinear programming formulation that jointly optimized the prices and seat allocation based on a latent choice model. A multinomial logit model was used to predict the passenger choices. Wang et al. [12] presented a two-stage stochastic formulation incorporating passenger choice for determining optimal seat allocations. The resulting policies were compared in simulations. Zhao and Zhao [13] considered the problem of optimal seat allocation on multiple train lines under customer choice model. They solved the problem using a heuristic algorithm and demonstrated their results on Beijing–Shanghai High Speed Rail (HSR). Zhai et al. [14] optimized seat allocation for multiple lines serving the same set of origin and destinations in order to maximize the passenger throughput. This study does not directly pertain to RM.

The survey of the literature on RM in railways reveals that:

1. Train sizing, in terms of number of coaches in each fare class, has not been considered as a decision variable. This aspect is particular to railways and not airlines. For instance, the operator can choose to vary the number of first- and second-class coaches for a particular train line based on the demand.
2. Logit-based models are primarily used for modelling price sensitivity when optimizing jointly for prices and seat allocations. However, such models are estimated from data and can introduce errors in the results of the optimization. Further, the use of such models introduces nonlinearity in the problem and makes them intractable.
3. Heuristic algorithms such as particle swarm optimization, simulated annealing are the dominant solution approaches. These algorithms claim to be able to find good feasible solutions. The algorithms do not provide an estimate of the optimality gap making it difficult to assess the quality of the solution. Hence, practitioners resort to running these algorithms for a long time in the hope of finding better solutions.

In this paper, we consider the data-driven joint optimization of pricing, seat allocation, sizing to maximize revenue for a single fare class over multiple legs and train lines. In particular, we: (i) consider number of coaches as a decision variable; (ii) assume that price-demand function is provided as pairs of price and demand which are derived from user surveys. In particular, we do not estimate a logit-based model from this data; (iii) present an exact approach to solve the problem based on an integer linear programming (ILP) formulation. Note that the data-driven approach allows us to model the problem as an integer linear optimization instead of a nonlinear optimization problem. Further, the formulation ensures that the variables modelling seat allocations and number of coaches are integer valued which has been ignored in prior works, especially when the formulation becomes nonlinear. ILP formulations can be easily solved using solvers such as CPLEX Optimizer [15], Gurobi [16], and FICO Xpress Solver [17]. ILP solvers are robust and are guaranteed to find the optimum. Further, the solvers easily scale to the problem sizes of interest. The results in Section 4 demonstrate this.

2 NOTATION

The set of integers and non-negative integers are denoted by $\mathbb{Z}$ and $\mathbb{Z}_+$ respectively. For a positive integer m, $[m]$ and $[m]_0$ denote respectively the sets $\{1, \ldots, m\}$ and $\{0, \ldots, m\}$. Let $\mathcal{T}$ be the set of train lines, n_t be the number of trains operated on the line over a day and $S(t)$ be the set of stops that are served by the train line $t \in \mathcal{T}$. Let $S = \cup_{t\in\mathcal{T}} S(t)$ be the set of all stops that are served by the train lines. Further we assume that the set of all stations served by the train lines can be ordered so $S := [|S|]$ i.e. if $i, j \in S(t)$ with $i < j$ then train line t stops at i before j for all $t \in \mathcal{T} : i, j \in S(t)$. The minimum and maximum number of coaches that can be attached to the train line $t \in \mathcal{T}$ are $\underline{u}_t$ and $\overline{u}_t$ respectively. Denote by C^{max} the number of seats in a coach of the train. Let $c_{t,l}$ be the cost of operating l number of coaches on train line $t \in \mathcal{T}$. A trip is denoted as a pair of stations $(i,j) \in S(t) \times S(t)$ such that a passenger can travel from station $i \in S(t)$ to station $j \in S(t)$ on the train line t. The set $J(t) \subset S(t) \times S(t)$ denotes all possible trips on the train line $t \in \mathcal{T}$ i.e. $J(t) := \{(i,j) \mid i,j \in S(t), i < j\}$. Let, $\underline{x}_{t,ij}(t)$, $\overline{x}_{t,ij}(t) \in \mathbb{Z}$ denote the minimum and maximum number of seats for the trip $(i,j) \in J(t)$ on the train line $t \in \mathcal{T}$. Let $J := \cup_{t\in\mathcal{T}} J(t)$ be the set of all trips that can be taken using any of the train lines. For each trip $(i,j) \in J$ there exists a set $PD(i,j) := \{(p_{ij,1}, d_{ij,1}), \ldots, (p_{ij,K(i,j)}, d_{ij,K(i,j)})\}$, representing the price-daily demand pairs for the trip (i,j) with $p_{ij,k} > 0$, $d_{ij,k} \in \mathbb{Z}_+$ for $k \in [K(i,j)]$. Let $J(t,q) \subset J(t)$ for $q \in S(t)$ be defined as $J(t,q) := \{(i,j) \mid i \leq q < j, (i,j) \in J(t)\}$. The set $J(t,q)$ defines the set of trips that cover the leg of the train line t from station q. By definition, $J(t,q) = \emptyset$ if q is the last station on the train line t.

3 INTEGER LINEAR PROGRAMMING FORMULATION

We present an ILP formulation for the joint optimization of sizing, price and seat allocation for a single fare class, multi-leg RM over multiple train lines. The variables in the problem are

- $y_{t,l} \in \{0,1\}$ denote the choice of l coaches for $l = \underline{u}_t, \ldots, \overline{u}_t$ on train line $t \in \mathcal{T}$;
- $x_{t,ij} \in \mathbb{Z}_+$ denote the number of seats that are allocated for the trip $(i,j) \in J(t)$ on train line $t \in \mathcal{T}$;
- $s_{ijk} \in \mathbb{Z}_+$ denote the number of seats that are sold at the price $p_{ij,k}$ on the trip $(i,j) \in J$, $k \in [K(i,j)]$;
- $z_{ijk} \in \{0,1\}$ denote the choice of price $p_{ij,k}$ for the segment $(i,j) \in J$, $k \in [K(i,j)]$. Note that the realized demand at price $p_{ij,k}$ is $d_{ij,k}$ with $(p_{ij,k}, d_{ij,k}) \in PD(i,j)$.

The optimization problem for maximizing profit is

$$\max \ \sum_{(i,j)\in J} \sum_{k\in[K_{ij}]} p_{ij,k} s_{ijk} - \sum_{t\in\mathcal{T}} n_t \cdot \left(\sum_{l=\underline{u}_t}^{\overline{u}_t} c_{t,l} y_{t,l} \right), \tag{1}$$

$$\text{s.t.} \ \sum_{(i,j)\in J(t,q)} x_{t,ij} \leq C^{max} \cdot \sum_{l=\underline{u}_t}^{\overline{u}_t} l \cdot y_{t,l} \quad \forall\, q \in S(t), t \in \mathcal{T}, \tag{2}$$

$$\underline{x}_{t,ij} \leq x_{t,ij} \leq \overline{x}_{t,ij} \quad \forall (i,j) \in J(t), t \in \mathcal{T}, \tag{3}$$

$$\sum_{k\in[K_{ij}]} s_{ijk} \leq \sum_{t\in\mathcal{T}:(i,j)\in J(t)} n_t \cdot x_{t,ij} \quad \forall (i,j) \in J, \tag{4}$$

$$s_{ijk} \leq d_{ij,k} z_{ijk} \quad \forall (i,j) \in J, k \in [K_{ij}], \tag{5}$$

$$\sum_{k\in[K_{ij}]} z_{ijk} = 1 \quad \forall (i,j) \in J, \tag{6}$$

$$\sum_{l=\underline{u}_t}^{\overline{u}_t} y_{t,l} = 1 \quad \forall\, t \in \mathcal{T}, \tag{7}$$

$$y_{t,l} \in \{0,1\} \quad \forall l = \underline{u}, \dots, \overline{u}, t \in \mathcal{T}, \tag{8}$$

$$s_{ijk} \in \mathbb{Z}_+ \quad \forall (i,j) \in J, k \in [\mathrm{K}_{ij}], \tag{9}$$

$$x_{t,ij} \in \mathbb{Z}_+ \quad \forall (i,j) \in J(t), t \in \mathcal{T}, \tag{10}$$

$$z_{ijk} \in \{0,1\} \quad \forall (i,j) \in J, k \in [\mathrm{K}_{ij}]. \tag{11}$$

The first term in the objective function models the revenue that is obtained on each segment as the product of the price $p_{ij,k}$ and the number of seats that are sold s_{ijk}. The second term in the objective models the cost incurred in operating a certain number of coaches and introduces a trade-off between revenue and operating costs. Note that the formulation does not impose any assumptions on the functional form of c_l. Constraint in eqn (2) ensures that the total number of seats allocated to all of the trips in $J(t,q)$ that include the segment starting from $q \in S(t)$ of the train line $t \in \mathcal{T}$ does not exceed the total capacity of the train. Constraint in eqn (3) requires the number of seats allocated for trip $(i,j) \in J(t)$ on train line $t \in \mathcal{T}$ to be between the specified lower ($\underline{x}_{t,ij}$) and upper ($\overline{x}_{t,ij}$) limits. Constraint in eqn (4) limits the number of seats sold $\sum_{k\in[K_{ij}]} s_{ijk}$ for trip $(i,j) \in J$ to be upper bounded by the total seat allocation over all train lines serving that trip. Constraint in eqn (5) limits the number of seats sold s_{ijk} for trip $(i,j) \in J$ at price $p_{ij,k}$ by the demand $d_{ij,k}$. The equality in eqn (6) enforces that only one of the available price alternatives is chosen for each of the trips $(i,j) \in J$. Eqn (7) ensures that only of the binary variables $y_{t,l}$ takes a value of 1. The remaining constraints enforce the appropriate integrality of the variables. Note that the formulation does not provide the number of seats that are sold on each train line. However, this can be obtained by apportioning the number of seats sold ($\sum_{k\in[K_{ij}]} s_{ijk}$) to each train line such that the assignment does not exceed the available seats for the trip ($n_t \cdot x_{t,ij}$) on the train line.

The constraints in the formulation are all linear and the decision variables are binary or integer valued. Hence, our formulation belongs to the class of ILPs. A number of commercial software packages have been developed for the solution of ILPs, namely CPLEX Optimizer [15], Gurobi [16], and FICO Xpress Solver [17].

We conclude the section by noting the following. Suppose a logit model is available for price-demand elasticity. Then a set with pairs of price and demand as in $PD(i,j)$ can be obtained by discretizing the logit-curve. An ILP formulation, so obtained, can alleviate the nonlinearity that is inherent when using the logit model. The number of price discretizations considered will influence the optimal solution. We will study the effect of the number of discretization points in Section 4.4.

4 NUMERICAL EXAMPLE

In order to showcase the effectiveness of our research, this section introduces an application example targeting high-speed railways in Japan. High-speed railways in Japan are called Shinkansen, and is a major means of transportation in Japan. In our study, we focus on “Tokaido-Sanyo Shinkansen” which is the one of the most famous Shinkansen lines in Japan. There are five train types “Nozomi”, “Hikari”, “Kodama”, “Sakura” and “Mizuho” in Tokaido-Sanyo Shinkansen. We use data from a variety of sources to estimate the demand data on Nozomi, which is the fastest type in Tokaido-Sanyo Shinkansen and used by a great deal of people. In the following section, we describe our approach to deriving the data required for the formulation. Section 4.1 presents our criteria for selecting the stations that are included in the study. This limits the set of trips that we consider in the formulation. Section 4.2 describes our method for estimating the number of passengers who take Nozomi from each station over a day. Section 4.3 provides the setting of the parameters used in the optimization formulation of Section 3. Section 4.4 describes the results from solving the optimization formulation in Section 3 under different parameter settings.

4.1 Selection of criteria for target station

Nozomi is a type of Shinkansen that operates between Tokyo and Hakata, and Table 1 lists the stations at which it stops. There are 15 Nozomi stops in total, but Himeji, Fukuyama, Tokuyama, and Shin-Yamaguchi are excluded from the study as they are only served by a few services. In this research, when there are two stations in one prefecture, only one of them is selected as representative in order to estimate the demand by using the prefecture's mutual departure and arrival table [18]. For instance, Tokyo Station and Shinagawa Station are in Tokyo prefecture, while Kokura and Hakata are in Fukuoka prefecture. Therefore, Tokyo and Hakata Station are respectively used as a representative of Tokyo and Fukuoka prefectures. This is done primarily due to a lack of disaggregated ridership data by stations in a prefecture. The column titled "Target station" in Table 1 shows the stations targeted in this study.

Table 1: List of stations.

Station name	Nozomi stop station	Target station
Tokyo	✓	✓
Shinagawa[2]	✓	
Shin-Yokohama	✓	✓
Nagoya	✓	✓
Kyoto	✓	✓
Shin-Osaka	✓	✓
Shin-Kobe	✓	✓
Himeji	✓[1]	
Okayama	✓	✓
Fukuyama	✓[1]	
Hiroshima	✓	✓
Tokuyama	✓[1]	
Shin-Yamaguchi	✓[1]	
Kokura[2]	✓	
Hakata	✓	✓

[1]Only some trains stop.
[2]Not representative stations in each prefecture.

4.2 Estimating Shinkansen ridership

In this section, we describe our method for estimating the passenger traffic information. We describe the method for estimating the number of passengers boarding and deboarding the Shinkansen at each main station in a day in Section 4.2.1. Section 4.2.2 describes the method for estimating the flow of passengers between stations.

4.2.1 Number of passengers at each stations per day

West Japan Marketing Communications Inc. records the average number of passengers boarding Shinkansen every day from each main station [19]. We assume that the ratio of

Shinkansen passengers to other railway passengers is 3:7 if some stations' data also includes non-Shinkansen passengers. The second column in Table 2 shows the number of one way Shinkansen users of our target stations by using this data.

4.2.2 Number of passengers on different trips

In Section 4.2.1 we estimated the number of passengers riding the train from each station. However, we have to estimate the number of passengers moving between stations in order to obtain the demand for each trip. The Ministry of Land, Infrastructure, Transport and Tourism surveyed the number of passengers moving between prefectures by Shinkansen or limited express train [18]. We utilize this information to estimate passengers flow.

As shown in Table 1, we consider the stations Tokyo, Shin-Yokohama, Nagoya, Kyoto, Shin-Osaka, Shin-Kobe, Okayama, Hiroshima and Hakata. In addition, the prefectures in which these stations belong to are Tokyo, Kanagawa, Aichi, Kyoto, Osaka, Hyogo, Okayama, Hiroshima, and Fukuoka in that order. These prefectures are also our target prefectures in this study. We used "One-way / round-trip inter-prefecture inter-arrival table" made available by the Ministry of Land, Infrastructure, Transport and Tourism [18] to obtain the percentage flows of passengers listed in fourth column of Table 2. The fifth column is the result multiplying the number of Shinkansen passengers form each station (column 2) with flow percentages (fourth column). The fifth column provides the number daily Shinkansen passengers between different stations.

4.2.3 Estimation for Nozomi passengers

In this section, we estimate the number of passengers using the Nozomi train (sixth column in Table 2) from the total number of Shinkansen passengers (fifth column). There are three types of Shinkansen, Nozomi, Hikari and Kodama from Tokyo. In addition, there are five types of Shinkansen, Nozomi, Hikari, Kodama, Sakura and Mizuho from Shin-Osaka. In order to estimate the number of Nozomi passengers, we consider the proportion of Nozomi departing from Tokyo to Hakata and from Shin-Osaka to Hakata. According to the JR Outing Net [20], about 62% of the Shinkansen from Tokyo is of Nozomi type, and about 50% of the Shinkansen from Shin-Osaka is of Nozomi type. Assuming that the number of passengers in Nozomi follows this ratio, we can estimate Nozomi passengers in sixth column of Table 2. The Nozomi consists of three different fare classes – business, reserved and non-reserved. We only focus on one fare class, reserved seat class. Assuming that 66% of Nozomi's passengers is reserved seat passengers, we can estimate reserved seat passengers in seventh column of Table 2.

4.3 Problem parameters

In this section we describe the values for the different parameters in the formulation provided in Section 3.

- Train lines: We consider 7 trains lines ($\mathcal{T} = \{1, \dots, 7\}$) as specified in Table 3. The number of trains that run on each of the train lines (n_t) is provided in the fourth column of Table 3. The cost for operating a coach on each of the train lines is provided in the fifth column of Table 3. The cost of operating l coaches on line t ($c_{t,l}$) is assumed to be l times the cost listed in the fifth column of Table 3. The cost for operating a coach consists of: (i) Electric cost (18.75 YEN/kilometer), (ii) Maintenance cost (YEN 10,000 per day), and (iii) Cleaning fee as Nozomi's operating expenses (YEN 1000 per train). Thus, the operating cost per coach varies by the train line. The coach capacity (C^{max}) is assumed to be 87.

Table 2: Information on number of passengers on different trips.

Departure	Total passengers	Arrival	Flow percentage[3]	Flow passengers[4]	Nozomi passengers[4]	Reserved seat passengers[4]
Tokyo	132,900	Shin-Yokohama	2.24	2,981	1,848	1,220
		Nagoya	17.57	23,350	14,477	9,555
		Kyoto	10.12	13,443	8,335	5,501
		Shin-Osaka	21.97	29,193	18,100	11,946
		Shin-Kobe	10.14	13,471	8,352	5,512
		Okayama	3.71	4,928	3,055	2,017
		Hiroshima	6.49	8,622	5,346	3,528
		Hakata	1.98	2,634	1,633	1,078
Shin-Yokohama	32,700	Nagoya	13.16	4,305	2,669	1,762
		Kyoto	11.59	3,789	2,349	1,551
		Shin-Osaka	18.39	6,013	3,728	2,460
		Shin-Kobe	9.00	2,943	1,825	1,204
		Okayama	3.33	1,090	676	446
		Hiroshima	6.04	1,976	1,225	809
		Hakata	2.11	689	427	282
Nagoya	62,670	Kyoto	4.23	2,650	1,643	1,084
		Shin-Osaka	12.76	7,998	4,959	3,273
		Shin-Kobe	4.46	2,792	1,731	1,143
		Okayama	1.86	1,165	722	477
		Hiroshima	3.74	2,342	1,452	958
		Hakata	4.25	2,661	1,650	1,089

Table 2: Continued.

Departure	Total passengers	Arrival	Flow percentage[3]	Flow passengers[4]	Nozomi passengers[4]	Reserved seat passengers[4]
Kyoto	37,600	Shin-Osaka	5.56	2,089	1,295	855
		Shin-Kobe	5.07	1,905	1,181	779
		Okayama	1.87	704	437	288
		Hiroshima	2.39	897	556	367
		Hakata	5.22	1,964	1,218	804
Shin-Osaka	79,300	Shin-Kobe	3.70	2,933	1,466	968
		Okayama	4.11	3,257	1,629	1,075
		Hiroshima	6.48	5,141	2,571	1,697
		Hakata	7.66	6,074	3,037	2,004
Shin-Kobe	9,300	Okayama	6.86	638	319	210
		Hiroshima	6.84	636	318	210
		Hakata	7.24	674	337	222
Okayama	20,160	Hiroshima	8.62	1739	869	574
		Hakata	2.98	601	300	198
Hiroshima	22,380	Hakata	9.86	2206	1,103	728

[3]Two decimal places display; [4]Integer display.

- Set of stations: The set of stations under consideration are provided in the third column of Table 1. The ordering of the stations is the sequence in which the stations are visited. Table 3 specifies the origin and destination terminals for each train line. The train line is assumed to stop at all the intermediate stations. For instance, the Tokyo-Shin-Osaka train line (Route no 1 in Table 3) stops at Tokyo, Shin-Yokohama, Nagoya, Kyoto, Shin-Osaka, i.e. stops at all intermediate stations mentioned in the third column of Table 2. Using this sets $S(t), J(t), J(t,q)$ can all be inferred.
- Price-demand function: We use the demand data inferred in the seventh column of Table 2 and the current fares for these trips (provided in Table 4) to estimate a price-demand logit function. Logit price-response curve for all $(i,j) \in J$

$$d_{ij}(p_{ij}) = C_{ij}e^{-(a_{ij}+b_{ij}p_{ij})}/1 + e^{-(a_{ij}+b_{ij}p_{ij})}. \qquad (12)$$

  The parameter b_{ij} is set to 0.001 for all $(i,j) \in J$. The demand $(\hat{d}_{ij})$ in the seventh column of Table 2 is observed at the price $(\hat{p}_{ij})$ stated in Table 4. We choose $a_{ij} = -b_{ij}\hat{p}_{ij}$ and hence, $C_{ij} = 2\hat{d}_{ij}$. We will discretize this logit curve at a set of prices to obtain our price-demand pairs. The price is discretized at intervals of Δp_{ij} as $\{\hat{p}_{ij} + k \cdot \Delta p_{ij}\}$ for $k = -n_{disc}, \ldots, n_{disc}$ to obtain $K_{ij} = (2n_{disc} + 1)$ prices. The demand function in eqn (12) is evaluated at the discretized prices to yield $(2n_{disc} + 1)$ price-demand pairs. We employ this approach since we do not have access to the raw data. But this suffices to illustrate the computational efficiency of our approach. The values of $\Delta p_{ij}, n_{disc}$ are provided in the next section.
- Lower and upper limit on each trip: We do not impose the constraint in eqn (3). In other words, $\underline{x}_{t,ij} = 0$ and $\overline{x}_{t,ij} = \infty$ for all $t \in \mathcal{T}, (i,j) \in J(t)$. This is done due to the lack of data on this parameter.

4.4 Results

We will begin this section by first studying the effect on operating costs, revenue and profits from having number of coaches and price to be a decision variable in the optimization. In the rest of the section the price interval parameter (Δp_{ij}) is set to $(0.01 \cdot \hat{p}_{ij})$, i.e. 1% of the nominal price in Table 4. With this setting the price for each trip will be varied within $\pm n_{disc}$% of $\hat{p}_{ij}$. We will vary the number of price points (n_{disc}) in the following. We consider the following four different realizations of the optimization formulation described in Section 3:

- Baseline : In this realization, we fix the number of coaches on each train line to be 10 $(\underline{u}_t = \overline{u}_t = 10)$. The parameter n_{disc} is set to 1. This implies that price for each trip is fixed to the value listed in Table 4. Hence, number of coaches and prices are not optimized. In this setting, the cost of operation is a fixed quantity since the number of coaches cannot be varied. The optimization algorithm finds a feasible allocation of seats according to the demand on each trip as specified in Table 2.
- FixedPrices: In this realization, we continue to use the parameter n_{disc} is set to 1. We allow the number of coaches on each train line to be a decision variable. The lower and upper limits for each train line is set as $\underline{u}_t = 6, \overline{u}_t = 16$. By varying the number of coaches in the optimization it is possible to reduce the operating costs and hence, increase profits. This realization allows to study the effect of variation in train sizes when prices are fixed.

Table 3: Nozomi's travel route.

Route no	Departure	Arrival	Trains per day	Cost per coach (YEN)
1	Tokyo	Shin-Osaka	56	20,656.25
2		Okayama	4	23675.00
3		Hiroshima	14	26,393.75
4		Hakata	33	31,043.75
5	Shin-Osaka	Okayama	4	14,000.00
6		Hiroshima	14	16,718.75
7		Hakata	37	21,368.75

Table 4: The fare table.

Departure	Arrival	Fare (YEN)
Tokyo	Shin-Yokohama	2,960
	Nagoya	11,090
	Kyoto	13,910
	Shin-Osaka	14,450
	Shin-Kobe	15,100
	Okayama	17,340
	Hiroshima	19,080
	Hakata	22,950
Shin-Yokohama	Nagoya	10,450
	Kyoto	13,250
	Shin-Osaka	14,120
	Shin-Kobe	14,770
	Okayama	17,010
	Hiroshima	18,760
	Hakata	22,630
Nagoya	Kyoto	5,800
	Shin-Osaka	6,560
	Shin-Kobe	8,180
	Okayama	11,290
	Hiroshima	14,230
	Hakata	18,540
Kyoto	Shin-Osaka	3,020
	Shin-Kobe	3,650
	Okayama	7,850
	Hiroshima	11,410
	Hakata	16,060
Shin-Osaka	Shin-Kobe	3,100
	Okayama	6,230
	Hiroshima	10,440
	Hakata	15,310
Shin-Kobe	Okayama	5,800
	Hiroshima	10,220
	Hakata	14,990
Okayama	Hiroshima	6,230
	Hakata	12,710
Hiroshima	Hakata	9,150

- FixedCoaches-5%: In this realization, we fix the number of coaches as specified in Baseline ($\underline{u}_t = \overline{u}_t = 10$). However, the price is allowed to vary on each trip. We use $n_{disc} = 5$ so the price for each trip is varied within $\pm 5\%$ of $\hat{p}_{ij}$ to yield $K_{ij} = 11$ price-demand pairs for each trip.
- Optimized-5%: In this realization, the number of coaches on each train line and the price are decision variables in the optimization as proposed in this paper. The prices on each trip are allowed to vary within $\pm 5\%$ of $\hat{p}_{ij}$ as in FixedCoaches-5%.

The resulting optimization problems for each of the four realizations are solved using Gurobi 9.0 [16]. In all instances, the solver took less than 0.1 second on a standard desktop to obtain the optimal solutions. Further, the solution is obtained at the root node of the tree. In all cases, the number of seats that are sold on each trip is equal to the demand on that trip. This is to be expected when the capacity that is available on each trip is in excess of the demand. An inspection of the solution for each of these cases also reveals that no seats are allocated to certain trips on some of the train lines. The number of coaches on each train line, total cost of operation, revenue, and profit is listed in Table 5. The number of coaches in Table 5 are listed in the order of the train lines specified in Table 3.

Table 5: Comparison of different realizations of the optimization formulation.

	Number of coaches	Operating cost (YEN)	Revenue (YEN)	Profit (YEN)
Baseline	10,10,10,10,10,10,10	1,538,562.50	868,245,230.00	866,706,668.00
Fixed Prices	6,6,6,6,6,6,6	923,137.50	868,245,230.00	867,322,092.50
FixedCoaches-5%	10,10,10,10,10,10,10	1,538,562.50	1,103,365,179.00	1,101,826,616.00
Optimized-5%	6,9,7,7,6,6,6	1,046,650.00	1,103,365,178.50	1,102,318,528.50

Note that in both Baseline and FixedPrices the demand for each trip is identical. Further, the number of seats sold is equal to the demand on the trips on all train lines. Hence, the revenue obtained from ticket sales is identical (see Table 5). However, using the freedom to vary the number of coaches on the train lines enables the optimizer to find a solution where fewer number of coaches are utilised and hence, lower operating costs. As a result, FixedPrices is able to attain a profit improvement of 0.07% over the Baseline. FixedCoaches-5%, which allows price variation while keeping the number of coaches fixed, is able to obtain a 27.2% increase in profit over the Baseline. In FixedCoaches-5% the price for all trips is reduced by 5% in order to increase the demand. The increased number of tickets contributes to the increase in revenues and profits. Finally, Optimized-5% is able to further improve the profit margin by reducing the operating costs. Note that the revenue in the cases of FixedCoaches-5% and Optimized-5% are identical even though fewer coaches are allocated to the train lines in Optimized-5%. The optimal prices identified in FixedCoached-5% and Optimized-5% are identical and hence, the demand that manifests are also identical in both cases. This reinforces the remark made earlier that the number of seats available in Baseline exceeds the demand.

The number of coaches on certain train lines is set to the lower limit ($\underline{u}_t$) at the optimal solution for Optimized-5% (refer Table 5). This seems to suggest that some train lines may not be necessary. To test this hypothesis, we modified the lower bound on the number of coaches to $\underline{u}_t = 0$ for all $t \in \mathcal{T}$. With this changed bound, we identify an optimal solution where the number of coaches on the lines are 13,0,0,6,0,0,0. In other words, 13 coaches are assigned to Tokyo–Shin-Osaka Nozomi which runs 56 times per day and six coaches are assigned to Tokyo–Hakata Nozomi which runs 33 times per day. All other train lines are eliminated. Despite the fewer train lines, the revenue is identical to that reported for Optimized-5% in Table 5. However, the operating cost is 527,506 which amounts to a reduction of 50% from the operating costs for Optimized-5% in Table 5. This result provides further evidence that indeed the capacity in Baseline far exceeds the demand.

In the next set of experiments, we study the scalability of our approach when the number of price-demand pairs are increased. As noted earlier, our intention is to work directly with the price-demand data pairs instead of the estimated logit function. The approach of discretizing the prices to obtain the price-demand is advocated in the absence of such raw-data. We simulate the case of increased price-demand pairs by increasing the number of price discretization points (n_{disc}). Table 6 present results from optimizing the number of coaches and prices for different number of discretization of the price $n_{disc} \in \{5,10,20,40\}$. These approaches are respectively called Optimized-5%, Optimized-10%, Optimized-20% and Optimized-40%.

Table 6: Effect of number of price discretization on the solution.

	Number of coaches	Operating cost (YEN)	Revenue (YEN)	Profit (YEN)
Optimized-5%	6,9,7,7,6,6,6	1,046,650	1,103,365,178.50	1,102,318,528.50
Optimized-10%	9,6,6,9,6,6,6	1,078,237.50	1,248,444,333	1,247,366,095.50
Optimized-20%	10,6,6,11,6,6,6	1,150,593.75	1,308,000,398.20	1,306,849,804.45
Optimized-40%	10,6,6,11,6,6,6	1,264,262.50	1,308,123,101.90	1,306,972,508.15

Table 6 shows that the profit improves as additional price discretizations are considered. Optimized-20% identifies an optimal solution with 50.8% improvement in profit over Baseline. The improvement in profit for Optimized-40% is negligible when compared with Optimized-20%. The size of ILPs increases as the n_{disc} is increased. Optimized-5% has 983 variables and 869 constraints while Optimized-40% has 6023 variables and 5909 constraints. Optimized-40% is solved in 0.2 seconds. Hence, solving ILPs with large number of price-demand pairs does not present a computational bottleneck for the proposed approach. Fig. 1 plots the percentage reduction in the prices for the different trips in the solution obtained from Optimized-10%, Optimized-20% and Optimized-40%. The prices are uniformly reduced by 10% for all trips in the solution for Optimized-10%. The solution to Optimized-20% exploits the difference in the price-demand elasticities for the trips to obtain an improved solution. The price for all trips are reduced by at least 13%. The trips Nagoya–Kyoto (#16 on x-axis of Fig. 1), Nagoya–Shin-Osaka (#17 on x-axis of Fig. 1), Nagoya–Shin-Kobe (#18 on x-axis of Fig. 1), Kyoto–Okayama (#24 on x-axis of Fig. 1), Shin-Osaka–Okayama (#28 on x-axis of Fig. 1), Shin-Kobe–Okayama (#31 on x-axis of Fig. 1), Okayama–Hiroshima (#34 on x-axis of Fig. 1) and Okayama–Hakata (#36 on x-axis of Fig. 1) all see a price reduction of 20% in the solution to Optimized-20%. In the solution to Optimized-40% the price reduction is identical to that for Optimized-20% for most trips. The difference occurs for those trips that reduced the price by 20% in the solution to Optimized-20%. These trips are further reduced by 1–2% in the solution to Optimized-40%.

5 CONCLUSION

We propose a data-driven approach for revenue maximization optimization by determining jointly the train sizes and prices for a single fare class over multiple legs and train lines. The resulting optimization problem belongs to the class of ILPs. The ILP is shown to be

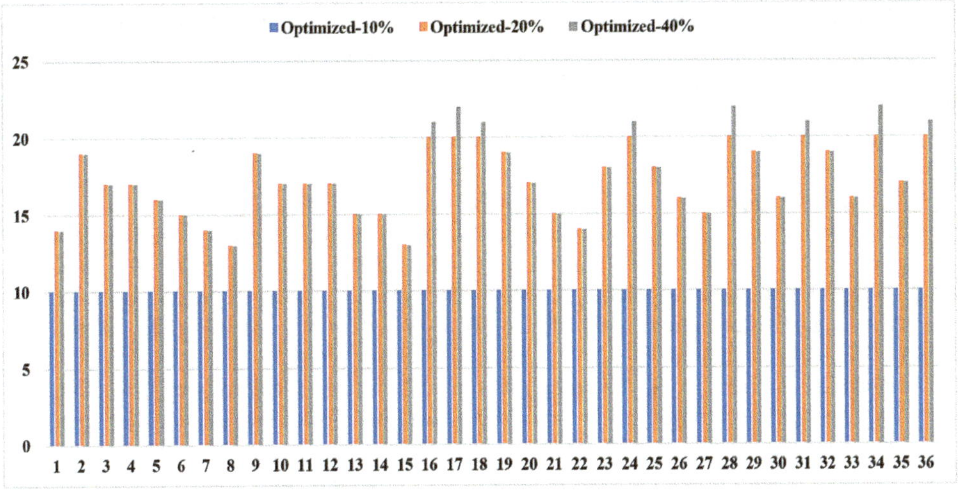

Figure 1: Plot of percentage reduction in price at the optimal solution for the different trips (trips on the x-axis are numbered in the order listed in Table 4).

computationally tractable through a numerical example. The numerical experiments demonstrate that the proposed approach is scalable in the number of price-demand pairs and is effective in improving the revenue by selling increased number of tickets at lower prices. In contrast to existing approaches for joint price and seat allocation that are nonlinear and ignore integral nature of decision variables, our approach shows that joint price and seat allocations can be effectively solved using an ILP formulation.

ACKNOWLEDGEMENT

The second author acknowledges discussions with Professor David Bergman on the connection of discretized logit curves with the data-driven approach.

REFERENCES

[1] Glover, F., Glover, R., Lorenzo, J. & McMillan, C., The passengers-mix problem in the scheduled airlines. *Interfaces*, **12**, pp. 73–79, 1982.
[2] Belobaba, P.P., Airline yield management: An overview of seat inventory control. *Transportation Science*, **21**, pp. 63–73, 1987.
[3] Davis, P., Airline ties profitability to yield management. *SIAM News – Mathematics that Counts*, **27**, 1994.
[4] Weatherford, L.R. & Bodily, S.E., A taxonomy and research overview of perishable-asset revenue management: Yield management, overbooking and pricing. *Operations Research*, **40**(5), pp. 831–844, 1992.
[5] McGill, J.G. & van Ryzin, G., Revenue management: Research overview and prospects. *Transportation Science*, **33**(2), pp. 233–256, 1999.
[6] Talluri, K.T. & van Ryzin, G.J., *The Theory and Practice of Revenue Management*, Kluwer Academic Press: New York, 2004.
[7] Strasser, S., The effect of yield management on railroads. *Transportation Quarterly*, **50**, pp. 47–55, 1990.

[8] Ciancimino, A., Inzerillo, G., Lucidi, S. & Palagi, L., A mathematical programming approach for the solution of the railway yield management problem. *Transportation Science*, **33**(2), pp. 168–181, 1999.

[9] You, P.-Y., An efficient computational approach for railway booking problems. *European Journal of Operations Research*, **185**(2), pp. 811–824, 2008.

[10] Armstrong, A. & Meissner, J., Railway revenue management: Overview and models. Technical report, Lancaster University Management School, 2010.

[11] Hetrakul, P. & Cirillo, C., A latent class choice based model system for railway optimal pricing and seat allocation. *Transportation Research Part E*, **61**, pp. 68–83, 2014.

[12] Wang, X., Wang, H. & Zhang, Z., Stochastic seat allocation models for passenger rail transportation under customer choice. *Transportation Research Part E*, **96**, pp. 95–112, 2016.

[13] Zhao, X. & Zhao, P., A seat assignment model for high-speed railway ticket booking system with customer preference consideration. *Transportmetrica A: Transport Science*, **15**(2), pp. 776–806, 2019.

[14] Zhai, X., Zhao, J. & Chen, Q., Optimization of the assignment of tickets for railway networks with large passenger flows. *Proceedings of the Institution of Mechanical Engineers, Part F: Journal of Rail and Rapid Transport*, **232**(2), pp. 632–642, 2018.

[15] CPLEX Optimizer, www.ibm.com/analytics/cplex-optimizer. Accessed on: 5 Feb. 2020.

[16] Gurobi, www.gurobi.com/. Accessed on: 5 Feb. 2020.

[17] FICO Xpress Solver, www.fico.com/en/products/fico-xpress-optimization. Accessed on: 5 Feb. 2020.

[18] Ministry of Land, Infrastructure, Transport and Tourism, Survey of mainline passenger flow. www.mlit.go.jp/common/001193645.xlsx. Accessed on: 9 Feb. 2020.

[19] West Japan Marketing Communications Inc., www.jcomm.co.jp/transit/price/databook/pdf/19D_41.pdf. Accessed on: 9 Feb. 2020.

[20] JR Outing Net, www.jr-odekake.net/. Accessed on: 15 May 2019.

MODERN RAILWAYS: CONNECTING TRAIN CONTROL SYSTEMS WITH MOBILE AND SATCOM TELECOM NETWORKS

FRANCESCO RISPOLI
Hitachi Rail STS, Italy

ABSTRACT

The growth of cellular and satellite telecom systems with the introduction of 5G and broadband mobile satellites are creating a global and pervasive telecom infrastructure offering connectivity services in any place in the world. Furthermore, 5G will provide millisecond latencies and network "slicing" capabilities that are ideal to realize bespoke virtual networks. A new generation of small satellites in low earth orbits are being launched promising a lower cost and a much higher throughput with latency below 100 m/s. The European Railways Train Management System (ERTMS) utilizes, since its conception in the early 1990s, GSM-R as the legacy telecom network linking trains with the Radio Block Center. This technology will have to be substituted as it approaches its end of life and cannot satisfy increasing demand of connectivity for new railway applications. UIC has defined the user needs of the Future Railway Mobile Communication System (FRMCS), one of the ERTMS game changers and the Shift2Rail – X2Rail-1 partners have agreed on adopting the ACS (Adaptable Communication System) as the architecture for the successor of the GSM-R. This paper addresses new railway needs with the goal of sharing cellular and satcom public telecom networks. An open multi-bearer system, coherent with ACS and FRMCS is evaluated as an alternative or complement to a dedicated network. A number of possible scenarios with different line lengths and operating trains demonstrate important cost reductions when bundling cellular and satellite bearers to achieve the QoS. Side benefits such as the possible reduction of new spectrum for the FRMCS applications, less emissions of CO_2 and a faster time to operate are evaluated. Yet, the role of new high throughput satcom is underlined to satisfy the increasing demand of train connectivity and also to extend the coverage of the future 5G services in the railways ecosystem.

Keywords: ERTMS evolution, public telecom networks, satcom, 5G.

1 INTRODUCTION

All European countries are moving towards implementing ERTMS at different speed as pointed out by the report of the European Court of Auditors (ECA) [1]. In fact although European countries have agreed to deploy European Railways Train Management System (ERTMS) on their network the current low status of ERTMS deployment is probably explained by the perceived high investments and the lack of an individual business case. In this respect, one important cost element of the ERTMS is due to the deployment, operation and maintenance of the GSM-R network that coupled with ETCS, as constituents of ERTMS, provide an interoperable railway signalling and communication system for all the European railway networks operating in the 876–880 MHz (train-to-ground) and 921–925 MHz (ground-to-train) bands. Meanwhile, Future Railway Mobile Communication System (FRMCS) has been designated by UIC [2] as the successor of GSM-R, implying that both GSM-R and FRMCS will have to operate in parallel until a full migration to the FRMCS has been completed. The overall migration throughout Europe is expected to take place between 2022 and 2035 with scenarios depending on plans of each country to replace the current GSM-R infrastructure.

The Adaptable Communication System (ACS) is the selected new architecture based on the principle of bearer independence, overcoming the identification of a legacy technology

WIT Transactions on The Built Environment, Vol 199, © 2020 WIT Press
www.witpress.com, ISSN 1743-3509 (on-line)
doi:10.2495/CR200361

such as the GSM in the early 1990s [3]. ACS will be using packet switching/IP technologies as LTE, Satellite, Wi-Fi to comply with the growing FRMCS user needs. In this evolving scenario the choice of the telecom bearer becomes a matter of economical convenience opening the way also to use cellular and satellite telecom services. In that case a service-based model could be implemented as part of the ERTMS system delivery instead of building and operating a dedicated telecom network. The use of public networks, compatible with the ACS bearer-independent architecture, will bring a substantial reduction of costs during the operational life time of the ERTMS because the investments, licencing and operations are undertaken by cellular and satellite operators to keep their networks competitive with market needs. Furthermore is not hard to predict the impact of these public services as early adopter and promoter of future 5G connectivity for the ERTMS. Most satellite broadband providers are recognising the demand arising from the ERTMS evolution and the opportunity to provide a worldwide connectivity for the railways sector. Inmarsat, Thuraya and Iridium are delivering narrow-band connectivity at global level in the L-band, while Intelsat is using the Ku-band to provide higher throughput with planar phased array antennas on trains. New high-power geostationary satellites under development for mobile applications in Ku and Ka bands will further improve the bitrate and capacity. Yet in perspective, low earth orbiting mega-constellation satellites (LEO) will set unprecedented performance in terms of connectivity and latency globally. By end of 2020 more than 500 LEO satellites were launched from roughly 200 at the end of 2019. The LEO mega-constellations proponents have announced more than 16,000 satellites in the next 3–6 years, potentially doubling the about 8,700 satellites that have been launched in the last 50 years. Starlink, promoted by Elon Musk, CEO of SpaceX is targeting service in North America and Canada in 2020, rapidly expanding to near global coverage of the populated world by 2021. A common objective of satcom proponents is to bring affordable high-speed access to every citizen and institution thanks to economy of scale in constructing and launching low mass LEO satellites. This paper starts introducing the evolving telecom needs for the ERTMS that is a de-facto global standard solution for train signaling in the world. The ERTMS connectivity requirements are evaluated looking to future years when GSM-R will have to be replaced by FRMCS whose high level architecture has been defined by the 3GPP Release 15. The FRMCS, as shown in Fig. 1 is connected to 3GPP radio access networks as cellular networks, non 3GPP radio access networks as the satcom, other than wirelines.

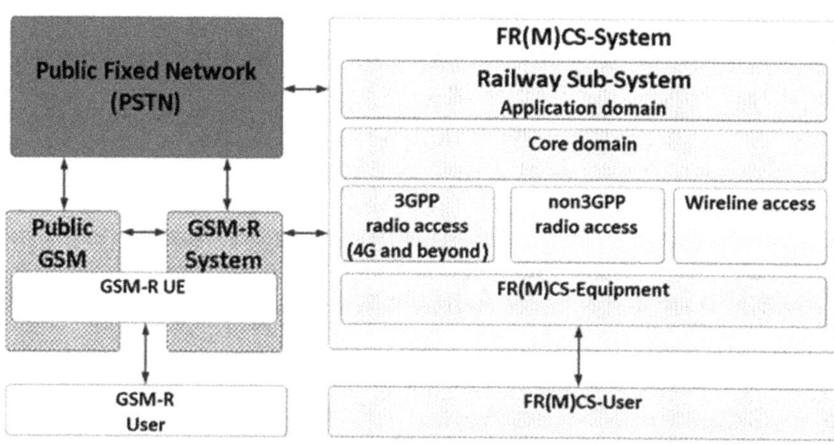

Figure 1: High level architecture of FRMCS.

The new requirements are then analyzed assuming a multi-bearer platform processing a combination of cellular and satcom bearers. The end results in terms of potential cost savings are shown in parametric form after having identified some realistic operational scenarios.

2 TRAFFIC CAPACITY EVOLUTION

The starting point is the ERTMS deployment status as indicated in Fig. 2 which clearly shows the global footprint of the ERTMS at end 2019 [4], [5]. Europe holds about 51% of tracksides and 69% of vehicles contracted. The globalization of the ERTMS is one of its strengths and to preserve this characteristic is important to look to different telecom networks to optimise the system according to regional constraints. For example in Australia, Hitachi Rail STS has deployed a TETRA network.

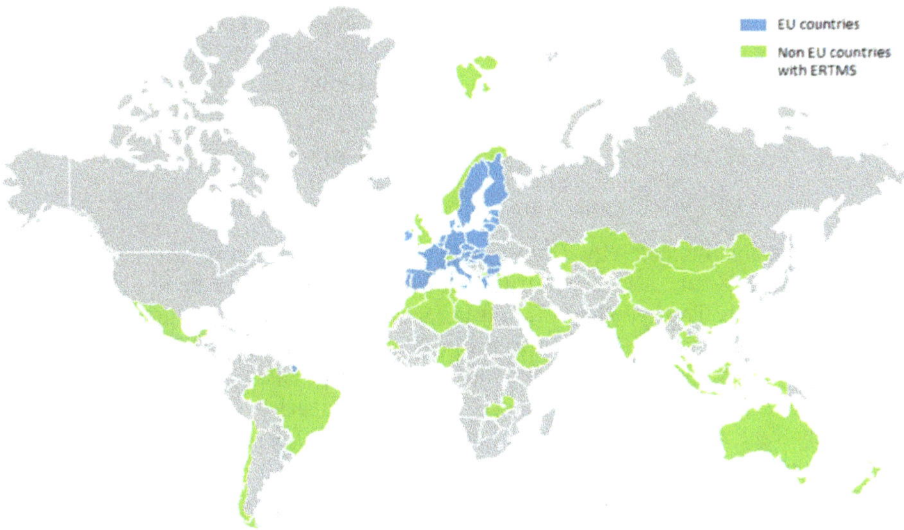

Figure 2: Status of ERTMS deployment/planned.

Looking in perspective, ERTMS will be evolving by adding new features which in turns increase the connectivity requirements. Here below there is a list of identified new applications that can be added to the ERTMS in a stepped manner starting from the basic services provided by the GSM-R and according to market needs [6], [7]:

- Current GSM-R critical applications: Voice (including emergency calls), ETCS L2;
- Train positioning;
- Automatic train operation (ATO) up to GoA2 (driver is still present in the cab);
- Remote control of engine;
- On-board intelligence and sensing (e.g. for ATO and object/person detection);
- Critical video (e.g. for remote control in shunting yards and in degraded mode);
- ETCS L3;
- ATO up to GoA4 (full automation, driver is not needed in the cab);
- Virtual coupling;
- Shunting voice;
- Other applications.

Table 1: Target traffic needs for each area type.

	DN link traffic per train (Mb/s)	UP link traffic per train (Mb/s)	Number of trains/km
Low density	4.38 (0.42 w/o video)	7.42 (3.46 w/o video)	0.33
High density			0.67
High speed			0.5

The foreseen aggregated traffic resulting from the above applications is indicated in Table 1 together with the number of trains/km for three type of lines.

The overall connectivity for the whole European trains (EU-28) is estimated assuming that the pick instantaneous number of trains (PINT) is 26,000, excluding high speed trains, and applying the traffic scenarios of Table 1. Even though PINT varies from country to country according to their traffic patterns, the forecasts indicated here below are weighted to get an upper bound of the connectivity with the actual train traffic. The results indicate a range between 310 Gbit/s (upper level with video) and 110 Gbit/s (lower level w/o video):

@ 11.8 Mbit/s per train (video included) = 310 Gbit/s;

@ 3.9 Mbit/s per train (video excluded) = 100 Gbit/s.

To guarantee this connectivity with a dedicated telecom network, the actual GSM-R spectrum is not sufficient also because the GSM-R network has to be active during the migration to FRMCS. The hypothesis is to using the extended 1.4 MHz allocated to GSM-R bringing the total spectrum to 2x5.6 MHz and reusing as much as possible the current radio network infrastructure. An additional spectrum of 10 MHz, possibly in the 1,900–1,910 MHz frequency band is also necessary. Alternatively, this extra spectrum would not be needed at all or partially if the approach is to sharing public telecom networks with commercial mobile and satellite networks. In this scenario, railways would either become users of public networks embedded into the ERTMS delivery or operate as a mobile virtual network operator (MVNO) interfacing with the ERTMS solution providers.

Whilst cellular mobile operators have capacity in excess to offer to the railways, satcom operators will be capable to fulfil this potential demand once new high throughput GEO and LEO orbiting satellites will be available. In spite of higher cost, satcom is essential to guarantee the QoS since satcom connectivity is pre-emptible for a specific service/area all the time. Fig. 3 gives the whole satcom connectivity as function of the share of satcom connectivity respect to the total connectivity required to account for different operational scenarios. A 5–10% satcom connectivity share is typical for railways where at least three cellular networks ensure the coverage, while higher percentages apply to rural railways with poor cellular coverage. However, a trade off to optimise the satcom usage should be made for each specific case. Another interesting scenario is to use satcom as intelligent back-up of a terrestrial railways network. In that case the network/base stations may use less equipment for redundancies with less stringent and costly service level agreements because traffic is routed to satcom in case of failures on ground network equipment.

3 REFERENCE TELECOM ARCHITECTURES

Multi bearer architectures with satcom have been tested in Sardinia in 2015 [8] and are currently employed in Australia on the AutoHaul system [9]. In the first case, Hitachi Rail STS has tested both a public-only telecom scenario and a scenario where satcom acted as

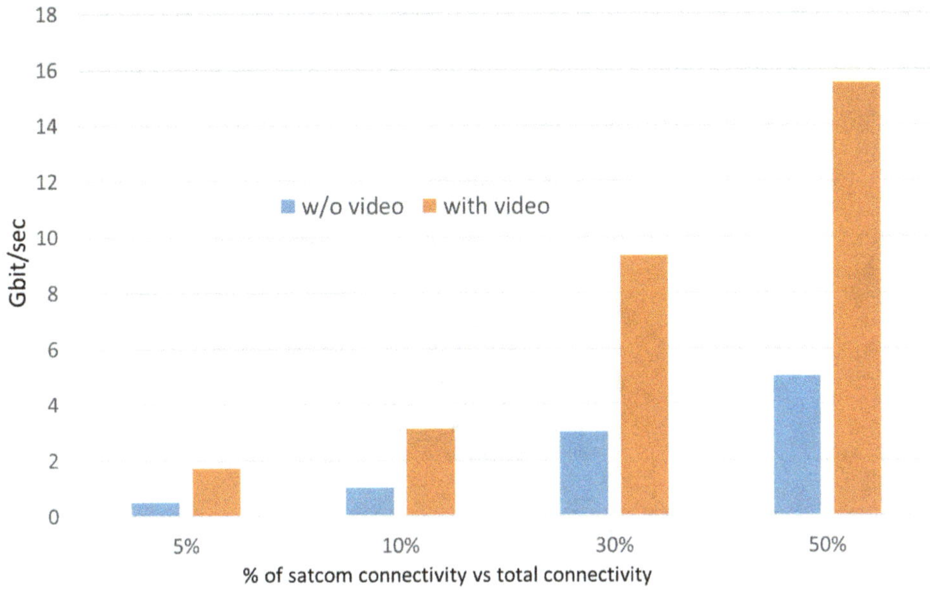

Figure 3: Satcom throughput versus different utilization scenarios.

back up to a TETRA network. The AutoHaul system uses the BGAN service of Inmarsat that is today a bench mark, having reached more than 500,000 km operating as the only telecom bearer for the Automatic Train Protection (ATP) and Automatic Train Operation (ATO) functions. Fig. 4 shows our roadmap, starting with the Sardinia test bed where we used the 2G/3G cellular networks of three mobile operators in addition to the BGAN satcom and a TETRA network. The second step is based on field trials with a dedicated train running on selected lines in north-west Italy using 3G/4G cellular connectivity and two Satcom bearers (BGAN, Thuraya). This platform is part of the satellite-based ERTMS pilot line Novara – Rho of RFI [10]. In this project a novel multi-link communication platform (MLCP) has been developed and tested. It is compliant with new European standard guaranteeing the backward compatibility with the GSM-R legacy networks and is ready for 5G and LEO satcom [11].

The last step entails the test of some of the new FRMCS applications over a 5G operational network. This activity is planned in the Radiolabs test bed at l'Aquila, one of the five cities in Italy with a 5G network to roll out pre-commercial services. These tests will be carried out in the urban and semi-urban environment using a vehicle equipped with 5G terminal [12]. The 5G test bed will be also used to evaluate the integration of satcom systems (including LEO) with terrestrial 5G network in order to prepare the ground for future terrestrial-satcom applications. In fact, differently from 4G networks the integration/interoperability of satellite systems in 5G terrestrial network has been explicitly considered and inserted in the 5G standardization roadmap. In particular, the main roles envisaged for satellite in the 5G networks are to provide:

- backhaul for the 5G access network;
- backup for the 5G network (useful in specific applications);
- differentiated/redundant communication paths (to be used with MP-TCP protocols).

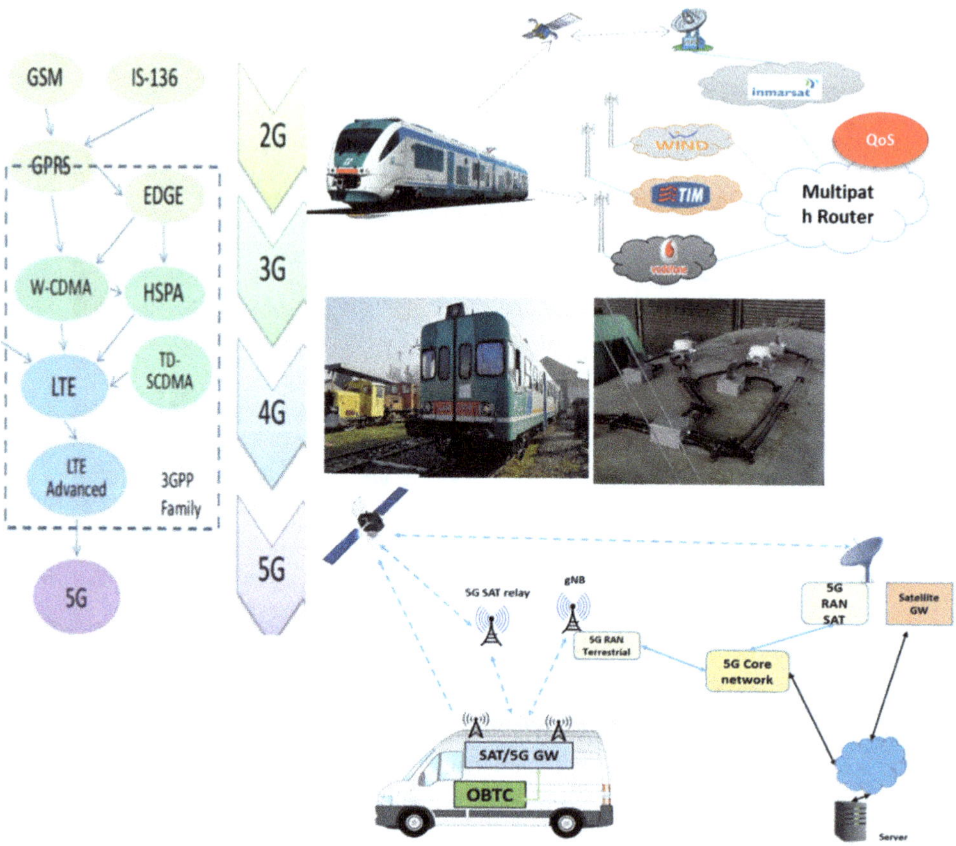

Figure 4: Phased approach to test public and satcom bearers.

The effort over the past 5 years and the growing consensus achieved so far are confirming the possibility to share public networks for ERTMS applications. The combination of cellular network with satcom, for prioritization and pre-emption, suggests a way to get a QoS for critical applications overcoming the difficulties (and costs) to imposing liabilities to MNO.

In the long term with the deployment of 5G networks this approach will become even more attractive using the network slicing capability to guarantee the QoS and satcom to extending the 5G footprint. Yet, resiliency and cyber security are important topics when using public access networks. In this respect a multi-bearer platform relying on physical separated networks is in principle more robust respect to a single network. Furthermore artificial intelligence techniques can be adopted to detect and mitigate possible interferences.

The concept of bearer independence has been selected by the Shift2Rail – X2Rail-1 project with the Adaptable Communication System (ACS) as the future railway communication infrastructure (Fig. 5). ACS provides an adaptable train-to-ground communications system usable for train control applications in all market segments, using traditional packet switching/IP technologies such as GPRS, EDGE, LTE, Satellite, Wi-Fi and other technologies. ACS is designed in accordance with the bearer independent principle (BIP) meaning that the ACS can interface with several and different communication bearers

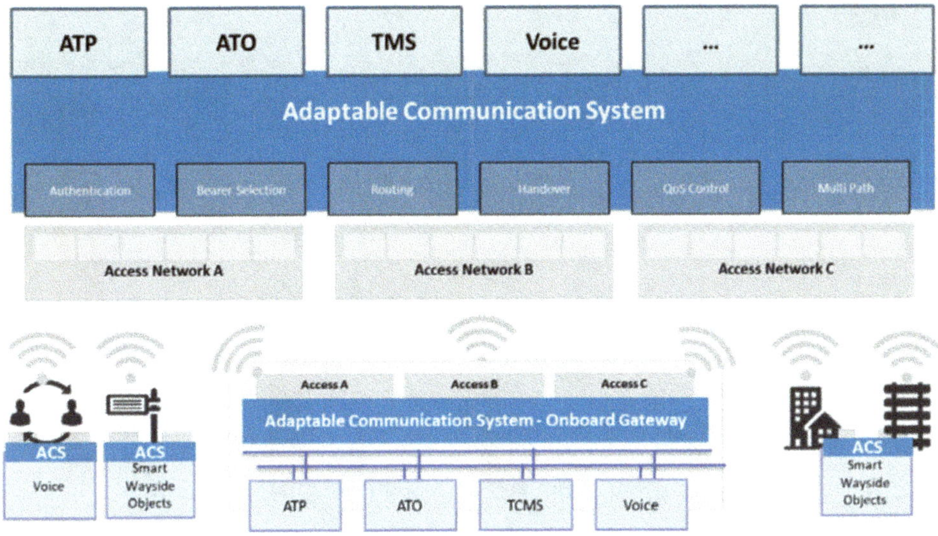

Figure 5: Adaptable Communication System (ACS).

(wired/wireless) to guarantee the end-to-end connectivity for railway applications. Interfacing of ACS with the bearers is made at IP level to fully implement the BIP concept. Communication of applications over the ACS can be provided over one or more IP-based bearers both alternative and/or traditional, public or private, without assuming that the bearer provides any capability beyond data transmission using the internet protocol (IP) [13].

Given the flexibility of ACS, is possible to design *end2end* connections by selecting proper bearers for the specific line and application. The economic benefits achievable with public bearers have been analyzed in a number of use cases by comparing the total cost of operations over 10 years with the costs of a dedicated telecom network. The business models are different: with public networks, the costs for a turnkey telecom service are proportional to the number of circulating trains; instead, for a dedicated network such as the GSM-R the costs including the installation of the network, the maintenance and operation depend on the length of the railways line. To help understanding the advantages of using public networks a mix between number of trains and railway lengths has been analyzed (Fig. 6). The benchmark is the cost of the GSM-R network, estimating an average value between a green field implementation and the upgrade of an existing network. Furthermore, it has been assumed that the *end2end* connectivity is achieved by using satcom for 20% and cellular networks for 80% of the train trip. The final results which must be considered preliminary due to the starting hypothesis, indicate a trend of significant cost savings with public networks in all the analyzed scenarios. On that basis a more accurate analysis should be done for each line by taking into account its characteristics.

A side-benefit when using public telecom networks is the reduction of CO_2 emissions. In fact, the electricity to feed H24 the base stations of a dedicated network is responsible for circa 60 to 100 kg of CO_2 emissions per day per km. Extrapolating this value to 30,000 km of regional lines in Europe, a potential saving of about 1.8–3 million kg of CO_2 emissions per day seems possible with public networks, assuming that 50% of the electricity is generated by renewable means.

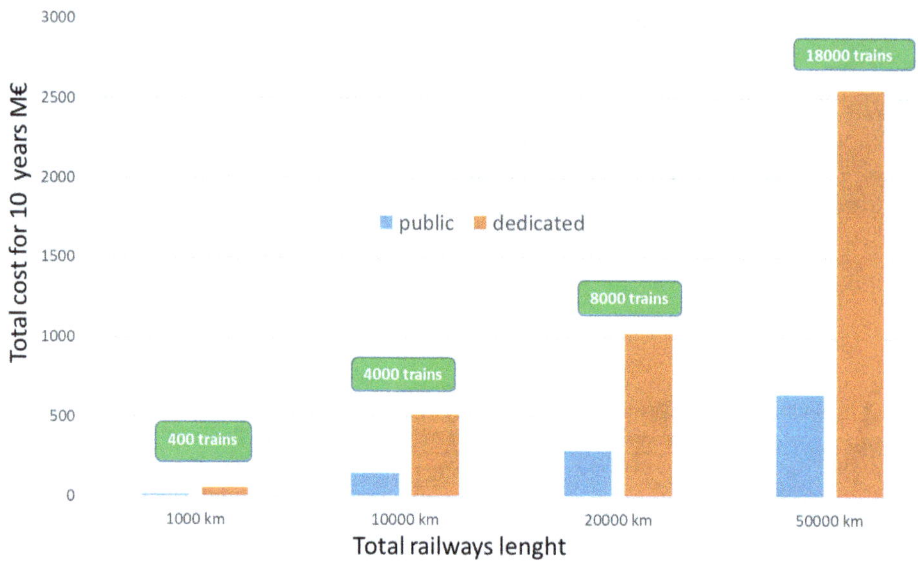

Figure 6: Estimated total cost of telecoms over 10 years.

4 FUTURE DEVELOPMENTS

Under TEN-T Regulation the European Member States are in principle obliged to deploy ERTMS on all the Core Network railways by 2030 (about 65,000 km) and 2050 for the Comprehensive networks (about 122,000 km). Yet, the fleet to be equipped by 2030 is estimated in the range of 27,500 to 38500 vehicles, compared to some 3,600 vehicles already equipped in Europe. In the frame of the ERTMS deployment, RFI has announced a plan in Italy to install ERTMS all over their lines by decommissioning the class B systems, thus becoming a full user of the ERTMS [1].

One of the great challenges ever faced by the rail sector is to satisfy the demand for communication during the migration from GSM-R to FRMCS and fulfilling at the same time an increase of train connectivity during the next years. A stepping stone is the evolution from circuit switched based communication systems towards IP-based communications, opening the way to a multitude of telecom bearers, instead of just one underlining technology as was the case of GSM-R.

To satisfy the increase of throughput, European Commission is evaluating an interim 1.4 MHz spectrum adjacent to the GSM-R band as well as the 1,900–1,910 MHz. Both solutions impose constraints since the new 900 MHz band – adjacent to those used by legacy mobile network operators – may need power restrictions to avoid adjacent channel interference, while the 1,900 MHz band will imply a higher number of base stations along the railways, thus increasing the costs.

A way to limit additional spectrum is to rely on public networks as described in this paper having recognised that in the railway environment, spectrum resources dedicated exclusively to railways may be insufficient, not available, difficult to coordinate with other users and perhaps not efficiently used since base stations are idle for most of the time being active just when train is within their footprint. Under these circumstances communication services offered by commercial mobile and satcom network operators should be considered strategic resources to alleviate these problems.

Sharing public networks would allow also to minimize on a national basis additional spectrum needs. From a technical point of view, cellular and satcom networks form a complementary mutually-sustainable bundle ensuring at the same time a whole coverage and QoS, as demonstrated with field tests in Italy. Resilience and availability, two peculiar characteristics can be achieved by the on-board multi-bearer processor on top of the standard bearer quality provided by the mobile network operators (MNO).

This aspect would entail further developments exploiting for example artificial intelligence with extensive field tests for training algorithms to optimize bearer selection along the railways lines. taking advantage of trains travelling frequently on the same lines.

Wi-Fi can also be used as additional bearer to off load the amount of data carried on cellular networks in proximity of stations where peak of capacity is generated by the simultaneous presence of many trains and satcom coverage may not be always ensured. In perspective, the development of 5G is set to revolutionize mobile services for railways, introducing in particular the network-slicing feature that allows to create a virtual network with guaranteed QoS. Network slicing – a breakthrough of 5G – is ideal to guarantee the railways specific connectivity needs and to exploiting also new train to train communications.

To meet the requirements of railways applications, 5G networks must ensure a coverage also outside main cities and high density areas where 5G networks are deployed at the beginning. In this respect, satcom are already evaluated at ETSI to foster the roll out of 5G service in unserved areas that cannot be covered by terrestrial 5G network (isolated/remote areas) and underserved areas (e.g. suburban/rural areas). In other words to upgrade the performance of limited 5G terrestrial networks in cost effective manner [16].

The mobile ecosystem for railways is evolving out of the traditional GSM-R supplier model and, other services for the passengers like web browsing, streaming, video surveillance, are boosting demand of connectivity in the Gbit range for the whole train. Such "backbone connectivity" on board trains travelling up to 300 km/h represents a real challenge also for the faster cell handovers due to a smaller cell size of the mobile broadband networks. The new paradigm of bearer independence backed by proper cybersecurity means is opening a wave of innovation in the railways sector. Similarly to aviation where a new satellite-based air–ground communication system (IRIS) is being developed for Air Traffic Management. (ATM). Actually pilots communicate with Air Traffic Control (ATC) by voice, over radio frequency channels suffering congestion, and with the projected doubling of aircraft in Europe's airspace by 2035 a shortage of communication may impact on air traffic, causing bottlenecks at airports. A study to evaluate the IRIS system for rail applications has been recently completed for ESA [16].

Synergies should be explored also with the automotive sector – by far the biggest user of vehicle to vehicle (V2V) and vehicle to infrastructure (V2I) technologies fulfilling safety requirements for connected and autonomous vehicles. These technologies are starting to be deployed along the roads to communicate with vehicles and in perspective could be shared with the railways being railways and roads generally close to each other. Yet satcom technologies for railways, already evaluated by ERA, are an important topic for ESA in supporting the introduction of the FRMCS in cooperation with ERA. The underlining objective is to foster the evolution of ERTMS with a building block architecture modular, upgradable and always connected [17], [18].

Whether railways communications will move gradually, endorsing new technologies will depend also by the mobile and satcom operators. Railways communication requirements have been predicted relying on large consensus for a steady migration to the FRMCS. It is almost certain that ERTMS competitivity will increase upon its ability to manage efficient, robust and high speed communication channels. Time is different from the early 1990s when

GSM-R has been selected and since then has contributed to the success of the ERTMS in the world. At that time, this choice was driven by a stepping innovation on mobile communications brought by the GSM, having had the merit to set a standard for ensuring the interoperability of ERTMS. These criteria still apply to the FRMCS, the successor of GSM-R, but with a technology-independent vision and more attention to the economical sustainability in stark contrast to past years. Interoperability will not be an issue with standard public mobile networks while satcom bearers would be selected to ensure as minimum a European coverage.

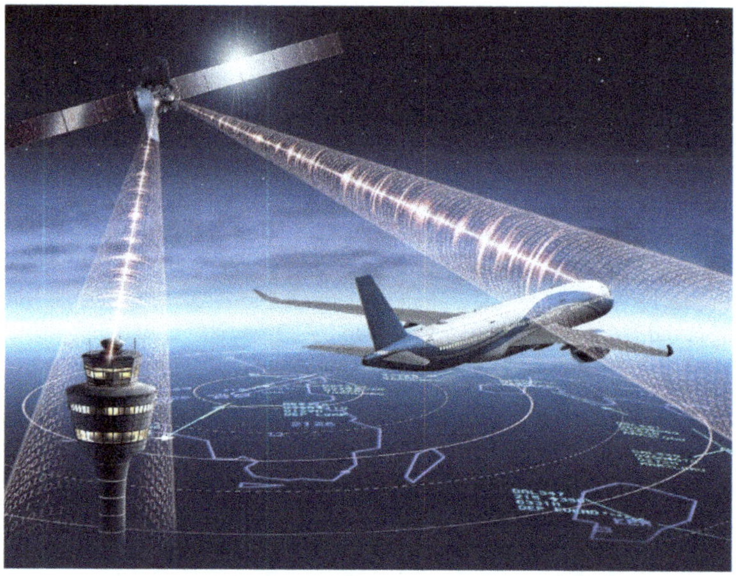

Figure 7: IRIS for air traffic control – ESA.

ACKNOWLEDGEMENT

Special gratitude is due to RFI and Trenitalia that are supporting the field tests within the projects co-funded by ASI (Agenzia Spaziale Italiana) through ESA and to Radiolabs for the use of the 5G test bed in l'Aquila.

REFERENCES

[1] European Court of Auditors, A single European rail traffic management system: Will the political choice ever become reality? Special report, 2017.
[2] FRMCS, UIC report. https://uic.org/rail-system/frmcs/.
[3] Shift2Rail-ACS, https://projects.shift2rail.org/s2r_ip2_n.aspx?p=X2RAIL-1.
[4] UNIFE, Report on ERTMS. www.unife.org/publication-press/publications/171-annual-report-2019-executive-summary.html.
[5] ERTMS, First Work Plan of the European Coordinator Matthias Ruete, May 2020. https://ec.europa.eu/transport/sites/transport/files/work_plan_ertms_2020.pdf.
[6] CEPT, Report 74 on spectrum needs.
[7] FW-ATwG 1903 v2.2.0/FM56(18)065: FRMCS Traffic Analysis.
[8] https://business.esa.int/projects/3insat.
[9] www.riotinto.com/news/releases/AutoHaul-successfully-deployed.

[10] ESA Space4Rail, Perspective of space utilisation in the Italian Railways, 13 Nov. 2019, Massimiliano Ciaffi, RFI Fabrizio Memmi, Trenitalia.
[11] https://artes.esa.int/projects/sat4train.
[12] www.radiolabs.it/en/radiolabs-unveils-the-initiative-emerge/.
[13] X2R1-WP3-D3.3-Communication system, Specification and technology guideline.
[14] RFI ERTMS Implementation plan in Italy: A system technological strategy, *CCRCC2019 ERTMS Conference*, Fabio Senesi.
[15] 3rd Generation Partnership Project (3GPP TM), Study on New Radio (NR) to support non-terrestrial networks, Release 15.
[16] ESA, IRIS for railway communications, Final Report, 2019.
[17] Final Report, ERA study on feasibility of SATCOM for railway communication.
[18] The ERA view on the future rail system, *CCRCC 2019*, Valenciennes Josef Doppelbauer, Executive Director.

Author index

WITPRESS ...for scientists by scientists

Urban Transport Systems

Edited by: ***G. PASSERINI****, Marche Polytechnic University, Italy, and* ***C. BORREGO****, University of Aveiro, Portugal*

Better urban transport systems are needed to achieve a healthier environment and as a result, a wide range of research has originated from many different countries. These studies highlight the importance of innovative systems, new approaches and original ideas, which need to be thoroughly tested and critically evaluated before they can be implemented in practice.

To address the need to solve important pollution problems the papers included in this book focus on the relationship with urban transport. There is also a growing need for integration with telecommunications systems and IT applications in order to improve safety, security and efficiency.

The variety of topics covered in this volume reflects the complex interaction of the urban transport systems with their environment and the need to establish integrated strategies. The aim is to arrive at optimal socio-economic solutions while reducing the negative environmental impacts of current transportation systems.

ISBN: 978-1-78466-369-8 eISBN: 978-1-78466-370-4
Published 2020 / 176pp

www.ingramcontent.com/pod-product-compliance
Lightning Source LLC
LaVergne TN
LVHW062322300926
849161LV00007B/1539